ROUTLEDGE HANDBOOK O DEMOCRACY

This handbook offers a comprehensive transdisciplinary examination of the research and practices that constitute the emerging research agenda in energy democracy.

With protests over fossil fuels and controversies over nuclear and renewable energy technologies, democratic ideals have contributed to an emerging social movement. Energy democracy captures this movement and addresses the issues of energy access, ownership, and participation at a time when there are expanding social, political, environmental, and economic demands on energy systems. This volume defines energy democracy as both a social movement and an academic area of study and examines it through a social science and humanities lens, explaining key concepts and reflecting state-of-the-art research. The collection is comprised of six parts:

1 Scalar Dimensions of Power and Governance in Energy Democracy
2 Discourses of Energy Democracy
3 Grassroots and Critical Modes of Action
4 Democratic and Participatory Principles
5 Energy Resource Tensions
6 Energy Democracies in Practice

The vision of this handbook is explicitly transdisciplinary and global, including contributions from interdisciplinary international scholars and practitioners. The *Routledge Handbook of Energy Democracy* will be the premier source for all students and researchers interested in the field of energy, including policy, politics, transitions, access, justice, and public participation.

Andrea M. Feldpausch-Parker is an associate professor of environmental and science communication at the State University of New York College of Environmental Science and Forestry (SUNY-ESF), USA.

Danielle Endres is a professor of communication and affiliated faculty in environmental humanities at the University of Utah, USA.

Tarla Rai Peterson is a professor of communication and affiliated faculty in environmental science and engineering at the University of Texas at El Paso, USA.

Stephanie L. Gomez is an assistant professor of communication studies at Western Washington University, USA.

"A jaw dropping, rich, and wondrously comprehensive treatment of the topic of energy democracy. A refreshing reminder than energy decisions, policies, and pathways have as much to do with politics and systems of political deliberation as they do hardware, infrastructure, or tariffs. For acts of energy consumption, investment or self-generation can be political statements alongside transactions in the marketplace or preferences for some technical criterion. This book offers a refreshing, urgent reminder of what is at stake—it is at once a sober diagnosis, a creative piece of scholarship, and a call for action."

Benjamin K. Sovacool, *Professor of Energy Policy, University of Sussex*

"This Handbook considers "energy democracy" as both a social movement and a terminological "composition" or way into important conversations about how technological innovation, new economic and political structures, and adaptive communication practices are all required to transform our broken relationship with the planet. Incredibly timely given recent events from Texas to India to around the globe!"

Stephen P. Depoe, *Professor and Head, Department of Communication, University of Cincinnati*

"Smart, comprehensive, and internationally authored, *Routledge Handbook of Energy Democracy* is an essential reference for scholars and climate activists alike in understanding the sociotechnical complexities of the energy transition now occurring and the urgent choices the climate crisis is demanding of us."

Robert Cox, *Professor Emeritus, University of North Carolina at Chapel Hill*

"A groundbreaking and highly recommended intervention that challenges taken-for-granted assumptions that energy transition necessarily delivers more sustainable futures. Contributors interrogate up-and-downstream aspects of energy assemblages, exploring new technologies and articulating participatory alternatives in the context of resource constraints and climate crisis. This collection is a must for exploring just transition."

Majia H. Nadesan, *Professor of Communication, Arizona State University*

"The intersection of energy, environmental, and security concerns creates urgent problems requiring collaborative solutions. This exciting volume provides a rich and ambitious overview of democratic concepts and practices that can empower scholars and activists in transforming the disastrous trends currently created by technocratic, neo-colonial, and corporate-capitalist control of energy systems."

Bryan C. Taylor, *Professor of Communication, University of Colorado Boulder*

ROUTLEDGE HANDBOOK OF ENERGY DEMOCRACY

Edited by
Andrea M. Feldpausch-Parker, Danielle Endres, Tarla Rai Peterson and Stephanie L. Gomez

Routledge
Taylor & Francis Group
LONDON AND NEW YORK

earthscan
from Routledge

First published 2022
by Routledge
2 Park Square, Milton Park, Abingdon, Oxon OX14 4RN

and by Routledge
605 Third Avenue, New York, NY 10158

Routledge is an imprint of the Taylor & Francis Group, an informa business

British Library Cataloguing-in-Publication Data
A catalogue record for this book is available from the British Library

Library of Congress Cataloging-in-Publication Data
Names: Feldpausch-Parker, Andrea M., editor.
Title: Routledge handbook of energy democracy / edited by Andrea M. Feldpausch-Parker, Danielle Endres, Tarla Rai Peterson and Stephanie L. Gomez.
Other titles: Handbook of energy democracy
Description: Abingdon, Oxon ; New York, NY : Routledge, 2022. | Includes bibliographical references.
Identifiers: LCCN 2021029051 (print) | LCCN 2021029052 (ebook)
Subjects: LCSH: Energy policy—Citizen participation. | Energy security—Social aspects | Energy consumption—Social asepcts. | Energy conservation—Social aspects. | Energy development—Social aspects.
Classification: LCC HD9502.A2 R677 2022 (print) | LCC HD9502.A2 (ebook) | DDC 333.79—dc23
LC record available at https://lccn.loc.gov/2021029051
LC ebook record available at https://lccn.loc.gov/2021029052

ISBN: 978-1-138-39225-0 (hbk)
ISBN: 978-1-032-13052-1 (pbk)
ISBN: 978-0-429-40230-2 (ebk)

DOI: 10.4324/9780429402302

Typeset in Bembo
by Apex CoVantage, LLC

CONTENTS

FIGURES

TABLES

CONTRIBUTORS

Elizabeth Allen is a research associate at Northeastern University's Global Resilience Institute. She holds a PhD in Environmental and Natural Resource Sciences from Washington State University. Her research focuses on food and agriculture systems, energy, and water resources management and policy decision making and applications of climate change impacts models. Her community engagement work centers on participatory approaches to resilience and sustainability assessments for community and economic development.

Kristin L. Arola is an associate professor in the Department of Writing, Rhetoric, and American Cultures at Michigan State University. She also holds affiliate appointments in the American Indian and Indigenous Studies and Digital Humanities programs. Arola's research and teaching focus on the intersections between technical writing, pedagogy, and American Indian studies. Her work is rooted in Anishinaabe, specifically Ojibwe, community practices, histories, and futures.

Jake Barnes is a researcher at the Environmental Change Institute, University of Oxford. His research interests include the politics and sociology of innovation and change at community through city to national systems. He has worked with local governments, community groups, and social enterprises to reflect, learn, and pursue societal transformations. His research is interdisciplinary and problem orientated, covering two broad areas: participation, actors, and agency in system transformations; and the governance of system change.

Susana Batel is a researcher at the Center for Social Research and Intervention of the University Institute of Lisbon. Her research has been adopting a critical and interdisciplinary perspective to examine the relationships between people, the territory, and the climate crisis, specifically around energy transformations toward carbon neutrality and associated social justice and political participation issues. She has published in journals like *Antipode* and *Energy Research & Social Science* and is coeditor of *Papers on Social Representations*.

Katelind Batill-Bigler is a marketing director at Patel, Greene & Associates, LLC. She graduated with dual master's degrees in Environmental Studies from the State University of New York College of Environmental Science and Forestry (SUNY-ESF) and in Public Relations from the S.I. Newhouse School of Public Communications at Syracuse University. Batill-Bigler enjoys spending time with her infant daughter, husband, and pup.

Sarah Becker is an associate researcher in the interdisciplinary Understanding Risk Research Group at Cardiff University. Overall, her research focuses on public perceptions of environmental issues with a particular interest in social justice. Her current research explores public views of heat decarbonization in Great Britain. In her previous research she has also examined public perceptions of climate change and how to pay for energy transitions.

Sören Becker is a researcher and lecturer at the University of Bonn, Germany. As a human geographer, he specializes in urban governance and strategies regarding transitions to more sustainable and climate-friendly infrastructures. His work on energy democracy was published in several academic journals within Human Geography and Sustainability Science. He also is coeditor of the first German-language university textbook on energy geographies.

Douglas Bessette is an assistant professor in the Department of Community Sustainability at Michigan State University. He conducts research and teaches courses in sustainable energy systems, energy transitions, and community energy development. Doug also develops and deploys structured decision-making frameworks, employs value-focused thinking, and investigates the trade-offs of renewable energy development in rural communities in the United States. Doug earned a PhD in Geography from the University of Calgary in Alberta, Canada.

Dominic Boyer teaches at Rice University where he also served as founding director of the Center for Energy and Environmental Research in the Human Sciences. His most recent book is *Hyposubjects*, an experimental collaboration with Timothy Morton. With Cymene Howe, he made a documentary film about Iceland's first major glacier (Okjökull) lost to climate change, *Not OK: A Little Movie about a Small Glacier at the End of the World* (2018). In August 2019, together with Icelandic collaborators, they installed a memorial to Okjökull's passing, an event that attracted media attention from around the world. He is currently pursuing research on electric futures across the world.

Ed Brown is research director of the Modern Energy Cooking Services (MECS) Program and a professor of Global Energy Challenges at Loughborough University. His interests lie in the fields of energy access and low-carbon energy transitions, questions of transparency, governance and corruption, globalization, and the financial needs of the poor. Ed cochairs the UK Low Carbon Energy for Development Network, building bridges between the divergent branches of academia working on energy and international development issues.

Matthew J. Burke explores the sociopolitical and ecological dimensions of the renewable energy transition. Matthew is a postdoctoral associate at the Rubenstein School of Environment and Natural Resources and Gund Institute for Environment at the University of Vermont, in collaboration with the Leadership for the Ecozoic project. Matthew completed a PhD in Renewable Resources—Environment at McGill University, examining energy democracy in theory and practice.

Ben Campbell is an environmental anthropologist with research interests in sustainability, in the cultural politics of indigenous agro-ecology and forests in the Himalayas, especially Nepal. Since climate change has had an increasing impact on Himalayan livelihoods, energy becomes a critical concern, including how to translate the concept into part-mechanized worlds far from modern infrastructural ideals. He is a fellow of the Durham Energy Institute and cochairs the UK Low Carbon Energy for Development Network.

Anabela Carvalho (PhD, University College London) is an associate professor at the Department of Communication Sciences of the University of Minho, Portugal. Her research focuses on various forms of environment, science, and political communication with a particular emphasis on climate change, as well as connections to public engagement. Anabela Carvalho is currently associate editor of *Frontiers in Science and Environment Communication*. She is also director of the PhD Programme on Communication Studies: Technology, Culture and Society.

Gabriel Chan is an assistant professor at the University of Minnesota's Humphrey School of Public Affairs. His research and engagement examines public policy to inform how technological innovation can enable a more rapid and equitable energy transition in domains such as community solar, consumer-owned electric utilities, and national innovation policy. He holds a PhD in Public Policy from Harvard University.

Jon Cloke is the national manager of the Low Carbon Energy for Development Network (LCEDN, www.lcedn.com) and also works with Modern Energy Cooking Services (https://mecs.org.uk/) and UK Innovate's Energy Catalyst. Previous low-carbon transitions work Jon has been involved in includes cofacilitating for USES (Understanding Sustainable Energy Solutions) and on two USES projects, Renewable Energy and Decentralization (READ, http://thereadproject.co.uk/) and Solar Nanogrids (SONG, http://songproject.co.uk/).

Brian Cozen is an assistant professor of environmental communication at California State University, Fresno (USA). His research examines the communication strategies involved when people make sense of and argue for social practices related to energy and mobility. He enjoys learning soup recipes and watching comedies, both versatile interests during a pandemic.

Laurence L. Delina is an assistant professor at the Division of Environment and Sustainability at The Hong Kong University of Science and Technology. His research interests are on rapid mitigation of climate change through accelerating sustainable and just sustainability transitions particularly in developing Asia. Laurence was a Philippine *Balik* Scientist, a Rachel Carson Fellow, and a Visiting Fellow at Harvard Kennedy School.

Christina Demski is a senior lecturer in the Understanding Risk Research Group at Cardiff University. She is also deputy director of the Centre for Climate Change and Social Transformations. Her wide range of research examines public responses to emerging sociotechnical issues with a focus on public views of environmental and energy issues. This includes, for example, public perceptions of whole energy system transformation and associated costs, acceptance of energy storage technologies and perceptions of climate change and low-carbon futures.

Danielle Endres is a professor of communication and affiliated faculty in environmental humanities at the University of Utah. Her research focuses on the rhetoric of science/environmental controversies including nuclear waste siting, climate justice, and energy democracy. Her research is guided by principles of environmental justice and decolonization and often focuses on how underrepresented groups engage in science and environmental controversies. Endres enjoys walking in the mountains, live music, reading, and being with her partner and two kids.

Andrea M. Feldpausch-Parker is an associate professor of environmental and science communication at the State University of New York College of Environmental Science and Forestry (SUNY-ESF). Her research is interdisciplinary in nature, focusing on science, environmental, and energy communication with interests in natural resources conservation through communication and collaboration. Much of her research addresses public participation in environmental decision making and social movements. Feldpausch-Parker loves spending time with her two kids and husband, crafting, playing in nature, and reading.

Mary Finley-Brook (PhD from the University of Texas at Austin) is an associate professor of geography and the environment at the University of Richmond. Her research focuses on natural resource management, land tenure, regional integration, climate change mitigation, Indigenous peoples, and carbon markets. Her geographic areas of interest are Latin America and the Caribbean.

Andrew Fiss is an assistant professor of technical and professional communication at Michigan Technological University, where he directs the Program in Scientific and Technical Communication and collaborates on grants with colleagues in engineering, writing studies, and environmental science. He has published in journals devoted to technical communication, science, technology, and society (STS), history, and education. His new book is *Performing Math: A History of Communication and Anxiety in the American Mathematics Classroom* (2021).

Valoree S. Gagnon serves as the director for University–Indigenous Community Partnerships in the Great Lakes Research Center at Michigan Technological University. Gagnon's expertise in environmental policy, food sovereignty, and community-engaged research focuses on the sociocultural impacts of environmental degradation, particularly on Indigenous communities. Her research, teaching, and service center on elevating Indigenous knowledge, facilitating equitable research practice and design, and guiding partnerships that prioritize the restoration of land and life in the Great Lakes region.

Stephanie L. Gomez is an assistant professor in the Department of Communication Studies at Western Washington University (USA). Her research focuses on critical media studies, critical/cultural studies, and rhetoric. Her research is guided by principles of social justice and equity, particularly regarding the representation of marginalized groups in the media. She enjoys reading, ballroom dancing, walking along Bellingham Bay, and spending time with her dog, Estrella.

Nihit Goyal is an assistant professor at the faculty of Technology, Policy and Management, Delft University of Technology. His research program is centered on analyzing and informing the governance of a sustainable energy transition with a focus on comparative public policy and computational social science.

Matthew Grimley practices engaged scholarship with electric utilities, cooperatives, nonprofits, and other community-based organizations at the hearts of their local energy transitions. Matthew is a research fellow at the Hubert H. Humphrey School of Public Affairs at the University of Minnesota, Twin Cities. He completed his MS and BA degrees at the University of Minnesota, Twin Cities, in Science, Technology, and Environmental Policy and English, respectively.

Kathleen E. Halvorsen is university professor of Natural Resource Policy and associate vice president for Research Development at Michigan Technological University. As a researcher, she is a policy scientist who studies the adoption of climate change mitigation strategies, including energy efficiency and renewable energy. She leads a large, convergent, and transdisciplinary team of Michigan researchers and Tribal Nation and nongovernmental organization staff studying solutions to barriers impeding transitions to local and Tribal renewable energy implementation.

Nicolas Hernandez is a PhD student at the Department of Communication at the University of Utah. He holds an MA in Communication from the University of Texas at El Paso. His research interests lie at the intersection of environmental communication, energy communication, environmental (in)justice, sociotechnical transitions, energy democracy, and decolonial theory. He has published original research in *Frontiers in Communication*.

Ana Horta is a researcher at the Institute of Social Sciences, University of Lisbon (Portugal). She holds a PhD in Sociology, and her current research interests include social practices and sociocultural aspects related to household energy consumption, energy transition, energy poverty, renewable energies, and media discourses on energy issues and climate change. She is one of the coordinators of the Environment and Society Section of the Portuguese Sociological Association.

Cristi Choat Horton is an assistant professor in the Department of Communication Studies at Tarleton State University, Stephenville, Texas. Her research focuses on how communication, policy, and democratic processes interconnect and impact policy options that contribute to sustainability, in particular to the development and deployment of low-carbon energy technologies.

Cymene Howe is a professor of Anthropology at Rice University and author of *Intimate Activism* (Duke 2013), *Ecologics: Wind and Power in the Anthropocene* (Duke 2019), and coeditor of the *Johns Hopkins Guide to Theory* and *The Anthropocene Unseen: A Lexicon* (Punctum 2020). She currently researches the dynamics between human populations and bodies of ice in the Arctic and the effects of sea level rise in coastal cities around the world.

Michael Howlett FRSC is Burnaby Mountain Professor and Canada Research Chair (Tier 1) in the Department of Political Science at Simon Fraser University in Vancouver, British Columbia, Canada. He specializes in resource and environmental policy making.

Taylor N. Johnson is an assistant professor of race, gender, and sexuality studies at the University of Wisconsin La Crosse. Her research is positioned at the intersections of Native American and Indigenous Studies and environmental justice. Her work primarily focuses on identifying and challenging settler colonialism in the context of US environmental decision-making processes.

Tatiana Kasperski is a research fellow at Pompeu Fabra University (UPF), Barcelona. Her research focuses on the history and politics of nuclear energy, in particular on memory, heritage, and forms of public engagement with nuclear energy in former Soviet spaces. More recently she has been working on a global nuclear environmental history.

William J. Kinsella (PhD, Rutgers University) is a professor emeritus in the Department of Communication at North Carolina State University (USA). His research and teaching examine the intersections of environmental and energy communication, organizational communication, and science and technology studies. His work on nuclear energy communication has encompassed the contexts of nuclear fusion, environmental remediation across the US nuclear weapons complex, and commercial nuclear energy in US and global contexts.

Olga Kuchinskaya is an associate professor of communication at the University of Pittsburgh. She is the author of *The Politics of Invisibility: Public Knowledge about Radiation Health Effects after Chernobyl* (MIT 2014). Her research has also focused on digital infrastructures that support the production of public knowledge about understudied health conditions and environmental risks and on pro-democracy technological design in the context of anti-regime social movements.

Stephanie Lenhart is a senior research associate at the Energy Policy Institute and faculty in the School of Public Service at Boise State University. Her research focuses on the governance of electricity systems and energy transitions. Her work examines how institutional designs, stakeholder participation, and the negotiation of authority are changing in response to a changing climate, new technologies, and increasingly diverse policy goals.

Julie L. MacArthur is an associate professor and Canada Research Chair in Reimagining Capitalism at Royal Roads University. She is the author of *Empowering Electricity: Co-operatives, Sustainability and Power Sector Reform in Canada* (UBC Press 2016) and a wide range of publications on energy democracy, participatory environmental governance, gendered employment in energy transitions, and comparative energy policy. She currently chairs the international WISER (women and inclusivity in sustainable energy research) network.

Elise Matz is a consultant who works at the intersection of government and industry to help communities build wealth in the transition to renewable energy. Since 2019, she has served on Michigan's Utility Consumer Participation Board, which makes grants to groups that represent residential consumers in the utility rate-making process. Prior to that, Elise served as a regional director for US Sen. Gary Peters. She holds a BA from the University of Michigan.

Meaghan McKasy (PhD, University of Utah) is an assistant professor of public relations in the Department of Communication at Utah Valley University. Her research focuses on strategic messaging, information processing, and attitude formation in science and environmental communication.

Laura Schmitt Olabisi is an associate professor in the Department of Community Sustainability at Michigan State University. She is an ecologist and a participatory systems modeler, working directly with stakeholders to build models that foster adaptive learning about the dynamics of coupled human–natural systems. Laura holds a doctoral degree from the State University of New York College of Environmental Science and Forestry. She was an AAAS Leshner Leadership Institute Public Engagement Fellow in 2018–2019.

Rebecca Ong is an assistant professor in the Department of Chemical Engineering at Michigan Technological University. Her research is on sustainable development of lignocellulosic bioenergy and bio-based products and waste plastic upcycling. Rebecca earned a PhD in Chemical Engineering from Michigan State University.

Lional Orama Exclusa holds a PhD in electrical engineering from Rensselaer Polytechnic Institute in New York. He is a professor at the Department of Electrical Engineering at the University of Puerto Rico–Mayaguez where he does research in the areas of electrical discharges and electrical transients and teaches courses in general engineering and electric power. He is also a cofounder and member of Puerto Rico's National Institute for Energy and Island Sustainability.

Cecilio Ortiz García holds a PhD in Public Administration and Environmental Policy from Arizona State University. He is a cofounder and member of Puerto Rico's National Institute for Energy and Island Sustainability. He is the chair of the Department of Public Affairs and a professor of Public Administration at the University of Texas at Rio Grande Valley.

Jennifer Peeples is a professor of communication studies at Utah State University. Her work uses environmental and visual rhetoric to examine discourses of energy transition and the construction of toxicity. In addition to journal articles and book chapters, she is the coauthor/coeditor of three books, including *Under Pressure: Coal Industry Rhetoric and Neoliberalism* (2016), and *Voice and Environmental Communication* (2014).

Marla Pérez-Lugo is an environmental sociologist with interests in disasters, environmental justice, and public participation in environmental and energy decision-making processes. She holds a PhD in Sociology from Rutgers, The State University of New Jersey. For nineteen years, she was a professor at the Department of Social Sciences at the University of Puerto Rico–Mayaguez and a cofounder and member of the Puerto Rico's National Institute for Energy and Island Sustainability's Steering Committee. She is currently a full professor of sociology and disaster studies at the Department of Sociology at the University of Texas at Rio Grande Valley.

Tarla Rai Peterson is a professor of communication and affiliated faculty in Environmental Science and Engineering at University of Texas at El Paso. Her research examines how intersections between communication and democracy enable and constrain science and environmental policy. She seeks to facilitate the emergence of an inclusive and nurturing community for Earth's citizens, while critiquing the normativity of that goal.

Chelsea Schelly is an associate professor of sociology in the Department of Social Sciences at Michigan Technological University. Her work explores how the sociotechnological systems used to support residential dwelling shape the organization of social life and conceptions of human–nature relationships. She researches and writes about a wide array of alternative socio-technological systems.

Patrick Scherhaufer is a researcher and senior lecturer at the Institute of Forest, Environmental, and Natural Resource Policy at the University of Natural Resources and Life Sciences, Vienna (BOKU). His research focuses on public policy, science, technology and environmental politics, participatory governance and democracy, knowledge integration, and inter- and transdisciplinarity. He has published in journals like *Energy Policy*, *Energy Research & Social Science*, *Futures and Systemic Practice*, and *Action Research*.

Jen Schneider is a professor in the School of Public Service at Boise State University. She writes about scientific and technological controversies, with a focus on the communication of energy transitions, environmental and social movements, and stakeholder engagement processes. Jen's PhD is in Cultural Studies, and she is coauthor of the books *Under Pressure: Coal Industry Rhetoric and Neoliberalism* and *The Joy of Science: Seven Principles for Scientists Seeking Happiness, Harmony, and Success*.

Jennie C. Stephens is a professor and director of School of Public Policy & Urban Affairs at Northeastern, and she is the author of the book *Diversifying Power: Why We Need Antiracist, Feminist Leadership on Climate and Energy* (Island Press 2020). Trained at Harvard and Caltech, she is an educator, social justice advocate, and internationally recognized expert on renewable transformation, energy democracy, climate justice, and gender in energy innovation.

M. Derya Tarhan is a postdoctoral fellow at the University of Toronto. Tarhan's work focuses on the theory and practice of the social and solidarity economy (SSE), community development, and socioenvironmental change. In his research, Tarhan focused on the impact, limitations, and potential of SSE organizations in enacting socioenvironmental change, with a keen eye on popular democratic participation and social justice. His further interests include economic democracy, critical theory, science and technology studies, and social learning.

Elizabeth Wilson is the founding director of the Arthur L. Irving Institute for Energy and Society and a professor of Environmental Studies at Dartmouth. Her research examines the interactions between technologies, decision making, policies, and institutions shaping energy system transitions in different contexts. She holds a doctorate in Engineering and Public Policy from Carnegie Mellon University and was selected as a Leopold Leadership Fellow in 2011 and an Andrew Carnegie Fellow in 2015 and spent 2016–2017 visiting the Danish Technical University.

Sara K. Yeo (PhD, University of Wisconsin–Madison) is an associate professor in the Department of Communication and a faculty affiliate with the Global Change and Sustainability Center and Environmental Humanities program at the University of Utah. Her research interests include public opinion of STEM issues, emotion, humor, and information seeking and processing. Dr. Yeo also holds an MS in Oceanography (University of Hawaii at Manoa), which has informed her research agenda.

ACKNOWLEDGMENTS

Andrea wants to thank her pandemic crew and family, Israel, Imogen, and Alaric, for all of their love and support.

Danielle wants to thank Wayne, Julie, Rosa, and Owen for their unwavering support.

Tarla is thankful for the support and ideas from her family, especially Gwen, Aspen, Zane, Sage, Greta, and Rose.

Stephanie wants to thank her family, especially Estrella, for their love and support.

We want to thank our colleague and friend, Leah Sprain, for the many brainstorm sessions and contributions she made in early stages of this project.

The National Science Foundation (SES 1655192), University of Utah College of Humanities, National Communication Association, and University of Colorado BoulderTalks provided funding and support for the Energy Democracy Symposium (2017). The symposium set the foundation for our work on energy democracy and the creation of this handbook.

1

ENERGY DEMOCRACY

An introduction

Andrea M. Feldpausch-Parker and Danielle Endres

As we are writing this chapter, the state of Texas in the United States is in a state of energy disaster. A winter storm intensified by climate change, exposed underlying structural problems in the state's dependence on fossil fuels, natural gas infrastructure, and an electricity grid designed without consideration of climate change–intensified bouts of extreme weather. This example is just the latest in a series of "unnatural disasters" (de Onís, 2018a) facing the world as a result of the ongoing climate crisis. It joins with the impacts that Hurricanes Irma and Maria had on Puerto Rico's aging energy system, reliant on fossil fuel imports (de Onís, 2018b), Superstorm Sandy's devastation to the US East Coast (Feldpausch-Parker et al., 2018), and wildfires in the United States, Australia, and the Amazon. Rather than a dystopian future scenario, climate disaster is happening now and affecting people all over the globe. Climate change acts as a threat multiplier, meaning that we have not seen the end of unnatural disasters and climate chaos.

In the midst of the ongoing climate crisis, energy transition is one of the most important issues facing local, national, and international communities. Although climate change often overshadows energy in public discourse, the ability to affect drastic transformations in the way we conceive of, plan, and use energy is crucial to any efforts to address climate change and create sustainable, equitable, and resilient futures. When viewing energy transition as a socio-technical phenomenon, focus shifts from determining technical feasibility of particular energy technologies to engaging with the messiness and complexity of social, political, and cultural contestation that must be navigated to actually implement energy system changes at multiple scales. Energy systems will continue to change in the coming years; the consequential question is *how* they will change. Protests over the continued use of fossil fuels; controversies over where (and whether) to locate solar, wind, hydro, and nuclear energy sites; and activist calls for a just transition remind us that while energy transition is inevitable, the contours of that transition remain uncertain and contested.

Working from the basic premise that an energy transition rooted in democratic principles is our best hope of achieving just, equitable, and culturally appropriate solutions across the many scales of decision making about energy, this handbook unpacks the relationships between energy and democracy in the context of the climate crisis and energy transition. Energy democracy (ED) first emerged as a term used by activists to call for greater levels of participation in decision making about energy transitions, including more localized control over energy production and consumption, distributional and procedural justice in decision making, and promotion

DOI: 10.4324/9780429402302-1

of renewable energy sources (e.g., Giancatarino, 2012). To work toward ED—a sociotechnical energy transition infused with democratic practices and ideals—requires that scholars and practitioners engage and experiment with new forms of participation, relations of power, practices of justice, and configurations of energy technologies. As such, there is a profound need to devote scholarly attention to understanding and developing theoretically informed democratic approaches to energy transitions.

Over the several years since we hosted a symposium in 2017 in Salt Lake City, Utah, to build a research agenda for ED, research in this area has grown exponentially (e.g., Burke & Stephens, 2017; Feldpausch-Parker et al., 2019; Hess, 2018; Szulecki & Overland, 2020; van Veelen & van der Horst, 2018). While Reinig and Sprain (2016) note that ED is not sufficiently addressed in interdisciplinary energy scholarship, four years later Szulecki and Overland (2020) characterize ED research as a nascent interdisciplinary field, with a growth spurt in publications starting in 2017. This growing area of research provides many opportunities to contribute to understanding ED as movement, concept, and practice through varied theoretical, conceptual, and empirical lenses.

Our vision is to offer a comprehensive transdisciplinary examination of the research and practices that constitute ED, or, as Chilvers and Pallett (2018) argue, the multiple energy democracies. While no one handbook can offer a fully comprehensive rendering of an emerging research area, we deliberately curated this collection to offer readers an introduction to ED that represents the many disciplines, regions of the world, and forms of social scientific and humanities research that make up this interdisciplinary area of study. Entries from leading international scholars and practitioners highlight various facets of ED and span a variety of theoretical, conceptual, critical, and empirical forms of research. The handbook includes six parts: scalar dimensions, discourses, grassroots and critical modes of action, democratic and participatory principles, energy resource tensions, and energy democracies in practice. Chapters across these parts explain key concepts, reflect state-of-the-art research, and elaborate on the broad range of actors, democratic values, democratic functions, and governance sites that are involved in ED. The handbook contributes to growing efforts to study examples of energy democracies in practice (e.g., Becker et al., 2020; Bloem et al., 2021; Gunderson & Yun, 2021; Hess, 2019; MacEwen & Evensen, 2021; Morris & Jungjohann, 2016; Stephens et al., 2018; Williams & Sovacool, 2020), as well as to engage with theoretical and conceptual perspectives toward ED (e.g., Burke & Stephens, 2017; Cantarero, 2020; Feldpausch-Parker et al., 2019; Sorman et al., 2020; Szulecki & Overland, 2020). The chapters in this handbook not only extend a growing field's understanding of ED but also, in recognition of its roots in social activism, contribute to the ongoing ED movement by offering research-informed insights into best practices and lessons learned from energy democracies in practice.

In the remainder of this introduction, we begin by defining ED as a composition of energy and democracy. Then we will introduce a framework for examining ED at the intersections of justice, participation, power, and technology. This is followed by a preview of the parts of the handbook and a brief conclusion.

Energy democracy as composition

Energy democracy is a composition, or a putting together, of energy and democracy for a specific sociotechnical purpose: a democratic energy transition in the face of the climate crisis. Bruno Latour (2010) offers the concept of *composition* to highlight that "things have to be put together (Latin, *componere*) while retaining their heterogeneity" (pp. 473–474). In response to growing ecological crises, Latour offers composition as an alternative to the forms of critique and deconstruction that commonly guide academic inquiry. Focusing primarily on putting things together turns attention to building alternatives and solutions that can have impact on addressing real-world problems, as opposed to the tendency of critique and deconstruction to stop at identifying problems. Composition works on two levels as a framing tool for this handbook's focus on ED: (1) It focuses attention on how seemingly disparate things—like energy and democracy—can be put together in a variety of ways, or *energy democracies* (Chilvers & Pallett, 2018); (2) it focuses attention on developing research programs that go beyond identifying what is wrong by offering immediate, yet thoughtful compositions of new solutions, imaginaries, and futures that respond to the many exigencies faced by society, such as the climate crisis. Composing is difficult but may provide an elemental framework for the energy exigencies at hand. In keeping with the obligation to put together while retaining heterogeneity, we will begin by defining energy and democracy.

Energy

While there are many ways of thinking about energy—from capacity to act, to spiritual essence, to power, to vigor—here energy refers to the forms of power, such as electricity, that are used to enable human technologies. Endres et al. (2016) define energy as:

> power that may be used to operate the infrastructures of the human-built environment. Humans derive that power from resources such as fossil fuels, solar, wind, hydroelectric, nuclear, biofuels, and geothermal sources that are extracted and harnessed, prepared, and distributed in a cycle of energy production. We use the term *energy resources* to discuss sources of energy, *energy production* to describe the cradle-to-grave process whereby energy is supplied to human-built infrastructures, and energy consumption to refer to the processes wherein people use energy resources to power infrastructure, technology, and other activities.
>
> *(p. 420)*

In this way, energy is an essential but often unseen and not reflected upon component of human society (e.g., Szeman & Boyer, 2017). Yet, while some form of energy is foundational to human society, there are significant choices to be made about which energy resources and forms of production and consumption to use. With the alarming present and future realities of climate change and inequities in the distribution of harms, we find ourselves in a time of energy transition with a complex matrix of contestation over what choices to make across scales of participation and governance.

Inherent to our definition of energy is the notion that energy resources, production, consumption, and transition work at the intersection of the technical and social. Energy, then, is also a composition. While many still overwhelmingly consider energy as a technical phenomenon that requires the work of innovative scientists and engineers, it would be a mistake to ignore its social aspects. As Sovacool and Dworkin (2014) note, energy is the:

> sociotechnical system in place to convert energy fuels and carriers into services—thus not just technology or hardware such as power plants and pipelines, but also other elements of the "fuel cycle" such as coalmines [sic] and oil wells in addition to the institutions and agencies, such as electric utilities or transnational corporations,

> that manage the system, as well as the households and enterprises that consume or put that energy to work.
>
> (pp. 7–8)

Furthermore, "energy transitions are complicated by the histories, cultures, practices, and existing spatial relations between sites of production and sites of consumption" (Curley, 2018, p. 58). Developing a technology does not guarantee that society will use it or that publics and policy makers will accept scientific evidence as valid. Put simply: Technology alone is not sufficient for implementation of that technology. Indeed, energy is a quintessential example of the inextricable intersections between technoscience and society (e.g., Jones, 2013; Laird, 2013; Sovacool & Brossmann, 2013). This handbook, therefore, resists viewing the social, political, and cultural elements of energy as merely contextual, instead seeing them as essential starting points for investigation into energy and energy transitions.

As a sociotechnical phenomenon, energy is the subject of research from a range of disciplinary perspectives, from sciences and engineering to humanities and social sciences. This handbook's engagement with energy encompasses the full range of social scientific and humanities perspectives, including our disciplinary grounding in energy communication (Cozen et al., 2018; Endres et al., 2016) and energy humanities (Szeman & Boyer, 2017; Wenzel, 2016), as well as the rich sets of interdisciplinary energy studies that are seen in journals such as *Energy Research & Social Science* and collaborations such as the Energy Impacts Research Coordination Network (see EnergyImpacts.org).

Democracy

Democracy is notoriously difficult to define. Communication scholars often think about democracy as what McGee (1980) called an ideograph, or an abstract ideological concept that is presented as self-evident across a variety of contexts to justify action. People may think they know what democracy means, and people living and working in democratic nations certainly hear repeated appeals to democracy as a social good. Yet those appeals rely less on a precise definition than on values, ideologies, and specific contexts. It should be no surprise to readers that democracy has been theorized, debated, and developed over thousands of years and across multiple disciplines. Scholarly literatures identify numerous models, forms, practices, and ideals of democracy. The purpose of our handbook is not to wade into the intricacies of scholarly debate about democracy, though many of our authors draw from various intellectual traditions and conversations about democracy. Rather, we offer a basic definition of democracy as a starting point.

Democracy is fundamentally "a method of group decision making characterized by a kind of equality among the participants at an essential stage of the collective decision making" (Christiano, 2018). This definition focuses on democracy as a process of collective decision making that can occur within a variety of groups ranging from a family to a community to a business to a state or national government. With this definition, the forms of decision making and participation can also vary: "It may involve direct participation of the members of a society in deciding on the laws and policies of the society or it may involve the participation of those members in selecting representatives to make the decisions" (Christiano, 2018). Mouffe's (2005) perspective on democracy assumes that decision making operates at the confluence of dialogue and rhetoric, where attempts to achieve mutual agreement include both the dissemination of persuasive messages (rhetoric) and intersubjective meaning-making (dialogue) processes among participants. Her theory highlights the concepts of *democratic paradox* and *agonism*, while also recognizing the very real requirement of moving beyond agonism to development, deployment, and administration of policies that convey sufficient public legitimacy to enable deliberation. The goal is to achieve (always temporary) consensus on procedure, without necessarily resolving the interest- and value-based conflicts between citizens, who will likely continue disagreeing and will hopefully continue arguing with one another. Only through this process of contestation of ideas is society able to enact changes in democratic practice that respond to contemporary needs.

Yet democracy is not only a process. Any democratic process is also interconnected with values and ideals. One value is the active and frequent participation by all of those who will be affected by a decision. Others are equity as an ideal that can be achieved by promoting justice among participants and a faith in the ability of a collective to compromise and make decisions with the best interests of all participants in mind. Democratic process is not easy. Indeed, it is a radical ideal that is often not met but is strived for in response to powerful interests and tyrannical leaders. With this definition, then, democracy is not affiliated with a particular political party, government, or nation-state. Democracy as a practice is possible on a variety of scales of decision making, ranging from the local to global, making possible the pursuit and practice of democracy in nondemocratic states.

Democracy is not unconnected with technology (e.g., Collins & Evans, 2002; Hamlett, 2003; Latour, 2004; Mercer, 1998). Indeed, Winner (1986) and Sclove (1995) argue for democratization of technological decision making so that publics, such as emergent ED movements, can be more involved in technological decision making. Jasanoff (2005) importantly highlights that any analysis of the interface between democracy and technology requires a consideration of the civic epistemologies at play across nations. Democracy, therefore, is not one thing in practice. It is a process and set of ideals that may look different when applied to different technologies in different contexts with different participants.

Energy democracy

As framed in this handbook, ED is a composition, a joining together of energy with democracy. Winner (1986) warns against viewing any one energy technology as intrinsically democratic, so the goal of ED movements and research is not to find the most democratic energy option. Rather, ED, in the sense of composition, simply puts the two terms together to encourage that energy decisions are as democratic as possible. Yet this is no easy task and involves a variety of different conceptions of ED. From a compositionist approach, there are many possible energy democracies that emerge within particular contexts, places, and configurations.

The rise of a social movement and area of scholarship

ED as a social movement is roughly a decade old, gaining traction in the early 2010s and gaining attention from a diversity of social science and humanities scholars shortly thereafter. As Szulecki and Overland (2020) note, energy democracy is "linked to the expanded deployment of distributed and small-scale renewable sources" and "the growing politicization of energy governance and climate policy" (p. 1). Feldpausch-Parker et al. (2019) define ED as "an emergent social movement that reimagines energy consumers as prosumers . . . who are involved in decisions at every stage, from energy production through consumption" (p. 2). The movement was borne out of valuing localized autonomy over energy resources and operations, democratic decision making, resisting various forms of environmental injustice, and promoting a just transition (e.g., Angel, 2016; Fairchild & Weinrub, 2017; Giancatarino, 2012; Sweeney, 2014). Angel

(2016) gave shape to the sentiments of this movement in a report from an international workshop on ED held in Amsterdam, The Netherlands, in February 2016 (roughly a year and a half prior to our research symposium in July 2017, as previously noted).

> From energy access to climate justice and from anti-privatisation to workers' rights, people across the world are taking back power over the energy sector, kicking-back against the rule of the market and reimagining how energy might be produced, distributed and used. For many (but not all) movements involved in struggles around energy, the concept of energy democracy is proving increasingly useful as a means of bringing together disparate but clearly linked causes under a shared discourse and, possibly, something of a common agenda.
>
> There is no singular understanding of the call for energy democracy. The term clearly evokes a desire for collective control over the energy sector, counterposed with the dominant neoliberal culture of marketisation, individualisation and corporate control. Energy democracy is concerned with shifting power over all aspects of the sector—from production to distribution and supply, from finance to technology and knowledge—to energy users and workers. Movements deploying the concept of energy democracy also demand a socially just energy system, meaning universal access, fair prices and secure, unionised and well-paid jobs. They want an energy system that works in the public interest, with the profit motive giving way to social and environmental goals. And they seek a transition from high to low carbon energy sources, ultimately meaning a world powered by 100 percent renewable energy.
>
> *(p. 3)*

ED, then, includes several key attributes: A broad range of actors, democratic values, democratic functions, and governance sites as related to energy transition. It is upheld by various processes that support democratic functions, such as public engagement as sense making around socio-technical systems (Einsiedel et al., 2013), making issues of power and competing values open to discussion and resolution (Sharpe et al., 2016), and enabling the conflict and contestation inherent in the redistribution of resources (Lawhon & Murphy, 2012). ED raises questions about what aspects of the energy system ought to be governed by democratic norms, which can be answered by mapping the range of sites of decision making, voice, and agency. Unlike transition management studies that often focus on solely technical or economic policy changes, ED considers a wide range of potential governance sites and stakeholders from energy production to everyday practices.

ED is primarily a movement of local energy practitioners. Proponents view it as a way ordinary people respond to their communities' most critical needs, especially those related to climate change and other environmental injustices (Climate Justice Alliance, n.d.). Expanding social demands on energy systems beyond access, reliability, and affordability to include a suite of environmental and economic benefits has fundamentally altered the configuration of who can directly influence shifts in energy systems. As Sweeney (2014) states:

> A timely and equitable energy transition can occur only with greater energy democracy, which requires that workers, communities, and the public at large have a real voice in decision making, and that the anarchy of liberalized energy markets is replaced with a comprehensive and planned approach.
>
> *(p. 217)*

ED is both a social movement that calls for more democratic decision making about the energy transitions needed to address the climate crisis and an academic area of study (Feldpausch-Parker et al., 2019).

ED is inclusive of considerations of energy justice (Finley-Brook et al., 2018; Sovacool et al., 2017), a just transition (McCauley & Heffron, 2018; Na'puti et al., 2018; Swilling & Annecke, 2012), and energy coloniality (de Onís, 2018b), as is demonstrated by chapters in this volume. More broadly, ED also allows for considerations of climate justice (Pezzullo & de Onís, 2018; Schlosberg & Collins, 2014). We do not offer ED as a replacement for research or activism centered on these terminologies. Rather, we propose that ED is broad and inclusive enough to think with these concepts toward a democratic energy transition built on the pillars of justice, power, participation, and technology.

As noted, ED is fundamentally a sociotechnical process of energy transition. Transition management presumes that emerging technologies influence and are influenced by social context as they progress through the phases of predevelopment, takeoff, breakthrough, and stabilization (Loorbach, 2010). Despite a growing sense of urgency, diffusion of low-carbon energy technologies has been slow and uncertain, and many obstacles have emerged (Stephens & Zwaan, 2005; Wilson & Stephens, 2009). Research exploring the challenges of energy technology diffusion has generally focused on the economic and technical aspects (Lawhon & Murphy, 2012). Drawing from research on the sociotechnical dimensions of energy technologies, ED considers a broader range of actors and social influences than just the economic and technical. ED contributes to a theory of sociotechnical energy transitions that recognizes that social change and technological change are fundamentally interrelated (Geels & Schot, 2007).

Energy democracy frameworks

As scholars attempt to capture the momentum of a social movement and its theoretical and practical implications, some, including ourselves, have attempted to provide a framework for an ED research agenda. An early framework from Kunze and Becker (2014) evolved from ED movements in Europe, pulling from case studies in Germany, Spain, the United Kingdom, France, Hungary, Italy, and Belgium. This framework includes democratization, property, surplus value production, and ecology. The authors take a postgrowth stance for which people and planet are prioritized above capitalism, citing the elimination of energy and fuel poverty as well as the conservation of biodiversity as examples. Focusing on efforts in North America, Burke and Stephens (2017) instead focus on the use of policy instruments and goal attainment in ED movements, noting that multiple goals are often sought from such efforts. Hess (2018) relatedly takes on a multicoalition perspective to energy transitions, whereas Szulecki (2018) breaks ED into: "popular sovereignty, participatory governance and civic ownership, and operationalizing with relevant indicators" (p. 21). The Climate Justice Alliance even assembled a list of ten principles for ED in an effort to guide the movement, including human rights, self-determination, energy as a commons, just transitions, energy use, community governance, diversity of scale, reclaiming relationship, acknowledging, acting on, and repairing historical harms, and rights of nature (Center for Earth, Energy & Democracy & CJA Energy Democracy Working Group, 2015, p. 6). These are just a handful of frameworks adding to this nascent body of literature that is both attempting to analyze energy system reimagining efforts and to contribute to articulating the principles undergirding those efforts. ED frameworks demonstrate the hope of shaping energy systems so they are more equitable and better able to deal with social and environmental challenges such as the climate crisis. In the following sections, we will outline our framework as an attempt to create a transdisciplinary research agenda.

Justice-participation-power framework

Feldpausch-Parker et al. (2019) synthesized and analyzed extant interdisciplinary scholarship on ED and proposed a framework for future research based on the conceptual pillars of *justice*, *participation*, and *power*. While these components may not be the only elements at play, their presence in various configurations can be seen throughout the literature and movement as fundamental to democratic energy transitions. Highlighting justice, participation, and power does not ignore the importance of technology, economics, and governance in energy transitions but shifts the organizing focus of energy transitions to ensuring adherence to democratic principles and practices in energy decision making. This framework offers a heuristic for understanding sociotechnical energy transition, "enabling examination of theoretical models, empirical examples of ongoing struggles over energy, and practical recommendations for communities engaged in promoting energy democracy" (Feldpausch-Parker et al., 2019, p. 3).

Justice in ED builds from social and environmental justice movements and discourses of climate justice, energy justice, and just transitions. As the Climate Justice Alliance (2018) notes on their website, "Transition is inevitable. Justice is Not." Similarly, no one energy technology is guaranteed to promote justice and democracy in all locations and contexts. Activists and scholars promote justice by illuminating and attempting to address issues of inequality and marginalization in energy processes, including inequalities of access to affordable energy, impacts from energy systems, and access to energy decision-making processes about energy infrastructure (e.g., Cha, 2020; Healy & Barry, 2017; Newell & Mulvaney, 2013). "The solution to resolving the Energy Trilemma"—economics, politics and environment—is justice and equity (Heffron et al., 2015, p. 168). Heffron et al. (2015) go on to argue that "this represents a move away from solely having economic thinking drive policy aims" (p. 169). A just transition sees economics from the lens of ensuring justice for workers and localities. Furthermore, while justice is traditionally conceived of as a human construct, it should also be applied to the more-than-human world to incorporate ecological systems into energy transition (Feldpausch-Parker et al., forthcoming; Whyte, 2016).

Participation pulls heavily from participation and democracy literatures to focus on how individual and collective actors participate in energy decision making. Engagement may include a variety of democratic practices, both within and outside formal decision-making processes sanctioned by governance institutions. Therefore, this category goes beyond the traditional practices of public comment periods and public hearings to include a broad swath of participatory actions (Feldpausch-Parker et al., 2019). Sociotechnical transitions research provides some examples where participation can include ownership and management of energy technologies by (re)imagining ratepayers "as prosumer, or innovators, designers, and analysts who are involved in decisions at every stage" (Feldpausch-Parker et al., 2019, p. 2). Social movements are also key to shaping energy politics and serve as an example of forms of participation that exceed official processes (Cozen et al., 2018; Endres et al., 2009; Pezzullo, 2007). Regardless of the form of participation, ED involves challenges to traditional technocratic or decide-announce-defend (DAD) forms of participation in environmental decision making (Depoe et al., 2004; Hendry, 2004; Hunt et al., 2019). Justice is relational to participation in that just forms of participation involve the genuine inclusion of all relevant stakeholders. For this to be fully realized, energy decision-making processes must expand beyond overreliance on technocratic and economic knowledges that privilege expert voices with institutional power to also include local, public, and Indigenous knowledges that engage with or have the capacity to engage with social and ecological systems in everyday life (Endres, 2009; Fischer, 2000; Kimmerer, 2013; Kinsella, 2004).

Power in this framework is focused on the ability to act (or not act) to elicit change. Mouffe (2000) notes that power is often unequally distributed in decision-making processes, leaving some actors to wield greater influence than others. This is very much the case in energy systems decision making. Burke and Stephens (2017) argue that "central to an energy democracy agenda is a shift of power through democratic public and social ownership of the energy sector and a reversal of privatization and corporate control" (p. 38). Power is multidirectional—it can be a form of control from an official position, or it can be resistive. In the case of the latter, power depends "on the hope that activism, grassroots democratic organizing, local governing structures, and public participation have the power to make changes in the status quo and possibly change existing hierarchies and relationships" (Feldpausch-Parker et al., 2019, p. 5). Understanding controlling and resistive forces of power can, therefore, provide an opening for reconstituting the understanding of energy transition via ecological processes, sociotechnical arrangements, and the societal consequences. Taken together, justice, participation, and power represent a set of intersecting nodes within ED that allow for a view of sociotechnical energy transition that centers the power of people to challenge inequitable structures and make change.

An expanded justice-participation-power-technology (JPPT) framework

Although the presence of energy technologies is assumed in the justice-participation-power framework, other ED frameworks include more explicit discussion of the technical and infrastructural components of energy transitions within a social context (see Kunze & Becker, 2014). As the body of literature on ED continues to grow and grapple with the complexities associated with socially and environmentally equitable energy systems, we have refined our framework to include technology as one of the nodes.

Research exploring the challenges of energy technology diffusion and implementation has generally focused on economic and technical solutions to technology deployment (Lawhon & Murphy, 2012). Yet, as we have argued, we cannot resolve sociotechnical problems like the climate crisis and energy transition without focusing on the intersecting societal, historic, cultural, and ecological factors. In this vein, Feldpausch-Parker et al. (2019) call to shift the focus of energy transition research "from viewing the sociopolitical elements as context to seeing them as key starting points for investigation of sustainable energy transitions" (p. 3). In addition to centering the sociopolitical in our understanding of energy technologies, the JPPT framework retains the centrality of actors and social influences other than economics and technology, while also exploring how economics and technology play into the composition of ED. This enables viewing energy technology transition through the lenses of justice, participation, and power, therefore reorienting the range of considerations, voices, and values available for analysis. The JPPT framework contributes to viewing sociotechnical energy transitions as matters of social change that are inseparable from justice, participation, and power. As such, the intersections of justice, participation, power, and technology offer a way to analyze the components of energy transition using democratic principles as the primary lens.

This expansion of our framework explicitly captures the roles played by technology and technological innovation in a just and equitable energy transition. What sets our framework apart from others is its continued focus on democratic theory and practice—via justice, participation, and people power—that we view as essential to the ED movement and scholarship. Practicing democracy requires conditions for productive interaction within communities of similar values and interests, between diverse communities, and at interfaces with decision makers.

This focus on democracy as the primary lens for energy transition brings its own limitations, particularly its narrowed focus on justice, participation, power, and technology. Szulecki and Overland (2020) acknowledge the difficulties of developing one overarching framework for ED movements and research programs. Instead, they provide three "ideal-typical understandings of energy democracy" including "a process," "an outcome of decarbonisation," and "a normative goal" (p. 2). Sorman et al.'s (2020) framework includes (1) the concept of energy as something that is socially and politically constructed; (2) the political, which captures the discourses of energy security versus energy sovereignty; and (3) the people, which addresses political polarization and post-truth energy politics. While the JPPT framework has less breadth than these frameworks, its strength is in its attention to how energy technology systems can be democratized via considerations of justice, participation, and power. Our framework also allows for comparisons of democratic constructs between case studies in locations with different social pressures and cultural considerations. Thus the JPPT framework provides a terministic screen (Burke, 1966), or lens, through which to highlight practices, that enables an energy transition that values just, participatory, and equitable power relations.

Scope of the handbook

Our vision for this handbook is to offer a transdisciplinary and international examination of the research and practices that constitute the emerging research agenda in ED. The handbook addresses ED as both a social movement and an academic area of study. It includes entries from leading scholars and practitioners that highlight various facets of ED theory and practice. The chapters include scholars and practitioners not only within communication, our primary affiliation, but across the humanities and social sciences. Moreover, the handbook engages with global energy issues through choice of both examples and international scholars. In sum, this handbook serves as a key source for students and researchers who want to teach about or engage in a research program on ED.

Part overviews

The handbook is divided into six parts that attempt to encompass different aspects of and scholarly lenses for examining ED. **Scalar Dimensions of Power and Governance in Energy Democracy** introduces dimensions of ED at different scales of governance, recognizing and working through dynamics of power. Chapters in this part demonstrate how ED is scale dependent. **Discourses of Energy Democracy** showcases prevalent discourses of ED that construct particular visions of energy transition. Together, the chapters in this part reveal patterns and tensions in how discourses of and about ED differently construct meanings and coordinate energy transition. **Grassroots and Critical Modes of Action** reviews grassroots and critical strategies for enacting ED. The chapters in this part provide a set of options that can operate singly or in combination to foster energy democracies in particular locales. **Democratic and Participatory Principles of Energy Democracy** introduces principles and constructs that characterize ED. Chapters in this part theorize and analyze a wide variety of democratic practices in many contexts and formats. **Energy Resource Tensions** introduces different types of energy resources (hydro, solar, wind, natural gas from fracking, and nuclear) and the tensions that exist in their continued or contested use. Chapters present these tensions through empirical and critical study. **Energy Democracies in Practice** introduces strategies and imaginaries for enacting ED through coalitions, projects, organizations, and movements that provide insight into doing the practical work of building ED. Chapters in this part offer analysis and description of both currently used and possible practices of ED.

Through these six parts, the handbook offers entries that explain key concepts, reflect state-of-the-art research, and build the research agenda for ED. Each section has an introductory chapter that outlines section contents, four chapters that engage with the section topic, and a response chapter that showcases section themes and points to future opportunities for research and practice.

Concluding thoughts

The research in the handbook illustrates the complexity of sociotechnical energy transitions; theorizes the principles of ED; analyzes on-the-ground energy democracies; illustrates the range of actors, scales, sites, and practices involved in ED; and assists practitioners in engaging in efforts to democratically participate in decisions regarding energy. This strategic integration of transdisciplinary and international research enhances and encourages future opportunities for ED research. The broader significance lies in its contribution to a better understanding of the composition of energy transitions and democratic world making.

This book is a composition in multiple senses of the word. Its subject matter requires a composition of energy and democracy into energy democracies. It is a composition of multiple disciplines. It is a composition of scales, locations, concepts, practices, and more. Rather than collapse differences, these compositions focus on the temporary "putting-together" for the purpose of analytic, theoretical, or practical clarity. This handbook is also a composition in Latour's (2010) second sense, as an intervention into an ongoing sociotechnical challenge facing society. The chapters in this book give guidance for policy makers, activists, and others seeking a democratic energy transition. As such, the handbook aligns with engaged and praxis-based research practices (Endres et al., 2008; Sismondo, 2008). We offer this composition with the hope that it spurs more research and action toward a just, equitable, and democratic energy transition that can begin to scale back human contributions to the climate crisis.

References

Angel, J. (2016). *Towards energy democracy: Discussions and outcomes from an international workshop*. www.tni.org/en/publication/towards-energy-democracy

Becker, S., Angel, J., & Naumann, M. (2020). Energy democracy as the right to the city: Urban energy struggles in Berlin and London. *Environment and Planning A: Economy and Space*, *52*(6), 1093–1111.

Bloem, S., Swilling, M., & Koranteng, K. (2021). Taking energy democracy to the streets: Socio-technical learning, institutional dynamism, and integration in South African community energy projects. *Energy Research & Social Science*, *72*(101906), 1–14. https://doi.org/10.1016/j.erss.2020.101906

Burke, K. (1966). *Language as symbolic action: Essays on life, literature, and method*. University of California Press.

Burke, M. J., & Stephens, J. C. (2017). Energy democracy: Goals and policy instruments for sociotechnical transitions. *Energy Research & Social Science*, *33*, 35–48. https://doi.org/10.1016/j.erss.2017.09.024

Cantarero, M. M. V. (2020). Of renewable energy, energy democracy, and sustainable development: A roadmap to accelerate the energy transition in developing countries. *Energy Research & Social Science*, *70*(101716), 1–15. https://doi.org/10.1016/j.erss.2020.101716

Center for Earth, Energy & Democracy, & CJA Energy Democracy Working Group. (2015). *Ten principles for energy democracy*. https://climatejusticealliance.org/workgroup/energy-democracy/

Cha, J. M. (2020). A just transition for whom? Politics, contestation, and social identity in the disruption of coal in the Powder River Basin. *Energy Research & Social Science*, *69*(101657), 1–9. https://doi.org/10.1016/j.erss.2020.101657

Chilvers, J., & Pallett, H. (2018). Energy democracies and publics in the making: A relational agenda for research and practice. *Frontiers in Communication*, *3*, 1–16. https://doi.org/10.3389/fcomm.2018.00014
Christiano, T. (2018, Fall). Democracy. In E. N. Zalta (Ed.), *The Stanford encyclopedia of philosophy*. https://plato.stanford.edu/archives/fall2018/entries/democracy/
Climate Justice Alliance. (2018). *Just transition*. https://climatejusticealliance.org/just-transition/
Climate Justice Alliance. (n.d.). *Energy democracy*. https://climatejusticealliance.org/workgroup/energy-democracy/
Collins, H. M., & Evans, R. (2002). The third wave of science studies: Studies of expertise and experience. *Social Studies of Science*, *32*(2), 235–296. https://doi.org/10.1177/0306312702032002003
Cozen, B., Endres, D., Peterson, T. R., Horton, C., & Barnett, J. T. (2018). Energy communication: Theory and praxis towards a sustainable energy future. *Environmental Communication*, *12*(3), 289–294. https://doi.org/10.1080/17524032.2017.1398176
Curley, A. (2018). A failed green future: Navajo Green Jobs and energy "transition" in the Navajo Nation. *Geoforum*, *88*, 57–65. https://doi.org/10.1016/j.geoforum.2017.11.012
de Onís, C. M. (2018a). Energy colonialism powers the ongoing unnatural disaster in Puerto Rico. *Frontiers in Communication*, *3*, 1–5. https://doi.org/10.3389/fcomm.2018.00002
de Onís, C. M. (2018b). Fueling and delinking from energy coloniality in Puerto Rico. *Journal of Applied Communication Research*, *46*(5), 535–560. https://doi.org/10.1080/00909882.2018.1529418
Depoe, S. P., Delicath, J. W., & Elsenbeer, M. A. (Eds.). (2004). *Communication and public participation in environmental decision making*. SUNY Press.
Einsiedel, E. F., Boyd, A. D., Medlock, J., & Ashworth, P. (2013). Assessing socio-technical mindsets: Public deliberations on carbon capture and storage in the context of energy sources and climate change. *Energy Policy*, *53*, 149–158. https://doi.org/10.1016/j.enpol.2012.10.042
Endres, D. (2009). Science and public participation: An analysis of public scientific argument in the Yucca Mountain controversy. *Environmental Communication: A Journal of Nature and Culture*, *3*(1), 49–75. https://doi.org/10.1080/17524030802704369
Endres, D., Cozen, B., Barnett, J. T., O'Byrne, M., & Peterson, T. R. (2016). Communicating energy in a climate (of) crisis. In E. Cohen (Ed.), *Communication yearbook* (Vol. 40, pp. 419–447). Routledge.
Endres, D., Sprain, L., & Peterson, T. R. (2008). The imperative of praxis-based environmental communication research: Suggestions from the step it up 2007 national research project. *Environmental Communication*, *2*(2), 237–245. https://doi.org/10.1080/17524030802141794
Endres, D., Sprain, L., & Peterson, T. R. (2009). *Social movement to address climate change: Local steps for global action*. Cambria Press.
Fairchild, D., & Weinrub, A. (Eds.). (2017). *Energy democracy: Advancing equity in clean energy solutions*. Island Press.
Feldpausch-Parker, A. M., Endres, D., & Peterson, T. R. (2019). Editorial: A research agenda for energy democracy. *Frontiers in Communication*, *4*, 1–8. https://doi.org/10.3389/fcomm.2019.00053
Feldpausch-Parker, A., Endres, D., & Peterson, T. R. (forthcoming). Energy democracy's relationship to ecology. In M. Nadesan, M. J. Pasqualetti, & J. Keahey (Eds.), *Democratizing energy*. Elsevier.
Feldpausch-Parker, A. M., Peterson, T. R., Stephens, J. C., & Wilson, E. J. (2018). Smart grid electricity system planning and climate disruptions: A review of climate and energy discourse post-Superstorm Sandy. *Renewable and Sustainable Energy Reviews*, *82*, 1961–1968. https://doi.org/10.1016/j.rser.2017.06.015
Finley-Brook, M., Williams, T. L., Caron-Sheppard, J. A., & Jaromin, M. K. (2018). Critical energy justice in US natural gas infrastructuring. *Energy Research & Social Science*, *41*, 176–190. https://doi.org/10.1016/j.erss.2018.04.019
Fischer, F. (2000). *Citizens, experts, and the environment: The politics of local knowledge*. Duke University Press.
Geels, F. W., & Schot, J. (2007). Typology of sociotechnical transition pathways. *Research Policy*, *36*(3), 399–417. https://doi.org/10.1016/j.respol.2007.01.003
Giancatarino, A. (2012). *Energy democracy: Supporting community innovation*. www.Centerforsocialinclusion.Org/New-Report-Energy-Democracy-Supporting-Community-Innovation
Gunderson, R., & Yun, S. J. (2021). Building energy democracy to mend ecological and epistemic rifts: An environmental sociological examination of Seoul's one less nuclear power plant initiative. *Energy Research & Social Science*, *72*(101884), 1–9. https://doi.org/10.1016/j.erss.2020.101884
Hamlett, P. W. (2003). Technology theory and deliberative democracy. *Science, Technology & Human Values*, *28*(1), 112–140. https://doi.org/10.1177/0162243902238498

Healy, N., & Barry, J. (2017). Politicizing energy justice and energy system transitions: Fossil fuel divestment and a 'just transition'. *Energy Policy, 108*, 451–459. https://doi.org/10.1016/j.enpol.2017.06.014
Heffron, R. J., McCauley, D., & Sovacool, B. K. (2015). Resolving society's energy trilemma through the energy justice metric. *Energy Policy, 87*, 168–176. https://doi.org/10.1016/j.enpol.2015.08.033
Hendry, J. (2004). Decide, announce, defend: Turning the NEPA process into an advocacy tool rather than a decision-making tool. In S. P. Depoe, J. W. Delicath, & M. F. A. Elsenbeer (Eds.), *Communication and public participation in environmental decision making* (pp. 99–112). SUNY Press.
Hess, D. J. (2018). Social movements and energy democracy: Types and processes of mobilization. *Frontiers in Energy Research, 6*, 1–4. https://doi.org/10.3389/fenrg.2018.00135
Hess, D. J. (2019). Coalitions, framing, and the politics of energy transitions: Local democracy and community choice in California. *Energy Research & Social Science, 50*, 38–50. https://doi.org/10.1016/j.erss.2018.11.013
Hunt, K. P., Walker, G. B., & Depoe, S. P. (Eds.). (2019). *Breaking boundaries: Innovative practices in environmental communication and public participation*. SUNY Press.
Jasanoff, S. (2005). *Designs on mature: Science and democracy in Europe and the United States*. Princeton University Press.
Jones, C. F. (2013). Building more just energy infrastructure: Lessons from the past. *Science as Culture, 22*(2), 157–163. https://doi.org/10.1080/09505431.2013.786991
Kimmerer, R. W. (2013). *Braiding sweetgrass: Indigenous wisdom, scientific knowledge and the teachings of plants*. Milkweed Editions.
Kinsella, W. J. (2004). Public expertise: A foundation for citizen participation in energy and environmental decisions. In S. P. Depoe, J. W. Delicath, & M. F. A. Elsenbeer (Eds.), *Communication and public participation in environmental decision making* (pp. 83–98). SUNY Press.
Kunze, C., & Becker, S. (2014). *Energy democracy in Europe: A survey and outlook*. Rosa Luxemburg Foundation.
Laird, F. N. (2013). Against transitions? Uncovering conflicts in changing energy systems. *Science as Culture, 22*(2), 149–156. https://doi.org/10.1080/09505431.2013.786992
Latour, B. (2004). *Politics of nature: How to bring the sciences into democracy*. Harvard University Press.
Latour, B. (2010). An attempt at a "compositionist manifesto". *New Literary History, 41*(3), 471–490. http://muse.jhu.edu/journals/new_literary_history/v041/41.3.latour.html
Lawhon, M., & Murphy, J. T. (2012). Socio-technical regimes and sustainability transitions: Insights from political ecology. *Progress in Human Geography, 36*(3), 354–378. https://doi.org/10.1177/0309132511427960
Loorbach, D. (2010). Transition management for sustainable development: A prescriptive, complexity-based governance framework. *Governance, 23*(1), 161–183. https://doi.org/10.1111/j.1468-0491.2009.01471.x
MacEwen, M., & Evensen, D. (2021). Mind the gap: Accounting for equitable participation and energy democracy in Kenya. *Energy Research & Social Science, 71*(101843), 1–9. https://doi.org/10.1016/j.erss.2020.101843
McCauley, D., & Heffron, R. (2018). Just transition: Integrating climate, energy and environmental justice. *Energy Policy, 119*, 1–7. https://doi.org/10.1016/j.enpol.2018.04.014
McGee, M. C. (1980). The "ideograph": A link between rhetoric and ideology. *Quarterly Journal of Speech, 66*(1), 1–16.
Mercer, D. (1998). Science, technology and democracy on the STS agenda. *Prometheus, 16*(1), 81–91. https://doi.org/10.1080/08109029808629255
Morris, C., & Jungjohann, A. (2016). *Energy democracy: Germany's energiewende to renewables*. Palgrave Macmillan.
Mouffe, C. (2000). *The democratic paradox*. Verso.
Mouffe, C. (2005). *On the political*. Psychology Press.
Na'puti, T. R., Pezzullo, P. C., Sprain, L., & Reinig, L. (2018). Engaging publics through climate math: Lessons from Boulder's 2016 climate action plan. *Journal of Argumentation in Context*, 7(3), 316–346. https://doi.org/10.1075/jaic.18020.nap
Newell, P., & Mulvaney, D. (2013). The political economy of the 'just transition'. *The Geographical Journal, 179*(2), 132–140. https://doi.org/10.1111/geoj.12008
Pezzullo, P. C. (2007). *Toxic tourism: Rhetorics of pollution, travel, and environmental justice*. University of Alabama Press.

Pezzullo, P. C., & de Onís, C. M. (2018). Rethinking rhetorical field methods on a precarious planet. *Communication Monographs, 85*(1), 103–122. https://doi.org/10.1080/03637751.2017.1336780
Reinig, L., & Sprain, L. (2016). Cultural discourses of public engagement: Insights for energy system transformation. In J. Goodwin (Ed.), *Confronting the challenges of public participation in environmental, planning, and health decision-making* (pp. 167–188). Iowa State University. https://doi.org/10.31274/sciencecommunication-180809-13
Schlosberg, D., & Collins, L. B. (2014). From environmental to climate justice: Climate change and the discourse of environmental justice. *Wiley Interdisciplinary Reviews: Climate Change, 5*(3), 359–374.
Sclove, R. (1995). *Democracy and technology*. Guilford Press.
Sharpe, B., Hodgson, A., Leicester, G., Lyon, A., & Fazey, I. (2016). Three horizons: A pathways practice for transformation. *Ecology and Society, 21*(2). https://doi.org/10.5751/ES-08388-210247
Sismondo, S. (2008). Science and technology studies and an engaged program. In E. J. Hackett, O. Amsterdamska, M. Lynch, & J. Wajcman (Eds.), *The handbook of science and technology studies* (pp. 13–31). MIT Press.
Sorman, A. H., Turhan, E., & Rosas-Casals, M. (2020). Democratizing energy, energizing democracy: Central dimensions surfacing in the debate. *Frontiers in Energy Research, 8*, 1–7. https://doi.org/10.3389/fenrg.2020.499888
Sovacool, B. K., & Brossmann, B. (2013). Fantastic futures and three American energy transitions. *Science as Culture, 22*(2), 204–212. https://doi.org/10.1080/09505431.2013.786999
Sovacool, B. K., Burke, M., Baker, L., Kotikalapudi, C. K., & Wlokas, H. (2017). New frontiers and conceptual frameworks for energy justice. *Energy Policy, 105*, 677–691. https://doi.org/10.1016/j.enpol.2017.03.005
Sovacool, B. K., & Dworkin, M. H. (2014). *Global energy justice*. Cambridge University Press.
Stephens, J. C., Burke, M. J., Gibian, B., Jordi, E., & Watts, R. (2018). Operationalizing energy democracy: Challenges and opportunities in Vermont's renewable energy transformation. *Frontiers in Communication, 3*, 1–12. https://doi.org/10.3389/fcomm.2018.00043
Stephens, J. C., & Zwaan, B. V. D. (2005). The case for carbon capture and storage. *Issues in Science and Technology*. https://issues.org/stephens/
Sweeney, S. (2014). Working toward energy democracy. In Worldwatch Institute (Ed.), *State of the world 2014: Governing for sustainability* (pp. 215–227). Island Press.
Swilling, M., & Annecke, E. (2012). *Just transitions: Explorations of sustainability in an unfair world*. UCT Press.
Szeman, I., & Boyer, D. (Eds.). (2017). *Energy humanities: An anthology*. Johns Hopkins University Press.
Szulecki, K. (2018). Conceptualizing energy democracy. *Environmental Politics, 27*(1), 21–41. https://doi.org/10.1080/09644016.2017.1387294
Szulecki, K., & Overland, I. (2020). Energy democracy as a process, an outcome and a goal: A conceptual review. *Energy Research & Social Science, 69*(101768), 1–14. https://doi.org/10.1016/j.erss.2020.101768
van Veelen, B., & van der Horst, D. (2018). What is energy democracy? Connecting social science energy research and political theory. *Energy Research & Social Science, 46*, 19–28. https://doi.org/10.1016/j.erss.2018.06.010
Wenzel, J. (2016). Taking stock of energy humanities. *Reviews in Cultural Theory, 6*(3), 30–34. http://reviewsinculture.com/archive/volume-6-issue-3/
Whyte, K. (2016). Indigenous experience, environmental justice and settler colonialism. In B. E. Bannon (Ed.), *Nature and experience: Phenomenology and the environment* (pp. 157–174). Rowman & Littlefield.
Williams, L., & Sovacool, B. K. (2020). Energy democracy, dissent and discourse in the party politics of shale gas in the United Kingdom. *Environmental Politics, 29*(7), 1239–1263. https://doi.org/10.1080/09644016.2020.1740555
Wilson, E. J., & Stephens, J. C. (2009). Wind deployment in the United States: States, resources, policy, and discourse. *Environmental Science & Technology, 43*(24), 9063–9070. https://doi.org/10.1021/es900802s
Winner, L. (1986). *The whale and the reactor: A search for limits in an age of high technology*. University of Chicago Press.

PART I

Scalar dimensions of power and governance in energy democracy

2

SCALAR DIMENSIONS OF POWER AND GOVERNANCE IN ENERGY DEMOCRACY

Introduction

Andrea M. Feldpausch-Parker

The first part of this handbook is focused on different scales of governance and how power dynamics play out within and between levels. Authors in this section explore these different levels and the players involved in interpretations of democratic practice, which Chilvers and Pallett (2018) argue is better described as energy democracies because of the "ever-multiplying diversity of ways in which citizens participate in energy systems" (p. 1). These multiple energy democracies are the result of different practices across geographies, technologies, societies, and cultures, leading scholars like Szulecki and Overland (2020) to propose "ideal-typical understandings of energy democracy," including democratic process, decarbonization, and the presence of a normative goal (p. 2).

Though not often explicitly discussed, dimensions of scale have been a consistent focus in energy democracy literature. Energy democracy as a social movement and as an area of scholarship has many examples of local- to subnational-level (e.g., state or province) case studies. The justification for this often has to do with not only the location of the energy resource or other technological considerations but also social and cultural considerations such as grassroots efforts to implement solar or a state attempting to meet renewable energy goals. It also has to do with receptivity to such efforts to be more inclusive of energy stakeholders in decision making. This more localized focus is therefore the product of both technical and social complexities. For example, Stephens et al. (2018) state in their analysis of energy transitions to renewables in the state of Vermont in the northeastern United States that "to facilitate this transition, *progressive cross-sectoral coalitions of Vermonters* are working toward various social, political, and institutional innovations that can be viewed as examples of operationalizing energy democracy goals" (p. 2; emphasis added). They later note that Vermont's "unique history and structure of local energy governance" played a role in providing a space for democratic practice in energy systems (p. 7).

As previously noted, scaling up energy democracy practices often involves a corresponding scaling up in complexity. This complexity, while often difficult to navigate and to ensure inclusivity in stakeholder participation, also has some benefits due to the polycentric nature of energy governance. Carlisle and Gruby (2019), who define polycentric governance as "a complex form of governance with multiple centers of decision-making, each of which operates with some degree of autonomy," argue that there are actually a handful of advantages to these complex

DOI: 10.4324/9780429402302-3

systems (p. 928). These advantages include enhanced adaptive capacity, good institutional fit, and risk mitigation/redundancy. These are due to the polycentric governance characteristics of (1) semiautonomous, overlapping decision-making institutions and (2) the ability to act in concert with other institutions or with understandings of other institutions through "processes of cooperation, competition, conflict, and conflict resolution" (p. 946). This also allows for greater potential for public participation opportunities because of the increased number of institutions involved in decision making, many of whom mandate public engagement in some form. This, however, is not a guarantee that such decision authorities will embrace the tenets of energy democracy, such as inclusivity and equality, when engaging the public. Späth and Rohracher (2010) provide a good example of scaled-up energy governance in their description of energy regions in Austria. These energy regions seek to implement shared "guiding visions" across regional levels of governance. They go on to state:

> Development and social propagation of such visions are inherently political and contested processes involving much strategizing and anticipation of conflict. We describe particular discursive strategies applied in niches—such as the combination and translation of sentiments into localised visions and demonstrations of feasibility. These strategies can be understood as systematic attempts to support discursive shifts at regime level by means of local activities, and aim to modify rather durable power structures.
>
> *(p. 449)*

This description brings together Szulecki and Overland's (2020) understanding of energy democracy through descriptions of democratic practice and normative goals with Carlisle and Gruby's (2019) model for the commons involving polycentric governance.

The following chapters of this section will take us from international energy governance down to the regional level. Though local to subnational examples of energy democracy are not included in this section, we have plenty of case studies at this level represented throughout the book (see Chapter 7 for a reference table of chapters organized by scale).

Starting at the international level in Chapter 3, Carvalho and Horta address both opportunities and challenges of democratic energy governance in the European Union. They focus specifically on a European project called the European Energy Union and its attempt to create common climate and energy policy, noting its struggles with inclusive participation, even after decades of promotion, but also opportunities for transformation. This chapter provides the context for global energy governance while examining the confluence of energy regulation and climate action.

Chapter 4 continues the international-level energy governance focus with national comparisons between European (Denmark and Germany specifically) and US electric cooperatives. Lenhart, Chan, Grimley, and Wilson provide a comprehensive look at institutional relationships and regulatory frameworks in both parts of the globe. They outline the different contexts in which energy systems developed in both places, noting some similarities but also some critical differences that influence opportunities and constraints for energy democracy in practice. In this exploration of coevolution of electric cooperatives, the chapter reflects on the four interdependencies of community identity, collective decision making, interdependent infrastructure, and market and regulatory frameworks.

Endres and Johnson address the challenges of Indigenous Native American participation in energy governance in the United States in this subnational- to national-level study described in Chapter 5. They examine issues of Indigenous governance, sovereignty, and settler colonialism and how these factors influence democratic practice in energy decision making. Endres and

Johnson explore the rhetorical dimensions of the Honor the Earth campaign, an Indigenous environmental justice organization, in their fight against the Dakota Access Pipeline and other energy issues. This case study addresses not only an attempt to influence energy policy but also an attempt at Native American self-determination and cultural sovereignty.

In Chapter 6, the final chapter of this section, Goyal and Howlett present a cross-scalar analysis of energy democracy and energy transitions, using a suite of international examples. In this chapter, they propose the use of Kingdon's (1995) multiple streams framework to explore the important questions of what, how, and where energy systems are being democratized. This framework addresses different actors (technical, epistemic, instrument, and advocacy) and their participation in the various "streams" of technology, problem, policy, and politics. These streams serve as the activities or processes involved in energy transitions. Goyal and Howlett make the argument that this framework is advantageous in its ability to discern relationships between structures, agency, and outcomes across institutions and geographies, hence its local to global utility.

References

Carlisle, K., & Gruby, R. L. (2019). Polycentric systems of governance: A theoretical model for the commons. *Policy Studies Journal*, *47*(4), 927–952. https://doi.org/10.1111/psj.12212

Chilvers, J., & Pallett, H. (2018). Energy democracies and publics in the making: A relational agenda for research and practice. *Frontiers in Communication*, *3*, 14. https://doi.org/10.3389/fcomm.2018.00014

Kingdon, J. W. (1995). *Agendas, alternatives, and public policies* (2nd ed.). HarperCollins College Publishers.

Späth, P., & Rohracher, H. (2010). 'Energy regions': The transformative power of regional discourses on socio-technical futures. *Research Policy*, *39*(4), 449–458. https://doi.org/10.1016/j.respol.2010.01.017

Stephens, J. C., Burke, M. J., Gibian, B., Jordi, E., & Watts, R. (2018). Operationalizing energy democracy: Challenges and opportunities in Vermont's renewable energy transformation. *Frontiers in Communication*, *4*. https://doi.org/10.3389/fcomm.2018.00043

Szulecki, K., & Overland, I. (2020). Energy democracy as a process, an outcome and a goal: A conceptual review. *Energy Research & Social Science*, *69*, 101768. https://doi.org/10.1016/j.erss.2020.101768

3

INTERNATIONAL ENERGY GOVERNANCE

Opportunities and challenges for democratic politics

Anabela Carvalho and Ana Horta

Introduction

Energy is critical to all societies and in the last few decades, energy-related issues have become more complex and more interconnected. Worldwide, policy makers and citizens are faced with an "energy trilemma"—addressing climate change, energy access, and energy security (e.g., Falkner, 2014). Decarbonization and decentralization are important trends in many regions of the world often in tandem with yet another "D" trend—digitalization. This intricate set of challenges is linked to the global, regional, national, and local scales and involves a variety of agents, such as governments, inter-governmental bodies, local authorities, business organizations, and nongovernmental organizations. Given the stakes and an increased awareness of these challenges across the world, energy governance has become hugely important to understand.

Energy governance refers to the actors and processes that shape decisions in the energy field and, crucially, to their relations. It is a key issue for the next few decades and a rapidly emerging research domain. Whereas in 2009 Florini and Sovacool argued that energy governance had "received almost no scholarly attention, despite the extraordinary importance of energy in current international affairs" (p. 5240), the subsequent decade saw a number of relevant studies and publications, several of which connect with "sustainability transitions" and "energy transitions" (e.g., Araújo, 2014; Karlsson-Vinkhuyzen et al., 2012; Van de Graaf & Colgan, 2016). Nonetheless, many aspects of energy governance remain under-researched. The social sciences, including critical and interpretive approaches, can offer important contributions to understanding and making decisions related to energy politics and governance.

This chapter focuses on energy governance at the international level through the lens of energy democracy by asking how current and novel forms of governance embed issues pertaining to power, participation and justice (cf. Feldpausch-Parker et al., 2019). More specifically, it analyzes the governance of the European Energy Union, a European project to create a common climate and energy policy, and identifies opportunities and limitations for energy democracy. Inclusive participation in the politics of energy and climate change is a pillar of future social sustainability. However, and in spite of decades of rhetorical promotion, public participation has garnered little attention and investment from states around the world (cf. Carvalho & Gupta,

 DOI: 10.4324/9780429402302-4

2012). We explore possibilities for the European Union serving as a beacon of transformation toward pluralism and democracy in energy governance.

The first section of this chapter looks at the global context for energy governance and major issues that must be addressed. The second section critically reviews extant research and outlines conclusions and gaps. Finally, we turn to the European project of an Energy Union and examine the Regulation on the Governance of the Energy Union and Climate Action (European Parliament and Council of the European Union, 2018).

Critical issues calling for multilateral governance

The growing use of the concept of governance has been linked to processes of neoliberalism and globalization, such as financial deregulation, trade liberalization, and the expansion of global production networks, which since the 1970s have eroded the traditional role of nation-states (Van de Graaf & Colgan, 2016). The liberalization of international oil markets in the aftermath of the oil shocks of the 1970s, driven more by market forces than by national governments (Goldthau & Witte, 2010), put into question processes of international governance in the energy field. This has been further emphasized by challenges such as increasing global energy demand, the security of energy supply due to geopolitical instability in several regions sourcing fossil fuels, climate change, and sustainability in general. There is no question that many energy issues are related to the provision of public goods and the regulation of externalities that cross borders and thus require international governance (Florini & Sovacool, 2009).

In spite of this, while there has been intergovernmental cooperation on climate change, international energy regulation has been integrated only at a rudimentary level (Vogler, 2018). A variety of entities have been identified as active in the international governance of energy. With various profiles and status, they range from international institutions to nonformal groupings and nongovernmental organizations (NGOs). Notably, the most influential international institution in the energy field has been the International Energy Agency, which was established in the wake of the 1973 oil crisis and is composed by members of the Organization for Economic Co-operation and Development (OECD). Also operating at the intergovernmental level are "fuel-driven institutions," such as the powerful Organization of the Petroleum Exporting Countries (OPEC), and "clubs," such as the Major Economies Forum on Energy and Climate (MEF), launched in 2009 (Bazilian et al., 2014). The Renewable Energy and Energy Efficiency Partnership (REEEP) is an international NGO that was set up by a group of regulators, banks, corporations, and NGOs in 2004. Country groups such as the G7 (G8 prior to the exclusion of Russia) and global and regional development banks are types of international structures that have also often taken stands on and/or financed energy policies.

Although several international organizations and various partnerships and initiatives are shaping the field of energy, globally there has been a lack of coordination among governments regarding energy issues, and not even the International Energy Agency, which has received the most attention, can be considered close to what would be a World Energy Organization (Lesage, van de Graaf & Westphal, 2010; van de Graaf & Colgan, 2016). As put by Karlsson-Vinkhuyzen et al. (2012), the pursuit of global governance of energy has been almost a taboo in international relations. This seems to be due to the facts that energy is crucial for socioeconomic development, on the one hand, and that the most widely used energy source in the world—oil—has its main reserves located in geopolitically sensitive areas, on the other. Therefore, energy became a critical military issue and a matter of national security. Thus governments have generally been reluctant to share power over this domain at the international level.

In the current socioenvironmental context, several critical issues call for multilateral governance. In the last few decades, it has become evident that climate change mitigation requires moving away from fossil fuels such as coal, oil, and natural gas as the main sources of energy. This would require major increases in the production capacity and uptake of renewable energy, as well as in energy efficiency. This "energy transition," as it has been widely termed, ought to occur very quickly, particularly in countries with the highest emissions of greenhouse gases. In order to fulfill the goals of the Paris Agreement (United Nations, 2015) of staying below a 2°C increase in global average temperature, global emissions would have to be cut by 80–90% (in relation to the baseline year of 1990) by 2050 and about half of that decrease ought to happen by 2030.

The shift to renewable energy sources poses other sustainability challenges as technologies for renewable energy, such as large hydropower dams and lithium batteries, can themselves have significant environmental impacts. Whereas the political goal is hegemonically framed as "decarbonization," there are other aspects to consider, namely the different technological pathways available and their impacts at various levels (besides social, economic, and other possible changes to achieve greenhouse gas emissions reductions).

Internationally, there are major inequities in historical and present responsibilities for greenhouse gas generation, a matter that must be accounted for in planning for future policies. Large parts of the world population face severe difficulties in accessing energy. There are problems of physical availability but also of affordability and quality (i.e., reliability, and environmental and health impacts of energy sources). Energy poverty can be defined as the "lack of access to affordable and high-quality energy services" (Bazilian et al., 2014, p. 217) and has impacts for lighting, heating, refrigeration, cooking, and transportation, inter alia. It is a grave problem in Sub-Saharan Africa, parts of Asia and Latin America, and it is also felt among the poorer in the "West." Sustainable Energy for All (2020) is an international initiative launched by then UN Secretary General Ban Ki-Moon aimed at achieving universal energy access, enhancing energy efficiency, and increasing the use of renewable energy.

Justice issues in energy governance continue to get insufficient attention from both decision makers and academics. For instance, as Bazilian et al. (2014) point out, the ways in which governance policies and practices at various scales address energy poverty are clearly under-researched. The few studies that exist show that energy poverty and the energy poor continue to be marginalized in global energy policies, including under the UN. Munro et al. (2016) maintain that framings of Sustainable Development Goal 7 take a technomanagerial approach to global energy distribution that does not account properly for political and economic power issues. Pointing to the dominance of a "modernist" epistemology, they argue that "for justice in energy policy to be realized holistically, there is a need to question *how* our knowledge of energy 'problems' have emerged to avoid epistemologically autarchic policy positions" (p. 635) and call for the combination of different knowledges and experiences.

That brings up the question of participation. A widely used concept in political discourse and in academic research, "participation" expresses concerns with the limitations of representative democracy and with the ability of policy makers to, alone, adequately address and reflect the expectations and views of different sectors of society. Opportunities for public participation can increase pluralism in debates and decision-making processes and hence further the democratic quality of political institutions. Research on global environmental governance has shown that the inclusion of civic society increases popular legitimacy (Bernauer & Gampfer, 2013). However, processes of public participation can also function as a mere appearance of inclusivity with legitimization gains for power holders; in other words, they can work as a simulation of democracy (Blühdorn, 2013) in the field of energy as in any others.

To sum up, guaranteeing a continuous supply of energy is a fundamental goal for national governments, but a number of issues, such as environmental problems and political instability in some parts of the world, can threaten energy security. In the last few decades, the growth of energy demand and associated infrastructure, the imperatives of environmental and social sustainability, changes in political systems and processes, and fast developing technological issues have posed new challenges and call for democratic and reflexive modes of governance. However, national governments are not capable of addressing these critical issues alone (e.g., Dubash & Florini, 2011). Cooperation and coordination are required at—and across—various scales and certainly beyond national borders.

Extant research in international energy governance

Following changes in governing processes in recent decades, the concept of governance has been increasingly used in a variety of disciplines. However, as it stems from diverse focuses and literatures, it has been given different meanings ranging from the "good governance" notion in the field of economic development to broader forms of organization without the involvement of governments. Across different approaches, governance tends to be viewed as pluricentric; organized in networks of interdependent actors; based on processes of negotiation (rather than traditional power structures); vulnerable to risks and uncertainties regarding cooperation; and often normative (van Kersbergen & van Waarden, 2004). A common concern behind uses of the concept is with how shifts in the role of the state are related to interactions between public and private sectors and how this impacts the delivery of public services. Despite its fluid character, the concept helps in understanding changes in forms of power and the roles of multiple actors in the construction of public policies and democracy.

One of the shifts in governance often discussed in research is an upward shift, from nation-states to international or intergovernmental institutions. Manifestly, the development of economic globalization has been accompanied by a proliferation of international organizations, networks, and agencies that have influenced the regulation of global activity in recent decades. The emergence of those institutions has contributed to erosion of the traditional autonomy of states. This is the case of the European Union, a new form of multilevel governance that has influenced transnational forms of governance in other parts of the world (Prado, 2007).

In order to understand contemporary global change, international relations scholars began using the concept of global governance in the 1990s (Hewson & Sinclair, 1999). That concept would also enter the literature on energy issues, and while early writings on global energy governance were produced by energy security specialists and global governance specialists, current challenges related to energy transitions have brought a renewed and wider interest to the topic. However, research on international energy governance has been mostly descriptive and under-theorized (Van de Graaf & Colgan, 2016). This is particularly the case for matters of participation and justice, which are critical to democratic politics.

Presenting the notion of global energy governance as "international collective action efforts undertaken to manage and distribute energy resources and provide energy services," Florini and Sovacool (2009, p. 5239) have argued that international energy markets need appropriate governance at both national and international levels in order to face global challenges in the energy field, such as the regulation of energy prices, investment in future energy supply or addressing environmental impacts (including climate change), as put forth in the previous section. Global energy governance mechanisms would help develop common rules and coordinate efforts within an overarching organization with inclusive membership, instead of the continued adoption of piecemeal and often contradictory measures by a fragmented set of players. However, the institutional architecture that has structured energy markets and the rules that have been in place for global energy governance have been considered insufficient and inadequate by some analysts (e.g., Goldthau & Witte, 2009, 2010). Scholars have also questioned the focus of global energy policy debates on energy security, since by emphasizing the issues related to access to energy, states tend to narrow current energy challenges to a national security perspective (Cherp et al., 2011; Goldthau & Witte, 2009, 2010; Lesage, van de Graaf & Westphal, 2010).

In the analysis of Dubash and Florini (2011), four objectives require global governance: ensuring energy supply security; fighting energy poverty; seeking to internalize the environmental externalities of energy, especially climate change; and seeking domestic good governance and fighting corruption related to energy. After examining the energy challenges and the major players addressing global energy governance, in particular the rising multipolarity of the global energy system, with growing economies such as China and India increasing energy demand and others, such as Russia and Brazil, emerging as energy exporters, Lesage, van de Graaf & Westphal (2010)) argued that it is necessary to create a global sustainable energy regime, one that addresses supply security, sustainability, and economic efficiency. In their view, such regime would allow for the integration of these new key players in governance mechanisms (from which they are excluded) that would provide a site for dialogue and cooperation at the global energy policy level. An example of how global governance could help address current sustainability challenges is the encouragement of the adoption of progressive energy policies without fear of harming the economy by losing national competitiveness. The European Union Energy Union Strategy, launched in 2015, is a pioneering project aimed at managing conjoint issues at a supranational level. It has five "mutually-reinforcing and closely interrelated dimensions": "energy security, solidarity and trust; a fully integrated European energy market; energy efficiency contributing to moderation of demand; decarbonising the economy; and Research, Innovation and Competitiveness" (European Commission, 2015).

As physical, political, and technological realities evolve in the energy field, new questions emerge for research. Dominant strands of research on international energy governance have been locked in a view of the power dynamics over global energy issues that is focused on leadership, while the development of globalization and the energy system have accentuated the relevance of networks and flows, thus calling for approaches that can account for higher levels of complexity with regard to the exercise of power. Critically analyzing the then emergent literature on global energy governance, Cherp et al. (2011) found it reductionist regarding its interpretations and focus of analysis on "who" governs (or should govern) energy. Alternatively, they claimed that more fundamental questions would be "what" and "how" it should be governed in energy and proposed a conceptual framework that departs from the usual approach that analyzes governance institutions and mechanisms in isolation from each other. Assuming that current energy challenges are highly interlinked and energy governance systems are complex, they claimed that no single institution or regime can achieve global energy governance, instead recommending a polycentric governance system. With a conceptual framework based on the notions of systems, complexity, and transitions, they depicted the global energy governance landscape as "dynamic arenas coevolving with energy systems and challenges" and argued that "to address energy challenges effectively, these arenas need to become more strongly interlinked and eventually converge to focus on commonly shared global energy goals," as most of those challenges would lead to systemic changes in how energy is produced, transformed, and consumed (Cherp et al., 2011, p. 77).

With the complexity inherent to energy transitions, many changes are unpredictable. One way for governance mechanisms to cope with uncertainty and complexity in energy transitions would be to use softer forms of coordination, such as flows of knowledge and information, as they facilitate flexibility and adaptation to evolving conditions (Cherp et al., 2011). An example of such soft modes of governance might be what is being implemented in the European Union through setting common objectives, pooling knowledge, monitoring, and accountability, in tune with the "soft power" resulting from its multilevel decision-making processes, as analyzed in the following section of this chapter. Such an approach calls into question the interconnections among power, participation, and knowledge.

The role of knowledge in global energy governance is starting to be investigated in an emergent strand of research. As argued by Roehrkasten (2018), global governance literature has missed the fact that since groups of actors involved in global governance are heterogeneous and have very diverse backgrounds and views, an important way of exercising power is through the creation and structuring of knowledge and social understandings, since this influences which ideas prevail. Indeed, from this point of view, for international cooperation to be possible, a minimum of shared understandings is necessary and must therefore be pursued by governance actors. This is especially relevant regarding contested ideas. As in the case of renewable energies, ideas may be subject to political struggles and to different understandings of problems and solutions, thus complicating transnational cooperation (Roehrkasten, 2014). Nonetheless, energy democracy can only be materialized if there are spaces and conditions that allow for those struggles and for differences in idea(l)s to be made explicit.

Epistemic approaches to global governance often build on international relations research about knowledge (see Allan, 2018), especially on Michel Foucault's broad view of political power as an imbrication of power relations and knowledge. Along these lines, Adler and Bernstein (2005) have argued that power involves the productive capacity of defining the global order of things and of developing formal and informal institutions that can fixate meanings necessary for global governance. In their perspective, "research on global governance must begin with the background knowledge that people share and selectively attach to material reality" (Adler & Bernstein, 2005, p. 295). Background knowledge in this sense refers to an encompassing concept of episteme: "intersubjective knowledge that adopts the form of human dispositions and practices" (Adler & Bernstein, 2005, p. 296), used by human beings to make sense of the world. Democratic forms of energy governance thus call for a politics and scholarship that give careful consideration to diversity in knowledge and values.

The power–knowledge nexus is at the core of governance procedures and structures from their inception. Voß and Freeman (2016) have contended that it is critical to understand the production and mobilization of ways of knowing about governance. From their point of view, the interactions and practices by which knowledge is formed shape perceptions of "what is rational and effective or unreasonable and futile for governments to do," thus providing "the basic framework of political reality, an ontology within which political action can be rationalized and performed" (Voß & Freeman, 2016, p. 5). Based on framing theory, Eriksson and Reischl (2019) have analyzed how technocratic organizations such as the Intergovernmental Panel on Climate Change (IPCC) and the International Energy Agency (IEA) get authority and legitimacy from scientific discourse to influence climate and energy policy agendas at the global level. They exert "epistemic power," a highly relevant form of power since "it is much easier for policy actors to change the way they define problems than it is to change interests and institutions" (Eriksson & Reischl, 2019, p. 68). Thus, even if an organization like the IEA is still focused on the world's energy demand, by including climate change in its agenda, it provides an opportunity for a change toward addressing environmental concerns.

Voß and Freeman (2016) claim that it is in transnational governance that epistemic authority is most advanced, especially through the use of monitoring mechanisms, benchmarks, or guidelines. The European Union has been building a new collective political order based precisely on a "kind of knowledge about governance" whereby "policy monitoring and reporting mechanisms work to establish shared models and metrics for representing and comparatively evaluating governing activities in different countries" (Voß & Freeman, 2016, p. 5). This innovative model, which has also been named experimentalist governance or the "open method of coordination," thus consists in achieving a political order by steering member states through comparisons and knowledge instead of "hard law" (p. 5). This framework also shapes possibilities for engagement or resistance by other actors. As power shifts from national governments to other forms of governance, rights and opportunities for the production, management, and regulation of knowledge become even more relevant. Hence it is important to ask specific questions such as "what forms of participation are being used in energy decisions, are extant forms of participation sufficient, and are local communities and relevant stakeholders (both human and non-human) involved in decision-making" (Feldpausch-Parker et al., 2019, p. 4). These are crucial matters at various scales, including the supranational one involved in the European Energy Union.

The case of Europe's energy union

For a number of years, the European Union has been portraying itself as a front-runner in international climate politics and has developed ambitious plans with regard to energy and climate policies for the next decades. The centerpiece of this project is the Energy Union (European Commission, 2015), an important political innovation in its holistic approach, in the integration of energy and climate goals, and in breaking the silos that have characterized past approaches to energy issues. However, it poses important challenges in terms of coordination among Member States and with other scales of governance.

At the end of 2015, the European Parliament called for the governance mechanism of the Energy Union to be "ambitious, reliable, transparent, democratic and fully inclusive of Parliament and to ensure that the 2030 climate and energy targets are achieved" (European Parliament, 2015). Under the United Nations Framework Convention on Climate Change (UNFCCC) and the Paris Agreement, Europe has committed to a 40% reduction in greenhouse gas emissions by 2030 and aims for a 27% increase in renewable energy and a 27% improvement in energy efficiency in the same period. Its goal for 2050 is of an 85–90% reduction in greenhouse gas emissions.

The new governance mechanism for the Energy Union was agreed under "Regulation (EU) 2018/1999 of the European Parliament and of the Council of 11 December 2018 on the Governance of the Energy Union and Climate Action" (European Parliament & Council of the European Union, 2018), henceforth "the Governance Regulation" or simply "the Regulation." The Regulation determines that Member States produce ten-yearly integrated national energy and climate plans (NECPs) and set up a complex reporting and monitoring process, with the Commission issuing recommendations on implementation. It is a mode of soft governance clearly anchored in the exercise of epistemic power, as previously discussed (Cherp et al., 2011).

The European Commission (2015) has stated that its vision of the Energy Union puts "citizens at its core," and in the negotiations over the governance mechanism, the European Parliament demanded that a multilevel dialogue platform be created allowing citizens to submit their views on climate and energy policies at the EU or national level. In this section, we will analyze the Governance Regulation focusing on its democratic potential and shortages, especially concerning public participation.

A morphologic analysis of The Regulation reveals that participation has a significant salience in the document. Up front, art. 1 states that the "governance mechanism" created by the

Regulation "ensures effective opportunities for the public to participate in the preparation" of national plans. Article 3 determines that the integrated national energy and climate plans that Member States must produce shall include "an overview of the process followed for establishing the integrated national energy and climate plan consisting of an executive summary, *a description of the public consultation and involvement of stakeholders and their results*, and of regional cooperation with other Member States in preparing the plan" (our emphasis). The Regulation has one article determining the realization of a "public consultation" (art. 10) and another one requiring a "multilevel climate and energy dialogue" (art. 11). In 77 pages, the words "participation," "participate" (excluding references to member states' participation), and derivatives are employed 18 times. "Consultation(s)" (excluding consultations of member states and other government-level consultations), "consult," and derivatives appear 15 times. "Dialogue" appears nine times. In contrast, there is only one (indirect) reference to democracy, one reference to citizens (although multiple to "the public"), and three references to "civil society." The term "involvement" (of stakeholders/social partners) is used three times. All of this suggests that the European Commission intends to promote inclusivity in the governance of energy and climate and to monitor processes of participation. It is clear that the Commission aims, at the very least, to award visibility to those matters.

A finer, critical-interpretive analysis points to both reproduction of norms (and practices) that are customary in top-down public participation *and* to elements of progressive innovation in provisions for involvement of society. Based on prior studies and our own research experience with public participation exercises, we will first point out continuities in the Governance Regulation that are problematic. First, as in other legal/regulatory instruments on public participation, there is a significant degree of vagueness and lack of clarity in the Regulation. For instance, it stipulates that "when carrying out public consultations . . . Member States should aim to ensure equal participation," but what is meant by "equal participation" is not specified, thus leaving much room to the interpretation of Member States' administrative bodies in charge of those processes. Similarly, ensuring "effective public participation" is mentioned as an important goal, but what this means is unclear. Article 2 offers "definitions" of a long list of practices, such as "quality control" and "regional cooperation"; of agents such as "implementing public authority" and "energy service provider"; of material matters, such as "biomass" and "fossil fuel"; and many other aspects. Contrastingly, it features no definition of anything or anyone related to public participation.

In comparison, the Aarhus Convention speaks of "the public" and "the public concerned," which it defines as:

> the public affected or likely to be affected by, or having an interest in, the environmental decision-making; for the purposes of this definition, non-governmental organizations promoting environmental protection and meeting any requirements under national law shall be deemed to have an interest.

Second, the Regulation does not seem to offer sufficient safeguards for crucial elements and conditions for public participation. Research has shown that knowledge and information are critical to public participation processes and that those issues are often not appropriately addressed. Several factors related to information influence "access" of the public to participation (Senecah, 2004). The means employed to inform the public that a consultation will take place are one of them (Fernandes-Jesus et al., 2019). The Regulation's provisions do not guarantee appropriate diffusion in terms of outlets to be used: the preamble states solely that the public should be "informed by public notices or other appropriate means such as electronic media." Knowledge about and understanding of the matters at stake and possible alternatives are obviously another crucial factor for (non)participation in public consultations. The Regulation states that the public should "be able to access all relevant documents." However, those documents are often very long and technical, and the majority of citizens do not have the skills, time, and/or disposition to read and digest them, which means that they will not be able to argue in ways that are fully grounded on or supported by those documents. This can easily lead to the future dismissal of those pronouncements in public consultation reports invoking those very reasons, as seen in past cases (Carvalho et al., 2016).

Time is also an important factor shaping access to public consultations. Critically, the Regulation is vague about the length of time of information dissemination (an important condition, as previously suggested) and about the duration of public consultations insofar as it refers to "reasonable timeframes allowing sufficient time for the public to be informed, to participate and express its views." Empirical research has shown that (many) citizens feel that the time normally planned for public participation in consultations is not sufficient (Fernandes-Jesus et al., 2019).

As argued, some aspects of the Governance Regulation have the potential to contribute positively and innovatively to democratic policy making. They concern the moments of public participation, the long-term nature of some of the policies under consultation, the requirement of plurality in scenarios that are put out for consultation, and the creation of an instance for multilevel dialogue on climate and energy issues.

Article 10 stipulates that each Member State:

> shall ensure that the public is given early and effective opportunities to participate in the preparation of the draft integrated national energy and climate plan—as regards the plans for the 2021 to 2030 period, in the preparation of the final plan well before its adoption—as well as of the long-term strategies referred to in Article 15.

The key aspect here concerns the stage of the policy-making process when public participation opportunities are to be created: it is said that public participation should happen "early," i.e., in the preparation of the draft plan. Exactly when in the policy-making process public consultations take place can prevent any meaningful influence over such process. Building on long standing recommendations for "upstream" public engagement and participation, references to "early public participation" can be found in legal/regulatory documents at least since the Aarhus Convention.[1] But the truth is that, in practice, public consultations have mostly continued to be treated as a sort of "end of pipe" procedure that is implemented when policy plans are no longer amenable to significant change (e.g., Hendry, 2004). The way the European Union has formulated this requirement could push for change in this respect.

It is also worth highlighting that the EU has determined that the "long-term (i.e. for 2050) strategies" for energy and climate that Member States are required to prepare also ought to undergo a public consultation. Typically, public participation exercises have only been carried out about matters of immediate effect and where there is no scope for deliberation between alternatives.

An express requirement of plurality in the pathways presented to the public is a beneficial addition toward democratizing policy making. In contrast with political discourses implicitly or explicitly suggesting that "there is no alternative," the Regulation refers to a discussion of "different scenarios [or "options" as formulated in the preamble of the Regulation] envisaged for energy and climate policies" (art. 11). Plural knowledges and values could find more room in this mode of governance—at least in debating processes prior to decision making—than is often the case with conventional processes for public participation. Finally, the Governance Regulation calls for each Member State to establish:

> a multilevel climate and energy dialogue [said to be "permanent" in the preamble, our note] pursuant to national rules, in which local authorities, civil society organisations, business community, investors and other relevant stakeholders and the general public are able actively to engage and discuss the different scenarios envisaged for energy and climate policies, including for the long term, and review progress, unless it already has a structure which serves the same purpose.
>
> *(art. 11)*

In theory, this resembles the global Talanoa process that the UNFCCC has been trying to put in place aiming to create a "process of inclusive, participatory and transparent dialogue" (UNFCCC, 2019). Multilevel dialogues building on Talanoa have been trialed in various places around the world. For instance, in May 2018, the City of Quito, Ecuador, hosted a Cities and Regions Talanoa Dialogue that "brought together representatives of local and national government in Ecuador along with key national and international climate stakeholders" (ICLEI, 2018). Hence, a structure and a practice of "permanent" interaction and debate among the social actors listed in the Regulation could be very positive (with "permanent" here to be understood as continuingly available or possible to activate at any time).

In spite of this, there are reasons to suspect that not much will change in terms of the actual involvement of society in most Member States. First, mechanisms and structures for "dialogue" may remain inappropriate or insufficient. The preamble of the Regulation states that the "dialogue may take place by means of any national structure, such as a website, public consultation platform or another interactive communication tool." There is a risk that some Member States will do little in terms of promoting effective and continuous debate via such structures, which in any case may be too "thin" as platforms of dialogue. Until the time of writing, there were records of some good practices in this domain in some countries (LIFE Planup, 2019), but several countries had not developed appropriate means for the kind of interaction envisaged in the Regulation.

Second, there may not be effective practices of inclusiveness on the side of governments. A report produced by Energycities (the European association of cities in energy transition) (Saïller, 2019) has exposed that most Member States did not consult cities/local authorities in the preparation of their Draft NECPs (all 28 Member States submitted their Draft NECPs by the end of 2018). In a clear expression of the epistemic stakes and struggles in energy governance, the report recommended creating specific consultation processes for local authorities and integrating their climate and energy good practices in the national strategies. A number of exemplary cases were described.

A different—but equally relevant—set of questions concern whether social actors are interested in participating in policy-making processes and whether they possess the conditions and/or resources for doing so. Citizens and civil society organizations may have some legal power of expressing their voices but may consider that they do not have the capacity to influence processes. Problems of trust and the perceived (non)relevance of issues for individuals and community organizations are also frequent barriers to participation.

In the face of all this, there is a risk that under the Regulation, public participation exercises continue to function mainly as state-sponsored opportunities for the reinforcement of power, as suggested by multiple studies on agency capture by the corporate sector. In consultations conducted by the Agency for the Cooperation of Energy Regulators (ACER), the European Agency in the field of energy, over 95% of participations were from business (85.9% from regulated business and 10% from other business) (Beyers & Arras, 2019). Cooperatives are a particular form of economic operator with a participatory mode of functioning and have been touted as key actors for energy democracy. In a study of Renewable Energy Cooperatives (RECs) in Germany (the country with the largest number of RECs), Tosun et al. (2019) have concluded that they are "less willing to participate in energy governance at the EU/transnational level than at the national and especially the subnational level" (p. 45). They also found that it was more likely for RECs that were already involved in governance processes to participate in future ones than for those without such experience, leading them to conclude that involvement in energy governance is affected by path dependence. Again, this suggests that there may be more continuities than transformations in public participation in the governance of the Energy Union.

All of this speaks much to the issues of power, participation, and justice that are central to energy democracy. Besides the procedural justice aspects discussed here, the Regulation makes provisions relevant to the analysis of other dimensions of justice in energy governance, especially via requirements for states to monitor and act upon energy poverty issues (see art. 3, for instance), but due to limitations of space those will not be detailed here.

Final considerations

This chapter has looked at the project of the European Energy Union as a critical test case for international energy governance along the values of energy democracy. Issues of power and participation, which, as we have argued, are intricately connected with knowledge/epistemic matters, are at the core of the governance of the Energy Union. We have examined how they have been formulated in the Governance Regulation and may—or may not—be enacted.

The engagement of civil society organizations has demonstrably positive effects in energy transitions (García & Khandke, 2019). If conducted under the principles of transparency, clarity, and openness, civil society engagement and participation can contribute to improved decisions, wider acceptance, and greater implementation. However, as Stirling (2008) has put it, public participation exercises often have opening-up and closing-down effects. Forms of participation initiated from the top (e.g., by official agencies) tend to follow standards that may not be appropriate or effective for the various parts of societies or for the domain or issue at stake.

Although the European Commission has put forth a vision of the Energy Union "where citizens take ownership of the energy transition" (European Commission, 2015, p. 2), the Regulation that sets up its governance is riddled with vagueness, opening the door for Member States to reenact traditional logics and formats of public participation. Moreover, there is already evidence that "old" problems and limitations in this domain are getting reproduced. In Marquardt's (2017) terms, the Regulation offers new resources of regulatory power to different groups and sectors in society (with special highlight to the multilevel dialogue requirement), but there seem to be no changes in the capacity dimension of power, which refers to social actors' ability to effectively use their own power resources, and most likely to structural power.

Although it constitutes a crucial opportunity to cultivate a continuing practice of dialogic policy making across Europe and to deepen the democratic nature of energy governance, the Energy Union Governance Regulation faces challenges repeatedly found at other scales. Drawing on a range of energy and sustainability programs in the United States and the UK, Peters et al. (2014) have argued that a "key challenge lies in local government providing sufficiently compelling reasons to participate, coupled with the development of flexible strategies capable of resonating effectively with the diversity of needs and priorities that characterize modern lifestyles" (p. 374). Across most of the continent, the European Union and its institutions tend to be perceived as distant and removed from citizens' and communities' pressing needs and problems. Motivating civil society to participate in EU-led climate and energy issues will require imaginative forms of communication and engagement.

On a different note, there are optimistic views about this governance project. In terms of participation formats and arrangements, a joint position paper by CEMR, Climate Alliance, Energy Cities, Eurocities, and ICLEI (2018), "representing thousands of local governments throughout the EU" as well as think tank E3G, argues that:

> different approaches already exist, from France's focus on broad consultation; to the Netherlands' formal agreements between different levels of governance; or Sweden's capacity building efforts and can be adapted to different countries' cultures and levels of decentralization. . . . These examples demonstrate that the European Parliament's idea of establishing permanent "dialogue platforms" is not only feasible but—where already existing—it also provides excellent outcomes.
>
> *(p. 5)*

Whether or not these expressions of hope on the democratic potential of the Energy Union will hold will be decisive for its near- and mid-term future governance.

Note

1 Aarhus Convention: "Each Party shall provide for early public participation, when all options are open and effective public participation can take place."

References

Adler, E., & Bernstein, S. (2005). Knowledge in power: The epistemic construction of global governance. In M. Barnett & R. Duvall (Eds.), *Power in global governance* (pp. 294–318). Cambridge University Press. https://doi.org/10.1017/CBO9780511491207

Allan, B. B. (2018). From subjects to objects: Knowledge in international relations theory. *European Journal of International Relations*, *24*(4), 841–864. https://doi.org/10.1177/1354066117741529

Araújo, K. (2014). The emerging field of energy transitions: Progress, challenges, and opportunities. *Energy Research and Social Science*, *1*, 112–121. https://doi.org/10.1016/j.erss.2014.03.002

Bazilian, M., Nakhooda, S., & de Graaf, T. (2014). Energy governance and poverty. *Energy Research and Social Science*, *1*, 217–225. https://doi.org/10.1016/j.erss.2014.03.006

Bernauer, T., & Gampfer, R. (2013). Effects of civil society involvement on popular legitimacy of global environmental governance. *Global Environmental Change*, *23*, 439–449. https://doi.org/10.1016/j.gloenvcha.2013.01.001

Beyers, J., & Arras, S. (2019). Who feeds information to regulators? Stakeholder diversity in European Union regulatory agency consultations. *Journal of Public Policy*, *40*(4), 1–26. https://doi.org/10.1017/S0143814X19000126

Blühdorn, I. (2013). The governance of unsustainability: Ecology and democracy after the post-democratic turn. *Environmental Politics*, *22*(1), 16–36. https://doi.org/10.1080/09644016.2013.755005

Carvalho, A., & Gupta, J. (2012). State commitment to promoting public participation: The UNFCCC and citizens' roles in climate change politics. In A. Carvalho & T. R. Peterson (Eds.), *Climate change politics: Communication and public engagement* (pp. 123–161). Cambria Press.

Carvalho, A., Pinto-Coelho, Z., & Seixas, E. (2016). Listening to the public—enacting power: Citizen access, standing and influence in public participation discourses. *Journal of Environmental Policy and Planning*, *21*(5), 563–576. https://doi.org/10.1080/1523908X.2016.1149772

CEMR, Climate Alliance, E3G, Energycities, Eurocities, & ICLEI. (2018, March 21). *Building on local climate and energy experience in energy union governance.* Joint Paper. https://energy-cities.eu/policy/building-on-local-climate-and-energy-experience-in-energy-union-governance/

Cherp, A., Jewell, J., & Goldthau, A. (2011). Governing global energy: Systems, transitions, complexity. *Global Policy, 2*(1), 75–88. https://doi.org/10.1111/j.1758-5899.2010.00059.x

Dubash, N., & Florini, A. (2011). Mapping global energy governance. *Global Policy, 2*(1), 6–18. https://doi.org/10.1111/j.1758-5899.2011.00119.x

Eriksson, J., & Reischl, G. (2019). Worlds apart, worlds together: Converging and diverging frames in climate and energy governance. *Globalizations, 16*(1), 67–82. https://doi.org/10.1080/14747731.2018.1518864

European Commission. (2015). *Communication from the commission to the European Parliament, the council, the European economic and social committee, the committee of the regions and the European investment bank—a framework strategy for a resilient energy union with a forward-looking climate change policy* (COM/2015/080 final). European Commission. https://eur-lex.europa.eu/legal-content/EN/TXT/?uri=COM%3A2015%3A80%3AFIN

European Parliament. (2015). *Motion for a European parliament resolution on towards a European energy union* (2015/2113(INI). www.europarl.europa.eu/doceo/document/A-8-2015-0341_EN.html

European Parliament & Council of the European Union (2018). Regulation (EU) 2018/1999 of the European Parliament and of the Council of 11 December 2018 on the governance of the Energy Union and climate action. *Official Journal of the European Union L32.* https://eur-lex.europa.eu/legal-content/EN/TXT/PDF/?uri=CELEX:32018R1999&from=EN

Falkner, R. (2014). Global environmental politics and energy: Mapping the research agenda. *Energy Research & Social Science, 1*, 188–197. https://doi.org/10.1016/j.erss.2014.03.008

Feldpausch-Parker, A., Endres, D., & Peterson, T. R. (2019). Editorial: A research agenda for energy democracy. *Frontiers in Communication, 4*, 1–8. https://doi.org/10.3389/fcomm.2019.00053

Fernandes-Jesus, M., Seixas, E. C., & Carvalho, A. (2019). Beyond the hindrances: Experiences in public consultations and the possibility of ethics and relevance in participation. *Comunicação e Sociedade, 36*, 57–80. https://doi.org/10.17231/comsoc.36(2019).2345

Florini, A., & Sovacool, B. K. (2009). Who governs energy? The challenges facing global energy governance. *Energy Policy, 37*, 5239–5248. https://doi.org/10.1016/j.enpol.2009.07.039

García, I., & Khandke, D. (2019). *Cities and civil society as allies for the energy transition in the EU and US: Lessons for successful civil society engagement.* Report No. 19. The German Marshall Fund of the United States. https://energy-cities.eu/wp-content/uploads/2019/09/Energy-Allies_report_July2019.pdf

Goldthau, A., & Witte, J. M. (2009). Back to the future or forward to the past? Strengthening markets and rules for effective global energy governance. *Institutional Affairs, 85*(2), 373–390. https://doi.org/10.1111/j.1468-2346.2009.00798.x

Goldthau, A., & Witte, J. M. (2010). *Global energy governance: The new rules of the game.* Global Public Policy Institute.

Hendry, J. (2004). Decide, announce, defend: Turning the NEPA process into an advocacy tool rather than a decision-making tool. In S. P. Depoe, J. W. Delicath, & M. A. Elsenbeer (Eds.), *Communication and public participation in environmental decision-making* (pp. 99–112). SUNY Press.

Hewson, M., & Sinclair, T. J. (Eds.). (1999). The emergence of global governance theory. In *Approaches to global governance theory* (pp. 3–22). SUNY Press.

ICLEI. (2018). *Quito is driving multilevel climate action through Talanoa dialogue.* http://old.iclei.org/index.php?id=100&tx_ttnews%5Btt_news%5D=7574&cHash=b5d6ec54c10a41475aedda22e1f0cdfa

Karlsson-Vinkhuyzen, S., Jolland, N., & Staudt, L. (2012). Global governance for sustainable energy: The contribution of a global public goods approach. *Ecological Economics, 83*, 11–18. https://doi.org/10.1016/j.ecolecon.2012.08.009

Lesage, D., van de Graaf, T., & Westphal, K. (2010). *Global energy governance in a multipolar world.* Routledge.

LIFE PlanUp. (2019). *Report on good practices in energy and climate governance.* https://energy-cities.eu/publication/life-planup-report-on-good-practices-in-energy-and-climate-governance/

Marquardt, J. (2017). Conceptualizing power in multi-level climate governance. *Journal of Cleaner Production, 154*, 167–175. https://doi.org/10.1016/j.jclepro.2017.03.176

Munro, P., van der Horst, G., & Healy, S. (2016). Energy justice for all? Rethinking sustainable development goal 7 through struggles over traditional energy practices in Sierra Leone. *Energy Policy, 105*, 635–641. https://doi.org/10.1016/j.enpol.2017.01.038

Peters, M., Fudge, S., & Hoffman, S. F. (2014). The persistent challenge of encouraging public participation in the low-carbon transition. *Carbon Management, 4*(4), 373–375. https://doi.org/10.4155/cmt.13.32

Prado, C. (2007). *Global multi-level governance: European and East Asian leadership*. Nations University Press.

Roehrkasten, S. (2014). *Global governance on renewable energy: Contrasting the ideas of the German and the Brazilian Governments*. Springer. https://doi.org/10.1007/978-3-658-10480-1

Roehrkasten, S. (2018). Are we on the cusp of a global renewable energy transition? In D. Davidson & M. Gross (Eds.), *The Oxford handbook of energy and society* (pp. 483–501). Oxford University Press.

Saïller, A. (2019). *National energy and climate plans fail to acknowledge cities leading role in the European energy transition*. http://energy-cities.eu/publication/national-energy-and-climate-plans-fail-to-acknowledge-cities-leading-role-in-the-european-energy-transition/

Senecah, S. L. (2004). The trinity of voice: The role of practical theory in planning and evaluating the effectiveness of environmental participatory processes. In S. P. Depoe, J. W. Delicath, & M. F. Elsenbeer (Eds.), *Communication and public participation in environmental decision making* (pp. 13–33). SUNY Press.

Stirling, A. (2008). 'Opening up' and 'closing down': Power, participation, and pluralism in the social appraisal of technology. *Science, Technology, & Human Values, 33*(2), 262–294. https://doi.org/10.1177/0162243907311265

Sustainable Energy for All. (2020). *Sustainable energy for all: About us*. www.seforall.org/about-us

Tosun, J., Zöckle, L., & Rilling, B. (2019). What drives the participation of renewable energy cooperatives in European energy governance? *Politics and Governance*, 7(1), 45–59. https://doi.org/10.17645/pag.v7i1.1782

UNFCCC. (2019). *2018 Talanoa dialogue platform*. https://unfccc.int/process-and-meetings/the-paris-agreement/the-paris-agreement/2018-talanoa-dialogue-platform

United Nations. (2015). *Paris agreement*. https://unfccc.int/files/essential_background/convention/application/pdf/english_paris_agreement.pdf

Van de Graaf, T., & Colgan, J. (2016). Global energy governance: A review and research agenda. *Palgrave Communications, 2*, 15047. https://doi.org/10.1057/palcomms.2015.47

van Kersbergen, K., & van Waarden, F. (2004). 'Governance' as a bridge between disciplines: Cross-disciplinary inspiration regarding shifts in governance and problems of governability, accountability and legitimacy. *European Journal of Political Research, 43*(2), 143–171. https://doi.org/10.1111/j.1475-6765.2004.00149.x

Voß, J. P., & Freeman, R. (2016). Introduction: Knowing governance. In J. P. Voß & R. Freeman (Eds.), *Knowing governance: The epistemic construction of political order* (pp. 1–33). Palgrave Macmillan. https://doi.org/10.1057/9781137514509

Vogler, J. (2018). Energy, climate change, and global governance: The 2015 Paris agreement perspective. In D. Davidson & M. Gross (Eds.), *The Oxford handbook of energy and society* (pp. 15–30). Oxford University Press.

4

COMPARING AND CONTRASTING THE INSTITUTIONAL RELATIONSHIPS, REGULATORY FRAMEWORKS, AND ENERGY SYSTEM GOVERNANCE OF EUROPEAN AND US ELECTRIC COOPERATIVES

Stephanie Lenhart, Gabriel Chan, Matthew Grimley, and Elizabeth Wilson

Energy democracy movements promote energy system transitions in support of broad environmental and social goals. The movements support actions that reduce environmental impacts to make energy systems more sustainable, to restructure economic and political institutions so as to reduce disparities across communities and subpopulations, and to create agency and opportunity to enable the prosperity of the full breadth of society. Advocates for these movements seek to amplify and translate grassroots support for greater individual agency and community control of energy systems.

Yet energy systems are composed of long-lived technologies and institutions that have coevolved to form complex interdependent infrastructures governed through multiple authorities with overlapping jurisdictions. For example, coal-fired power plants in the United States typically have 50-year lifespans and are regulated simultaneously by multiple state and federal authorities with jurisdiction over electricity operations and environmental impacts. At the same time, these plants can sell power into multistate regional electricity markets, can be owned and operated by electric utilities that serve customers across state lines, and often seek siting approval from local county or municipal governments. These polycentric relationships institutionalize the authority of actors at different scales of governance and shape how they interact with and influence one another (Andersson & Ostrom, 2008). Energy democracy can therefore be viewed as a challenge to the institutional arrangements of the incumbent fossil fuel regime that concentrates economic and political power at scales and in hierarchical structures that limit individual and community agency (Colgan & Keohane, 2012). Against powerful incumbents,

 DOI: 10.4324/9780429402302-5

energy democracy frames decentralized energy ownership and control as integral to enabling low-carbon energy transitions, more equitable outcomes, and new forms of governance that empower individuals and communities (Szulecki, 2018; van Veelen & van der Horst, 2018).

Renewable distributed energy resources (DERs) are seen as technological priorities for energy democracy due to their smaller scale and potential for decentralized ownership and control (Burke & Stephens, 2018). Unlike traditional electric generation through large-scale fossil fuels, DERs, such as rooftop and community-scale solar, load-control devices, energy efficiency, and geothermal heat pumps, can be owned and controlled by individuals or communities. Yet this decentralization is not inevitable, and possible technology configurations for DERs can concentrate or distribute power at different scales through a wide range of ownership and participation mechanisms. This, in turn, suggests that if energy democracy is to realize its promises, the movement must necessarily grapple with how more decentralized ownership and control models integrate with the multiscalar contexts of energy systems and the structures and actors that mediate power (electrical, financial, and political) across scales.

In this chapter, we explore the important cross-scale structures in energy systems that create interdependencies among energy consumers, communities, energy service providers, regulators, and policy makers. We then introduce a framework for understanding issues of scale in energy systems and illustrate how concepts of energy democracy are being advanced through organizational relationships that cross scales. We apply this framework through a focused examination of electric cooperatives in Europe and the United States. These organizations were established with many of the same principles that are advanced by energy democracy movements.

Energy democracy and electric cooperatives

While energy democracy movements can be found across all energy sectors, we focus on the electric sector where certain organizations have extensive experience implementing democratic principles within highly polycentric systems. We focus on the formation, development, and operation of electric cooperatives, an organizational form that has been broadly adopted over many decades in Europe and the United States. As organizations founded on democratic member control, equality, equity, and solidarity, cooperatives embody many energy democracy goals, making them relevant cases for understanding energy democracy movements in theory and practice. Cooperatives theoretically further the building of a participatory democratic society by "promoting interests of neglected issues and population, cultivating democratic skill sets, and monitoring socio-political processes across scales" (Taylor, 2015, p. 156).

Electric cooperatives in Europe and the United States were essential to rural electrification in the first half of the twentieth century. These early electric cooperatives are founded on well established ideals of cooperative governance that can be traced to social movements in the 1840s (Boland, 2017; Grimley, 2019). They also represent some of the first examples of cooperatives formed to provide a service of general interest to all members of a community rather than to a specific group of users (Mori, 2014). The cooperative values and principles that evolved from this early experimentation and in response to social disadvantage provide a model for pluralistic governance (Mori, 2014; Taylor, 2015).

A cooperative is a distinct business model in which the organization is owned by members or users rather than investors (Hansmann, 1996). Thus the organization is oriented to serving the needs of owner-members rather than private investors. In an idealized cooperative model, ownership—rather than size of investment, confers decision-making power through the concept of "one member, one vote," and confers rights to residual profits distributed based on volume of transactions (i.e., patronage). Around the world and across many sectors, cooperatives are guided by the same set of core principles (International Cooperative Alliance, n.d.). These shared principles include open and voluntary membership, democratic member control, economic participation, autonomy and independence, education, cooperation among cooperatives, and concern for community. However, fully adhering to cooperative principles is challenging. Research documents the many and varied ways in which cooperatives deviate from these principles and the ideals of pluralistic governance (Monaci & Caselli, 2005; Taylor, 2015; Yildiz et al., 2015). In practice, electric cooperatives must negotiate multiple sources of authority and are shaped by relationships across scales that govern ownership and operation of energy technologies.

Providing electricity is complex: It is generated, distributed, and consumed through nested systems that largely coevolved with centralized generation, strong economies of scale, vertically integrated business models, top-down government policy and financial guarantees, and a quasi-judicial approach to regulation. Mirroring the complex, interdependent structure of the electric grid that integrates large and long-lived private capital assets (most of generation, transmission, and distribution), shared-access assets (most of transmission), and highly regulated markets and operations (most of wholesale markets and dispatch decisions), actors governing electricity systems are required to coordinate across institutional scales. To understand systems with this type of polycentric governance, it is important to consider the ways in which local organizations and actors are embedded within larger sociotechnical systems and the relationships among actors at different scales (Andersson & Ostrom, 2008; Ostrom, 2005).

In this chapter, we explore the extent to which the coevolution of electric cooperatives and other interdependent organizations across scales has advanced or hindered progress toward the goals of energy democracy. We do this by examining differences across cooperatives on four intra- and interscale (horizontal and vertical) interdependencies (see Figure 4.1):

1. Linkages among members that form a cooperative based on a particular ***community identity***.
2. Linkages between the community and the cooperative that define ***collective decision making*** through membership and participation.
3. Linkages across energy service providers that embed cooperatives within an ***interdependent infrastructure***.
4. Linkages to ***market and regulatory frameworks*** that constrain and enable what is possible for a cooperative.

Diversity of electric cooperatives: Denmark, Germany, and the United States

Around the world, electric cooperatives demonstrate a diversity of legal and organizational forms. Yet what defines a cooperative is a commitment to core cooperative principles, as outlined by the International Cooperative Alliance (International Cooperative Alliance, n.d.). In this section, we provide examples of electric cooperatives in three countries experienced with this model: Denmark, Germany, and the United States. We choose these cases not as representative examples of cooperatives in their respective countries but to illustrate the breadth of organizational forms electric cooperatives can take and to exemplify the cross-scale challenges of implementing the goals of energy democracy. Table 4.1 summarizes the scalar dimensions of cooperative functions, as illustrated schematically in Figure 4.1, across the three cases.

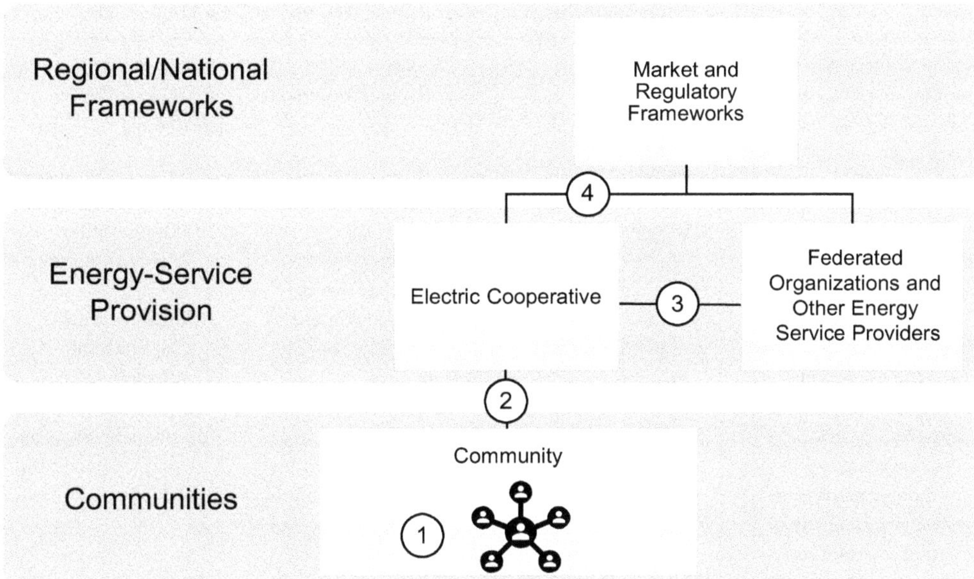

Figure 4.1 *Electricity system inter- and intrascale interdependencies.* Cooperatives are embedded in an electricity system that is built around three scales: communities, energy-service provision, and regional/national frameworks. These scales are defined by the many different mechanisms and relationships that form to deliver electricity service.

Denmark: Middelgrunden Cooperative

In Denmark, new wind energy cooperatives emerged in the late 1970s alongside long established distribution cooperatives (Bauwens et al., 2016; Eikeland & Inderberg, 2016; Mey & Diesendorf, 2018; Wierling et al., 2018). Denmark has been a leader in renewable cooperatives, reaching a peak of 931 wind cooperatives in 1999 (Wierling et al., 2018). Before 1999, Denmark restricted ownership of wind turbines to local electricity consumers living or registered in geographic proximity to the turbine (Bauwens et al., 2016). Then, in the late 1990s in conjunction with market restructuring, reforms removed this exclusive commitment to local communities and opened ownership to private actors and any individual regardless of geographic proximity. In the 2009 Energy Act, the local ownership model was revived through a requirement that the local population be given priority rights to acquire 20% ownership of new wind power plants (Eikeland & Inderberg, 2016). This policy opens the possibility of joint ownership while retaining a commitment to local control.

Of the many wind cooperatives in Denmark, the Middelgrunden Cooperative, established in 1996, is the first offshore wind project based on sale of shares to local electricity consumers. The project's 20 offshore 2-MW wind turbines are located 3.5 km east of the Copenhagen harbor. Rather than being fully owned and operated by a cooperative of local residents, Middelgrunden was established through joint ownership with the local municipal electric utility owning ten of the turbines and cooperative members collectively owning the other ten turbines. The municipal utility subsequently merged with other companies to become part of Ørsted, a privately

Table 4.1 Electricity system scale interdependencies in Denmark, Germany, and the United States

Inter- and Intrascale interdependencies among actors and institutional arrangements	*Denmark: Middelgrunden*	*Germany: UrStrom/Bürgerwerke eG*	*United States: Kit Carson/Basin Electric*
Generation resources and participation	40-MW wind capacity 8,500 investors	26-MW renewable capacity 300 members/ 15,000 members	38-MW solar capacity 29,000 members/ 3 million members
1 Community identity: foundational principles and the expressed concern for the community that forms a shared sense of representativeness among individuals within a community	Local involvement in planning and financing Social buy-in, voluntary economic participation	Renewable energy instead of coal or nuclear Individual involvement in energy transitions, voluntary economic participation	Affordable, reliable, and renewable Community-developed solar goal Membership required based on geographic service area
2 Collective decision making: rules and norms within the cooperative that establish practices for ownership and participation	Joint ownership by individuals, private developers, and municipal utilities One-member/one-vote Profits distributed by shares	Individual ownership One-member/one-vote Profits distributed by shares	Community ownership and/or specific programs One-member/one-vote Profits distributed by patronage
3 Interdependent infrastructure: relationships among cooperatives and other energy service providers based on cooperation, education, and information sharing	Generation only Joint ownership of generation with municipal or private distribution utilities Joint education, research, lobbying, and wholesale trading	Generation only Retail marketing and distribution services from jointly owned cooperatives Joint education, research, lobbying, and wholesale trading	Distribution utility Generation and transmission services through jointly owned cooperatives Joint education, research, lobbying, and wholesale trading
4 Market and regulatory frameworks: infrastructure, regulatory frameworks, or market rules that establish linkages between local cooperative autonomy and energy system institutions	Cooperative and municipal ownership promoted by a policy framework defining ownership by locality Nationally/regionally determined monetizable value of DERs National/regional balancing of fairness with opportunity to participate	Cooperative ownership promoted but not defined by locality Nationally/regionally determined monetizable value of DERs National/regional balancing of fairness with opportunity to participate	Cooperative ownership defined by service area Locally determined monetizable value of DERs for cooperative members Local balancing of fairness with opportunity to participate

traded company, with a majority of state-controlled ownership. Ørsted offers renewable energy and is an offshore wind developer. Then, in 2018 the municipal utility repurchased its share of the project. The cooperative has more than 8,500 members that mostly—but not exclusively—live in or around Copenhagen. A small number of members live outside of Denmark. As co-owners, cooperative members make investment decisions and negotiate terms with operators of larger electric networks (Hoeschele, 2017; Larsen et al., 2005). The partners believe the project would not have been built without the public support generated by the cooperative and local ownership (Pahl, 2007).

In a more recent example of the joint ownership model, a private project developer organized a local group in an industrial neighborhood of Copenhagen to create the Prøvestenen Wind Cooperative. Launched in 2010, the project's three onshore 2-MW wind turbines are jointly owned by the local municipal electric utility and cooperative members. The cooperative serves to engage local community members, and the project was designed to support Copenhagen's target of being fossil fuel–free by 2025 (De Sotto, 2014).

Germany: UrStrom eG Cooperative

In Germany, electric cooperatives reemerged in the late 2000s as a common organizational form to promote customer-sited solar, expanding from fewer than 50 to more than 800 cooperatives by 2015 (Klagge & Meister, 2018; Yildiz et al., 2015). The UrStrom eG Cooperative operates 15 photovoltaic systems mounted on the roofs of public and industrial buildings in the Rheinhessen/Nahe region of Germany. It was founded in 2010 with the goal of contributing to the energy transition in the city of Mainz and the surrounding region. The cooperative generates about 984,000 kWh per year, approximately enough power for 220 four-person households, and it also offers biogas for heating and cooking and electric car sharing services. The cooperative has more than 300 members who exchange views on new projects and developments in the field of renewable energy (Clean Energy Wire, 2018; UrStrom, n.d.).

UrStrom is a founding member of the Bürgerwerke eG Cooperative, a network of 92 cooperatives with over 15,000 members and more than 400 decentralized power plants with 26 MW of capacity, making it the largest association of energy cooperatives in Germany. Described as "everywhere regional," the cooperatives are located throughout Germany. Bürgerwerke eG is a nonprofit retail electricity marketer owned by the member cooperatives to promote the sale of renewable energy from local cooperatives to customers across all of Germany. It aims to supply as much electricity as possible from the facilities of the member cooperatives in order to keep the added value from electricity purchases in local communities, but it also uses electricity from the Töging hydroelectric plant to meet demand (Bürgerwerke, n.d.).

In 2014, Germany changed the Renewable Energy Sources Act from offering renewable generators long-term contracts at fixed prices above the retail rate to requiring facilities with installed capacity in excess of 500 kW to market electricity directly in the wholesale market. As a result, renewable generation, which is supported by local democratic practice, must now engage in market structure at a regional scale. With the shift to auctions, UrStrom is considering whether to merge with other cooperatives to build new projects that compete in auctions or to pursue smaller projects that use electricity locally and has reconsidered previous plans to provide wind energy (Chandrashekhar, 2018).

United States: Kit Carson Electric Cooperative

In contrast to the more recent proliferation of cooperatives in Europe, in the United States, many electric cooperatives have continuously served local communities for nearly 100 years (Youngblood, 1995). Today more than 900 cooperatives supply power to 13% of US electricity customers and provide exclusive service in more than half of the nation's geographic area (National Rural Electric Cooperative Association, n.d.; US Energy Information Administration, 2019).

Kit Carson Electric Cooperative (KCEC) is a distribution utility that has served communities in northern New Mexico since 1944. It provides electricity to approximately 23,000 members in a three-county service area and also offers propane and internet services. As a distribution utility, KCEC provides retail service and owns electricity distribution infrastructure but has not historically generated a significant share of its own power. Instead, since the early 1950s, KCEC, together with 42 other distribution utilities, has owned and received power from the Tri-State Generation and Transmission (G&T) Association. To enable the financibility of its long-lived and jointly owned infrastructure, Tri-State provides power through contracts that generally limit local self-generation to 5% of each distribution cooperative's total power supply.

Tri-State also procures about one-tenth of its energy from Basin Electric Power Cooperative, adding a third tier to KCEC's power supply network (Tri-State Generation and Transmission Association, 2015). Basin Electric Power Cooperative is one of several "super G&Ts" formed through collaboration among G&T co-ops to generate and transmit wholesale bulk electricity. More than 60% of Tri-State's electricity is produced from fossil fuels, and its electricity rates have risen 106% from 2000 to 2016 (Cates & Feaster, 2019).

In 2002, KCEC took the lead in a community initiative to become the "solar capital of the world," and in 2016 it set a goal of achieving 100% daytime solar by 2022 (an estimated 34% of total load) (Kit Carson Electric Cooperative, n.d.). This community goal, amid rising rates and a perceived lack of transparency from Tri-State, led KCEC to negotiate a $37 million buyout of obligations under a power supply contract with Tri-State that was slated to run until 2040. This buyout was enabled by a ten-year contract with Guzman Renewable Energy Partners, a private energy trading and marketing firm, offering clean energy at a cost projected to lower electric rates compared to terms offered by Tri-State—even incorporating the buyout cost (Cates & Feaster, 2019; Logan, 2016). To reach its goals, KCEC has finalized agreements for 38 MW of solar and 15 MW of storage (Ludt, 2019).

Electric cooperatives in practice

To explore how the ideals of cooperative principles work in practice, we compare the selected examples of electric cooperatives in Europe and the United States across the four dimensions introduced in Figure 4.1: community identity, collective decision making, interdependent infrastructure, and market and regulatory frameworks.

Community identity

At the community-scale we focus attention on the relationships among cooperative member-owners and on how the cooperative principles of *concern for community* and *voluntary and open membership* form a collective identity. Cooperatives express concern for community through their goals and purposes. In Denmark and Germany, electric cooperatives have been central to the renewable energy transition, and many also have explicit ideological goals of promoting community economic development, public acceptance of new technologies, and social cohesion (Huybrechts & Haugh, 2018; Huybrechts & Mertens, 2014). For example, the Middelgrunden Cooperative in Denmark brought together individuals who believed in a vision of large-scale offshore wind and Danish national leadership in this emerging technology. The cooperative has been described as forming a "public trust" that carefully considers environmental and local community concerns, and it is credited with solving potential conflicts and generating social acceptance of the offshore location (Copenhagen Environment and Energy Office, 2003; Larsen et al., 2005). The UrStrom eG Cooperative in Germany draws together individuals for the purpose of "getting out of coal" and offers several options to "take the energy transition in your own hands" (UrStrom, n.d.). Membership in these Danish and German cooperatives is voluntary and is achieved through investment or contribution of assets. Overall, a large share of Danish and German cooperatives define themselves as communities of interest and are explicitly oriented toward social impact (Huybrechts & Haugh, 2018; Huybrechts & Mertens, 2014; Klagge & Meister, 2018; Mey et al., 2016; Mori, 2014).

In contrast, in the United States, many electric cooperatives largely retain a commitment to the goals of reliability and affordability that were foundational during early rural electrification. The initial purpose of US electric cooperatives was centered around providing electricity access and helping members serve themselves (Beecher, 2013). For example, KCEC has only relatively recently identified solar energy goals, but it continues to define itself as a community of place, serving the needs of—and balancing the interests of—all local members. Unlike most European electric cooperative, US electric cooperatives have defined exclusive service areas; if a consumer lives in a cooperative's territory, then the cooperative is the only utility that provides service, and consumers become owner-members through purchase of electricity service. As small-scale renewable generation and other DERs have become more widely available and affordable, a growing number of US cooperatives are responding to their geographically defined communities and providing voluntary programs, like community-shared solar, that distribute economic and political power. These programs have different pricing schemes but do not alter the patronage basis for distributing residual profits. Electric cooperatives have been among the early adopters of community-shared solar in the United States (Chan et al., 2018); however, these offerings and the required investment for participation vary across cooperatives and are dependent on the balance of interests within the preestablished local service areas.

In many parts of the world, cooperatives have enlarged their ideal of how to serve their communities. They have embraced broader societal benefits and are less tightly tied to producing member benefits (Mori, 2014). In Denmark and Germany, cooperative membership is voluntary and defined around social benefits. In the United States, cooperative membership is not voluntary but is defined by a geographic place and requires more explicit balancing of member and social benefits. As old and new cooperatives begin to explicitly define their community not only as a community of place but also—or instead—as a community of interest (e.g., transitioning to renewable energy), this shift raises questions of fairness and participation. For example, if cooperatives act on behalf of a willing coalition, could their actions create exclusive benefits that leave out portions of the public or even cause harm to others?

Collective decision making

At the cooperative scale, we focus attention on the relationships between the cooperative and member-owners and on how the cooperative principles of *economic participation* and *democratic member control* create interdependencies between individuals and the cooperative.

According to cooperative principles, ownership confers membership. Moreover, membership in a cooperative translates to voice in its governance and the right to residual profits distributed according to use (i.e., patronage) rather than investment. Approaches to ownership are important in distinguishing cooperatives from other firms. Yet in practice, cooperative ownership is quite varied. In Denmark, each member has one vote regardless of shares owned, but, profits are distributed based on a member's shares (Bolinger, 2001). For example, one of the Middelgrunden Cooperative wind turbines is deemed a "children's wind turbine" because members have had their children vote on their behalf, with one vote for each member and profits distributed based on shares (Hoeschele, 2017; Larsen et al., 2005). In Denmark, cooperative membership is also defined by locality (Bauwens et al., 2016). Individual ownership of the Middelgrunden Cooperative was initially restricted to people living or working in Copenhagen or the neighboring municipalities. Local residents and organizations owned half of the project, and the local municipal electric utility owned the other half. The Prøvestenen Cooperative is another example of joint local ownership, with the developer initiating a cooperative in order to meet the 20% local ownership requirement. Thus in Denmark cooperative ownership and control are a mix of private, municipal, local community, and outside actors. These boundaries have been renegotiated over time and shaped by national legislation that has sought to balance the benefits of local ownership, including public acceptance of projects, with policies to promote commercialization and market restructuring.

In Germany, the concept of cooperative membership is not closely tied to location. For example, the UrStrom eG Cooperative's voluntary and open membership has attracted more than 300 owner-members from Mainz and across the region to invest in renewable energy. The Bürgerwerke eG Cooperative, a network of 92 generation cooperatives, is a competitive electricity retailer that actively markets 100% renewable electricity to customers nationwide regardless of membership in local cooperatives (UrStrom, n.d.). Members of these cooperatives do not necessarily need to be customers; they can join just as investing owners. They exert collective agency as investors, through the exercise of voting on a one-member/one-vote basis, and receive a return based on shares rather than patronage.

In the United States, electric cooperatives have an obligation to serve defined geographic service area. For example, KCEC is required to provide electricity in parts of three counties, six municipalities, and two Native American pueblos in North New Mexico. Given this regulatory framework, US electric cooperative members exert collective agency on cooperative decision making through voting in periodic board elections on a one-member/one-vote basis and other community participation opportunities. As the case of KCEC demonstrates, the cooperative structure allowed members to exercise democratic control to deploy more solar energy and alter KCEC's power supply relationship with Tri-State. However, as will be discussed, these changes required a shared governance decision among cooperatives engaged in joint ownership of infrastructure. So, while KCEC's internal democratic structure precipitated the cooperative's exit from Tri-State, this democratic control was constrained and had to also be approved at a higher governance level.

Electric cooperatives are engaging in experimentation with different technology scales and different models of ownership and control. In Denmark and Germany, owner-members with the interest and resources exert individual agency by choosing an electric cooperative retail provider and have the opportunity to exert collective agency on a one-member/one-vote basis with residual profits distributed by shares. In the United States, all owner-members receiving service have the opportunity to exert collective agency on a one-member/one-vote basis, profits are distributed according to patronage, and opportunities to exert individual agency through separate investments are dependent on the offerings of the cooperative.

Interdependent infrastructure

At the scale of energy service provision, we examine the relationships between the electric cooperative and other energy service providers. To meet the needs of their owner-members, cooperatives form relationships with other cooperatives, link with external services providers, and are part of highly interdependent infrastructures and systems. Some electric cooperatives provide only generation, others provide distribution, others provide generation and transmission services, and there are many other combinations (US Department of Energy, 2017; Table A-5). The cooperative principle of *cooperation among cooperatives* orients electric cooperatives toward creating these federated systems. This allows cooperatives to retain their local autonomy and community identity, while pursuing favorable economies of scale from collective decision making and investment in shared infrastructure (see also Taylor, 2015). Yet cooperatives vary in how and where they develop these nested institutional relationships.

In Europe, modern electric cooperatives are primarily formed as a way for individuals to jointly own renewable generation (Eikeland & Inderberg, 2016; Yildiz et al., 2015). Although some cooperatives operate distribution networks (e.g., EWS in Germany) and some include other sectors, such as heating, most are small organizations that own wind turbines or solar photovoltaic panels and have no paid staff (Brummer, 2018). For example, the Prøvestenen Cooperative in Denmark and UrStrom eG in Germany are generation-only cooperatives that do not own distribution infrastructure (although the Danish wind project is jointly owned with a municipal electric distribution utility). In Germany, 70% of all energy cooperatives own or invest in power generation, only 1% rely on fossil fuels, and none rely on nuclear fuel (Yildiz et al., 2015). These cooperatives leverage member financing, often supporting projects that would not produce sufficient profit to attract investment funds, and they usually sell electricity to utilities rather than to individual members or enter into joint ownership arrangements with utilities. For example, the UrStrom eG generation cooperative relies on Bürgerwerke eG, a network of 92 cooperatives, for retail marketing services. Thus these cooperatives integrate with the electricity grid infrastructure through long-term purchase contracts, with renewable energy auctions, or by marketing electricity into wholesale electricity markets.

European electric cooperatives have nested relationships to advance the interests of their organizational form at the European Union level and collectively provide shared services. The UrStrom eG and Bürgerwerke eG cooperatives in Germany and the Middelgrunden Cooperative in Denmark are members of a European federation of more than 1,500 renewable energy cooperatives. This network promotes renewable energy and seeks to provide information technology, financing, and forecasting services to new and existing cooperatives (REScoop.eu, n.d.). Cooperatives in Denmark, Germany, and other European countries have also established trading companies, like Vindenergi Danmark, that allow small and medium-sized electric cooperatives to reduce their risks and transaction costs while sharing infrastructure and expertise for wholesale market trading (Bauwens et al., 2016).

In the United States, 93% of electric cooperatives are distribution utilities that don't own generation but rather obtain power from G&T cooperatives, federal power marketing administrations, or, in some cases, investor-owned utilities (Lazar, 2016; National Rural Electric Cooperative Association, 2019). US electric cooperatives coevolved with fossil fuel generation and developed nested institutional relationships that enabled small distribution cooperatives to take advantage of the favorable economies of scale of centralized generation. For example, prior to its exit in 2016, KCEC received services through long-term contracts with Tri-State Cooperative, which owns some large generation and also purchases wholesale power from Basin Cooperative, a cooperative of G&T cooperatives. Basin, directly and through 141 member cooperatives, serves 3 million member-consumers and has a generation capacity that is 68% fossil fuels (Basin Electric Power Cooperative, 2017). Basin is comparable in total sales and number of customers served to some of the largest investor-owned utilities in the United States (US Energy Information Administration, 2019). The governance and ownership of these different cooperatives is linked, with distribution cooperatives (e.g., KCEC) sharing ownership and voting control of G&T cooperatives (e.g., Tri-State) with other distribution cooperatives. Similarly, G&T cooperatives share ownership and control of super G&T cooperatives (e.g., Basin). When KCEC moved to terminate its power supply contract with Tri-State, all 44 of Tri-State's member cooperatives voted on the fairness of the negotiated $37 million exit fee. In October 2015, the proposed exit fee passed unanimously (Cates & Feaster, 2019). Similar to cooperatives in Europe, US electric cooperatives have also formed joint service companies for research and planning, retail marketing, and wholesale energy trading (e.g., the Alliance for Cooperative Energy Services) (Rocha, 2017).

Electric cooperatives rely on federated organizational forms to achieve economies of scale while maintaining local autonomy. Similarities can be seen across Denmark, Germany, and the United States in how electric cooperatives jointly pursue education, research, lobbying, and wholesale market trading. However, differences can also be seen between these countries in how cooperatives integrate into the architecture of the grid, define relationships with third-party vendors, and share information and management technologies. Electric cooperatives rely on nested relationships for effective energy service delivery, but these relationships also create constraints in their ability to make decisions due to the interdependence of technologies, infrastructure, contracts, and other service relationships.

Market and regulatory frameworks

This dimension of a cooperative's organizational interdependence is about how it is situated in the infrastructure and technology of the electricity transmission grid, regional market rules, and regulatory frameworks. Viable choices about local generation investments and operating practices are shaped by market rules and regulatory requirements at the regional and national scale (Lenhart et al., 2016; Stafford & Wilson, 2016). Although cooperatives are guided by the principle of *autonomy and independence*, these external structures shape what is possible.

Market restructuring and renewable energy policies differ across Denmark, Germany, and the United States. In 1999, Denmark restructured the electricity sector and promoted the commercialization of wind power by giving customers the ability to choose their retail electricity provider and removing local ownership restrictions on wind projects (Bauwens et al., 2016; Wierling et al., 2018). These market and policy changes coincided with engineering improvements for larger turbine sizes, investment by large companies, and a decline in cooperative ownership. However, in the face of rising local resistance to wind installations and stagnation in the wind energy sector, Denmark reinstated a partial local ownership requirement in 2009 (Eikeland & Inderberg, 2016). Despite this effort, the total number of wind cooperatives in Denmark has declined dramatically, with most of the loss in medium-sized cooperatives with 5 to 50 members and in areas with low wind yield (Wierling et al., 2018). The Middelgrunden Cooperative marks this transition from small-scale, onshore wind generation with local ownership and control to larger-scale offshore generation technology with joint municipal and local ownership. Over the intervening years, the project has been transformed. At one time, it was jointly owned by one of the largest private wind developers in the world, but more recently, this share of the project has been repurchased by the original municipal utility owner (Renews, 2018).

In Germany, the growth of cooperatives has been facilitated by policies aimed at expanding renewable electricity generation. From 2000 to 2011, Germany implemented fixed prices for renewable generation under a feed-in tariff and priority purchasing requirements, revised the cooperative incorporation laws, and accelerated the shift away from nuclear power. The UrStrom eG cooperative has leveraged these incentives to expand rooftop solar photovoltaics in an urban area and to compete for customers within liberalized electricity markets. However, revisions to the Renewable Energy Resources Act in 2012, 2014, and 2017, implementing a phaseout of fixed pricing under the feed-in tariff and the introduction of competitive auctions for solar, have contributed to UrStrom eG reconsidering how to position itself for the future. And across the country, there has been a leveling off in the formation of new electric cooperatives (Klagge & Meister, 2018; Wierling et al., 2018).

In contrast to the market liberalization that has driven change in European cooperatives, the majority of the United States lacks retail electricity choice, and electric cooperatives generally do not compete for retail customers.[1] Furthermore, in most jurisdictions, cooperatives are exempt from—or face less restrictive requirements under—many state regulations and mandates that cover investor-owned utilities. Instead, electric cooperatives largely retain local authority over retail rates and can determine the price paid to distributed renewable energy generators. For example, KCEC has been compensating rooftop solar PV generators for energy sold back to the grid at the retail rate of electricity under a net-metering policy but has recently stated that this unfairly benefits those who can invest in rooftop solar and creates social justice concerns (Cantwell, 2020). These local debates within the cooperative demonstrate an important level of independence that KCEC has from the dominant market and regulatory frameworks in the United States.

Decisions made by national, regional, and local regulators and grid operators determine the value of local generation and other system assets (e.g., the outcome of competitive auctions or whether or not a particular renewable generating source is dispatched). The interdependencies between how electric cooperatives that are embedded in federated systems deliver energy services and the institutional frameworks for markets and regulation have implications for economic value, system reliability, and the potential expansion of renewable generation and distributed energy resources.

Discussion

Energy democracy has multiple goals, including sustainability, fairness, and empowerment. It also emphasizes the potential for decentralized decision making as a necessary means to achieving these ends. However, the ideal of community ownership and the electric cooperative form have not been critically explored in much of the energy democracy movement or literature. Due to the complex, polycentric, and embedded natures of electricity systems, decentralizing ownership alone is not sufficient for achieving the goals of energy democracy. Institutional arrangements operating at other scales also have critical roles to play in self-organizing regimes (Zinck & Newen, 2008). Electricity systems involve expensive and long-lived infrastructure, complex regulatory frameworks across multiple agencies, decades-long service contracts and financial arrangements, actor relationships that span multiple interconnected scales, and enhanced needs for coordination (Goldthau, 2014). Decentralization of the system at one level restructures these relationships and can redistribute economic and political power, but in most cases, cross-scale interdependencies remain. Thus simultaneously making progress toward energy democracy's multiple goals requires situating decentralized control within the multiscale, polycentric governance context of the energy system.

The examples of variations in cooperative organizational forms across Denmark, Germany, and the United States highlight how decentralized control is situated within different multiscalar contexts and results in distinct local institutional coherences. These examples illustrate how differences in nested institutional relationships and actions across intra- and interscale interdependencies have important implications for energy democracy.

Cooperatives interpret community in different ways. Like many cooperatives in Denmark and Germany, Middelgrunden and UrStrom are defined by ideological goals and as communities of interest. They have explicit aims to promote the welfare of nonmembers through transformation of the energy system. Thus the concerns of the cooperative broaden attention to shifting economic and political power, in alignment with the social benefit goals of energy democracy. In contrast, the electric cooperatives in the United States have a long tradition of providing access to affordable and reliable electric service to all members and are defined as communities of place. Most often their explicit aim is serving their local communities, not furthering the broader public interest beyond their service territory. Some, like KCEC, have explicitly chosen to pursue goals that align more closely with energy democracy goals. Yet having established contractual and institutional relationships decades ago in the pursuit of affordability, KCEC has faced constraints as it seeks to meet its members' interests, requiring it to address systemic institutional barriers. The expanded purposes of electric cooperatives to include social benefits, either newly formed in Denmark and Germany or, to a more limited extent, evolving in the United States, has been described as "something radically new" (Mori, 2014).

The mechanisms of collective ownership, economic participation, and democratic control represent cross-scale linkages between cooperatives and communities. In contrast to traditional cooperative principles, under which voluntary participation and patronage incentivize use and limit conflict, the ways in which economic participation translates to voice and rights to profits in practice are varied. For example, in response to changing technological and market conditions, Denmark has connected ownership to a member's locality and accepted a mixed-ownership model across community, public, and private entities, whereas, in Germany, ownership is not closely connected to a locality and is held by individual members. Economic participation in UrStrom eG is voluntary, open to individuals across the region, does not depend on use, and permits members to be eligible for distributions based on shares. This form of economic participation creates the risk of excluding certain populations and the risk of external conflicts between owner-members and users. The experience in the United States is different, reflecting an exclusive service obligation, rather than voluntary participation. Ownership in KCEC does not require an additional investment beyond cost of service, and all owners are considered members eligible for patronage distributions. This form of economic participation requires an internal negotiation among members and their elected representatives about the appropriateness of new programs or voluntary offerings like KCEC's net-metering policy.

Cooperatives are embedded in interdependent infrastructures and can affect change across scales in less obvious ways than other business models. The mechanisms and relationships that form around community-based cooperatives situate them within federated systems of shared services, shared obligations, and require negotiations of fairness with evolving missions within and across different scales. Within federated electricity systems, institutional relationships enable utilities to innovate and achieve more than they could achieve alone, but they also constrain actions through shared obligations. These interdependencies require negotiation among cooperatives that collectively share investment in services and infrastructure. For example, if a US distribution cooperative, like KCEC, increases local renewable generation, this action displaces their use of older fossil fuel generation and shifts the financial obligation for the fossil fuel generation to the other cooperatives that are joint owners of the asset. In the case of KCEC, the exit fee negotiated with Tri-State was rationalized on the impact of the cost shift that KCEC's exit would impose on the other Tri-State member cooperatives with the intent to "make them whole."

Electricity grid operating rules, power market constructs, and state and federal regulatory frameworks simultaneously enable and constrain the local autonomy and control of cooperatives, implying the need for coordination across scales. The examples of Denmark and Germany illustrate how market liberalization and national policy can significantly influence the opportunities and challenges of local actors seeking to engage with DERs to create environmental, economic, and social benefits. In the United States, the self-regulation of electric cooperatives may provide opportunities for innovation and experimentation with DERs, but the potential to change the system depends on balancing the needs of all community owner-members and renegotiating long established institutional relationships, financial obligations, and existing contracts.

Scale places energy transitions within particular contexts, relations, and dynamics (Coenen et al., 2012). From this perspective, scale and boundary-spanning become critical considerations for understanding the implications and potential of energy democracy. The challenges of multiscale governance are both relational and emergent, requiring ongoing negotiation as actors seek to further unique positions in polycentric systems. In other words, "actors construct scales as they seek to look after their own interests within the networks most salient to them" (Coenen et al., 2012, p. 975). A situated scale is built around the many different mechanisms and relationships that emerge through the networking of actors working in cooperation or in competition to construct collective understandings and mechanisms to deliver electricity service (Coenen et al., 2012; Nelson-Marsh, 2017).

Energy democracy has focused on grassroots efforts to decentralize ownership and control to the smallest possible scale. However, interdependencies across infrastructure and institutions continue to create opportunities and constraints. Careful attention to interdependencies could potentially enable greater energy democracy by identifying opportunities to strengthen shared services and public goods, as well as the constraints from structures that concentrate economic and political power. The cooperatives highlighted as examples in this chapter illustrate that decentralizing political and economic power requires strategic action within interdependencies that promote the goals of energy democracy (e.g., shared education and research provided by the REScoop Network in Europe) and breaking the interdependencies that cause constraints (e.g., long-lived power contracts in KCEC).

Conclusion

Furthering energy democracy requires changes at multiple scales and is more complex than the adoption of novel distributed energy technologies by grassroots organizations and community-level actors. This chapter is intended to advance the understanding of how energy democracy movements are playing out in practice—and might play out in the future. We focus on electric cooperatives, new and old, to provide important empirical evidence of how the cooperative organizational form deviates from ideals of pluralistic, democratic governance as it integrates with existing structures, actors, and practices. The electricity system is governed by overlapping jurisdictions and polycentric relationships that span multiple scales. For the ideals of energy democracy to be realized, initiatives must grow and adapt within this scaffolding to allow for contextualization, experimentation, and innovation of these complex systems. This chapter demonstrates a critical need to understand scales and interdependencies as energy democracy movements continue to experiment with bottom-up efforts and seek to identify how to contribute to system change that not only limits environmental impacts but also distributes economic and political power.

Note

1 Only 13 states and the District of Columbia offer retail choice, and even in these states, most electric cooperatives have opted out, meaning that if an electric customer wants grid-supplied electricity service in a cooperative's service territory, they typically must become owner-members of the cooperative (US Energy Information Administration, 2018).

References

Andersson, K. P., & Ostrom, E. (2008). Analyzing decentralized resource regimes from a polycentric perspective. *Policy Sciences*, *41*(1), 71–93. https://doi.org/10.1007/sl

Basin Electric Power Cooperative. (2017). *Built to serve: Basin electric power cooperative 2017 annual report.* www.basinelectric.com/news-center/publications/annual-report

Bauwens, T., Gotchev, B., & Holstenkamp, L. (2016). What drives the development of community energy in Europe? The case of wind power cooperatives. *Energy Research & Social Science Social Science*, *13*, 136–147. https://doi.org/10.1016/j.erss.2015.12.016

Beecher, W. G. (2013). Is it time to revoke the tax-exempt status of rural electric cooperatives? *Washington and Lee Journal of Energy, Climate, and the Environment*, *5*(1), 221. http://scholarlycommons.law.wlu.edu/jece/vol5/%0Aiss1/7

Boland, M. (2017). *An introduction to cooperation and mutualism*. University of Minnesota Libraries Publishing. https://doi.org/10.24926/9781946135384

Bolinger, M. (2001, May). *Community wind power ownership schemes in Europe and their relevance to the United States*. Lawrence Berkeley National Laboratories. http:/eetd.lbl.gov/EA/EMP/

Brummer, V. (2018). Of expertise, social capital, and democracy: Assessing the organizational governance and decision-making in German renewable energy cooperatives. *Energy Research & Social Science*, *37*(Supplement C), 111–121. https://doi.org/10.1016/j.erss.2017.09.039

Bürgerwerke. (n.d.). *Bürgerwerke Energie in Gemeinschaft*. https://buergerwerke.de/

Burke, M. J., & Stephens, J. C. (2018). Political power and renewable energy futures: A critical review. *Energy Research & Social Science*, *35*, 78–93. https://doi.org/10.1016/j.erss.2017.10.018

Cantwell, D. (2020, January 16). Kit Carson electric CEO: Rooftop solar not paying its fair share. *Taos News*. www.taosnews.com/stories/kit-carson-electric-ceo-rooftop-solar-not-paying-its-fair-share,61764

Cates, K., & Feaster, S. (2019, April). *Case study: How Kit Carson electric engineered a cost-effective coal exit*. Institute for Energy Economics and Financial Analysis. https://ieefa.org/wp-content/uploads/2019/04/How-Kit-Carson-Electric-Engineered-a-Cost-Effective-Coal-Exit_April-2019.pdf

Chan, G., Grimley, M., Arnold, E., Evans, I., Herbers, J., Hoffman, M., Ihde, B., Mazumder, P., Morgan, J., Neuman, N., & Streitz, R. (2018, March). *Community shared solar in Minnesota: Learning from the first 300 megawatts*. Center for Science, Technology, and Environmental Policy, Humphrey School of Public Affairs, University of Minnesota. www.hhh.umn.edu/step-updates/community-shared-solar-minnesota-learning-first-300-megawatts

Chandrashekhar, V. S. (2018, June 12). As subsidies wane, market forces drive the growth of renewables. *Yale Environment 360*. https://e360.yale.edu/features/as-subsidies-wane-market-forces-drive-the-growth-of-renewables

Clean Energy Wire. (2018). *Investors, drivers, troublemakers—citizens and the local powers of the Energiewende*. www.cleanenergywire.org/sites/default/files/files/event/investors-troublemakers-citizens-local-powers-energiewende/investorsdriverstroublemakers-clew-media-workshop-programme-22-24042018-incl-schedule-i-updated-1804.pdf

Coenen, L., Benneworth, P., & Truffer, B. (2012). Toward a spatial perspective on sustainability transitions. *Research Policy*, *41*(6), 968–979. https://doi.org/10.1016/j.respol.2012.02.014

Colgan, J. D., & Keohane, R. O. (2012). Punctuated equilibrium in the energy regime complex. *Review of International Organizations*, 117–143. https://doi.org/10.1007/s11558-011-9130-9

Copenhagen Environment and Energy Office. (2003). *The Middelgrunden offshore wind farm: A popular initiative*. Copenhagen Environment and Energy Office. https://doi.org/ISBN 87-986690-3-6

De Sotto, J. (2014). *Management of uncertainty and ambiguity in wind power projects* [Unpublished master's thesis, Technical University of Denmark].

Eikeland, P. O., & Inderberg, T. H. J. (2016). Energy system transformation and long-term interest constellations in Denmark: Can agency beat structure? *Energy Research & Social Science, 11*, 164–173. https://doi.org/10.1016/j.erss.2015.09.008

Goldthau, A. (2014). Rethinking the governance of energy infrastructure: Scale, decentralization and polycentrism. *Energy Research & Social Science, 1*, 134–140. https://doi.org/10.1016/j.erss.2014.02.009

Grimley, M. (2019). *'Cooperative is an oxymoron': The polycentric energy transition of Midwestern electric cooperatives to load management technologies, 1940s to present* (Order No. 13886811). ProQuest Dissertations & Theses Global (2281953253). http://libproxy.boisestate.edu/login?url=https://search.proquest.com/docview/2281953253?accountid=9649

Hansmann, H. (1996). *The ownership of enterprise*. Harvard University Press.

Hoeschele, W. (2017, June 12). *Danish energy cooperative lets consumers collectively build wind turbines*. www.shareable.net/danish-energy-cooperative-lets-consumers-collectively-build-wind-turbines/

Huybrechts, B., & Haugh, H. (2018). The roles of networks in institutionalizing new hybrid organizational forms: Insights from the European renewable energy cooperative network. *Organization Studies, 39*(8), 1085–1108. https://doi.org/10.1177/0170840617717097

Huybrechts, B., & Mertens, S. (2014). The relevance of the cooperative model in the field of renewable energy. *Annals of Public and Cooperative Economics, 85*(2), 193–212. https://doi.org/10.1111/apce.12038

International Cooperative Alliance. (n.d.). *Cooperative values, identity and principles*. www.ica.coop/en/cooperatives/cooperative-identity

Kit Carson Electric Cooperative. (n.d.). https://kitcarson.com/

Klagge, B., & Meister, T. (2018). Energy cooperatives in Germany—an example of successful alternative economies? *Local Environment, 23*(7), 697–716. https://doi.org/10.1080/13549839.2018.1436045

Larsen, J. H. M., Soerensen, H. C., Christiansen, E., & Naef, S. (2005, October 26–28). *Experiences from Middelgrunden 40 MW offshore wind farm*. Copenhagen Offshore Wind. https://energytransitionkorea.org/sites/default/files/2019-07/5.%20Copenhagen%20Offshore%20Middelgrunden.pdf

Lazar, J. (2016). *Electricity regulation in the US: A guide* (2nd ed.). The Regulatory Assistance Project. www.raponline.org/knowledge-center/electricityregulation-in-the-us-a-guide-2

Lenhart, S., Nelson-Marsh, N., Wilson, E. J., & Solan, D. (2016). Electricity governance and the Western energy imbalance market in the United States: The necessity of interorganizational collaboration. *Energy Research & Social Science, 19*, 94–107. http://doi.org/10.1016/j.erss.2016.05.015

Logan, J. R. (2016, June 30). Kit Carson CEO Reyes says Tri-State break has two big advantages. *The Taos News*. www.taosnews.com/stories/kit-carson-ceo-reyes-says-tri-state-break-has-two-big-advantages,23584

Ludt, B. (2019, November 7). New Mexico electric cooperative will soon supply customers 100% through solar. *Solar Power World*. www.solarpowerworldonline.com/2019/11/new-mexico-electric-cooperative-will-soon-supply-customers-through-solar/

Mey, F., & Diesendorf, M. (2018). Who owns an energy transition? Strategic action fields and community wind energy in Denmark. *Energy Research & Social Science, 35*, 108–117. https://doi.org/10.1016/j.erss.2017.10.044

Mey, F., Diesendorf, M., & MacGill, I. (2016). Can local government play a greater role for community renewable energy? A case study from Australia. *Energy Research & Social Science, 21*, 33–43. https://doi.org/10.1016/j.erss.2016.06.019

Monaci, M., & Caselli, M. (2005). Blurred discourses: How market isomorphism constrains and enables collective action in civil society. *Global Networks, 5*(1), 49–69. ISSN 1470-2266.

Mori, P. A. (2014). Community and cooperation: The evolution of cooperatives towards new models of citizens' democratic participation in public services provision. *Annals of Public and Cooperative Economics, 85*(3), 327–352. https://doi.org/10.1111/apce.12045

National Rural Electric Cooperative Association. (2019). *Co-op facts & figures*. www.electric.coop/wp-content/uploads/2019/10/Co-op-Facts-and-Figures_WEB_PAGES_10-11-19.pdf

National Rural Electric Cooperative Association. (n.d.). *Powering America*. www.electric.coop/our-mission/we-are-americas-electric-cooperatives/

Nelson-Marsh, N. (2017). Boundary spanning. In C. R. Scott, L. K. Lewis, J. R. Barker, J. Keyton, T. Kuhn, & P. K. Turner (Eds.), *International encyclopedia of organizational communication*. Wiley Publishers.

Ostrom, E. (2005). *Understanding institutional diversity*. Princeton University Press.

Pahl, G. (2007). *The citizen-powered energy handbook: Community solutions to a global crisis*. Chelsea Green Publishing Company.

Renews. (2018, November 14). Danes take back the Middelgrunden. *ReNews.Biz*. https://renews.biz/50044/danes-take-back-the-middelgrunden/

REScoop.eu. (n.d.). *Welcom to REScoop.eu*. www.rescoop.eu/

Rocha, V. (2017). *A co-op voice in the wholesale market*. www.cooperative.com/remagazine/articles/Pages/Celebrating-75-Years-of-NRECA.aspx

Stafford, B. A., & Wilson, E. J. (2016). Winds of change in energy systems: Policy implementation, technology deployment, and regional transmission organizations in the United States. *Energy Research & Social Science*, *21*, 222–236. https://doi.org/10.1016/j.erss.2016.08.001

Szulecki, K. (2018). Conceptualizing energy democracy. *Environmental Politics*, *27*(1), 21–41. https://doi.org/10.1080/09644016.2017.1387294

Taylor, K. (2015). Learning from the co-operative institutional model: How to enhance organizational robustness of third sector organizations with more pluralistic forms of governance. *Administrative Sciences*, *5*(3), 148–164. https://doi.org/10.3390/admsci5030148

Tri-State Generation and Transmission Association. (2015). *Integrated resource plan/electric resource plan*. http://doi.org/9771682584003-32963

UrStrom. (n.d.). *UrStrom BürgerEnergieGenossenschaft Mainz eG*. www.urstrom.de/

US Department of Energy. (2017, January). *Transforming the nation's electricity system: The second installment of the quadrennial energy review*. www.energy.gov/sites/prod/files/2017/02/f34/Quadrennial%20Energy%20Review-Second%20Installment%20%28Full%20Report%29.pdf

US Energy Information Administration. (2018, November 8). Electricity residential retail choice participation has declined since 2014 peak. *Today in Energy*. www.eia.gov/todayinenergy/detail.php?id=37452

US Energy Information Administration. (2019). *Annual electric generator report database (Form 861). Final data file for 2018*. (Dataset).US EIA.

van Veelen, B., & van der Horst, D. (2018). What is energy democracy? Connecting social science energy research and political theory. *Energy Research & Social Science*, *46*, 19–28. https://doi.org/10.1016/j.erss.2018.06.010

Wierling, A., Schwanitz, V. J., Zeiß, J. P., Bout, C., Candelise, C., Gilcrease, W., & Gregg, J. S. (2018). Statistical evidence on the role of energy cooperatives for the energy transition in European countries. *Sustainability (Switzerland)*, *10*(9). https://doi.org/10.3390/su10093339

Yildiz, Ö., Rommel, J., Debor, S., Holstenkamp, L., Mey, F., Müller, J. R., Radtke, J., & Rognli, J. (2015). Renewable energy cooperatives as gatekeepers or facilitators? Recent developments in Germany and a multidisciplinary research agenda. *Energy Research & Social Science*, *6*, 59–73. https://doi.org/10.1016/j.erss.2014.12.001

Youngblood, J. A. (1995). Alive and well, the rural electrification act preempts state condemnation law: City of Morgan City v. South Louisiana electric cooperative. *Energy Law Journal*, *16*(489), 491–492.

Zinck, A., & Newen, A. (2008). Classifying emotion: A developmental account. *Synthese*, *161*(1), 1–25. https://doi.org/10.1007/sl

5

ENERGY DEMOCRACY AT THE SCALE OF INDIGENOUS GOVERNANCE

Indigenous Native American struggles for democracy, justice, and decolonization

Danielle Endres and Taylor N. Johnson

We are in a period of energy transition. From protest over the Dakota Access Pipeline to activist calls for a just energy transition to controversies over community solar projects, energy transition is becoming one of the most important sociotechnical challenges that faces society. This challenge is made even more acute by the pressing ecological, cultural, and economic exigencies of the ongoing climate crisis. Narratives coming from many voices demand and make possible an energy transition that reconfigures our relationship with energy, that democratizes energy through participation in energy decision making, and that infuses principles of justice and equity into our collective future. Energy transition is already happening, but there are still opportunities to shape the transition and even to achieve a radical transformation in energy systems across a variety of scales. This is the goal of grassroots energy democracy activists. The emerging energy democracy movement has seen some important wins in this ongoing struggle: wins by Keystone XL protesters; Western Shoshone, Southern Paiute, and Skull Valley Goshute efforts to prevent nuclear energy waste storage on their lands; the Diné nation's ban on uranium mining; and *Honor the Earth*'s creation of Tribal Administrative Hearings in the absence of a sufficient governmental public participation process about the Dakota Access Pipeline. These enactments of more democratized and just forms of energy decision making constitute energy democracy.

Energy democracy is simultaneously a just and participatory ideal to which many communities strive, a process for making decisions about energy futures, and a set of practices that seek to shift power away from energy corporations to local communities, away from fossil fuels toward alternatives. Energy democracy is committed to more equitable distribution of harms and benefits across the entire life cycle of energy extraction, production, and consumption. In the past several years, there has been a slow but steady move within interdisciplinary energy studies to theorize, analyze, and collaborate with energy democracy practitioners. In an editorial essay, Feldpausch-Parker et al. (2019) lay out the justice, participation, and power framework for energy democracy, asserting that energy democracy theoretically lies at the intersections of these components while energy democracies in practice embody differing combinations of these elements.

DOI: 10.4324/9780429402302-6

In this chapter, we analyze energy democracy at the scale of Indigenous Native American governance and offer examples of how energy democracy is being practiced at this level. Actions by "water protectors" to stop the Dakota Access Pipeline provide a recent and visible example; yet the tentacles connecting Indigenous people and Nations in the United States and Canada to energy issues extend well beyond the Dakota Access Pipeline. Indigenous Nations, people, and organizations are engaged in protesting pipelines, resisting future fossil fuel development on Indigenous lands, banning fracking, promoting Indigenous government public participation processes, addressing sexual violence in extraction zones, and developing solar and wind projects on Indigenous lands.

This chapter begins by unpacking the relationships between Indigenous Nations, democracy, and governance. Then we explain energy democracy with an eye toward its relationship to Indigenous Nations. We use the justice, participation, and power framework to examine the enactments of energy democracy by Indigenous people and Nations. Next we offer a brief rhetorical analysis of the tactics that *Honor the Earth*, a North American Indigenous environmental justice organization, uses to engage in and support energy democracy at the scale of Indigenous governance. This analysis will examine how *Honor the Earth*'s rhetoric both enacts and challenges the principles, processes, and practices of energy democracy. The chapter concludes by considering the possibilities of energy democracy as a framework for a just response to the climate crisis for Indigenous Nations, many of which are disproportionately harmed by the impacts of the climate crisis.

Decolonizing democracy

Our focus on Indigenous governance in this chapter builds on a growing body of scholarship in Native American and Indigenous Studies (NAIS) centered on the specificities of settler colonialism experienced by Indigenous nations. This literature calls for scholars to attend to the particular forms of violence experienced by Indigenous communities, for this chapter particularly in the United States, and to recognize and celebrate the unique tactics through which Indigenous people engage in survivance, resurgence, and decolonization. This rich literature theorizes decolonization in ways that radically interrupt forms of critical theory and political theory, which insufficiently acknowledge and incorporate Indigenous governance, sovereignty, and ongoing structures of settler colonialism (Byrd, 2011; Simpson, 2014). To do so necessitates divesting from "settler futurities" that drive many progressive movements and reinvesting in Indigenous futurities (Tuck & Gaztambide-Fernández, 2013; Tuck & Yang, 2012). Decolonization in the context of Indigenous governance demands the rejection of settler institutions (such as the settler state) that constrain self-determination in favor of investing in sovereign Indigenous governance and decision making. Decolonization, then, is defined as "the repatriation of Indigenous land and life," which entails both the returning of land bases to Indigenous peoples and the (re)assertion of Indigenous epistemologies, systems of governance, and forms of relationality that have been marginalized through settler colonialism (Tuck & Yang, 2012, p. 1). Settler colonialism, which we understand as "a structure not an event," is a process of invasion in which non-Indigenous people arrive with the goal of replacing Indigenous people and producing a society that normalizes and legitimizes non-Indigenous presence and dominance in stolen Indigenous territories (Wolfe, 2006, p. 388).

Settler colonialism, from this perspective, cannot be disentangled from conceptions of American democracy, both because the production and maintenance of the United States nation-state is an inherently settler colonial project and because the ideals of American democracy are tied up with settler dominance. As Dahl (2018) writes, "[D]ispossession was not an unfortunate by-product of modern democracy, nor was settler colonial ideology an entirely separate political tradition from democratic thought. The two surged alongside each other and reinforced each other in their historical development" (p. 5). He argues that the core ideals that define American democracy—"pluralism, the rule of law, social equality, and popular sovereignty"—are rooted in processes of colonial agricultural settlement that cannot be separated from colonial dispossession (p. 2). Furthermore, Goldstein (2008) examines American resistance to the UN Declaration on the Rights of Indigenous Peoples, which posits that Indigenous self-determination "would fatally undermine U.S. democracy . . . because 'collective rights'—an enduring dilemma within the discourse of liberal legalism—supposedly threaten majority rule with the tyranny of a minority" (p. 837). This logic frames Indigenous people as a racial minority (rather than as peoples with distinct and inalienable claims to territory) who must assimilate to the will of an amorphous "people" whose collective will is the cornerstone of democratic governance in the United States (Byrd, 2011; Na'puti, 2019).

Given the role of settler colonialism in the development and continued practice of American democracy, we would be remiss to discuss Indigenous enactments of energy democracy or energy democracies without acknowledging the ways that democracy itself is a fraught term in the context of Indigenous governance and decolonization. This is not to say, however, that democracy is incompatible with decolonization. Indeed, many of the practices that make up American democracy are rooted in Indigenous governance and are compatible with Indigenous governance. Democracy, in our definition, is not synonymous with the American political system. Rather, it is a means of equitable collective decision making (Christiano, 2018). The Iroquois Confederacy, Cherokee Nation, Penacook Federation, Wabanaki Confederacy, and Powhatan Confederacy, among others, have been enacting forms of democratic governance since well before the advent of the United States, which could not have emerged without the models of democracy provided by Indigenous Nations (Begaye, 2008). Yet American democracy can be incompatible with decolonization in, for example, the constraints placed on Indigenous sovereignty when Indigenous Nations are framed as "domestic dependent" nations, the rhetorical positioning of Indigenous peoples' dual citizenship, and the enactment of Congress's plenary power (Bruyneel, 2007; Wilkins & Stark, 2010). By highlighting the paradoxical nature of American democracy as simultaneously reliant on Indigenous governance for its emergence and inextricable from settler colonialism, we assert that the practice of energy democracy at the scale of Indigenous governance necessarily involves understanding democracy in a broader sense than the practice of democracy in the United States political system. We argue that energy democracy, as a movement and research agenda, must attend to these complexities in any consideration of energy democracy at the scale of Indigenous governance.

What is energy democracy?

Energy democracy works at the intersections of energy systems decision making and democratic principles and practices for decision making.

Energy

Energy is "power that may be used to operate the infrastructures of the human-built environment. Humans derive that power from resources such as fossil fuels, solar, wind, hydroelectric, nuclear, biofuels, and geothermal sources that are extracted and harnessed, prepared, and distributed in a cycle of energy production" (Endres et al., 2016, p. 420). We are bracketing out other uses of the word "energy," such as strength, vitality, the capacity to be active, a spiritual force, or an embodied affect (e.g., Cozen, 2018), and narrowing consideration to those resources that humans use to fuel the technologies that support our lifestyles.

We find ourselves in a time of great changes in energy resources, production, and consumption, spurred on by ecological contexts, climate crisis, innovation, growing human populations, geopolitics, and sociotechnical imaginaries about the good life (for more on sociotechnical imaginaries, see Jasanoff & Kim, 2009). Quite simply, energy systems will change in the coming years. Many people still overwhelmingly consider energy transition as a technical phenomenon that requires the work of innovative scientists and engineers. Yet while the technoscientific aspects of energy are certainly important, we cannot ignore the sociopolitical aspects of energy resources, production, and consumption (Laird, 2013). Developing or advancing a technology does not guarantee that society will use it or that publics and policy makers will accept scientific evidence. Yet conceptual frameworks for energy transition often inadequately account for political dynamics, public engagement, grassroots civil society, and other sociopolitical aspects (Chilvers & Longhurst, 2016; Grin et al., 2010; Lawhon & Murphy, 2012). Building from this effort, we need to move away from viewing the sociopolitical elements as context to seeing them as an essential starting point for investigation into energy transitions and the inevitable decision making needed (Feldpausch-Parker et al., 2019). Ultimately, the impending energy transition offers a significant opportunity experiment with new forms of participation in decision making while attending to the intricacies and complexities of relations of power and forms of justice.

Democracy

The question of how to define democracy—as concept, ideal, form, practice, and model—is a perennial question in scholarly literature, making it difficult to narrow to one definition. For our purposes, we adopt the *Stanford Encyclopedia of Philosophy*'s definition of democracy as "a method of group decision making characterized by a kind of equality among the participants at an essential stage of the collective decision making" (Christiano, 2018). This definition aligns with the concept of participatory democracy that prizes active citizen participation from the bottom up, equity, and justice. Although Eberly (2002) warns of "delegitimation of participatory democracy in the United States" resulting from the growth of neoliberalism, this is a critique of how democracy is practiced in the United States (p. 290). We argue that participatory democracy also holds the potential for democratizing energy transition, particularly in the case of Indigenous governments.

Although not always immediately apparent, democracy is closely related to technology (Jasanoff, 2011; Latour, 2004). Explicit democratization of technological decision making can allow publics, social movements, and other stakeholders to be more involved in decision making that involves technology, in this case energy technology (Sclove, 1995; Winner, 1986). This allows for decision making about energy transition to be less technocratic and more in line with the principles of participatory democracy. Further, analysis of the interface between democracy and technology requires a consideration of the civic epistemologies, values, and cultural differences at play across and within nations (Jasanoff, 2011).

Energy democracy

Energy democracy describes efforts to make decision making about energy systems as democratic as possible. The goal of energy democracy is not to find the most democratic form of energy technology (Winner, 1986). Indeed, there is no single energy technology—even if we limit our purview to renewables—that will satisfy the needs of all communities. Rather, the goal is to think critically and inventively about the interrelations between energy and democracy in both broad and particular contexts. This is not an easy task and can involve a variety of different ways of conceiving of energy democracy. Yet, this task is necessary when we consider the alternative. As Schneider and Peeples (2018) note, the prevailing system of Energy Dominance in the United States "emphasizes central control, fossil fuels, swift decision making that favors private industry, and short-term profits" (p. 9).

Feldpausch-Parker et al. (2019) argue that *energy democracy* refers to an "emergent social movement that re-imagines energy consumers as prosumers, or innovators, designers, owners, and analysts who are involved in decisions at every stage, from production through consumption" (p. 2). Energy democracy includes a broad range of actors, democratic values, sites of participation, forms of justice, and power dynamics (Burke & Stephens, 2017; Fairchild & Weinrub, 2017; Szulecki & Overland, 2020). Importantly, Chilvers and Pallett (2018) remind us to focus not on one definition of energy democracy but on "multiple diverse *energy democracies* which intermingle in wider systems" (p. 3). We must also reserve space for the possibility of energy democracy taking hold in nondemocratic countries (Delina, 2018). In this chapter, we are interested in those energy democracies that are emerging and already happening in Indigenous Nations.

Justice, participation, and power within energy democracy

We use Feldpausch-Parker et al.'s (2019) justice, participation, and power framework to analyze enactments of energy democracy at the Indigenous governance scale. Justice, as a key component of participatory democracy, centers marginalization, inequality, and privilege in any decision making about energy framework. The justice component affirms the significant role that social and environmental justice, climate and energy justice, and just transition concepts and practices have played in energy democracies and "encourages scholars to ask questions about, for example, who is served, what is the role of structural inequities" in energy decision making (Feldpausch-Parker et al., 2019, p. 3). Participation focuses on individual and collective actors' abilities to meaningfully engage in the decisions about energy that affect them via access to a variety of democratic processes and practices. These may include formal modes of public participation enacted by federal, state, local, and Indigenous governments, protests that make voices that have been silenced in formal participatory processes heard, and other means of making room for involvement of stakeholders who have been overlooked, silenced, or excluded from decision-making processes. Power, in this case, is not a synonym for energy but rather refers to "a relationship between human actors and their capacities to act or not freely" to elicit change within structures and systems of governance and within energy decision making in particular (Feldpausch-Parker et al., 2019, p. 4). Power is intrinsically related to conceptualizations of justice and participation but allows for focusing on the structures and practices that enable and constrain the agency of stakeholders to act in line with their interests. Energy democracy will involve a variety of shifts in power.

In the spirit of considering multiple energy democracies, this framework creates a heuristic that allows for examination of theoretical models, empirical examples of ongoing struggles over energy, and practical recommendations for communities engaged in promoting energy democracy that exist in the interplay between the elements, rather than determining an ideal configuration between these components. Taken together, justice, participation, and power are not simply words that appear frequently in the discourse of energy democracy advocates; they are necessary to the democratization of energy transition. Seeing energy democracy as being made up of a complex web of relationships among justice, participation, and power serves as a framework with which scholars can examine particular localized performances of energy democracy.

Energy democracy enacted at the indigenous governance scale

Informed publics need not look far to find examples of energy democracies in practice. Indigenous resistance to Tar Sands Oil Development, Keystone XL, and Dakota Access Pipeline protests, rooftop solar development in Salt Lake City and other cities, energy municipalization in Boulder, calling attention to the dirty cradle and grave of nuclear energy, and Puerto Rico's post-Katrina deliberation about energy infrastructure and futures are some of the many examples that pervade our screens (de Onís, 2018; Endres, 2009b; Johnson, 2019; Reinig & Sprain, 2016). Energy democracy advocates call attention to the public nature of energy decision making, seeking to democratize these decisions. Energy democracy advocates engage in both mundane and extraordinary enactments that seek to assert their right to participate in decisions that affect them, highlight and seek to redress injustices, and engage in a bottom-up redistribution of power.

We focus on the efforts of *Honor the Earth* to seed and support energy democracy enactments by Indigenous governments. *Honor the Earth* is an Indigenous Native American and First Nation environmental organization rooted in Anishanaabe lands (now called the United States and Canada), peoples, and values. Anishanaabe includes Odawa, Ojibwe, Potawatomi, Oji-Cree, and Algonquin nations. Led by executive director Winona LaDuke (Ojibwe of the White Earth Reservation in Minnesota) and cofounded by settler (non-Indigenous) performers, Indigo Girls, *Honor the Earth*'s mission is "to create awareness and support for Native environmental issues and to develop needed financial and political resources for the survival of sustainable Native communities" (Honor the Earth, n.d.a). *Honor the Earth* is an example of an unlikely alliance between Indigenous and settler (often white) environmentalists (Grossman, 2017). The organization is unique in its focus on not only raising awareness to outside communities about issues facing Indigenous Nations but also in building capacity for Indigenous people and Nations to engage in campaigns that enact sovereignty, self-determination, and governance by and for Indigenous communities (Honor the Earth, n.d.b). Energy is a major focus of *Honor the Earth*'s campaigns.

> The process of determining our destiny is at the core of our survival as Indigenous peoples. As tribal communities grow and we deepen our strategies and infrastructure for our Nations, it is essential for us to look at the world's economic and environmental realities in order to make critical decisions about our future. That means we must address issues such as climate change, peak oil and food insecurity. Food and energy consume huge portions of our tribal economies and must be considered in relation to tribal sovereignty and self-determination.
>
> *(Honor the Earth, n.d.j, preface)*

Energy, then, is not simply a context for *Honor the Earth*'s work. Rather, it is tied to the survival of healthy Indigenous communities and to the enactment of sovereignty and self-determination. We argue that the specific campaigns that *Honor the Earth* pursues are enactments of energy democracy at the scale of Indigenous governance. To support this argument, we analyzed a variety of *Honor the Earth*'s texts, including the website, social media posts, imagery, and reports.

Honor the Earth's tactics of energy democracy

In what follows, we highlight three enactments of energy democracy at the Indigenous governance scale. These enactments are both instrumental and consummatory, in the sense that they are geared at both enacting a particular goal (e.g., stopping the Dakota Access Pipeline) and at enacting Native American self-determination and cultural sovereignty (Endres, 2011; Lake, 1983). This mix of instrumental and consummatory enactment is evidenced in *Honor the Earth*'s statement of goals:

> Native people are in a pivotal position in this time and region. It is essential that we affirm principled and culturally-driven agency. That is to say that, tribal communities often conflicted over extraction as a result of a historic set of decisions forced upon us, are able to be essential agents of change in this time. Honor the Earth will work in the next two years, with first nations, Indigenous communities, and tribal governments to oppose extraction, support tribal regulatory push for environmental regulation, strengthen renewable energy and food systems work in our region, and create a curriculum and learning tool for tribal youth in Indigenous Economics.
>
> *(Honor the Earth, n.d.a)*

Through campaigns focused on sustainable forms of economic self-determination, energy installation on reservations, and participation in decision making, *Honor the Earth* seeks to not only critique dominant energy practices but also compose viable processes of energy democracy.

Sustainable economic self-determination

Honor the Earth views a transformed economy for Indigenous Nations as a key element in adapting to the climate crisis, engaging in decolonizing resistance, enacting self-determination, and localizing the energy transition. Within this campaign, "Honor the Earth is engaging in a two-year program of collaborative organizing on climate change, opposition to extreme [energy] extraction and laying the groundwork for restored Indigenous economies in Native American communities" (Honor the Earth, n.d.i). This campaign, which includes conducting economic assessments for Indigenous governments, creating local energy and food economies, supporting education in tribal colleges, and creating local energy jobs is highlighted as a means to support self-determination while contributing to an energy transformation away from fossil fuels. This relies on the energy resources that already exist in Indigenous nations.

Honor the Earth's economic transformation cannot be considered outside of ongoing processes of settler colonialism, struggles for self-determination, and sovereignty (Black, 2009; Endres, 2009a; Tuck & Yang, 2012; Whyte, 2016, 2017). As such, sustainable economic self-determination, a form of resistance to conquest, is also an enactment of sovereignty and self-determination. In a report called "Sustainable Tribal Economies: A Guide to Restoring Energy and Food Sovereignty in Native America," *Honor the Earth* notes that creating local economies with local energy production within Native American Nations is key to enacting sovereignty:

> In all cases, we are looking at the creation of local economies, using the resources available to each Indigenous community. We are hopeful that some of these strategies will not only be viable for tribal self-determination, but also, when appropriate, be a possible source of export revenues for tribal communities. Recovering and restoring local food and energy production requires a conscious transformation and set of

> technological and economic leaps for our communities. We must decide whether we want to determine our own future or lease it out for royalties. In the end, developing food and energy sovereignty is a means to determine our own destiny.
>
> *(Honor the Earth, n.d.j, preface)*

Economic self-determination is needed, in part, as a form of resistance to ongoing patterns of colonialism and dependency: "We cannot erase the process of economic colonization and the deliberate creation of dependency. But we can join with others and take action to reclaim our future" (Honor the Earth, n.d.j, p. 5). *Honor the Earth* then poses a challenge to Indigenous Nations to take control of their own decision-making processes. They argue that "decisions will either be made for us or we will make our own decisions about how to proceed in developing green economic opportunities and our future. By making our own decisions and taking action to establish and implement sustainable economic development, our tribal communities will exercise sovereignty and forge a green path for our coming generations" (Honor the Earth, n.d.j, p. 28). Economic self-determination enacts energy democracy through considerations of justice and power (colonization, cultivated dependencies, and the fossil fuel regime) with a participatory solution that returns democratic decision making to tribal governments.

Sustainable energy installation on tribal lands

In addition to their campaign to encourage economic self-determination for Indigenous Nations, *Honor the Earth* also engages in the project of energy democracy by promoting renewable energy on Indigenous lands. One of the authors first learned about this part of *Honor the Earth*'s campaigns when she visited her late friend Margene Bullcreek at the Skull Valley Reservation. Danielle noticed that Margene had solar panels installed on her home, and Margene told her that Winona LaDuke and *Honor the Earth* had come out to install it for her as a way to promote energy self-sufficiency at her remote home. Danielle soon found out that Bullcreek's home was just one of many *Honor the Earth* projects to bring renewable energy to particularly remote Indigenous reservations. *Honor the Earth* does not just argue for the use of renewable solar and wind energy on Native lands but actually funds and installs renewable energy projects and builds capacity through "advocacy groundwork, training, and technical support" for Native nations to be able do this work themselves (Honor the Earth, n.d.f). *Honor the Earth* positions this campaign as a mechanism to bring efficient and sustainable energy to Indigenous governments that need it while also developing knowledge and fostering energy justice.

> The reality is that building a renewable energy economy on Native lands (and restoring local, non-industrial food systems and foods themselves) will not only help mitigate the climate change crisis, but also address the poverty and social injustices that plague our communities. We will continue to oppose the fossil fuel and nuclear economy, with your help. And we will support our communities to restore the Indigenous knowledge, foods, and ways of living for the next generations. The technology and the wind and solar resources exist to transform a highly inefficient and exploitative energy production system into one that is safe and clean, a transformation that would signify an era of energy justice.
>
> *(Honor the Earth, n.d.b)*

Beyond the positive impacts of renewable energy projects for Indigenous lands, these projects can also be a way for Indigenous communities to atone for the role of Indigenous reservation lands in supporting fossil fuel and nuclear energy under a system of resource colonialism (Gedicks, 1994). According to Honor the Earth:

> In the US—the largest and most inefficient energy economy in the world—tribal communities have long supplied the raw materials for nuclear and coal plants, huge dam projects, and oil and gas development. These resources have been exploited to power far-off cities and towns, while we remain in the toxic shadow of theory lethal pollution and without our own sources of heat an electricity.
>
> *(n.d.j, preface)*

Installing renewable energy on Indigenous lands allows these communities to resist the "prominent false solutions that mirror the existing paradigm of energy conquest, and simply extend our reliance on a fossil fuel and nuclear economy" (Honor the Earth, n.d.j, p. 23). The installation of solar energy on Indigenous lands is an embodied and emplaced act of localized energy democracy.

Honor the Earth also integrates their sustainable energy installation campaign into the other work they do. *Honor the Earth* was a key participant in the Dakota Access Pipeline Protests, providing various forms of support to the activists while also standing with them on the front lines. Yet, in addition to fighting the Dakota Access Pipeline, *Honor the Earth* also worked with the Standing Rock Reservation government to address energy poverty.

> In addition to battling the pipeline, we also continue to work towards energy infrastructure for the Standing Rock Reservation that actually serves its people. After all, three years ago Debbie Dogskin, a Standing Rock resident, froze to death because she could not pay her propane bill. That is the reality here. We plan to install 20 solar thermal panels on tribal houses at Standing Rock, beginning to address fuel poverty on the reservation.
>
> *(Honor the Earth, n.d.d)*

This example indicates *Honor the Earth* does not see their campaigns separately but instead views fighting the fossil fuel industry hand in hand with bringing renewable energy to reservations. Supporting sustainable renewable energy installation on Indigenous lands is a form of energy democracy by being a part of the energy transition.

Installing renewable energy on Indigenous lands engages most heavily with the justice and power dimensions of the framework, yet it does not engage strongly with the participation dimension. *Honor the Earth*'s materials, while emphasizing the need to choose a new and just energy path that frees dependency from fossil fuels and US colonialism, do not emphasize the importance of Indigenous government decision making about whether renewable energy is best for the land under consideration. In other words, *Honor the Earth*'s messaging attempts to persuade Indigenous Nations on the importance of renewable energy, which would be enhanced by explicitly bringing in participatory processes within the Indigenous Nation prior to funding and building capacity to install solar and wind energy on the Nation's lands. Without the participatory element, this tactic risks imposing a new energy system on a community without a clear process for local community deliberation and participation in the decision.

Enacting participation in decision making

In a variety of campaigns, *Honor the Earth* encourages participatory democracy through protest, creating Indigenous government-sponsored public comment processes, and supporting Indigenous governments in making decisions to ban hydraulic fracturing (fracking). As previously noted, participation in energy democracy can take many forms depending on the situation. *Honor the Earth* highlights these forms of protest, public comment, and fracking bans as responses to unjust policies imposed by powerful federal and/or industry groups that have such limited venues for participation that these become the only options for participation. *Honor the Earth* regularly supports protest events as a way to encourage and promote the necessity of Indigenous people's participation in decisions that affect them. In addition to contributing to the Dakota Access Pipeline Protests, the *Honor the Earth* website lists a variety of pipeline projects for which protesting played a big role. Yet participation goes well beyond protest.

In response to perceived limitations in the Army Corp of Engineers' public participation process with regard to the Dakota Access Pipeline, *Honor the Earth* worked with the Standing Rock Sioux government to hold tribal government-sponsored public hearings (Johnson, 2019).

> We are also working closely with the Standing Rock Sioux Tribe as they hold a series of public hearings on the reservation, collecting public comment and expert testimony about the impacts of the pipeline—basically, doing the public outreach that the Army Corps never did.
>
> *(Honor the Earth, n.d.d)*

These public hearings offered a way that public comments could be "received and recorded to address public concerns for the environment and the potential impacts to the drinking water, tribal wildlife, natural resources, spiritual, historical and cultural sites of Standing Rock" (Honor the Earth, 2016). This public hearing is more than a symbolic alternative. As a sovereign nation, the Standing Rock Sioux Tribe can offer a governmental public participation process that simultaneously allows for comments to be heard but also reaffirms the Standing Rock Sioux as a sovereign government as opposed to being assimilated in the United States public, as often happens to Native Americans in public hearing processes (Endres, 2009a). In a similar example, in response to the Line 3 Oil Pipeline Project, the Minnesota Chippewa nation decided to prepare its own environmental impact statement as a form of participation in the decision-making process. *Honor the Earth* (n.d.g) notes:

> The Minnesota Chippewa Tribe is a federally recognized tribal government comprised of 6 Ojibwe/Chippewa Bands (Bois Forte, Fond du Lac, Grand Portage, Leech Lake, Mille Lacs, and White Earth) and their more than 42,000 members. A tribal EIS is a very rare thing in the context of big infrastructure projects. This is a bold and courageous assertion of self-determination.

This highlights that participation, in the case of Indigenous governments, can happen both within and outside of US federal government practices, and *Honor the Earth* presents the public processes started by Indigenous governments as crucial acts of self-determination that democratize the participation process in a way that affirms the sovereign status of Indigenous Nations. In doing so, *Honor the Earth* rejects the colonial mind-set that invests greater legitimacy in decision-making processes facilitated by the settler state, instead centering processes that are rooted in Indigenous self-determination.

A final way in which *Honor the Earth* enacts participation is in fracking bans adopted by the Turtle Mountain Chippewa and Haudenosaunee nations. According to Honor the Earth (n.d.e):

> The Turtle Mountain band of Chippewa passed a resolution banning fracking, along with the Haudenosaunee. For copies of the resolutions, and other resources on how to

> get your tribal council to ban fracking please visit the resources section on this website, or see below.

Indigenous government resolutions are a form of participation in energy decision making that seek to upend the power dynamics between the federal government and Indigenous people by offering a more just form of participation in decisions. Even if these public processes do not succeed in changing the outcomes of decision-making processes, they do serve an important role as enactments of self-determination.

A potential limitation in Honor the Earth*'s campaigns*

Thus far, our analysis has shown that *Honor the Earth*'s campaigns support performances of energy democracy at the scale of Indigenous governance. It is important, however, to reflect on a potential limitation in *Honor the Earth*'s campaigns that might undermine the principles of energy democracy. *Honor the Earth* is not an Indigenous government; it is a nonprofit group set up to support Indigenous communities, including Indigenous governments. *Honor the Earth*'s No Coal Goal on Diné (Navajo) and Crow lands campaign opposes Diné and Crow tribal governments' decisions to pursue continued development of fossil fuel energy on their reservation lands. Similar to Endres's (2015) work on the Ute Tribal Government's approval of the Utes' nickname, this example demonstrates the tensions that arise when arguing against the sovereign decision of an Indigenous Nation. Because of the absolute importance of sovereignty in Indigenous rights, engaging in an argument against the sovereign decision of another Nation could undermine a right at the very core of Indigenous decolonization. On the other hand, sovereignty can become dangerous when used to stifle dissent (Endres, 2015). In this case, intertribal conflicts between the Hopi and Diné about coal energy operate at the intersection of these tensions, as does *Honor the Earth* as a pro-Indigenous sovereignty group.

Honor the Earth attempts to address this dilemma in three interrelated ways. First, *Honor the Earth* argues that they are working to support Diné and Crow grassroots activists engaged in protest against their own Diné and Crow governments. For example:

> We believe it is possible to avert the purchase of the coal strip mine, close a power plant, and create and actualize a transition strategy to renewable solar power for the Navajo nation. Based on our long and illustrious history in working at Diné Bii Kaya [an activist group of Diné people], we would like to continue and deepen this work in partnership with grassroots organizers in 2013 and actualize this vision.
>
> *(Honor the Earth, n.d.h)*

Honor the Earth positions itself as an ally to the Diné activists, not an enemy of the Diné government. If any government's decisions are subject to dissent and protest, the act of protest is not inherently an act of delegitimizing Indigenous sovereignty. Second, *Honor the Earth* implies that the Diné tribal government's decision is complicit with colonialism by using sovereignty in a way that compounds the exploitation of their people. *Honor the Earth* (n.d.h) contends, "It appears that the Navajo nation, like other tribal governments, will seek to use its sovereign status to protect the fossil fuel industry and continue exploitation." Similarly, with regard to the Crow, *Honor the Earth* (n.d.c) argues, "Tribal sovereignty and tribal politics have kept the Crow largely out of the eye of the storm on fossil fuels extraction, as jobs are hard to come by, political allies are able to keep jobs and others are not, and the cloak of invisibility which covers tribal sovereignty is likely to be used to keep others out." This framing allows *Honor the Earth* to uphold sovereignty as a goal for Indigenous governments, while arguing against how the particular nation is enacting their sovereignty. Finally, *Honor the Earth* attributes the actions of the Diné and Crow governments to the conditions of colonialism that have created unfair policies and dependencies that are difficult to escape. They state:

> Tribal economics like Crow are skewed towards fossil fuels as a result of historic leasing policies by the Bureau of Indian Affairs, complemented by large amounts of fossil fuel influence, and a lack of re-indigenized and long-term self-sustainable development strategies. The problems are mirrored throughout Native America, and require a careful deconstruction, in order to solve. This is our work.
>
> *(Honor the Earth, n.d.c)*

Honor the Earth justifies its campaign against the Diné and Crow governments' decisions as a fight against Indigenous Nations who are not using their sovereignty responsibly but cannot be blamed for such because of colonialism. Yet, even while *Honor the Earth* attempts to negotiate this tension with regard to its struggles against Diné and Crow governments, the rhetorical consequence could be at odds with the principles of participation within these governments. This tension draws attention to a question that troubles energy democracy in multiple contexts: What happens when a democratic process comes to a conclusion that does not encourage a sustainable energy transition?

Conclusion

In this chapter, we have used the example of *Honor the Earth*'s campaigns to unpack practices of energy democracy at the scale of Indigenous governments. The most important consideration for any discussion of energy democracy for Indigenous people and communities, particularly in North America, is to recognize those people and communities as members of sovereign Nations. While the way that democracy is being practiced in the United States is incommensurable with Indigenous sovereignty and decolonization, democracy as an equitable and participatory process is certainly commensurable with Indigenous sovereignty and decolonization. *Honor the Earth*'s work with various Indigenous communities and governments exemplifies how justice, participation, and power interact to create localized enactments of energy democracies for Indigenous Nations as well as where the energy democracy framework reveals challenges.

Energy democracy for *Honor the Earth* is also inherently connected to health, self-determination, economics, food, fuel poverty, and other considerations. Indeed, recall that many of the quotations we presented from *Honor the Earth*'s campaigns seamlessly moved between energy and food sustainability or between the role of local energy and food production in sustainable economies. When we think about energy democracies at the scale of Indigenous governance, as opposed to perhaps viewing Indigenous people as stakeholders with the settler nations or as communities of protestors rather than citizens of Indigenous Nations, we are forced to confront the ways in which Indigenous governance is not just about energy systems, although that is certainly important, but also about the broader well-being of Indigenous communities. Because Indigenous governments juggle all the same issues that face any government, energy democracy has to exceed its focus on energy systems to consider how those systems may benefit or harm the sustainability of the entire nation.

In closing, we must acknowledge that a limitation of our chapter is its exclusive focus on Indigenous Nations that share boundaries with the United States and in some cases with Canada. Our perspective highlights the unique situations of Indigenous people and nations in North

America. To gain a fuller understanding of energy democracies at the scale of Indigenous governance, it is important to expand beyond the United States context to understand the forms and practices of Indigenous governance in other regions of the world, as well as acknowledging that many Indigenous communities are not recognized as governments. As such, a consideration of the relationship between energy democracy and Indigenous people and communities necessarily exceeds our focus on the scale of Indigenous governance. We welcome extensions of this conversation that attend to other vectors of this relationship.

References

Begaye, T. (2008). Modern democracy: The complexities behind appropriating indigenous models of governance and implementation. In N. K. Denzin, Y. S. Lincoln, & L. T. Smith (Eds.), *Handbook of critical and indigenous methodologies* (pp. 459–470). Sage Publications, Inc.

Black, J. E. (2009). Native resistive rhetoric and the decolonization of American Indian removal discourse. *Quarterly Journal of Speech, 95*(1), 66–88. https://doi.org/10.1080/00335630802621052

Bruyneel, K. (2007). *The third space of sovereignty: The postcolonial politics of US-Indigenous relations*. University of Minnesota Press.

Burke, M. J., & Stephens, J. C. (2017). Energy democracy: Goals and policy instruments for sociotechnical transitions. *Energy Research & Social Science, 33*(Supplement C), 35–48. https://doi.org/10.1016/j.erss.2017.09.024

Byrd, J. A. (2011). *The transit of empire: Indigenous critiques of colonialism*. University of Minnesota Press.

Chilvers, J., & Longhurst, N. (2016). Participation in transition(s): Reconceiving public engagements in energy transitions as co-produced, emergent and diverse. *Journal of Environmental Policy & Planning, 18*(5), 585–607. https://doi.org/10.1080/1523908X.2015.1110483

Chilvers, J., & Pallett, H. (2018). Energy democracies and publics in the making: A relational agenda for research and practice. *Frontiers in Communication, 3*, 14.

Christiano, T. (2018, Fall). Democracy. In E. N. Zalta (Ed.), *The Stanford encyclopedia of philosophy*. https://plato.stanford.edu/archives/fall2018/entries/democracy/

Cozen, B. (2018). Stabilizing energies: Intersections between energy promotion texts and rhetorical theory. In B. McGreavy, J. Wells, G. F. McHendry, Jr., & S. Senda-Cook (Eds.), *Tracing rhetoric and material life* (pp. 315–342). Springer.

Dahl, A. (2018). *Empire of the people: Settler colonialism and the foundations of modern democratic thought*. University Press of Kansas.

Delina, L. L. (2018). Can energy democracy thrive in a non-democracy? *Frontiers in Environmental Science, 6*. https://doi.org/10.3389/fenvs.2018.00005

de Onís, C. M. (2018). Energy colonialism powers the ongoing unnatural disaster in Puerto Rico. *Frontiers in Communication, 3*. https://doi.org/10.3389/fcomm.2018.00002

Eberly, R. A. (2002). Rhetoric and the anti-logos doughball: Teaching deliberating bodies the practices of participatory democracy. *Rhetoric and Public Affairs, 5*(2), 287–300.

Endres, D. (2009a). The rhetoric of nuclear colonialism: Rhetorical exclusion of American Indian arguments in the Yucca Mountain nuclear waste siting decision. *Communication and Critical/Cultural Studies, 6*(1), 39–60. https://doi.org/10.1080/14791420802632103

Endres, D. (2009b). From wasteland to waste site: The role of discourse in nuclear power's environmental injustices. *Local Environment, 14*(10), 917–937. https://doi.org/10.1080/13549830903244409

Endres, D. (2011). American Indian activism and audience: Rhetorical analysis of Leonard Peltier's response to denial of clemency. *Communication Reports, 24*(1), 1–11. https://doi.org/10.1080/08934215.2011.554624

Endres, D. (2015). American Indian permission for mascots: Resistance or complicity within rhetorical colonialism? *Rhetoric & Public Affairs, 18*(4), 649–689.

Endres, D., Cozen, B., Barnett, J. T., O'Byrne, M., & Peterson, T. R. (2016). Communicating energy in a climate (of) crisis. In *Communication yearbook* (Vol. 40, pp. 419–447). Routledge.

Fairchild, D., & Weinrub, A. (Eds.). (2017). *Energy democracy: Advancing equity in clean energy solutions*. Island Press.

Feldpausch-Parker, A. M., Endres, D., & Peterson, T. R. (2019). Editorial: A research agenda for energy democracy. *Frontiers in Communication, 4*. https://doi.org/10.3389/fcomm.2019.00053

Gedicks, A. (1994). *The new resource wars: Native and environmental struggles against multinational corporations*. Black Rose Books Ltd.

Goldstein, A. (2008). Where the nation takes place: Proprietary regimes, antistatism, and US settler colonialism. *South Atlantic Quarterly, 107*(4), 833–861. https://doi.org/10.1215/00382876-2008-019

Grin, J., Rotmans, J., & Schot, J. (2010). *Transitions to sustainable development: New directions in the study of long term transformative change*. Routledge.

Grossman, Z. (2017). *Unlikely alliances: Native nations and White communities join to defend rural lands*. University of Washington Press.

Honor the Earth. (2016, November 4). *Standing rock hearings to begin: Protect the lands, and water no DAPL*. www.honorearth.org/standingrockhearings

Honor the Earth. (n.d.a). *About us*. Retrieved June 11, 2018, from www.honorearth.org/about

Honor the Earth. (n.d.b). *Background*. Retrieved June 11, 2018, from www.honorearth.org/about_us

Honor the Earth. (n.d.c). *Crow coal*. Retrieved June 21, 2018, from www.honorearth.org/crow_coal

Honor the Earth. (n.d.d). *Dakota access pipeline*. Retrieved June 18, 2018, from www.honorearth.org/dapl

Honor the Earth. (n.d.e). *Frack no*. Retrieved June 21, 2018, from www.honorearth.org/frack_no

Honor the Earth. (n.d.f). *Indigenous economics for the seventh generation*. Retrieved June 21, 2018, from www.honorearth.org/indigenous_economics

Honor the Earth. (n.d.g). *Line 3*. Retrieved June 19, 2018, from www.honorearth.org/sandpiper_line_3_corridor

Honor the Earth. (n.d.h). *Navajo and Diné Bii Kiya*. Retrieved June 21, 2018, from www.honorearth.org/navajo_and_din_bii_kiya

Honor the Earth. (n.d.i). *Our campaigns*. Retrieved June 11, 2018, from www.honorearth.org/campaigns

Honor the Earth. (n.d.j). *Sustainable tribal economies: A guide to restoring energy and food sovereignty in Native America*. Honor the Earth.

Jasanoff, S. (2011). *Designs on nature: Science and democracy in Europe and the United States*. Princeton University Press.

Jasanoff, S., & Kim, S. H. (2009). Containing the atom: Sociotechnical imaginaries and nuclear power in the United States and South Korea. *Minerva, 47*(2), 119–146. https://doi.org/10.1007/s11024-009-9124-4

Johnson, T. N. (2019). The Dakota access pipeline and the breakdown of participatory processes in environmental decision-making. *Environmental Communication, 13*(3), 335–352. https://doi.org/10.1080/17524032.2019.1569544

Laird, F. N. (2013). Against transitions? Uncovering conflicts in changing energy systems. *Science as Culture, 22*(2), 149–156. https://doi.org/10.1080/09505431.2013.786992

Lake, R. A. (1983). Enacting red power: The consummatory function in Native American protest rhetoric. *Quarterly Journal of Speech, 69*(2), 127–142. https://doi.org/10.1080/00335638309383642

Latour, B. (2004). *Politics of nature: How to bring the sciences into democracy* (C. Porter, Trans.). Harvard University Press.

Lawhon, M., & Murphy, J. T. (2012). Socio-technical regimes and sustainability transitions: Insights from political ecology. *Progress in Human Geography, 36*(3), 354–378. https://doi.org/10.1177/0309132511427960

Na'puti, T. R. (2019). Speaking of indigeneity: Navigating genealogies against erasure and #RhetoricSoWhite. *Quarterly Journal of Speech, 105*(4), 495–501. https://doi.org/10.1080/00335630.2019.1669895

Reinig, L., & Sprain, L. (2016). *Cultural discourses of public engagement: Insights for energy system transformation*. Iowa State University Summer Symposium on Science Communication. https://lib.dr.iastate.edu/sciencecommunication/2016/proceedings/13

Schneider, J., & Peeples, J. (2018). The energy covenant: Energy dominance and the rhetoric of the aggrieved. *Frontiers in Communication, 3*. https://doi.org/10.3389/fcomm.2018.00005

Sclove, R. (1995). *Democracy and technology*. Guilford Press.

Simpson, A. (2014). *Mohawk interruptus: Political life across the borders of settler states*. Duke University Press.

Szulecki, K., & Overland, I. (2020). Energy democracy as a process, an outcome and a goal: A conceptual review. *Energy Research & Social Science, 69*, 101768. https://doi.org/10.1016/j.erss.2020.101768

Tuck, E., & Gaztambide-Fernández, R. A. (2013). Curriculum, replacement, and settler futurity. *Journal of Curriculum Theorizing, 29*(1), Article 1. https://journal.jctonline.org/index.php/jct/article/view/411

Tuck, E., & Yang, K. W. (2012). Decolonization is not a metaphor. *Decolonization: Indigeneity, Education & Society, 1*(1), 1–40.

Whyte, K. P. (2016). Indigenous experience, environmental justice and settler colonialism. In B. E. Bannon (Ed.), *Nature and experience: Phenomenology and the environment*. Rowman & Littlefield International. https://papers.ssrn.com/abstract=2770058

Whyte, K. P. (2017). The Dakota access pipeline, environmental injustice, and US colonialism. *Red Ink*, *19*(1), 154–169.

Wilkins, D. E., & Stark, H. K. (2010). *American Indian politics and the American political system* (3rd ed.). Rowman & Littlefield Publishers.

Winner, L. (1986). *The whale and the reactor: A search for limits in an age of high technology*. University of Chicago Press.

Wolfe, P. (2006). Settler colonialism and the elimination of the native. *Journal of Genocide Research*, *8*(4), 387–409. https://doi.org/10.1080/14623520601056240

6

CONCEPTUALIZING ENERGY DEMOCRACY USING THE MULTIPLE STREAMS FRAMEWORK

Actors, public participation, and scale in energy transitions

Nihit Goyal and Michael Howlett

Introduction

The growing realization that climate justice (equitable distribution of the burden of climate change), energy justice (a human rights perspective to the energy life cycle), and environmental justice (involvement of all citizens in environmental policy making) need to be considered holistically rather than in silos has led to a call for a "just transition" to a low-carbon economy (Heffron & McCauley, 2018). This transition lies at the heart of the idea of energy democracy, which emerged from a movement to "recapture" energy utilities and accelerate the transformation to renewable energy (Farrell, 2014).

A "just" transition will entail governing the synergies and trade-offs involved in shifting to low-carbon energy while catalyzing universal energy access and ensuring justice for those dependent on the fossil fuel economy (Newell & Mulvaney, 2013). Scholars have argued that this process will not suffice and that catalyzing it will require grassroots mobilization to not only support but also challenge "top-down" action (Evans & Phelan, 2016; Healy & Barry, 2017; Routledge et al., 2018). Consequently, energy democracy has received increasing attention as a model for governing a just energy system.

To analyze whether energy democracy can meet this goal, a better understanding of the concept and its relationship to energy transitions is necessary. While recent work has advanced research on the topic, the theoretical underpinnings of energy democracy remain vague (van Veelen & van der Horst, 2018). While for some it represents a continuous struggle, even beyond the electoral arena, against an established energy regime (Angel, 2016), for others it is an outcome of public participation in the ongoing transition (Vansintjan, 2015) toward an efficient, equitable, flexible, low-carbon, but also locally governed energy system (Farrell, 2014). Although one reason for this divergence in the notion of energy democracy is its origin in the gray literature, the emergent nature of transitions also leads to the possibility of multiple "energy democracies" (Chilvers & Pallett, 2018).

DOI: 10.4324/9780429402302-7

Consequently, numerous ways of classifying and disaggregating energy democracy have been proposed. Welton (2018), for example, has suggested thinking of energy democracy as consumer choice, ownership or control of energy infrastructure, and access to the decision-making process. Comparably, Szulecki (2018) has classified energy democracy based on civic ownership, popular sovereignty, and participatory governance. More broadly, Chilvers and Longhurst (2016) have proposed disaggregation based on subjects, objects, and procedural formats of energy democracy, and Burke (2018) has identified actors, collective action frames, discourses, and futures as the elements of energy democracy. Meanwhile, van Veelen and van der Horst (2018) have distinguished the rationale for pursuing energy democracy, the stakeholders involved, the types of engagement necessary, and the scale at which energy democracy can be achieved.

Although these typologies seem difficult to reconcile and synthesize, their analytical focus is easier to discern: democratize what (activities), by whom (actors), why (rationales), how (modes), where (scales), and to what effect (outcomes). In this chapter, we delve into three questions specifically: democratizing what, democratizing how, and democratizing where. We develop a conceptual framework for energy democracy to facilitate analytical inquiry along these dimensions of energy democracy and help examine whether and how they are related empirically. To do so, we use the multiple streams framework (MSF), which was originally proposed by Kingdon (1995) to examine agenda setting in policy making and has subsequently been extended to the entire policy process (Herweg et al., 2018; Howlett et al., 2015).

Recently, a variant of the MSF has been proposed to depict the different types of activities involved in sustainability transitions: technological development, problem framing, alternatives specification, and political decision making (Goyal & Howlett, 2019; Goyal et al., 2019; Goyal et al., 2021). Further, scholars have also analytically disaggregated the actors involved in different streams of the MSF (Goyal & Howlett, 2020; Mukherjee & Howlett, 2015). However, the role of the public in these activities and their relationship to different actors in transitions remain unclear. To address this gap, we delve into the modes of public participation within each stream of the MSF.

Further, we use the MSF for conceptualizing the interaction between different scales in energy transitions. While scale in the energy democracy literature is often assumed to be local (Cumbers et al., 2013; van Veelen & van der Horst, 2018), energy transitions are, however, cross-scalar processes (Calvert, 2015) that operate simultaneously at other scales—the subnational, the national, the regional, and the global (Becker & Naumann, 2017). We suggest that the MSF can depict these processes at each scale, and, therefore, the cross-scalar nature of the transition can be modeled through interaction among the streams in different scales or settings (Goyal, 2021).

The MSF is, therefore, appropriate for analyzing energy democracy as it can help analyze the relationship between structures, agency, and outcomes while accounting for path dependence and coevolution in public engagement (Chilvers & Pallett, 2018), variation in local context (van Veelen, 2018), different imperatives and objectives for energy democracy across countries (Becker & Naumann, 2017), and cross-scalar interaction between institutions as well as geographies (Becker & Naumann, 2017; Calvert, 2015; Cumbers et al., 2013).

Democratizing what? Actors and activities in sustainability transitions

In this section, we present a conceptualization of activities and actors in sustainability transitions based on a four-stream variant of the MSF. Kingdon (1995) suggested that the dynamics among three relatively independent "streams" explained how issues came to the agenda of the government: the problem stream, the policy stream, and the politics stream. The problem stream depicts elite perceptions of issues based on policy feedback, focusing events, and indicators. In contrast, the policy stream represents the evolution of policy ideas as they go through "mutations" and "recombinations" in the "policy primeval soup" leading to the creation of policy alternatives and proposals. Distinct from these, the politics stream moves based on changes in interest groups, party ideologies, and the public mood. The MSF has since been proposed as a generic representation of the entire policy process (Herweg et al., 2018; Howlett et al., 2015).

The problem, policy, and politics streams of the MSF can capture the high-level processes necessary for a sustainability transition: delegitimization of unsustainable technologies, institutions, and structures; creation of new governance arrangements to overcome path dependence; and political struggles to embed new technologies and governance arrangements in societies (Geels, 2011; Van Poeck et al., 2018). However, they do not adequately capture the development process of novel technologies. This shortcoming can be addressed by incorporating a technology stream in the framework (Elzen et al., 2011). The technology stream depicts the creation of technology alternatives through research, development, and deployment to represent relatively independent, but coevolutionary technological dynamics in sustainability transitions (Hoppmann et al., 2014). Building on work in policy studies and the transitions literature (Goyal & Howlett, 2018; Mukherjee & Howlett, 2015; Voß & Simons, 2014), we have proposed an analytical disaggregation of the actors involved in these four streams (Goyal & Howlett, 2019).

Technology constituencies and the technology stream

Technology constituencies "foster technological innovation and promote its diffusion amongst citizens, businesses, and governments" (Goyal & Howlett, 2018, p. 6). While generally associated with the development of a technology, technology constituencies may also refer to actors who foster the diffusion of sociotechnical alternatives built around specific technologies (Geels, 2012; Manders et al., 2018; Schwanen, 2015). These constituencies may include technologists, lobby groups, political actors, users, and civil society organizations. Actors within the constituency are brought together by a shared interest in promoting the technology (e.g., solar) and do not necessarily have a shared interest in the purpose(s) it serves. The technologies, cultural milieu, and actors comprising technology constituencies coevolve within a sociotechnical system (Molina, 1995; Staudenmaier, 1989).

Technology constituencies are primarily responsible for fostering technological innovation in sustainability transitions, through coordinated or distributed action (Farla et al., 2012; Fuenfschilling & Truffer, 2014; Garud & Karnøe, 2003). While these constituencies are generally expected to include actors supporting the technology, actors associated with other possibly competing technologies may nonetheless participate in them and provide resources for further development (Boon & Bakker, 2016) in exchange for other resources (Normann, 2017), as an image-building exercise, to engage in technological learning, to combine the new technology in some form with old technology, or to anchor technological disruption and lead the transition (Bakker et al., 2014; Bosman et al., 2014; Elzen et al., 2012; Manders et al., 2018; Markard & Truffer, 2008; Raven et al., 2016; Raven, 2006).

Epistemic communities and the problem stream

An epistemic community is "a network of professionals with recognized expertise and competence in a particular domain and an authoritative claim to policy-relevant knowledge within that domain or issue-area" (Haas, 1992, p. 3). The members of epistemic communities include scientists, academic experts, public sector officials, and other stakeholders with a common interpretation of science for public policy. Epistemic communities share normative beliefs regarding social action, epistemic beliefs regarding the phenomenon under investigation, causal beliefs emerging from their epistemological leanings, and a common policy enterprise (Haas, 1992). Thus knowledge regarding sustainability issues keeps members of these communities together despite their possibly diverse backgrounds (Mukherjee & Howlett, 2015). Illustratively, such communities have been found to influence problem framing and research priorities in areas such as climate change and oceans policy (Haas, 1992; Rudd, 2015).

In sustainability transitions, epistemic communities are involved in framing issues and defining problems (Gough & Shackley, 2001; Zito, 2001). They might also support these activities indirectly by facilitating debate and discussion regarding sustainability issues, generating policy-relevant evidence, supplying technical knowledge, articulating broad policy alternatives, and supporting vested interests (Adler & Haas, 1992; Dobson, 2019). These activities gain prominence in transitions due to the often complex and technical nature of the issues involved in addressing long-term, multigenerational issues that are connected to a wide range of anthropogenic activities (Underdal, 2010; Zito, 2018).

Instrument constituencies and the policy stream

Instrument constituencies are actors supporting "heterogeneous practices involved in the making of [policy] instruments: scientific theory building, data production, and publishing, political issue framing, agenda setting, coalition building, business development, marketing and lobbying, management of innovation networks, professional organization" (Voß & Simons, 2014, p. 737). Instrument constituencies usually comprise members from think tanks, consulting firms, academia, bureaucracies, businesses, and civil societies. These constituencies can emerge during the creation of new tools of governance and thrive in supporting expansion of the tool. Studies have found the presence of such constituencies in areas such as carbon emissions trading (Voß & Simons, 2014), the field of transition management in innovation policy studies (Voß, 2014), conservation trading (Mann & Simons, 2015), conditional cash transfers (Foli et al., 2018), social insurance and pension privatization (Béland & Howlett, 2016), public participation and deliberative democracy (Amelung & Grabner, 2017), and transportation (Perl & Burke, 2018).

In sustainability transitions, instrument constituencies facilitate the creation and diffusion of policy instruments, or tools of governance. Members and practices involved in the development of the policy instrument, in a coordinated or distributed fashion, come to be associated with the instrument constituency due to the twin expectations that the instrument will address a policy problem while also furthering their interests (Simons & Voß, 2018; Voß & Simons, 2014). They create a "supply push" for the instrument as a solution to various sustainability issues beyond its original policy context and into new policy areas and geographies, in the process creating resources to also possibly support the development of novel technologies.

Advocacy coalitions and the politics stream

Advocacy coalitions consist of "people from a variety of positions . . . who share a particular belief system—i.e., a set of basic values, causal assumptions, and problem perceptions—and who show a non-trivial degree of coordinated activity over time" (Sabatier, 1988, p. 139). Members of these coalitions can include elected politicians, political appointees and advisors, political parties, and lobbyists, among other stakeholders. Advocacy coalitions are held together by common beliefs in an issue area as well as a specific policy preference for addressing that issue (policy core beliefs), even as their broader values or worldviews (normative core beliefs) and their stands on on-the-ground calibrations of those policies (secondary aspects) may diverge. These coalitions remain relatively stable as their core beliefs do not change much over time. Advocacy coalitions have been found to exist in numerous policy areas across various jurisdictions around the world (Weible et al., 2009).

In sustainability transitions, advocacy coalitions participate in political competition and contestation against opposing coalitions to further the adoption and implementation of their policy preferences. They engage in matching specific sustainability issues to specific policy instruments and/or technological solutions (Harborne & Hendry, 2012; Vasseur et al., 2013). These coalitions typically engage in "instrumental" use of knowledge to support their policy preferences and undermine those of their competitors. Unlike the other three types of actors, they are also believed to be more localized and typically focus on one issue area in one jurisdiction.

These different types of actors and their key activities in sustainability transitions are summarized in Table 6.1. The claim is not that these actors are necessarily distinct in practice; individuals, groups, or organizations from one might also participate in another (Goyal & Howlett, 2019; Mukherjee & Howlett, 2015). Technology constituencies or instrument constituencies might, for example, work with advocacy coalitions to create traction for their policy or technology solutions in a specific jurisdiction. However, the analytical distinction serves to highlight the different activities in transitions, examine public participation, and link it to various transition outcomes.

Democratizing how? Public participation in energy transitions

While the MSF—and the conceptualization of agency within its streams—is useful for analyzing sustainability transitions, its application to energy democracy requires a better account of the possibilities for public participation within transition processes. In this section, we present an initial discussion on how public participation in energy transitions can be examined within this framework. We utilize the distinction by Angel (2017) on participation in and against the state but suggest that disaggregation using the multiple streams can capture activities "beyond the state" in which the public can engage. Such disaggregation can also provide a more nuanced account of which aspect of a sociotechnical system public participation is likely to influence.

Public participation in the technology stream

The democratization of the technology stream implies public participation in (or against) activities that foster technological innovation through research, development, or deployment. The most direct form of public participation in the technology stream can be in the form of a

Table 6.1 Types of agents and their roles in sustainability transitions (Goyal & Howlett, 2019)

Type of agent	*Reference*	*Primary role in sustainability transition*
Technology constituency	Goyal & Howlett, 2018	Technology development and diffusion
Epistemic community	Haas, 1992	Sustainability problem definition and framing
Instrument constituency	Voß & Simons, 2014	Policy instrument creation and diffusion
Advocacy coalition	Sabatier, 1988	Problem and solution matching and implementation

technology user. This is alluded to in the literature on energy democracy as consumer choice (Welton, 2018). The literature on transitions has also emphasized the importance of user perceptions and preferences for technological development and highlighted the role users can play in facilitating interactive learning for innovation (Kemp et al., 1998). Makonese and Bradnum (2017), for example, found that better communication between technologists and users increased the design of improved cookstoves and the users' sense of ownership in them in South Africa. As mentioned earlier, though, this is only one form of (limited) participation in the technology stream.

The public can play a more active role in the technology stream as well, for example as a "prosumer" through ownership or control over energy generation or distribution utilities. In the German energy transition, for example, "backyard-pioneers" were involved in supporting distributed renewable energy technologies through experimentation with biogas, solar, and wind (Morris & Jungjohann, 2016; Paul, 2018). Similarly, Karanasios and Parker (2018) observed a shift from hydro and small wind power driven by electricity utilities to solar energy driven by indigenous communities in Canada, whose participation in electricity generation induced experimentation and learning.

More broadly, participation in the technology stream can also be in the form of support for the technology through the policy process rather than, or possibly even in order to bring about, direct ownership or control. Hess (2018), for example, observed that one among several coalitions in the energy transition in New York focused on influencing the implementation of the Reforming the Energy Vision policy of the state to advocate for not only more control over energy decision making but also explicit support for renewable energy and energy efficiency technologies.

Participation in the technology stream can also be against a specific type of technology. The German Energiewende, for instance, had its roots not in support for renewable energy technologies but in antinuclear movements in the country dating back to the 1970s (Paul, 2018). Similarly, Hess (2018) found a distinct public coalition in the energy transition in New York that focused on opposition to natural gas and advocated for a moratorium on hydraulic fracturing technologies and the construction of natural gas infrastructure in the state. Further, support for one technology can be combined with explicit opposition to another technology. Sagaris (2018) found that, in Chile, cyclists constituted an organized citizen group that supported public transportation initiatives that complemented cycling but opposed initiatives that were perceived as threats to it.

Public participation in the problem stream

The democratization of the problem stream implies public participation in (or against) processes through which issues in the energy transition are framed. While this might be more significant at the community or the "local" level, the literature on policy making has highlighted the limited role the public often has in problem definition and framing at the subnational or national scale. Kingdon (1984) himself, for example, suggested that the problem stream was influenced by *elite* perceptions of societal issues and not necessarily those of the public. The deliberative turn in public policy was, in fact, a response to the exclusion of the citizen from policy making and policy analysis through the framing of expert knowledge as "value free" and rational (Fischer, 2007).

Public participation in epistemic communities might also be limited as these communities are networks of professionals with authoritative claim to knowledge regarding issues in a certain area. In an example to the contrary, however, Chilvers (2008) found that an "emergent multidisciplinary, multisectoral network of participatory appraisal experts" (p. 16) were involved in defining issues at the science–policy interface in environmental risk policy in the United Kingdom. The actors involved in the process included not only decision makers from the private and public sectors but also participatory practitioners, participatory researchers, and civil society organizations.

More commonly, public participation in the problem stream can counter epistemic communities and contribute to delegitimization of existing institutions and structures that reinforce epistemic authority. A key activity here might involve creating awareness about the problem or connecting issues for the public (Boasson & Wettestad, 2014; Kalafatis et al., 2015; Knaggård, 2016; Meijerink & Huitema, 2010). Blanchet (2015), for example, argued that grassroots initiatives in Berlin increased citizen awareness about energy-related issues and created an alternate vision for the energy system. Using a focusing event to reframe the problem (Mallett & Cherniak, 2018), as was done by the civil society in Germany in the wake of the Fukushima accident (Paul, 2018), can be another mechanism through which the public can participate in the problem stream.

In focusing on discourse(s) and framing in the diffusion of community choice aggregation in California, Hess (2019) has provided another account of public participation in the problem stream (as much as the politics stream). He found that the procommunity choice aggregation coalition in California initially framed the issue as one of affordability but subsequently highlighted its rationale as job creation and energy democracy. More broadly, the literature on citizen science has shed light on public participation in creating issue awareness (Dickinson et al., 2012; Goodchild, 2007), and the literature on social movements has examined another route to public participation in problem framing (Bennett & Segerberg, 2012; Schlosberg, 2004).

Public participation in the policy stream

Democratization of the policy stream involves public participation in alternatives specification and, subsequently, possibly policy implementation. An indirect role of the public in the policy stream is influencing the survival of ideas in the policy stream based on their value acceptability (Kingdon, 1984). A more direct role of the lay public in the policy stream is, as in the case of the problem stream, not automatic as policy expertise might act as a barrier for participation (Crow, 2010). Voß and Simons (2018), for example, highlighted how policy experimentation can create a cycle wherein policy expertise and political authority reinforce one another. Similarly, Fitch-Roy et al. (2019) highlighted that the demand for knowledge about a policy instrument can result in the supply of specialists on that instrument, with further implementation of the instrument augmenting expertise among the select community.

A positive example of public participation in an instrument constituency can, once again, be seen in the previously mentioned study by Hess (2018). The participation, however, did not occur directly but more *through* environmental and public interest organizations, as is generally expected in the literature on instrument constituencies. The constituency involved in the energy transition in New York pushed for an in-state cap-and-trade program for greenhouse gas emissions and contributed to its design. The involvement of this constituency is primarily in the policy stream, and not in the politics stream, from the description of its activities: "The primary strategy of the coalition was participation in the complex rule-making process; thus, this mobilization did not involve street protest or similar repertoires of action" (Hess, 2018, p. 181).

Procedural policy instruments can facilitate public participation in the policy stream. Fiorino (1990), for example, argued that instruments such as public hearings, surveys, negotiated rule making, and citizen panels could democratize environmental risk decision making. Fremeth et al. (2014) also found that "consumer advocate participation" in regulatory fora in the United States led to fewer electricity rate reviews, lower return on equity, and lower residential rates. However, it is important to examine how these tools work in practice to determine who is participating in the policy stream. Newman et al. (2004), for example, found that numerous constraints affected public participation in deliberative forums of collaborative governance. In examining the application of crowdsourcing to policy making, another potential tool for public engagement, Taeihagh (2017) found that its use was largely limited to agenda setting and policy evaluation and did not extend to policy formulation itself.

Recently, codesign—the concerted use of process, principles, and practical tools to facilitate public engagement in designing policy solutions (Blomkamp, 2018)—has received attention in the literature. Rotmann (2017), for example, has described the use of storytelling by the International Environment Agency to promote stakeholder collaboration and codesign for behavioral policies. Ansell et al. (2017) have argued that deliberations amongst "upstream and downstream actors" during policy design can results in better policy implementation outcomes. Blomkamp (2018) has, however, found that while codesign can lead to solutions that are innovative, efficient, and responsive, its application in practice has been patchy thus far. MacArthur (2016) has in fact cautioned about the capacity and design challenges in conducting public engagement as they can otherwise be ineffective and further erode public confidence in the government.

Public participation in the politics stream

Public participation can also result from and in turn influence involvement in political processes. Where such processes are issue-based and "democratic," the public can influence energy transitions through referenda. Otherwise, public participation can also occur more indirectly. Kingdon (1984), for example, posited that public "mood" can influence agenda setting, while Sabatier (1988) indicated that public opinion can influence the activities of advocacy coalitions or the government directly. The influence of public opinion can be even more significant when issue salience is high (Burstein, 2003). More actively, the public might itself participate in or against advocacy coalitions.

In pluralist societies such as the United States, interest groups represent a traditional forum for public participation in the politics stream. In theory these groups can be both business-based and mass-based, but Gilens and Page (2014), for example, found that groups involving economic elites and business had significant impact on policy making in the United States, while groups representing average citizens had negligible impact. Moreover, Hendriks (2002) even highlighted tensions between different forms of participation: The author found that interest groups did not necessarily support or participate in newer mechanisms for public participation, such as citizenship juries.

A more positive account of public engagement in the politics stream in case of the German Energiewende has been provided by Paul (2018). She observes that the civil society registered its preferences through activism, civil disobedience, and significant political lobbying from the antinuclear movement, the environmental movement, and the peace movement. This mobilization of the public was key as far as implementing the decision on nuclear phaseout was concerned. Angela Merkel did not go through with her initial strategy of reversing this decision as she "would have had to sail against the wind of public opinion" after the Fukushima accident (Paul, 2018, p. 5).

In another example of public participation in the politics stream, Cervas and Giancatarino (2017) highlighted the role of women representing working-class communities of color in California who worked as members of the California Environmental Justice Alliance (CEJA) in successfully advocating for the Multifamily Affordable Housing Solar Roofs Program in 2015. After an effort to enact a similar bill in 2011 had failed to clear the Senate due to opposition from fossil fuel industry lobbyists, the movement allied with a member of the state assembly, the California Solar Energy Industries Association, and with Everyday Energy to achieve a more favorable outcome in 2015.

Democratizing where? Interaction between scales in energy transitions

As mentioned earlier, energy democracy can simultaneously occur at several different interdependent scales, from the "local" to the global. The specific scale or scales involved in energy transitions depend on the issues that are significant and the institutional contexts in which they are relevant. In this section, we propose an approach to analyzing scalar relations as interactions among the multiple streams across jurisdictions or levels. The basic idea is that as the different settings or scales are interdependent, the streams in one setting or scale are subject to not only "internal" influences but also "external" influences from other settings or scales (Goyal, 2021; Lovell, 2016). This might be the result of diffusion processes (Simmons et al., 2006), possibly involving actors operating across settings or scales (Paterson et al., 2014).

Transition processes in one setting might influence the technology stream in another setting. Participation in the form of consumer choice might, for example, increase the market penetration of certain technologies, creating a virtuous cycle of learning, efficiency or performance improvement, and higher adoption. This can also increase the likelihood of technological diffusion to another setting. An influence might also occur through a more direct exchange among the public, especially in case of ownership or control over energy generation or distribution. Capellán-Pérez et al. (2018), for example, found that the challenges faced by renewable energy cooperatives in Spain due to an unfavorable policy environment led to the adoption of innovative participatory approaches and investment instruments, which subsequently diffused to other political and social levels through information sharing and learning among cooperatives.

Transition processes in one setting can also influence or be perceived to influence the problem stream as well. As mentioned earlier, the Fukushima accident in Japan and its influence on public opinion in Germany and the consequent stickiness of the antinuclear problem frame in the country is an example of such a spatial interaction in energy democracy. The dynamics of the diffusion of community choice aggregation in California provide another account of codependent changes in the problem stream, where the expansion of the initiative led to a change in problem framing from energy pricing to job creation and energy democracy to counter the mobilization of energy utilities against the community choice aggregation (Hess, 2018).

The policy stream can also be influenced by the energy transition in another setting or scale. While an instrument constituency is a more likely pathway for such an influence (Howlett et al., 2018; Saguin & Howlett, 2019; Voß & Simons, 2014), such a constituency can witness participation through a nongovernment organization. Pacheco-Vega (2015), illustratively, highlighted the role of civil society organizations in facilitating horizontal and vertical coordination in North America to bring about the diffusion of two procedural tools in environmental policy—the Pollutant Release and Transfer Registry and the Citizen Submission on Enforcement Matters—across Canada, Mexico, and the United States not through "political" activities but through knowledge sharing.

The policy process in a setting can also affect the politics stream in a different setting. The presence of common political parties (Butler et al., 2017), interest groups (Barrilleaux et al., 2015), or coalitions (Meckling, 2011) can create a pathway for this influence between different settings. While transnational civil society networks can also act as a pathway for influencing politics across different settings, in the case of climate justice at least, Derman (2014) found their impact between the European Union and the United States to be limited and argued for greater engagement of civil society within established fora such as the United Nations Framework Convention on Climate Change. Public opinion, though, has been found to be a driver for policy diffusion in case of morality policy (Mooney & Mei-Hsien, 1999) or antismoking policy (Pacheco, 2012) and might play such a role in transmitting the influence of energy democracy in one setting or scale on another.

Conclusion

While energy democracy has become popular as an instrument for realizing the vision of just transitions, the relationship between energy democracy as a process and its outcomes is not straightforward. For instance, although decentralized and distributed energy generation and public participation are believed to go hand in hand, studies have found that the former does not necessarily translate into the latter (Weinrub, 2014) and even that the latter does not always promote the former (Ajaz, 2019). It is, therefore, imperative to understand the different dimensions of energy democracy and empirically examine whether and under what conditions energy democracy leads to desirable energy outcomes. Despite inconsistencies in conceptualization and operationalization, recent literature on the topic has identified numerous important dimensions that define democratization—such as the stakeholders involved, their motivations for action, the mechanisms of participation, and the geographies of democratization—and influence its relationship with the futures of the transition.

In this chapter, we built on this research by conceptualizing energy democracy using a four-stream variant of the MSF. We focused on three issues in the existing literature which require clarification: the subject of participation, the mode of participation, and the scale of participation. Based on recent literature, we presented categorization of different actors—technology constituencies, epistemic communities, instrument constituencies, and advocacy coalitions—who lend agency to sustainability transitions. We proposed that the disaggregation of public participation in relation to these actors and their activities can facilitate a functional analysis of energy democracy and discussed various modes of public engagement in a transition. Finally, we posited that this conceptualization is also useful for examining the scalar dimension of energy democracy, as energy democracy at each scale can be depicted using the MSF and cross-scalar settings can be modeled as interactions between different streams at different scales. The proposed framework enables synthesis of analytical inquiry into questions such as democratizing what, democratizing how, and democratizing where and can, therefore, help identify the conditions and behaviors that enhance or undermine energy democracy. Moreover, it also permits examination of the relationship between these dimensions of energy democracy and its outcomes (Goyal, 2021) and can thereby help create scholarly knowledge to accelerate the energy transition.

References

Adler, E., & Haas, P. M. (1992). Conclusion: Epistemic communities, world order, and the creation of a reflective research program. *International Organization*, *46*(1), 367–390. www.jstor.org.libproxy1.nus.edu.sg/stable/2706960

Ajaz, W. (2019). Resilience, environmental concern, or energy democracy? A panel data analysis of microgrid adoption in the United States. *Energy Research and Social Science*, *49*, 26–35. https://doi.org/10.1016/j.erss.2018.10.027

Amelung, N., & Grabner, L. (2017). Making citizen panels a 'universal bestseller'. In L. Bherer, M. Gauthier, & L. Simard (Eds.), *The professionalization of public participation* (pp. 189–213). Routledge.
Angel, J. (2016). *Strategies of energy democracy*. Rosa Luxemburg Stiftung.
Angel, J. (2017). Towards an energy politics in-against-and-beyond the state: Berlin's struggle for energy democracy. *Antipode, 49*(3), 557–576. https://doi.org/10.1111/anti.12289
Ansell, C., Sørensen, E., & Torfing, J. (2017). Improving policy implementation through collaborative policymaking. *Policy & Politics, 45*(3), 467–486.
Bakker, S., Maat, K., & van Wee, B. (2014). Stakeholders interests, expectations, and strategies regarding the development and implementation of electric vehicles: The case of the Netherlands. *Transportation Research Part A: Policy and Practice, 66*, 52–64. https://doi.org/10.1016/j.tra.2014.04.018
Barrilleaux, C., Garrett, K. N., & Jansa, J. M. (2015). Interest group influence in policy diffusion networks. *State Politics and Policy Quarterly, 15*(3), 387–417. https://doi.org/10.1177/1532440015592776
Becker, S., & Naumann, M. (2017). Energy democracy: Mapping the debate on energy alternatives. *Geography Compass, 11*(8), Article e12321. https://doi.org/10.1111/gec3.12321
Béland, D., & Howlett, M. (2016). How solutions chase problems: Instrument constituencies in the policy process. *Governance, 29*(3), 393–409. https://doi.org/10.1111/gove.12179
Bennett, W. L., & Segerberg, A. (2012). The logic of connective action: Digital media and the personalization of contentious politics. *Information Communication & Society, 15*(5), 739–768. https://doi.org/10.1080/1369118x.2012.670661
Blanchet, T. (2015). Struggle over energy transition in Berlin: How do grassroots initiatives affect local energy policy-making? *Energy Policy, 78*, 246–254. https://doi.org/10.1016/j.enpol.2014.11.001
Blomkamp, E. (2018). The promise of co-design for public policy. *Australian Journal of Public Administration,* 77(4), 729–743. https://doi.org/10.1111/1467-8500.12310
Boasson, E. L., & Wettestad, J. (2014). Policy invention and entrepreneurship: Bankrolling the burying of carbon in the EU. *Global Environmental Change, 29*, 404–412. https://doi.org/10.1016/j.gloenvcha.2014.09.010
Boon, W. P. C., & Bakker, S. (2016). Learning to shield: Policy learning in socio-technical transitions. *Environmental Innovation and Societal Transitions, 18*, 181–200. http://doi.org/10.1016/j.eist.2015.06.003
Bosman, R., Loorbach, D., Frantzeskaki, N., & Pistorius, T. (2014). Discursive regime dynamics in the Dutch energy transition. *Environmental Innovation and Societal Transitions, 13*, 45–59. https://doi.org/10.1016/j.eist.2014.07.003
Burke, M. J. (2018). Shared yet contested: Energy democracy counter-narratives. *Frontiers in Communication, 3*(22). https://doi.org/10.3389/fcomm.2018.00022
Burstein, P. (2003). The impact of public opinion on public policy: A review and an agenda. *Political Research Quarterly, 56*(1), 29–40. https://doi.org/10.1177/106591290305600103
Butler, D. M., Volden, C., Dynes, A. M., & Shor, B. (2017). Ideology, learning, and policy diffusion: Experimental evidence. *American Journal of Political Science, 61*(1), 37–49. https://doi.org/10.1111/ajps.12213
Calvert, K. (2015). From 'energy geography' to 'energy geographies': Perspectives on a fertile academic borderland. *Progress in Human Geography, 40*(1), 105–125. https://doi.org/10.1177/0309132514566343
Capellán-Pérez, I., Campos-Celador, Á., & Terés-Zubiaga, J. (2018). Renewable energy cooperatives as an instrument towards the energy transition in Spain. *Energy Policy, 123*, 215–229. https://doi.org/10.1016/j.enpol.2018.08.064
Cervas, S., & Giancatarino, A. (2017). Energy democracy through local energy equity. In D. Fairchild & A. Weinrub (Eds.), *Energy democracy: Advancing equity in clean energy solutions* (pp. 57–75). Island Press. https://doi.org/10.5822/978-1-61091-852-7_4
Chilvers, J. (2008). Environmental risk, uncertainty, and participation: Mapping an emergent epistemic community. *Environment and Planning A: Economy and Space, 40*(12), 2990–3008. https://doi.org/10.1068/a39279
Chilvers, J., & Longhurst, N. (2016). Participation in transition(s): Reconceiving public engagements in energy transitions as co-produced, emergent and diverse. *Journal of Environmental Policy & Planning, 18*(5), 585–607. https://doi.org/10.1080/1523908X.2015.1110483
Chilvers, J., & Pallett, H. (2018). Energy democracies and publics in the making: A relational agenda for research and practice. *Frontiers in Communication, 3*(14). https://doi.org/10.3389/fcomm.2018.00014
Crow, D. A. (2010). Policy entrepreneurs, issue experts, and water rights policy change in Colorado. *Review of Policy Research, 27*(3), 299–315. https://doi.org/10.1111/j.1541-1338.2010.00443.x

Cumbers, A., Danson, M., Whittman, G., Morgan, G., & Callaghan, G. (2013). *Repossessing the future: A common weal strategy for community and democratic ownership of Scotland's energy resources*. The Jimmy Reid Foundation.

Derman, B. B. (2014). Climate governance, justice, and transnational civil society. *Climate Policy, 14*(1), 23–41. https://doi.org/10.1080/14693062.2014.849492

Dickinson, J. L., Shirk, J., Bonter, D., Bonney, R., Crain, R. L., Martin, J., Phillips, T., & Purcell, K. (2012). The current state of citizen science as a tool for ecological research and public engagement. *Frontiers in Ecology and the Environment, 10*(6), 291–297. https://doi.org/10.1890/110236

Dobson, J. (2019). Reinterpreting urban institutions for sustainability: How epistemic networks shape knowledge and logics. *Environmental Science & Policy, 92*, 133–140. https://doi.org/10.1016/j.envsci.2018.11.018

Elzen, B., Geels, F. W., Leeuwis, C., & van Mierlo, B. (2011). Normative contestation in transitions 'in the making': Animal welfare concerns and system innovation in pig husbandry. *Research Policy, 40*(2), 263–275. https://doi.org/10.1016/j.respol.2010.09.018

Elzen, B., Van Mierlo, B., & Leeuwis, C. (2012). Anchoring of innovations: Assessing Dutch efforts to harvest energy from glasshouses. *Environmental Innovation and Societal Transitions, 5*, 1–18. https://doi.org/10.1016/j.eist.2012.10.006

Evans, G., & Phelan, L. (2016). Transition to a post-carbon society: Linking environmental justice and just transition discourses. *Energy Policy, 99*, 329–339. https://doi.org/10.1016/j.enpol.2016.05.003

Farla, J., Markard, J., Raven, R., & Coenen, L. (2012). Sustainability transitions in the making: A closer look at actors, strategies and resources. *Technological Forecasting and Social Change, 79*(6), 991–998. https://doi.org/10.1016/j.techfore.2012.02.001

Farrell, J. (2014). Beyond utility 2.0 to energy democracy. *Institute for Local Self-Reliance, 10*.

Fiorino, D. J. (1990). Citizen participation and environmental risk: A survey of institutional mechanisms. *Science, Technology, & Human Values, 15*(2), 226–243. https://doi.org/10.1177/016224399001500204

Fischer, F. (2007). Deliberative policy analysis as practical reason: Integrating empirical and normative arguments. In F. Fischer, G. J. Miller, & M. S. Sidney (Eds.), *Handbook of public policy analysis: Theory, politics, and methods* (pp. 223–236). Routledge.

Fitch-Roy, O. W., Benson, D., & Woodman, B. (2019). Policy instrument supply and demand: How the renewable electricity auction took over the world. *Politics and Governance*, 7(1), 81–91. https://doi.org/10.17645/pag.v7i1.1581

Foli, R., Béland, D., & Fenwick, T. B. (2018). How instrument constituencies shape policy transfer: A case study from Ghana. *Policy and Society, 37*(1), 108–124. https://doi.org/10.1080/14494035.2017.1377987

Fremeth, A. R., Holburn, G. L. F., & Spiller, P. T. (2014). The impact of consumer advocates on regulatory policy in the electric utility sector. *Public Choice, 161*(1), 157–181. https://doi.org/10.1007/s11127-013-0145-z

Fuenfschilling, L., & Truffer, B. (2014). The structuration of socio-technical regimes: Conceptual foundations from institutional theory. *Research Policy, 43*(4), 772–791. https://doi.org/10.1016/j.respol.2013.10.010

Garud, R., & Karnøe, P. (2003). Bricolage versus breakthrough: Distributed and embedded agency in technology entrepreneurship. *Research Policy, 32*(2), 277–300. https://doi.org/10.1016/S0048-7333(02)00100-2

Geels, F. W. (2011). The multi-level perspective on sustainability transitions: Responses to seven criticisms. *Environmental Innovation and Societal Transitions, 1*(1), 24–40. https://doi.org/10.1016/j.eist.2011.02.002

Geels, F. W. (2012). A socio-technical analysis of low-carbon transitions: Introducing the multi-level perspective into transport studies. *Journal of Transport Geography, 24*, 471–482. https://doi.org/10.1016/j.jtrangeo.2012.01.021

Gilens, M., & Page, B. I. (2014). Testing theories of American politics: Elites, interest groups, and average citizens. *Perspectives on Politics, 12*(3), 564–581. https://doi.org/10.1017/S1537592714001595

Goodchild, M. F. (2007). Citizens as sensors: The world of volunteered geography. *GeoJournal, 69*(4), 211–221. https://doi.org/10.1007/s10708-007-9111-y

Gough, C., & Shackley, S. (2001). The respectable politics of climate change: The epistemic communities and NGOs. *International Affairs*, 77(2), 329–346. https://doi.org/10.1111/1468-2346.00195

Goyal, N. (2021), *Explaining Policy Success Using the Multiple Streams Framework: Political Success Despite Programmatic Failure of the Solar Energy Policy in Gujarat, India*. Politics Policy. https://doi-org.tudelft.idm.oclc.org/10.1111/polp.12426

Goyal, N., & Howlett, M. (2018). Technology and instrument constituencies as agents of innovation: Sustainability transitions and the governance of urban transport. *Energies*, *11*(5), 1198. www.mdpi.com/1996-1073/11/5/1198

Goyal, N., & Howlett, M. (2019). Who learns what in sustainability transitions? *Environmental Innovation and Societal Transitions*. https://doi.org/10.1016/j.eist.2019.09.002

Goyal, N., Howlett, M., & Chindarkar, N. (2019). Who coupled which stream(s)? Policy entrepreneurship and innovation in the energy—water nexus in Gujarat, India. *Public Administration and Development*, *40*, 49–54. https://doi.org/10.1002/pad.1855

Goyal, N., & Howlett, M. (2020). Making sense of the babble of policy-making: a general framework of the policy process. In *A Modern Guide to Public Policy*. Cheltenham, UK: Edward Elgar Publishing. doi: https://doi.org/10.4337/9781789904987.00012

Goyal, N. (2021). Policy Diffusion Through Multiple Streams: The (Non-)Adoption of Energy Conservation Building Code in India. *Policy Stud J*. https://doi-org.tudelft.idm.oclc.org/10.1111/psj.12415

Goyal, N., Howlett, M., & Taeihagh, A. (2021), Why and how does the regulation of emerging technologies occur? Explaining the adoption of the EU General Data Protection Regulation using the multiple streams framework. *Regulation & Governance*. https://doi-org.tudelft.idm.oclc.org/10.1111/rego.12387

Haas, P. M. (1992). Introduction: Epistemic communities and international policy coordination. *International Organization*, *46*(1), 1–35.

Harborne, P., & Hendry, C. (2012). Commercialising new energy technologies: Failure of the Japanese machine? *Technology Analysis & Strategic Management*, *24*(5), 497–510. https://doi.org/10.1080/09537325.2012.674671

Healy, N., & Barry, J. (2017). Politicizing energy justice and energy system transitions: Fossil fuel divestment and a 'just transition'. *Energy Policy*, *108*, 451–459. https://doi.org/10.1016/j.enpol.2017.06.014

Heffron, R. J., & McCauley, D. (2018). What is the 'just transition'? *Geoforum*, *88*, 74–77. https://doi.org/10.1016/j.geoforum.2017.11.016

Hendriks, C. (2002). Institutions of deliberative democratic processes and interest groups: Roles, tensions and incentives. *Australian Journal of Public Administration*, *61*(1), 64–75. https://doi.org/10.1111/1467-8500.00259

Herweg, N., Zahariadis, N., & Zohlnhöfer, R. (2018). The multiple streams framework: Foundations, refinements, and empirical applications. In C. M. Weible & P. A. Sabatier (Eds.), *Theories of the policy process* (pp. 17–53). Taylor and Francis.

Hess, D. J. (2018). Energy democracy and social movements: A multi-coalition perspective on the politics of sustainability transitions. *Energy Research and Social Science*, *40*, 177–189. https://doi.org/10.1016/j.erss.2018.01.003

Hess, D. J. (2019). Coalitions, framing, and the politics of energy transitions: Local democracy and community choice in California. *Energy Research and Social Science*, *50*, 38–50. https://doi.org/10.1016/j.erss.2018.11.013

Hoppmann, J., Huenteler, J., & Girod, B. (2014). Compulsive policy-making: The evolution of the German feed-in tariff system for solar photovoltaic power. *Research Policy*, *43*(8), 1422–1441. https://doi.org/10.1016/j.respol.2014.01.014

Howlett, M., McConnell, A., & Perl, A. (2015). Streams and stages: Reconciling Kingdon and policy process theory. *European Journal of Political Research*, *54*(3), 419–434. https://doi.org/10.1111/1475-6765.12064

Howlett, M., Ramesh, M., & Saguin, K. (2018). Diffusion of CCTs from Latin America to Asia: The Philippine 4Ps case. *Revista de Administracao Publica*, *52*(2), 264–284.

Kalafatis, S. E., Grace, A., & Gibbons, E. (2015). Making climate science accessible in Toledo: The linked boundary chain approach. *Climate Risk Management*, *9*, 30–40. https://doi.org/10.1016/j.crm.2015.04.003

Karanasios, K., & Parker, P. (2018). Tracking the transition to renewable electricity in remote indigenous communities in Canada. *Energy Policy*, *118*, 169–181. https://doi.org/10.1016/j.enpol.2018.03.032

Kemp, R., Schot, J., & Hoogma, R. (1998). Regime shifts to sustainability through processes of niche formation: The approach of strategic niche management. *Technology Analysis & Strategic Management*, *10*(2), 175–198. https://doi.org/10.1080/09537329808524310

Kingdon, J. W. (1984). *Agendas, alternatives, and public policies*. Little, Brown. http://nus.summon.serialssolutions.com/2.0.0/link/0/eLvHCXMwY2AQNtIz0EUrE8C3uxkFRxm7ehtHIdb-DAltSQE2lkBMDi9MRAxPGoNk6YMnKZAy6jsEyHJ4GgQ18CzMLE1PwxCuwqW9pDKz-voAfpwPigrSZAQ5HqCTdBBhbQ3gEhBqbUPBEGLUdg5KQkFusogKek88BHbIN4eSkKkNOl-

FUBXJAAzV7Eog4yba4izhy7QwHjoiEo8zH3GRmIMLMBeeqoEg0JaYpplsnmKgXGSiaGJoaWJZ-VqqabJxErB1k5pkkJJsKskgitUIKRzi0gxchpYWJpDevgwDaxowNabKgj0mBw4RAFGnW-w

Kingdon, J. W. (1995). *Agendas, alternatives, and public policies* (2nd ed.). HarperCollins College Publishers.

Knaggård, Å. (2016). Framing the problem: Knowledge brokers in the multiple streams approach. In R. Zohlnhöfer & F. W. Rüb (Eds.), *Decision-making under ambiguity and time constraints: Assessing the multiple-streams framework* (pp. 125–146). ECPR Press.

Lovell, H. (2016). The role of international policy transfer within the multiple streams approach: The case of smart electricity in Australia. *Public Administration, 94*(3), 754–768. https://doi.org/10.1111/padm.12259

MacArthur, J. L. (2016). Challenging public engagement: Participation, deliberation and power in renewable energy policy. *Journal of Environmental Studies and Sciences, 6*(3), 631–640. https://doi.org/10.1007/s13412-015-0328-7

Makonese, T., & Bradnum, C. M. S. (2017). Public participation in technological innovation: The case of the Tshulu stove development programme. *Journal of Energy in Southern Africa, 28*(1), 13–24. https://doi.org/10.17159/2413-3051/2017/v28i1a1379

Mallett, A., & Cherniak, D. (2018). Views from above: Policy entrepreneurship and climate policy change on electricity in the Canadian Arctic. *Regional Environmental Change, 18*(5), 1323–1336. https://doi.org/10.1007/s10113-018-1317-7

Manders, T. N., Wieczorek, A. J., & Verbong, G. P. J. (2018). Understanding smart mobility experiments in the Dutch automobility system: Who is involved and what do they promise? *Futures, 96*, 90–103. https://doi.org/10.1016/j.futures.2017.12.003

Mann, C., & Simons, A. (2015). Local emergence and international developments of conservation trading systems: Innovation dynamics and related problems. *Environmental Conservation, 42*(4), 325–334. https://doi.org/10.1017/S0376892914000381

Markard, J., & Truffer, B. (2008). Actor-oriented analysis of innovation systems: Exploring micro—meso level linkages in the case of stationary fuel cells. *Technology Analysis & Strategic Management, 20*(4), 443–464. https://doi.org/10.1080/09537320802141429

Meckling, J. (2011). *Carbon coalitions: Business, climate politics, and the rise of emissions trading*. MIT Press.

Meijerink, S., & Huitema, D. (2010). Policy entrepreneurs and change strategies: Lessons from sixteen case studies of water transitions around the globe. *Ecology and Society, 15*(2), 17. https://doi.org/10.5751/ES-03509-150221

Molina, A. H. (1995). Sociotechnical constituencies as processes of alignment: The rise of a large-scale European information technology initiative. *Technology in Society, 17*(4), 385–412. https://doi.org/10.1016/0160-791X(95)00016-K

Mooney, C. Z., & Mei-Hsien, L. (1999). The temporal diffusion of morality policy: The case of death penalty legislation in the American States. *Policy Studies Journal, 27*(4), 766–780.

Morris, C., & Jungjohann, A. (2016). *Energy democracy: Germany's Energiewende to renewables*. Palgrave Macmillan.

Mukherjee, I., & Howlett, M. (2015). Who is a stream? Epistemic communities, instrument constituencies and advocacy coalitions in public policy-making. *Politics and Governance, 3*(2), 65–75. https://doi.org/10.17645/pag.v3i2.290

Newell, P., & Mulvaney, D. (2013). The political economy of the 'just transition'. *The Geographical Journal, 179*(2), 132–140. https://doi.org/10.1111/geoj.12008

Newman, J., Barnes, M., Sullivan, H., & Knops, A. (2004). Public participation and collaborative governance. *Journal of Social Policy, 33*(2), 203–223. https://doi.org/10.1017/S0047279403007499

Normann, H. E. (2017). Policy networks in energy transitions: The cases of carbon capture and storage and offshore wind in Norway. *Technological Forecasting and Social Change, 118*, 80–93. https://doi.org/10.1016/j.techfore.2017.02.004

Pacheco, J. (2012). The social contagion model: Exploring the role of public opinion on the diffusion of antismoking legislation across the American states. *Journal of Politics, 74*(1), 187–202. https://doi.org/10.1017/S0022381611001241

Pacheco-Vega, R. (2015). Transnational environmental activism in North America: Wielding soft power through knowledge sharing? *Review of Policy Research, 32*(1), 146–162. https://doi.org/10.1111/ropr.12111

Paterson, M., Hoffmann, M., Betsill, M., & Bernstein, S. (2014). The micro foundations of policy diffusion toward complex global governance: An analysis of the transnational carbon emission trading network. *Comparative Political Studies*, *47*(3), 420–449. https://doi.org/10.1177/0010414013509575

Paul, F. C. (2018). Deep entanglements: History, space and (energy) struggle in the German Energiewende. *Geoforum*, *91*, 1–9. https://doi.org/10.1016/j.geoforum.2018.02.017

Perl, A., & Burke, M. I. (2018). Does institutional entrenchment shape instrument adjustment? Assessing instrument constituency influences on American and Australian motor fuel taxation. *Policy and Society*, *37*(1), 90–107. https://doi.org/10.1080/14494035.2018.1402527

Raven, R., Kern, F., Verhees, B., & Smith, A. (2016). Niche construction and empowerment through socio-political work. A meta-analysis of six low-carbon technology cases. *Environmental Innovation and Societal Transitions*, *18*, 164–180. https://doi.org/10.1016/j.eist.2015.02.002

Raven, R. P. J. M. (2006). Towards alternative trajectories? Reconfigurations in the Dutch electricity regime. *Research Policy*, *35*(4), 581–595. http://dx.doi.org/10.1016/j.respol.2006.02.001

Rotmann, S. (2017). 'Once upon a time. . .' Eliciting energy and behaviour change stories using a fairy tale story spine. *Energy Research & Social Science*, *31*, 303–310. https://doi.org/10.1016/j.erss.2017.06.033

Routledge, P., Cumbers, A., & Derickson, K. D. (2018). States of just transition: Realising climate justice through and against the state. *Geoforum*, *88*, 78–86. https://doi.org/10.1016/j.geoforum.2017.11.015

Rudd, M. A. (2015). Scientists' framing of the ocean science—policy interface. *Global Environmental Change*, *33*, 44–60. https://doi.org/10.1016/j.gloenvcha.2015.04.006

Sabatier, P. A. (1988). An advocacy coalition framework of policy change and the role of policy-oriented learning therein. *Policy Sciences*, *21*(2–3), 129–168. www.jstor.org.libproxy1.nus.edu.sg/stable/4532139

Sagaris, L. (2018). Citizen participation for sustainable transport: Lessons for change from Santiago and Temuco, Chile. *Research in Transportation Economics*, *69*, 402–410. https://doi.org/10.1016/j.retrec.2018.05.001

Saguin, K., & Howlett, M. (2019). Policy transfer and instrument constituency: Explaining the adoption of conditional cash transfer in the Philippines. *Asia & the Pacific Policy Studies*, *6*(3), 352–366. https://doi.org/10.1002/app5.280

Schlosberg, D. (2004). Reconceiving environmental justice: Global movements and political theories. *Environmental Politics*, *13*(3), 517–540. https://doi.org/10.1080/0964401042000229025

Schwanen, T. (2015). The bumpy road toward low-energy urban mobility: Case studies from two UK cities. *Sustainability (Switzerland)*, 7(6), 7086–7111. http://doi.org/10.3390/su7067086

Simmons, B. A., Dobbin, F., & Garrett, G. (2006). Introduction: The international diffusion of liberalism. *International Organization*, *60*(4), 781–810. www.jstor.org.libproxy1.nus.edu.sg/stable/3877847

Simons, A., & Voß, J. P. (2018). The concept of instrument constituencies: Accounting for dynamics and practices of knowing governance. *Policy and Society*, *37*(1), 14–35. https://doi.org/10.1080/14494035.2017.1375248

Staudenmaier, J. M. (1989). The politics of successful technologies. In S. H. Cutliffe & R. C. Post (Eds.), *In context: History and the history of technology: Essays in honor of Melvin Kranzberg* (pp. 150–171). Lehigh University Press.

Szulecki, K. (2018). Conceptualizing energy democracy. *Environmental Politics*, *27*(1), 21–41. https://doi.org/10.1080/09644016.2017.1387294

Taeihagh, A. (2017). Crowdsourcing: A new tool for policy-making? *Policy Sciences*, *50*(4), 629–647. https://doi.org/10.1007/s11077-017-9303-3

Underdal, A. (2010). Complexity and challenges of long-term environmental governance. *Global Environmental Change*, *20*(3), 386–393. https://doi.org/10.1016/j.gloenvcha.2010.02.005

Van Poeck, K., Östman, L., & Block, T. (2018). Opening up the black box of learning-by-doing in sustainability transitions. *Environmental Innovation and Societal Transitions*, *34*, 298–310. https://doi.org/10.1016/j.eist.2018.12.006

Vansintjan, D. (2015). The energy transition to energy democracy. *Report of the REScoop*, 20.

Vasseur, V., Kamp, L. M., & Negro, S. O. (2013). A comparative analysis of photovoltaic technological innovation systems including international dimensions: The cases of Japan and the Netherlands. *Journal of Cleaner Production*, *48*, 200–210. https://doi.org/10.1016/j.jclepro.2013.01.017

van Veelen, B. (2018). Negotiating energy democracy in practice: Governance processes in community energy projects. *Environmental Politics*, *27*(4), 644–665. https://doi.org/10.1080/09644016.2018.1427824

van Veelen, B., & van der Horst, D. (2018). What is energy democracy? Connecting social science energy research and political theory. *Energy Research & Social Science*, *46*, 19–28. https://doi.org/10.1016/j.erss.2018.06.010

Voß, J. P. (2014). Performative policy studies: Realizing 'transition management'. *Innovation: The European Journal of Social Science Research*, *27*(4), 317–343. https://doi.org/10.1080/13511610.2014.967666
Voß, J. P., & Simons, A. (2014). Instrument constituencies and the supply side of policy innovation: The social life of emissions trading. *Environmental Politics*, *23*(5), 735–754. https://doi.org/10.1080/09644016.2014.923625
Voß, J. P., & Simons, A. (2018). A novel understanding of experimentation in governance: Co-producing innovations between 'lab' and 'field'. *Policy Sciences*, *51*(2), 213–229. https://doi.org/10.1007/s11077-018-9313-9
Weible, C. M., Sabatier, P. A., & McQueen, K. (2009). Themes and variations: Taking stock of the advocacy coalition framework. *Policy Studies Journal*, *37*(1), 121–140. https://doi.org/10.1111/j.1541-0072.2008.00299.x
Weinrub, A. (2014). *Expressions of energy democracy: Perspectives on an emerging movement*. Local Clean Energy Alliance.
Welton, S. (2018). Grasping for energy democracy. *Michigan Law Review*, *116*(4), 581–644.
Zito, A. R. (2001). Epistemic communities, collective entrepreneurship and European integration. *Journal of European Public Policy*, *8*(4), 585–603. https://doi.org/10.1080/13501760110064401
Zito, A. R. (2018). Instrument constituencies and epistemic community theory. *Policy and Society*, *37*(1), 36–58. https://doi.org/10.1080/14494035.2017.1416929

7
PART I RESPONSE

Andrea M. Feldpausch-Parker

Introduction

Much of energy democracy practice and its corresponding research happens at the local or subnational level (i.e., state or province) (e.g., Bloem et al., 2021; MacEwen & Evensen, 2021; Stephens et al., 2018). Scholars, however, including those featured in this part of the handbook, have been examining efforts at scaling up these ideas, both from a theoretical perspective (Chapter 6) as well as in practice (Chapters 3–5). Chapters in this section showcase the struggles and opportunities of such attempts, recognizing that scalar dimensions of energy governance tend to correlate with complexity in people, resources, and technology. By focusing on the people side of complexity, these chapters are able to address both social and cultural aspects of energy transitions that go beyond just economics, politics, and technology (Feldpausch-Parker et al., 2019). These include but are not limited to exploring governance structures, power dynamics, participatory structures, identities, and issues of colonialism. Resources and technologies are acknowledged as playing a role in setting these contexts, thus giving a more comprehensive picture of dynamic situations.

Themes

Chapters in this part of the handbook probe at the interplay between participation, power, and justice, with two of the four implementing Feldpausch-Parker et al.'s (2019) energy democracy framework. For example, Carvalho and Horta point out in Chapter 3 that even with statements promoting transparency and greater opportunities for public participation, processes aimed at engaging relevant publics by the Energy Union's Governance Regulation are still vague and ill defined. Additionally, the authors point to energy sector governance falling back on past norms and practices such as top-down decision making, even with attempts at progressive innovation. They argue that this could be addressed with more concrete definitions for engagement as well as more sufficient safeguards for public participation.

In Chapter 4, Lenhart, Chan, Grimley, and Wilson discuss the pathways taken by electric cooperatives in Denmark, Germany, and the United States. Using case studies from each country, their findings focus on a strong ethic of participation and distribution of decision-making power. They argue that this is due to the fact that electric cooperatives already embody many

DOI: 10.4324/9780429402302-8

aspects of energy democracy because they are member controlled and focus on equality, equity, and solidarity. Some of the big differences between the European nations and the United States were that renewables were at the center of Denmark and Germany's efforts, whereas the United States was more focused on reliability and affordability of electricity, which could come from both fossil fuels and renewables. They also found that the three different countries used different participatory models for owner membership as well as different market and regulatory frameworks. As a final point, they argue that decentralizing ownership alone will not guarantee energy democracy goals but that restructuring relationships across scales in order to redistribute power is also needed, especially when dealing with the complex and polycentric nature of energy systems.

Endres and Johnson put the concept of justice front and center in Chapter 5, with their focus on Indigenous governance and the case study of the *Honor the Earth* campaign and the Dakota Access Pipeline. By first addressing the need to decolonize democracy through the rejection of settler institutions, in this case the American political system, they argue that democracy should instead be seen as equitable collective decision making. This, however, also brings to the fore the issue of sovereignty and its lack of recognition by American democracy and resultant marginalization of Indigenous voices in energy systems governance. In their chapter, they focus on three enactments of energy democracy practiced by the *Honor the Earth* campaign as a representation of the Indigenous governance scale, which includes sustainable economic self-determination, sustainable energy installation on Tribal lands, and enacting participation in decision making. This chapter therefore makes the point that governance complexities surrounding colonialism, sovereignty, and marginalization must be addressed in energy democracy movements and scholarship.

Unlike as in the other chapters, Goyal and Howlett (Chapter 6) offer a conceptual framework to work across scales of energy governance using examples from across Europe. This framework, known as the multiple streams framework (MSF), addresses the important questions of what, how, and where energy systems are democratized. Interestingly, these questions also align with our energy democracy framework of justice, power, public participation, and technology outlined in Chapter 1. For example, Goyal and Howlett focus heavily on public participation as the instrument (the "how" if you will) of stakeholder engagement in energy decision making across governance levels. Power and justice within this framework is captured by the "who" which is composed of communities and constituencies. And finally, the "where" captures the interactions between scales as well as technology itself. Though not intentional on the part of the chapter authors, they were also able to demonstrate convergence between the frameworks used to analyze and contribute to energy democracy practice.

Future directions

As energy democracy practice as well as its scholarship continue to grow, more examples at the national and international levels will hopefully start to be more prominent in the literature. This is especially important for the Global South as there are only a handful of such examples to date, and most of them are either at the local level (see Bloem et al., 2021; MacEwen & Evensen, 2021; Chapter 24) or describe a partnership between Global North and Global South nations, with efforts again still occurring at local levels (see Chapter 35). Table 7.1 provides a list of handbook chapters with their dominant level of scale identified. Even with considerable efforts to ensure a diversity of scholarship by scale, geographic location, government type, and level of national development, it was especially difficult to capture energy research in the Global South (Chapters 24, 29, and 35 only). This is compounded by whether or not democracy is

Table 7.1 Energy democracy chapters by scale (level of governance approximated using dominant scales discussed in the text)

Chapter	*Level of governance*
Chapter 3. International energy governance: opportunities and challenges for democratic politics	International
Chapter 4. Comparing and contrasting the institutional relationships, regulatory frameworks, and energy system governance of European and US Electric cooperatives	National to international
Chapter 5. Energy democracy at the scale of Indigenous governance: Indigenous Native American struggles for democracy, justice, and decolonization	National
Chapter 6. Conceptualizing energy democracy using the multiple streams framework: actors, public participation, and scale in energy transitions	Cross-scalar
Chapter 9. Energy security: from security of supply to public participation	Local to national
Chapter 10. The premise and the promise: energy poverty, capabilities, and the language of moral commitments	International
Chapter 11. A brief excursion into the many scales and voices of renewable energy colonialism	National to international
Chapter 12. Energy dominance	National
Chapter 15. The state or the citizens for energy democracy? Municipal and cooperative models in the German energy transition	National
Chapter 16. Institutionalizing energy democracy: the promises and pitfalls of electricity cooperative development	International
Chapter 17. A feminist lens on energy democracy: redistributing power and resisting oppression through renewable transformation	Cross-scalar
Chapter 18. Energy commons and alternatives to enclosures of sunshine and wind	Local
Chapter 21. Splitting (over) the atom: nuclear energy and democratic conflict	National
Chapter 22. Public participation and energy system transformations	Local
Chapter 23. The complex relations between justice and participation in collaborative planning processes for a renewable energy transition	Local to subnational
Chapter 24. Participation in nondemocracies: rural Thailand as a site of energy democracy	National
Chapter 27. Energy democracy, nuclear power, and participatory knowledge production about radiation risks	National
Chapter 28. A fracked society: multistate media analysis of hydraulic fracturing in the United States	Subnational to national
Chapter 29. Latin American hydropower sacrifice zones	National to international
Chapter 30. Postcards from the future: a case study in Hawaii's transition to wind and solar energy	Subnational
Chapter 33. Carbon-neutral pledges: public opinions, opportunities, and challenges for energy democracy	Local
Chapter 34. Beyond the ivory tower: exploring the role of universities toward sustainable energy transitions in postdisaster environments	Subnational
Chapter 35. Low-carbon energy democracy in the Global South?	International
Chapter 36. Energy democracy in practice: centering energy sovereignty in rural communities and Tribal Nations	Local to national

the form of government practiced in particular Global South countries and, even if a country is a democracy, democratic processes may not be carried out (see Chapter 29). This has led to the editors of this handbook questioning: Are energy democracies a thing of circumstance and privilege, or do such movements have the ability to permeate all levels and types of governance? Only time, effort, and future research will tell.

References

Bloem, S., Swilling, M., & Koranteng, K. (2021). Taking energy democracy to the streets: Socio-technical learning, institutional dynamism, and integration in South African community energy projects. *Energy Research & Social Science*, *72*. https://doi.org/10.1016/j.erss.2020.101906

Feldpausch-Parker, A. M., Endres, D., & Peterson, T. R. (2019). Editorial: A research agenda for energy democracy. *Frontiers in Communication*, *4*. https://doi.org/10.3389/fcomm.2019.00053

MacEwen, M., & Evensen, D. (2021). Mind the gap: Accounting for equitable participation and energy democracy in Kenya. *Energy Research & Social Science*, *71*. https://doi.org/10.1016/j.erss.2020.101843

Stephens, J. C., Burke, M. J., Gibian, B., Jordi, E., & Watts, R. (2018). Operationalizing energy democracy: Challenges and opportunities in Vermont's renewable energy transformation. *Frontiers in Communication*, *3*, 43.

PART II

Discourses of energy democracy

8

DISCOURSES OF ENERGY DEMOCRACY

Introduction

Stephanie L. Gomez and Danielle Endres

Part II of the handbook focuses on the importance of discourses for energy democracy research and practice by highlighting particular keywords—such as energy dominance and energy poverty—that come into discussions about energy transition. These keywords reflect prevalent discourses about energy that contribute to constructing particular visions of energy transition and reveal how it can be framed in a variety of ways. Thus each chapter focuses on a different keyword, or discursive frame, in order to look at energy democracy from myriad angles. As a whole, this part of the handbook reveals themes, disjunctures, and debates about and within energy democracy and its practice.

The keywords presented in these chapters are windows into broader discourses about the roles of energy in society. Discourse is more than discrete instances of talking or meaning-making. Following Foucault's (1972) theorization, discourse is linked with the production of knowledge, subject positions, and enactments of power in both restrictive and productive ways. Discourses are ways of structuring, disciplining, and viewing the world. Discourses function in the background of the way energy issues are communicated, conceived, and deliberated. In a sense, discourses function similarly to frames, or what Kenneth Burke (1966) has called terministic screens. Viewing discourses as terministic screens allows for analysis of how each discourse selects, reflects, and deflects different aspects of a phenomenon. In this case, energy transition might be framed in a variety of different ways—for example, in relation to dominance, poverty, colonialism, justice, or security—with each frame highlighting certain elements of energy's relationship to society and obscuring others. As Feldpausch-Parker and Endres argue in the introduction chapter of this handbook (see Chapter 1), energy democracy itself frames energy transition to particularly emphasize the importance of democratic principles and practices in the transition.

The keywords in these chapters, then, are not dictionary definitions but words or phrases that are of particular importance or significance for understanding a sociocultural phenomenon (Burgett & Hendler, 2020). Inspired by Raymond Williams's (1985) work on cultural studies keywords, the terms unpacked in this section represent an inquiry into key discourses that are not static but ever changing and evolving to adapt to changing contexts and situations. The chapters on energy security (Chapter 9), energy poverty (Chapter 10), energy colonialism (Chapter 11), and energy dominance (Chapter 12) trace these keywords as discourses that frame energy's relation to society. Chapters focus on how the discourse emerged, how it

DOI: 10.4324/9780429402302-10

has evolved, and how it frames discussions of energy. The keywords covered in this part of the handbook do not represent all of the keywords that relate to energy democracy, but they do offer a model for how to engage with other keywords that frame energy democracy research and practice. Some other keywords—such as energy justice (Chapters 9 and 11), energy commons (Chapter 18), and low-carbon energy (Chapter 35)—are referenced in other chapters in other parts of the handbook.

The goal of this part of the handbook is not just to highlight the discursive dynamics of the four keywords covered in it but to highlight the importance of discourse and framing in our understandings of energy democracies. To put it simply: discourse and framing matter. Complex debates, controversies, and topics are conveyed through language and other symbol systems, which not only reflect social systems and values but also can play a role in constructing new systems. Discourse and framing hold the potential to make material changes and construct new ways of being and understanding.

In Chapter 9, Demski and Becker explore the discourses surrounding energy security, a highly contested and contextual term. They draw out various tensions surrounding energy security by providing an overview of the various ways in which it has been used and understood across a variety of contexts, including multiple publics and countries. They argue that while energy security varies from energy democracy in important ways (e.g., different emphases and scales), there are important synergies regarding how both perspectives understand publics. Ultimately, they underscore the importance of broadening the scope of energy democracy by imagining the possibilities of a more fully democratized energy system writ large.

In Chapter 10, Cozen also argues for imagined possibilities through an analysis of the term "energy poverty." Through three case studies, Cozen discusses various ways by which the alleviation of energy poverty is positioned as a moral imperative, which serves to normalize energy use as a public good. According to Cozen, this positioning of energy use as central to human flourishing forecloses on any possibilities for imagining collective action regarding energy systems and practices. By deconstructing this process, Cozen argues for the necessity of building toward a democratized energy future that bypasses traditional modes of energy production, distribution, and consumption.

Batel, in Chapter 11, takes up the term "renewable energy colonialism," drawing parallels between this idea and neo- and carbon colonialism. Using several case studies, Batel explores how renewable energy colonialism is deployed at different scales, including transnationally, internationally, and intranationally. Additionally, Batel investigates the ways that renewable energy colonialism is enacted through different voices, including governmental and corporate practices, the media, and everyday discourses. Through a recognition of the ways that everyday discourses shore up colonial practices and policies regarding energy use, Batel argues for the importance of decolonizing energy practices and systems.

In Chapter 12, the final chapter of this section, Schneider and Peeples similarly take up the idea of energy colonialism by focusing on "energy dominance," a term introduced by former President Donald Trump. They analyze energy dominance as a means of reinforcing the status quo and refusing other possibilities for more democratic energy systems. Through an exploration of various energy policies that use energy dominance as justification, Schneider and Peeples point out the ways that this type of energy colonialism reinforces notions of US exceptionalism and dominance across the world. Essentially, they draw out the ways that energy dominance is fundamentally opposed to the ideals of energy democracy.

References

Burgett, B., & Hendler, G. (2020). Keywords: An introduction. In B. Burgett & G. Hendler (Eds.), *Keywords for American cultural studies* (3rd ed., pp. 1–6). New York University Press.

Burke, K. (1966). *Language as symbolic action: Essays on life, literature, and method*. University of California Press.

Foucault, M. (1972). *The archaeology of knowledge & the discourse on language*. Pantheon Books.

Williams, R. (1985). *Keywords: A vocabulary of culture and society* (revised ed.). Oxford University Press.

9

ENERGY SECURITY

From security of supply to public participation

Christina Demski and Sarah Becker

Energy security is a multifaceted concept that continues to be an influential discourse in decision making within energy policy. It became particularly prominent during the oil crisis of the 1970s but has recently reemerged due to increased global energy demand, disruptions to gas supplies in Europe, increased dependence on imports from politically unstable regions, and the move toward decarbonizing energy systems (Cherp & Jewell, 2014; Chester, 2010). In its early conceptions, energy security was mostly concerned with the securing of supplies. However, in more recent discussions, energy security has become much more contested, and multifaceted (for reviews, see Ang et al., 2015; Chester, 2010; Winzer, 2012). The following sections will discuss the various dimensions and meanings associated with energy security, as well as exploring the related multiple public and expert perspectives. The chapter finishes with a reflection on linkages and contrasts between energy security and energy democracy discourses.

What is energy security?

On a basic level, energy security is concerned with "access" to energy and the services that it affords, for example heating/cooling, lighting, and cooking. Many texts begin with the International Energy Agency definition, which considers energy security in terms of "the uninterrupted availability of energy sources at an affordable price" (International Energy Agency, 2017). From this, it is evident that energy accessibility is critically about two things: reliability or uninterrupted supply of energy and affordability (Hughes, 2009). Furthermore, energy security is often discussed in relation to the factors that might compromise access to energy, as well as solutions that might minimize these risks.

First, threats to a continuous supply of energy may arise through inadequate infrastructure. Winzer (2012) identifies technical, human, and natural risks that may lead to unexpected shocks or disruption. Some of the most commonly cited concerns are about physical threats such as terrorism and natural disasters, as well as inadequate planning and investment (Ang et al., 2015). Indeed, in the UK and many other countries, the renewal of old infrastructure is a key policy driver (Demski et al., 2014). Other, more recently emerging, concerns include cyber-attacks on national energy infrastructure due to increased digitization (Onyeji et al., 2014) and the increasing use of renewable energy, making it more difficult to match demand with supply (Zeyringer et al., 2018).

DOI: 10.4324/9780429402302-11

Second, dependence on energy imports is another concern (Chester, 2010). Import dependence is of particular concern for nations that do not have sufficient resources to satisfy their own energy demand (e.g., Japan) (Kitamura & Managi, 2017; Yao et al., 2018) or where domestic sources have been depleted (e.g., North Sea oil and gas reserves, Demski et al., 2014). This leaves countries dependent on imports from more resource-rich regions. For example, much of Eastern Europe is heavily dependent on gas imports from Russia, while other European regions are increasingly reliant on imports from Middle Eastern countries (Chester, 2010). The import dependence per se is not the issue of concern for many industrialized countries, but it becomes salient if import dependence is associated with regions that are considered politically unstable or regions with which countries have not had substantial trade relationships in the past (Umbach, 2010). These risks associated with import dependence tend to be addressed and managed in two ways: emphasizing the need to further liberalize international energy markets and attempting to increase transparency within these markets (Chester, 2010; International Energy Agency, 2017). In addition, encouraging domestic energy production to increase self-sufficiency is also a favored option (Umbach, 2010). In Europe, for example, increasing renewable energy to decarbonize energy systems has the added benefit of reducing reliance on fossil fuel imports (European Commission, 2014).

Third, there is increased pressure on energy resources globally due to increased demand, especially from Asia (Chester, 2010; Umbach, 2010). Thus, ensuring sufficient supplies may be threatened if there are simply not enough resources to cover demand. Related to this is the concern of being too dependent on specific fuel types, especially considering fossil fuels are being depleted and increasingly expensive to extract. An often cited solution to these concerns is diversification, both in terms of fuel types and where energy is imported from (European Commission, 2014; Hughes, 2009). This is thought to increase resilience of a national energy system (i.e., if there is a supply disruption regarding one resource/import source, there are others to fall back on).

Finally, many of the previous discussions have focused on the supply side, but recent decades have seen a focus on the demand side as well. Energy efficiency and conservation are considered effective ways to reduce vulnerabilities associated with energy security (Ang et al., 2015; Cox, 2016; Hughes, 2009). For this reason, encouraging reductions in overall energy use is an important part of energy policy in Europe (European Commission, 2014).

Beyond security of supply

The previous section mostly discussed energy security in terms of securing energy supplies, which remains a core part of energy security discourses to date. However, other dimensions are becoming increasingly important, with the two most prominent ones being affordability and environmental impacts[1] (Ang et al., 2015).

Affordability is an established dimension of energy security, but it is also highly contested in definition and meaning. Energy security scholars often focus on energy prices, particularly absolute energy price levels as well as price volatility (Winzer, 2012). These issues are then closely linked to energy imports through exchange rates and a country's purchasing power (Ang et al., 2015) or ensuring that energy prices do not adversely affect the economic performance of a country (Chester, 2010). More recently, energy security scholars have also begun to consider aspects that go beyond economic indicators, to consider what affordability means at the household level, for example by including a focus on household energy prices and levels of fuel poverty (Ang et al., 2015). Nonetheless, the energy security literature is still relatively poorly connected to literature on fuel or energy poverty, which seeks to bring into focus more social aspects of energy, including issues of equity and welfare (Day et al., 2016; Oppenheim, 2016; Walker & Day, 2012).[2] For example, while the energy security literature tends to focus on price signals, other literature suggests that energy could, to some extent, be considered a public good (Demski et al., 2019; Sovacool, 2011; Stern & Aronson, 1984; Walker et al., 2016). This way of examining energy affordability accounts for energy being essential to live decent and healthy lives and argues that people's access to energy should not be compromised by their (in)ability to purchase it (Hughes, 2009). Having said that, some authors have advocated for conceiving energy security in a way that can better take into account these issues (Cherp & Jewell, 2014), for example by considering it in terms of having low vulnerabilities to vital energy systems that are understood in terms of energy services, not resources (Laldjebaev et al., 2018; Sovacool, 2011). These ways of assigning meaning to energy security emphasize the importance of energy for maintaining individual and societal welfare.

Advocates of broadening the concept of energy security to include aspects such as equity and welfare also make the case for including issues of sustainability and environmental protection such as air pollution, deforestation, oil spills, waste production, etc. (Ang et al., 2015). Climate change is also increasingly being incorporated into the discourse on energy security. While there is no dispute over the importance of producing and using energy in a way that is not harmful to environments, there are differing views on whether this should be considered an integral part of energy security. While some authors argue that both policy goals of securing supplies and environmental protection can and should be pursued in harmony, there can also be clear conflicts between them (Bazilian et al., 2011; Brown & Huntington, 2008; Duffield, 2009; Froggatt & Levi, 2009). Indeed, there is evidence to suggest that national energy policy decisions can favor one goal over another. For example, Rehner and McCauley (2016) conclude that Germany adopted an environmental justice rather than energy security stance in their nuclear phaseout policy, which will have significant consequences for securing energy supply in the long term. In contrast, it could be argued that the UK has seen a shift in energy policy focus toward prioritizing energy security separately from climate change (Cox, 2016). Finally, some authors suggest that energy security is already a complex concept, and it may become meaningless if it becomes a catchall term for all aspects to do with energy policy. In this regard, Winzer (2012) notes: "[E]nergy security has thus become an umbrella term for many different policy goals" (p. 36).

Whether we agree or disagree that energy security should incorporate environmental dimensions, one thing is clear: Energy security is complex and polysemic (Chester, 2010). And while the multidimensionality is one source of such complexity, further aspects need to be considered. In particular, the complexity increases when we consider what energy security means at multiple temporal and geographic scales, as well as the diverse fuels that make up "energy" and the services they provide. As Chester (2010) notes, "Energy security may be delineated through multiple dimensions and it takes on different specificities depending on the country (or continent), timeframe or energy source to which it is applied" (p. 893). Similarly, Jain (2010) suggests that "each country will face specific energy security concerns depending on its state of economic development, availability of energy resources, energy consumption pattern, and other related factors" (p. 2835). For example, some countries might focus more on import dependence in their understanding of energy security, such as some European nations being reliant on gas supply from Russia (Chester, 2010). The Government of India instead emphasizes the need to provide affordable and nonpolluting energy at the household level as a key energy security concern (Jain, 2010).

In terms of geographic scale, these quotations emphasize that much of the energy security discourse occurs at the international and national scales (as opposed to more local scales).

This is perhaps not surprising given that resources are unequally divided across the globe and energy provision is increasingly international and polycentric, involving actors and organizations beyond nation-states (Umbach, 2010). Simultaneously, much of energy policy and decision making is still conducted at the national level, even within regions that closely cooperate on cross-national matters such as the European Union (Brown et al., 2014). A quick review of the academic literature reveals a host of articles that focus on assessing the energy security of regions and countries and how to improve it (Jain, 2010; Kitamura & Managi, 2017; Kumar Singh, 2013; Laldjebaev et al., 2018; Rehner & McCauley, 2016). Much of these discussions focus on ensuring adequate infrastructure, securing supplies, and the associated geopolitics. One dimension of energy security that does, however, bring more local scales into view is that of affordability, especially when considered at the household level. Local scales are also more prominent in discussions of energy security in developing countries where large-scale infrastructures are not (yet) widespread and/or energy markets have not been liberalized (Jain, 2010).

An important conclusion to be drawn, then, is that energy security is not one specific thing and that specific meanings of energy security depend on the perspective taken. Further to this, important political and social actors also shape how energy is viewed, in addition to the physical, geographic, and system characteristics that inform particular meanings of energy security (Cox, 2016). Indeed, there is an emerging literature that attempts to capture the perspectives of different stakeholders such as policy makers, industry, and individuals. The next section will go on to examine some of these perspectives.

Public and other stakeholder perspectives on energy security

Public perceptions of energy security

Starting with what we know about public perspectives, much of the research that has been done involves quantitative surveys asking people to rate the importance or express their level of concern about a number of energy security–related issues. Most of this research has been carried out in the United States and Europe. The precise nature of these surveys varies, but they generally focus on aspects that might threaten supply and access to electricity/energy (e.g., Reiner, 2006).

In one of the first surveys conducted in the UK, we found that people expressed most concern about fuel prices being high and least concern about terrorist attacks on infrastructure. We also found high concern regarding importing fuels from far away, being dependent on other countries, and running out of traditional energy sources (i.e., fossil fuels). Less concern was expressed for more immediate personal threats to energy access such as rationing of energy and power cuts (Corner et al., 2011; Demski et al., 2014). The finding that people are concerned about the affordability of energy as well as dependence on energy imports and fossil fuels, while lower concern is expressed with regard to reliability of energy, has been relatively robust over time and across countries (Becker et al., 2016; DeCicco et al., 2015; Jones et al., 2017). Perhaps the strongest evidence to date comes from the 2016 European Social Survey (ESS), which surveyed 22 European countries (plus Israel). It also showed that the highest levels of worry were associated with the affordability of energy, followed by concerns about fossil fuel and import dependence. Relatively lower worry was associated with energy reliability, with most countries not being concerned about this aspect, especially in Northern and Western Europe (Demski et al., 2018).

The concern members of the public express in these studies serves to highlight energy security issues that are particularly salient currently. In this respect, affordability is paramount and most often the reason why people do not have access to energy services. While it could be argued that reliability is equally important, especially in societies where energy is integral to practices associated with comfort, convenience, hygiene, communication, and so on, many countries typically enjoy high levels of reliability. For example, Becker et al. (2016) calculate that in Germany the average person has experienced only 17 minutes of power cuts per year. In line with this, we found that members of the UK public noted the taken-for-granted nature of uninterrupted energy access[3] (Demski et al., 2019).

Some studies have also explicitly compared levels of concern about energy security and other environmental issues (DeCicco et al., 2015; Reiner, 2006). For example, the ESS findings show that, across most European countries, energy affordability attracts most worry followed by concern about climate change and then energy reliability. Only in some countries did climate change attract more worry than affordability—these being Denmark, Iceland, Switzerland, Austria, Hungary, Norway, Netherlands, and Sweden (Poortinga et al., 2018). Finally, when asked to make direct trade-offs between climate change and energy security goals, most respondents consider both equally important (Demski et al., 2014). In summary, energy affordability is consistently a predominant energy security concern for members of the public, especially in countries where energy reliability is relatively high. In most but not all cases, concerns about affordability trump concerns over climate change.

Energy security perceptions across groups and countries

The previous section discussed public perceptions of energy security, focusing on key findings and commonalities across studies and countries. There is, of course, also variability in perceptions, and understanding multiple perspectives is important. To this end, some studies have begun to examine differences across sociodemographic groups, although these are still relatively sparse. For example, Reiner (2006) found that concern about energy independence, that is, not being dependent on energy imports from other countries, was higher among participants that were male, older, or conservative. In line with this, Knox-Hayes et al. (2013) found that older respondents ascribed higher importance to a majority of energy security issues. In contrast to Reiner's study, they found that women, not men, expressed higher importance for almost all of the energy security issues included in the survey. Findings were mixed with regard to education and income. As such, while some similarities across studies can be found, conclusions should be interpreted with caution because studies vary widely in terms of the types of energy security questions that are asked.

Studies that examine differences across countries have also recently emerged (Demski et al., 2018; Jones et al., 2017; Knox-Hayes et al., 2013; Sovacool, 2016; Sovacool & Blyth, 2015). As discussed in the previous section, energy security takes on different meanings depending on the context in which it is examined; for example, countries that do not have their own energy resources may strongly emphasize the need for transparency within energy markets, while others may focus more heavily on providing universal access to energy supplies for its population if this has not yet been achieved. The national level is a particularly salient scale because many countries have their unique histories and strategies when it comes to energy resources, production, and use. It is therefore plausible that people's experiences of the energy system and energy security across countries differ and that differences in national context may affect people's concerns about energy security.

To this end, we used the ESS data to examine whether national context is able to explain differences in energy security concerns across the 23 surveyed countries (Demski et al., 2018). Relationships were found for indicators of energy-specific national context and people's expressed worry about a number of energy security dimensions. Most prominently, national electricity prices and fossil fuel consumption were positively related to concerns about affordability, reliability, and vulnerability of energy, meaning that these concerns were higher in countries that also had higher prices and fossil fuel consumption. In addition, higher net energy imports were positively related to concerns about import dependency, affordability, and fossil fuel dependency. While these findings provide initial evidence that the energy-specific context of a country informs people's perspectives on energy security, the study also examined national context that was not directly related to the energy system. Indeed, countries with higher economic and human well-being[4] were found to have lower energy security concerns on the main three dimensions measured—reliability, affordability, and vulnerability. This suggests that people's experience of energy security is related to more than just the energy context of a country but includes the wider societal conditions that exist.

A 2011 survey conducted in a different set of countries[5] made a similar suggestion (Sovacool, 2016; Sovacool et al., 2012). This survey differed from the ESS survey because it used a much broader definition of energy security, asking respondents about the importance of a large set of energy policy areas such as security of supply, environmental issues, and justice issues. The authors conclude that people's perceptions clearly varied nationally and suggest that this variability is linked to political and economic differences across countries. Knox-Hayes et al. (2013) similarly found that level of reliance on energy imports may explain higher concern with regard to the availability and affordability of energy.

It is important to note that this 2011 survey used a sample that was skewed toward people working in governmental, nongovernmental, and university organizations. In addition, there may also have been a relatively strong self-selection bias in favor of people who are interested in energy issues (Knox-Hayes et al., 2013). It is therefore possible that this sample contained a large number of respondents who have a higher level of expertise regarding energy policy than a more general population sample would. This does not make these perspectives any less valuable, but it is worthwhile to consider the differences that may exist between public and expert perspectives. In fact, there has been relatively little attention on expert perspectives to date.

Comparing public and expert perspectives

In one of the only studies to date that has examined expert energy security perspectives, Cox (2016) interviewed 22 UK energy experts,[6] revealing multiple competing and context-specific views rather than convergence. However, the authors did identify some agreements on the broad tenants of energy security. Experts thought policies should incorporate measures to help "respond" to threats and insecurities (e.g., through increasing flexibility in the system such as demand-side management and energy storage) instead of focusing on reducing the causes of insecurity. Experts also focused on the cobenefits of addressing energy security and other areas of energy policy (e.g., climate change) and agreed that political stability and long-term planning should be emphasized to enable adequate investment in critical infrastructure. In a separate study, Manley et al. (2013) surveyed 884 members of professional organizations related to energy, asking how the United States should prioritize energy policy across the goals of energy supply security, environment and climate, and economics and job creation. In line with the findings from public surveys, experts did not subscribe more or less importance to climate change, the economy or energy security; the majority favored policy making that is balanced across all three. The authors conclude that the survey responses "illustrate the challenges of attempting to articulate distinct, independent goals of energy policy" (p. 694). This echoes earlier arguments forwarded in the academic literature that energy security should not be considered distinct from environmental goals.

When comparing public and expert perspectives, one clear difference is the level of knowledge of the energy system and associated terminology. Multiple studies have found that people lack understanding of the wider energy system (Devine-Wright et al., 2010) and that most people's views are instead informed by their experiences of using electricity/gas/petrol or engaging with energy companies through energy bills (DeCicco et al., 2015). Multiple studies have also found that people are unfamiliar with the term "energy security" (DeCicco et al., 2015; Demski et al., 2014; Kim & Kim, 2015). The one group for whom this may be different is for households that have their own energy production systems, most often a solar-powered generation system. This group appears relatively more informed than the average household (Becker et al., 2016).

Finally, to date, only one study has directly compared public and expert perceptions of energy security. Blumer et al. (2015) found that Swiss energy experts tend to focus on the physical aspects, whereas Swiss energy users appear to have a much broader understanding of energy security; for example, people's discussions included considerations around equity and fairness. This is in line with the earlier discussed finding from the ESS data, which showed that people's concerns about energy security appear connected to a much wider range of issues beyond energy-specific ones, including the level of well-being enjoyed by residents of a particular country (Demski et al., 2018). These findings are also consistent with other research we conducted on public engagement with energy systems and futures. The research shows that people's perspectives are informed by a large number of values including those related to security but also issues in relation to equality and fairness. People expect approaches toward energy, and therefore energy security, to be considerate of all of these issues (Demski et al., 2015).

In conclusion, considering multiple perspectives on energy security opens up its definition and meaning to include issues beyond simply the security of energy supplies as it is traditionally conceived. Having discussed the energy security discourse, including its complexity and multiplicity, the remainder of this chapter will briefly consider how this connects with the emerging discourse on energy democracy. On the surface, energy security and energy democracy emphasize very different aspects within energy systems and their wider context. While the energy security discourse is more established and the energy democracy discourse is only emerging, both are dynamic concepts with multiple meanings and perspectives.

Energy security and energy democracy

One central aim of the energy democracy movement is to change the socioeconomic relations embedded in energy systems by encouraging greater public involvement and control. This appears to be most closely associated with participatory democracy and direct participation (van Veelen & van der Horst, 2018). As such, reconfigured energy systems and decision making are envisaged to include significant decentralization and local community/individual ownership. This implies a desired shift in power away from privatized systems and corporate control toward models of social ownership, for example, through self-organized, voluntary civil society organizations (Burke & Stephens, 2017).

On the one hand, social ownership and local control is in line with ideas of self-sufficiency and independence, which are also central to (some forms of) energy security discourse. In our research with publics, these ideas are evaluated positively and perhaps even discussed as something to be desired (Demski et al., 2015; Thomas et al., 2019). This preference for self-sufficiency relates to being independent from other countries and depleting fossil fuels but also extends beyond this. For example, there is emerging evidence that, at least in some countries, people would welcome a shift away from privatization of the energy system and that producing their own energy would enable them to do so (Thomas et al., 2019). Furthermore, visions of energy systems that are more inclusive, equitable, and low-carbon as forwarded by the energy democracy literature are also in line with public visions for the future of energy (Demski et al., 2015). As the previous section has discussed, people's sense of energy security and desires for energy transitions are about more than just securing energy resources and a reliable supply of energy (Demski et al., 2018; Knox-Hayes et al., 2013).

On the other hand, there are clear differences between energy security and energy democracy, most notably when it comes to scale. Energy security focuses heavily on international and national scales, while energy democracy tends to more often emphasize the local. Indeed, it could be argued that the energy security literature has not yet sufficiently explored what energy security means at more local levels. As such, the energy security discourse tends to focus on maintaining the status quo (i.e., large-scale energy systems that are centrally controlled and in which local communities or households are passive consumers of energy). The reverse criticism has been voiced in relation to the energy democracy literature—it has, to date, mostly neglected broader scales, in particular the historic role of the state as owner, investor, and developer of the national grid. It has been noted that this vast national energy infrastructure not only is implicated in the production and distribution of energy for use in communities and households but also indirectly supports a large array of essential goods and services that people rely on, e.g., food distribution (van Veelen & van der Horst, 2018).

Perhaps the starkest criticism of the energy democracy discourse to date is that the decentralization rhetoric can play into the hands of neoliberalization, where responsibility for previously collectively provided services is increasingly placed on individuals (DeFilippis et al., 2006; Newell et al., 2015; van Veelen & van der Horst, 2018). This is an argument that is particularly pertinent if we consider energy security to be a service that current national energy systems provide, at least in its basic conception of providing an uninterrupted supply of energy.[7] As the previous sections have shown, such continuous access to energy is something people have learned to rely on (Demski et al., 2019).

Guaranteeing a continuous supply of energy may be particularly challenging in the future, where energy systems are increasingly reliant on inflexible renewable energy production as opposed to more flexible fossil fuel power stations. As such, in a renewable energy world, additional solutions such as energy storage and/or demand-side management might be necessary to maintain some flexibility (Braff et al., 2016). In a future where locally owned, decentralized energy systems dominate, this also means that the responsibility of ensuring a continuous supply will be held at community level. This brings into focus a number of justice concerns. For example, there are limits on who will be able to afford new energy storage technologies and engage in shifting behavior and practices to ensure energy demand matches supply[8] (Forman, 2017; Mander et al., 2015). This concern about shifting responsibility for energy reliability onto individuals and communities is similar to issues echoed in other parts of the energy democracy literature. For example, ensuring equal participation in self-governance requires assets (e.g., roofs for solar panels), financial resources, time and knowledge, which not everyone may have (Faber & McCarthy, 2003; van Veelen & van der Horst, 2018). Indeed, in a recent study we asked diverse groups of people to deliberate different energy storage options, and participants raised precisely these issues in relation to community and individual ownership. Participants questioned how equality and fairness could be ensured in these circumstances, particularly making sure that vulnerable people would not simply end up with an unreliable energy supply (Thomas et al., 2019).

These arguments have led some authors to suggest that direct participation is not the only way energy democracy could be enhanced and that the state and its institutions should continue to play a role (Angel, 2017), for example through renationalization and remunicipilization (Chavez, 2015). As such, direct forms of participation and self-governance may be able to exist in parallel with other forms of democracy, including institutional and deliberative democracy (van Veelen & van der Horst, 2018). This would open up ways of democratizing the energy system beyond owning generation technology and be able to engage with actors at the regional, national, and international scales. An additional focus on these other types of democracy could ensure that the state and institutions could still play a role, thus avoiding a significant individualization of responsibility for aspects like reliability of energy supply. For example, the state may still own aspects of the energy system infrastructure but with citizens holding power and having a direct say over state institutions (Chavez, 2015; van Veelen & van der Horst, 2018). This would require a stronger focus on increasing transparency, accountability, deliberation, and conflict resolution (Burke & Stephens, 2017; van Veelen, 2018; van Veelen & van der Horst, 2018)—and would require innovative methods and procedures that can capture and consider a multiplicity of views and values for decision making (Burke & Stephens, 2017; Demski et al., 2015).

Conclusion

Energy security has traditionally been defined as securing of supplies, but more recent conceptualizations are much broader, including aspects such as affordability and at times issues of environmental protection, justice, and welfare. This chapter has revealed the multiple, dynamic, and varied perspectives that make up the energy security discourse, including those of publics and experts across multiple countries. Comparing the energy security discourse with the emerging discourse on energy democracy, it becomes clear that they tend to emphasize different aspects (e.g., investing in existing national infrastructure versus local ownership) and scales (e.g., national versus local). However, there are also synergies, especially when considering the dynamic and multiple meanings associated with energy security as understood by publics; for example, both emphasize the importance of self-sufficiency. A focus on energy security also emphasizes the importance of considering how continuous access to energy is ensured in a future where energy is supplied by renewable sources and predominantly owned by communities and individuals. Those advocating for greater energy democracy may therefore need to think beyond local scales and consider how the wider energy system can be more fully democratized.

Notes

1 For a comprehensive overview of different dimensions and how emphasis on different issues has changed over time, see Ang et al. (2015).
2 See Brian Cozen's Chapter 10 in this book for a more in-depth examination of energy poverty.
3 We would expect different perceptions in countries where power cuts are more common.
4 Measured in terms of per capita GDP and the Human Wellbeing Index (Demski et al., 2018).
5 Brazil, China, Germany, India, Kazakhstan, Japan, Papua New Guinea, Saudi Arabia, Singapore, United States
6 The study defined energy experts as people with expertise on energy security, who hold positions in research or policy and strategy at different types of organizations in the UK energy sector, including utility companies, NGOs, policy/regulation, and academia.
7 There are, of course, exceptions to this, for example if we consider remote rural locations that are not connected to electricity or gas grids.
8 For example, in our own research, people often talk about shift workers, people with health issues, children, etc.

References

Ang, B. W., Choong, W. L., & Ng, T. S. (2015). Energy security: Definitions, dimensions and indexes. *Renewable and Sustainable Energy Reviews*, *42*, 1077–1093. https://doi.org/10.1016/j.rser.2014.10.064

Angel, J. (2017). Towards an energy politics in-against-and-beyond the state: Berlin's struggle for energy democracy. *Antipode*, *49*(3), 557–576. https://doi.org/10.1111/anti.12289

Bazilian, M., Hobbs, B. F., Blyth, W., MacGill, I., & Howells, M. (2011). Interactions between energy security and climate change: A focus on developing countries. *Energy Policy*, *39*(6), 3750–3756. https://doi.org/10.1016/j.enpol.2011.04.003

Becker, S., Schober, D., & Wassermann, S. (2016). How to approach consumers' nonmonetary evaluation of electricity supply security? The case of Germany from a multidisciplinary perspective. *Utilities Policy*, *42*, 74–84. https://doi.org/10.1016/j.jup.2016.06.012

Blumer, Y. B., Moser, C., Patt, A., & Seidl, R. (2015). The precarious consensus on the importance of energy security: Contrasting views between Swiss energy users and experts. *Renewable and Sustainable Energy Reviews*, *52*, 927–936. https://doi.org/10.1016/j.rser.2015.07.081

Braff, W. A., Mueller, J. M., & Trancik, J. E. (2016). Value of storage technologies for wind and solar energy. *Nature Climate Change*, *6*(10), 964–969. https://doi.org/10.1038/nclimate3045

Brown, M. A., Wang, Y., Sovacool, B. K., & D'Agostino, A. L. (2014). Forty years of energy security trends: A comparative assessment of 22 industrialized countries. *Energy Research and Social Science*, *4*(C), 64–77. https://doi.org/10.1016/j.erss.2014.08.008

Brown, S. P. A., & Huntington, H. G. (2008). Energy security and climate change protection: Complementarity or tradeoff? *Energy Policy*, *36*(9), 3510–3513. https://doi.org/10.1016/j.enpol.2008.05.027

Burke, M. J., & Stephens, J. C. (2017). Energy democracy: Goals and policy instruments for sociotechnical transitions. *Energy Research and Social Science*, *33*, 35–48. https://doi.org/10.1016/j.erss.2017.09.024

Chavez, D. (2015). The meaning, relevance and scope of energy democracy. *Transnational Institute*. Retrieved from April 12, 2019, from www.tni.org/en/article/the-meaning-relevance-and-scope-of-energy-democracy

Cherp, A., & Jewell, J. (2014). The concept of energy security: Beyond the four as. *Energy Policy*, *75*, 415–421. https://doi.org/10.1016/j.enpol.2014.09.005

Chester, L. (2010). Conceptualising energy security and making explicit its polysemic nature. *Energy Policy*, *38*(2), 887–895. https://doi.org/10.1016/j.enpol.2009.10.039

Corner, A., Venables, D., Spence, A., Poortinga, W., Demski, C., & Pidgeon, N. (2011). Nuclear power, climate change and energy security: Exploring British public attitudes. *Energy Policy*, *39*(9). https://doi.org/10.1016/j.enpol.2011.06.037

Cox, E. (2016). Opening the black box of energy security: A study of conceptions of electricity security in the United Kingdom. *Energy Research and Social Science*, *21*, 1–11. https://doi.org/10.1016/j.erss.2016.06.020

Day, R., Walker, G., & Simcock, N. (2016). Conceptualising energy use and energy poverty using a capabilities framework. *Energy Policy*, *93*, 255–264. https://doi.org/10.1016/j.enpol.2016.03.019

DeCicco, J., Yan, T., Keusch, F., Muñoz, D. H., & Neidert, L. (2015). US consumer attitudes and expectations about energy. *Energy Policy*, *86*, 749–758. https://doi.org/10.1016/j.enpol.2015.08.022

DeFilippis, J., Fisher, R., & Shragge, E. (2006). Neither romance nor regulation: Re-evaluating community. *International Journal of Urban and Regional Research*, *30*(3), 673–689. https://doi.org/10.1111/j.1468-2427.2006.00680.x

Demski, C., Butler, C., Parkhill, K. A., Spence, A., & Pidgeon, N. F. (2015). Public values for energy system change. *Global Environmental Change*, *34*, 59–69. https://doi.org/10.1016/j.gloenvcha.2015.06.014

Demski, C., Poortinga, W., & Pidgeon, N. (2014). Exploring public perceptions of energy security risks in the UK. *Energy Policy*, *66*. https://doi.org/10.1016/j.enpol.2013.10.079

Demski, C., Poortinga, W., Whitmarsh, L., Böhm, G., Fisher, S., Steg, L., Umit, R., Jokinen, P., & Pohjolainen, P. (2018). National context is a key determinant of energy security concerns across Europe. *Nature Energy*, *3*(10), 882–888. https://doi.org/10.1038/s41560-018-0235-8

Demski, C., Thomas, G., Becker, S., Evensen, D., & Pidgeon, N. (2019). Acceptance of energy transitions and policies: Public conceptualisations of energy as a need and basic right in the United Kingdom. *Energy Research and Social Science*, *48*, 33–45. https://doi.org/10.1016/j.erss.2018.09.018

Devine-Wright, P., Devine-Wright, H., & Sherry-Brennan, F. (2010). Visible technologies, invisible organisations: An empirical study of public beliefs about electricity supply networks. *Energy Policy*, *38*(8), 4127–4134. https://doi.org/10.1016/j.enpol.2010.03.039

Duffield, J. S. (2009). Germany and energy security in the 2000s: Rise and fall of a policy issue? *Energy Policy*, *37*(11), 4284–4292. https://doi.org/10.1016/j.enpol.2009.05.021
European Commission. (2014). *Communication from the commission to the European Parliament and the Council: European energy security strategy*. European Commission.
Faber, D. R., & McCarthy, D. (2003). Neo-liberalism, globalization, and the struggle for ecological democracy: Linking sustainability and environmental justice. In J. Agyeman, R. Bullard, & B. Evans (Eds.), *Just sustainabilities: Development in an unequal world* (pp. 38–63). Earthscan.
Forman, A. (2017). Energy justice at the end of the wire: Enacting community energy and equity in Wales. *Energy Policy*, *107*, 649–657. https://doi.org/10.1016/j.enpol.2017.05.006
Froggatt, A., & Levi, M. A. (2009). Climate and energy security policies and measures: Synergies and conflicts. *International Affairs*, *85*(6), 1129–1141. https://doi.org/10.1111/j.1468-2346.2009.00853.x
Hughes, L. (2009). The four 'R's of energy security. *Energy Policy*, *37*(6), 2459–2461. https://doi.org/10.1016/j.enpol.2009.02.038
International Energy Agency. (2017). *World energy outlook 2017*. International Energy Agency. https://doi.org/10.1016/0301-4215(73)90024-4
Jain, G. (2010). Energy security issues at household level in India. *Energy Policy*, *38*(6), 2835–2845. https://doi.org/10.1016/j.enpol.2010.01.016
Jones, C. R., Kaklamanou, D., & Lazuras, L. (2017). Public perceptions of energy security in Greece and Turkey: Exploring the relevance of pro-environmental and pro-cultural orientations. *Energy Research and Social Science*, *28*, 17–28. https://doi.org/10.1016/j.erss.2017.04.002
Kim, J., & Kim, J. (2015). Korean public's perceptions on supply security of fossil fuels: A contingent valuation analysis. *Applied Energy*, *137*, 301–309. https://doi.org/10.1016/j.apenergy.2014.10.016
Kitamura, T., & Managi, S. (2017). Energy security and potential supply disruption: A case study in Japan. *Energy Policy*, *110*, 90–104. https://doi.org/10.1016/j.enpol.2017.08.008
Knox-Hayes, J., Brown, M. A., Sovacool, B. K., & Wang, Y. (2013). Understanding attitudes toward energy security: Results of a cross-national survey. *Global Environmental Change*, *23*(3), 609–622. https://doi.org/10.1016/j.gloenvcha.2013.02.003
Kumar Singh, B. (2013). South Asia energy security: Challenges and opportunities. *Energy Policy*, *63*, 458–468. https://doi.org/10.1016/j.enpol.2013.07.128
Laldjebaev, M., Morreale, S. J., Sovacool, B. K., & Kassam, K. A. S. (2018). Rethinking energy security and services in practice: National vulnerability and three energy pathways in Tajikistan. *Energy Policy*, *114*, 39–50. https://doi.org/10.1016/j.enpol.2017.11.058
Mander, S., Abi Ghanem, D., Belhomme, R., Delgado Espinos, I., Gonsalez Sainz-Maza, R., Kessels, K., & Lombardi, M. (2015). Socio-economic aspects of demand response. In A. Losi, P. Mancarella, & A. Vicino (Eds.), *Integration of demand response into the electricity chain: Challenges, opportunities and smart grid solutions* (pp. 215–240). ISTE.
Manley, D. K., Hines, V. A., Jordan, M. W., & Stoltz, R. E. (2013). A survey of energy policy priorities in the United States: Energy supply security, economics, and the environment. *Energy Policy*, *60*, 687–696. https://doi.org/10.1016/j.enpol.2013.04.061
Newell, P., Bulkeley, H., Turner, K., Shaw, C., Caney, S., Shove, E., & Pidgeon, N. (2015). Governance traps in climate change politics: Re-framing the debate in terms of responsibilities and rights. *Wiley Interdisciplinary Reviews: Climate Change*, *6*(6), 535–540. https://doi.org/10.1002/wcc.356
Onyeji, I., Bazilian, M., & Bronk, C. (2014). Cyber security and critical energy infrastructure. *Electricity Journal*, *27*(2), 52–60. https://doi.org/10.1016/j.tej.2014.01.011
Oppenheim, J. (2016). The United States regulatory compact and energy poverty. *Energy Research and Social Science*, *18*, 96–108. https://doi.org/10.1016/j.erss.2016.04.022
Poortinga, W., Fisher, S., Böhm, G., Steg, L., Whitmarsh, L., & Ogunbode, C. (2018). European attitudes to climate change and energy. *ESS Topline Results Series, Round 8*(9), 20.
Rehner, R., & McCauley, D. (2016). Security, justice and the energy crossroads: Assessing the implications of the nuclear phase-out in Germany. *Energy Policy*, *88*, 289–298. https://doi.org/10.1016/j.enpol.2015.10.038
Reiner, D. M. (2006). *2006 EPRG public opinion survey on energy security: Policy preferences and personal behaviour 2004* (pp. 1–40). www.eprg.group.cam.ac.uk/wp-content/uploads/2008/11/eprg0706.pdf
Sovacool, B. K. (2011). Security of energy services and uses within urban households. *Current Opinion in Environmental Sustainability*, *3*(4), 218–224. https://doi.org/10.1016/j.cosust.2011.06.004

Sovacool, B. K. (2016). Differing cultures of energy security: An international comparison of public perceptions. *Renewable and Sustainable Energy Reviews*, *55*, 811–822. https://doi.org/10.1016/j.rser.2015.10.144

Sovacool, B. K., & Blyth, P. L. (2015). Energy and environmental attitudes in the green state of Denmark: Implications for energy democracy, low carbon transitions, and energy literacy. *Environmental Science and Policy*, *54*, 304–315. https://doi.org/10.1016/j.envsci.2015.07.011

Sovacool, B. K., Valentine, S. V., Jain Bambawale, M., Brown, M. A., de Fátima Cardoso, T., Nurbek, S., Suleimenova, G., Li, J., Xu, Y., Jain, A., Alhajji, A. F., & Zubiri, A. (2012). Exploring propositions about perceptions of energy security: An international survey. *Environmental Science and Policy*, *16*, 44–64. https://doi.org/10.1016/j.envsci.2011.10.009

Stern, P. C., & Aronson, E. (1984). *Energy use: The human dimension*. The National Academies Press.

Thomas, G., Demski, C., & Pidgeon, N. (2019). Deliberating the public acceptability of energy storage in the UK. *Energy Policy*, *133*. https://doi.org/10.1016/j.enpol.2019.110908

Umbach, F. (2010). Global energy security and the implications for the EU. *Energy Policy*, *38*(3), 1229–1240. https://doi.org/10.1016/j.enpol.2009.01.010

van Veelen, B. (2018). Negotiating energy democracy in practice: Governance processes in community energy projects. *Environmental Politics*, *27*(4), 644–665. https://doi.org/10.1080/09644016.2018.1427824

van Veelen, B., & van der Horst, D. (2018). What is energy democracy? Connecting social science energy research and political theory. *Energy Research and Social Science*, *46*, 19–28. https://doi.org/10.1016/j.erss.2018.06.010

Walker, G., & Day, R. (2012). Fuel poverty as injustice: Integrating distribution, recognition and procedure in the struggle for affordable warmth. *Energy Policy*, *49*, 69–75. https://doi.org/10.1016/j.enpol.2012.01.044

Walker, G., Simcock, N., & Day, R. (2016). Necessary energy uses and a minimum standard of living in the United Kingdom: Energy justice or escalating expectations? *Energy Research and Social Science*, *18*, 129–138. https://doi.org/10.1016/j.erss.2016.02.007

Winzer, C. (2012). Conceptualizing energy security. *Energy Policy*, *46*, 36–48. https://doi.org/10.1016/j.enpol.2012.02.067

Yao, L., Shi, X., & Andrews-Speed, P. (2018). Conceptualization of energy security in resource-poor economies: The role of the nature of economy. *Energy Policy*, *114*, 394–402. https://doi.org/10.1016/j.enpol.2017.12.029

Zeyringer, M., Price, J., Fais, B., Li, P. H., & Sharp, E. (2018). Designing low-carbon power systems for Great Britain in 2050 that are robust to the spatiotemporal and inter-annual variability of weather. *Nature Energy*, *3*(5), 395–403. https://doi.org/10.1038/s41560-018-0128-x

10

THE PREMISE AND THE PROMISE

Energy poverty, capabilities, and the language of moral commitments

Brian Cozen

Energy poverty discourses consistently suggest further scarcity—not merely lack in infrastructure but in all the practices and values assumed through modern energy. Such discussions proffer light as a premise and a promise: access to modern energy as the necessary grounds (premise) for expanding human capacities and achieving human freedom (promise). This energy poverty discourse often draws upon the language of human development (e.g., Sen, 1999) to naturalize energy practice as the sole means for overcoming scarcity. These narratives constitute equivalencies, between ends and means, toward the proposition that increased energy consumption is a universal moral good and inevitable. Further, the phrase "energy poverty" itself presents problematic assumptions and prescriptions. "Poverty" encourages a view of the other as lacking not simply in access to services but in overall human potential. But even further, "energy" may be directive, as it discourages conceiving of alternatives to meeting human development ends outside of known relations to energy, environment, economy, and political participation.

This chapter examines the discursive process by which energy practices become legitimated as the inevitable means for expanding human development and freedoms. Overall, I argue that this discourse draws upon a human development paradigm to naturalize energy practices. Namely, a capabilities approach theorizes human development in terms of well-being and offers a conceptual framework to measure development via "ableness" (Dowding, 2006) or "adequate social opportunities" (Sen, 1999, p. 11), countering traditional economic measurements. This approach, largely drawing from the work of Nussbaum (2000, 2011) and Sen (1985, 1999, 2005), has been adapted to initiate and assess global development reforms in a variety of contexts, such as education (Boni & Walker, 2013), technology (Oosterlaken & van den Hoven, 2012), and food security (Hunt, 2018).

This chapter starts with how the capabilities approach has been applied to energy policy and use (Day et al., 2016) in order to make sense of the overlapping discourses of capabilities and energy poverty. I argue that, while capabilities can function as a beneficial heuristic for conceiving of transformational practices for achieving human well-being, the language of energy poverty diminishes this benefit. Therefore, studying "capabilities" as a trope illuminates how energy poverty issues are understood and acted upon by those promoting various interventions in governance, markets, and energy access. Namely, the language of capabilities naturalizes the link between energy and human potential. This discourse turns attention away from the process

DOI: 10.4324/9780429402302-12

of collective action, by which communities might reimagine systems, practices, and habits that bypass traditional modes of energy use to find alternative approaches to human flourishing. Therefore, as opposed to a capabilities approach to energy, it might be more fruitful to name "energetic capabilities," turning attention from energy poverty to energetic democratization. That is, promoting discourse on human energies, and the democratic mechanisms that enable new conceptions of human practice, can counter this tendency to reify energy practice as the necessary ground for human development.

The chapter proceeds as follows. First, I discuss the value of a capabilities approach to examine and assess energy poverty. Next, I explain how the term "capabilities," more than a neutral framework, also names a language strategy or trope for justifying energy practices. Then I offer three cases illustrating how "capabilities" discourse regarding energy poverty circulates and legitimates energy practice: one, an analysis of a United Nations campaign linking energy poverty, development, and social practices (see also Cozen, 2017); two, an illustration of pro-nuclear advocacy, which calls forth a capabilities perspective to construct discursive space for nuclear energy's expanded role; and three, a reflection on the GravityLight–Shell collaboration and its allusions to light's necessity for imaginative capacities. The conclusion considers how democratizing energy or nurturing democratic capabilities can constitute alternative visions, practices, habits, and structures for energy use.

Energy poverty through a capabilities framework

"Energy poverty" usually denotes issues of energy access in developing countries and, as such, contains certain premises, assumptions, and exigences. Different themes emerge when defining energy poverty, including, for one, naming energy poverty as a condition of lack, specifically in terms of modernity, and two, emphasizing a broader relationship between energy access and social capacities. Illustrative of the first theme, the *Dictionary of Energy* defines *energy poverty* as a "condition" of the underdeveloped, lacking in "modern energy technology for basic household needs" due to cost and/or infrastructure (Cleveland & Morris, 2015, p. 200). To intervene in this condition of energy poverty, Bouzarovski and Petrova (2015) point out that, in focusing on technology and infrastructure, such interventions often focus on top-down efforts that modernize domestic energy supply, such as grid expansion.

Other definitions offer broader associations between energy-as-means and the ends they serve. This focus sometimes seeks to muddle the developed/developing world distinction commonly found in energy poverty discussions, especially with the construct "fuel poverty" generally reserved to discussing heat affordability in developed countries (Li et al., 2014). Similar to Simcock and Petrova's (2017) argument that the developed/developing binary obscures addressing energy poverty as "a global-scale issue of human security" (p. 425), Bouzarovski (2018) attempts to account for all energy vulnerability research by defining *energy poverty* as "a set of domestic energy circumstances that do not allow for participating in the lifestyles, customs and activities that define membership of a society" (pp. 13–14). Emphasizing whether or not energy services are "sufficient for [a household's] social and material needs" (p. 1) links these services to culturally sensitive and subjectively defined requirements and, as such, may allow for greater complexity than linear assumptions that modernizing infrastructure initiates growth. This second theme, of access and social opportunity, is suggested by the 2000 World Energy Assessment (WEA) definition, which names *energy poverty* as an "absence of sufficient choice" in the provisions that "support economic and human development" (as cited in González-Eguino, 2015, p. 379). This definition suggests a negative correlation between energy poverty and human development, as well as hints at Sen's (1999) emphasis on choice and freedoms when assessing development. The capabilities approach to human development (e.g., Nussbaum, 2011; Sen, 1999) places emphasis on the ends, goals, or aims of development: less about base economic terms (like increased GDP or energy use) and more about increasing individuals' potential for realizing social opportunities.

Both González-Eguino (2015) and Day et al. (2016) specify the association of energy poverty to development frameworks, with the latter structuring their model around this relationship, arguing that a multidimensional approach to development offers a theoretically rich framework for energy poverty. This approach focuses on the social practices that energy affords, or the ends people seek or may potentially wish to seek, and not simply a focus on energy fuels, supplies, and services. Day et al. (2016) propose a definition of *energy poverty* in these terms: "An inability to realise essential capabilities as a direct or indirect result of insufficient access to affordable, reliable and safe energy services, and taking into account available reasonable alternative means of realizing these capabilities" (p. 260). The capabilities approach that underscores their framework has numerous benefits that inform a theoretical framework for energy poverty, including, as detailed here: identifying energy-related interventions across a spectrum; starting with the question of ends, or what energy is used for (Shove & Walker, 2014), not with what energy services need to be met (yet are not being met); and a consideration for alternative means, which can encourage conceiving of and realizing development ends beyond energy use.

Defining development in terms of "wider human flourishing, and on what people can achieve and do" (Day et al., 2016, p. 258) establishes "energy poverty" interventions across a spectrum, from fuel source to basic needs satisfied, and what people can do (or not do). Day et al. (2016, p. 260) visualize the energy poverty capabilities framework as a series of related elements, from left to right, that illustrate the ultimate outcomes of energy use:

Fuel source → power supply → energy service → secondary capabilities → basic capabilities

A fuel or energy source might be coal; the power supply could be an electric grid; a type of energy service might be refrigeration, with a secondary capability, storing food, and a basic capability, maintaining health. Ultimately, this framework assumes that people wish to maintain good health or at least have the capability to do so; various additional capabilities, like storing food, aid the potential for satisfying such basic conditions. Various means/services to store food include refrigeration. Interventions interested in weaning people off coal would seek an alternative fuel source that powers the grid that runs the refrigeration that holds the capacity to store food and help maintain health. This setup places emphasis on individuals' capacity to satisfy basic desired outcomes.

This capabilities model highlights how different elements related to energy and its social practices suggest different points of intervention. Goodwin (2010) similarly visualizes the "socially understood links" that underscore the cultural logics of automobility: "gasoline → cars → mobility → human flourishing" (p. 61). Goodwin argues that the mobility–flourishing link is historical or not pregiven, and so along with other links can be reimagined and reconstituted. Electric cars illustrate delinking cars from gasoline; alternative modes delink cars from mobility; and interventions in cultural norms would delink mobility from human flourishing, perhaps seen in downtown microliving. Conceptually, the Day et al. (2016) capabilities model also turns attention from the energy services required for human use and economic growth to open-ended questions about how social practice can enable the social ends desired (Shove & Walker, 2014) and, indeed, how it can be otherwise.

By focusing on the facilitation of capabilities, energy provisions can be understood as one possible means, among others, to enable social opportunities. As long as the social ends are (or have the potential to be) met, "energy poverty"–related interventions can prescribe either energy-centered means or alternatives. This conception of alternative means, or the second part of Day et al.'s (2016) definition of energy poverty, opens possible interventions that satisfy human development without necessarily expanding energy consumption, such as "supporting alternatives to energy services" that still fulfill secondary capabilities, or "shaping or shifting expectations, customs and practices" that influence how basic human needs and wants are imagined and realized (Day et al., 2016, p. 262). While grounded in a privileged position of choice, not actual precarity, Colin Beavan's *No Impact Man* project illustrates such alternatives. For the project, Beavan's family spent a year trying to live carbon-neutral in New York City. The project can be understood as an effort to constitute public models for green identities (DeLaure, 2011). This included testing out new power supplies to satisfy old needs, such as installing a solar panel to charge Beavan's laptop; meeting basic needs without a power supply (including failing at pot-in-pot refrigeration and so for them an unfeasible alternative); or rethinking basic conceptions of needs and desires (reflections the project continuously encouraged) (Gabbert & Schein, 2009). Such delinking of ends and means underscores the capabilities energy poverty definition, its final clause encouraging "alternative routes to realising capabilities. . . [and] not locking in assumptions about required energy services" (Day et al., 2016, pp. 260–261).

Ultimately, realizing alternative modes of thinking, being, and doing, expanding capabilities while minimizing energy use, can best manifest in the energetic capacities that emerge out of democratic engagement. This energetic democratization reflects Sen's (1999) ultimate focus, on the capacity for individuals to actively take part in and "effectively shape their own destiny and help each other" (p. 11). In the context of energy, this focus is less about energy provisions as the premise and promise of capabilities and more about the capacities of democratic mechanisms: creating the social opportunities to engage, from which energized publics might emerge (Chilvers & Longhurst, 2016). Circulating discourses that emphasize the "alternative means" component of a capabilities definition of energy poverty (Day et al., 2016) might help cultivate the types of participatory structures, constituted identities, and social norms that can simultaneously engender human well-being and diminish reliance on energy practices.

Energy poverty in terms of capabilities: moralizing energy access

A capabilities perspective does not simply define a theoretical framework but names a trope by which energy providers emphasize the moral authority of their provisions. Day et al.'s (2016) capabilities framework for energy poverty illuminates two aspects of and for energy discussions. On the one hand, it suggests a way to consider alternatives to energy consumption in satisfying basic needs, opportunities, and capacities. One sees this emphasis in the second part of Day et al.'s (2016) definition of energy poverty, as previously discussed. On the other hand, as elaborated in this section, it opens a space to examine, through a communication lens, how energy is framed in various discourses: not as simple means but as the starting point for human capabilities. In energy poverty discourses, energy sources, supplies, and services are rendered the ground or premise on which development-as-freedom (Sen, 1999) can begin. Paying attention to the discourse that circulates around energy poverty can clarify which aspects of the concept are "appropriated and prioritized" (Schneider et al., 2016, p. 142). This section illustrates how and to what ends discourse promoting energy poverty alleviation often maps onto the components of Day et al.'s (2016) framework.

As I illustrate here and elsewhere (Cozen, 2015, 2017, 2018), the discourse of energy poverty often frames energy access in terms of capabilities while, in calling forth the energy source or supply, reinforcing the links among elements of Day et al.'s (2016) framework. That is, energy poverty discourse often profits in essentialized conflations, legitimating fuel sources as a precondition for human achievement. The promise of energy poverty alleviation is in naming energy as the initial premise on which "expanding substantive freedoms" (Sen, 1999, p. 3) are built. While Sen focuses on ends over means, arguing for development paradigms to begin with freedoms as their ultimate aim, energy practice often takes on an assumed role not as the means for human development so much as the grounds by which such development might be attainable. Therefore, such discourses tend to reinforce the linking of energy provisions to a good life. Functioning as the premise of social opportunity, modern energy access holds out promise, and energy provisions (and providers) take on "moral authority" (Schneider et al., 2016, p. 107).

The material effects of such communication dynamics have been documented elsewhere. Schneider et al. (2016) point to coal companies using a similar linguistic slide. Schneider et al. (2016) argue that, beyond guiding scholarly constructs, different components of the WEA definition previously mentioned are "appropriated and prioritized differently in competing rhetorics surrounding energy poverty" (p. 142). Their analyses of coal industry communication illustrate such appropriation, including the "energy utopia" strategy by which promotional discourses champion coal for lifting people out of poverty. I examine the same strategy in other fossil fuel promotions, arguing that texts such as oil company advertisements communicate "energy" in a dual sense: on the one hand as human potential or an energetic spirit and on the other hand as energy resource systems (Cozen, 2018). Energy systems are depicted as making various activities possible—from hitting the road to find freedom, to cooking a family meal, to playing nighttime baseball—though not as deterministic in how such human vitality might manifest. That is, these ads use an "energy" trope to link human capabilities with fuel sources, infrastructures, and provisions, positioned as the necessary grounds by which human flourishing, in all its forms, might actualize. This same communicated logic transfers over to energy poverty discussions, in which speaking of others who lack modern energy services (the "condition" theme in energy poverty definitions) assumes a concurrent impossibility to fully actualize their human potential.

Energy access, in this moral reasoning, is the premise by which the promise of humans' energies begins, rendering any alternative means absent. Such moralism is not restricted to fossil fuel legitimacy campaigns, as Cross (2019) illustrates in the narrative framings of for-profit solar entrepreneurs. Further, the appropriated logics of "access" are not restricted to energy. Gordon (2018) argues, "It is not that inequities in food access don't exist, but that the language we use to make sense of food access can quickly be capitalized on by those on the outside, who often make decisions for, and not with" those involved (p. 126). The same claim can be made with "energy poverty" and questions of access. Studying "the language we use" turns attention to the circulating discourses that capitalize on associations to energy poverty, connecting energy service provisions to "wider human flourishing, and on what people can achieve and do" (Day et al., 2016, p. 258). The next section examines how this moralizing discourse materializes in three distinct but related cases.

Energy poverty as development problem: the UN–*Revolution* collaboration

Throughout the second and final season of the NBC television series, *Revolution* (2012–2014), creators collaborated with United Nations representatives interested in promoting, to a broad audience, energy access initiatives (e.g., Alliance for Clean Cookstoves). UN affiliates, while sharing experiences that influenced some of the fictional storylines, promoted the show's universe as not simply depicting a speculative world but as an accurate reflection of current energy poverty. UN blog posts, for example, emphasized how the series's depictions, such as the dangers of childbirth without electricity, reflect reality. In all, "the UN saw *Revolution*'s first season as representing certain problems that arise in areas that are 'energy poor' and as a platform that could help 'communicate the message . . . on energy poverty' to a wider audience" (UN Foundation, as cited in Cozen, 2017, p. 331). This section summarizes a prior analysis of the collaboration (Cozen, 2017), positioning the campaign, and the influence of the dystopian television show, as a discursive rendering of the capabilities framework (Day et al., 2016).

Revolution narrates a postapocalyptic world that suddenly lost electricity. The UN campaign encouraged those involved with the television series to turn attention from a world lost—specifically, a United States where technology stops working—to a world of lack: that what they depict reflects the global poor's lived realities. This added content encouraged audiences to reflect on what they depend upon in their own lives, to sense what it might be like to lose all that energy affords, and to transfer that reflection to others' experiences. Such reflections were particularly evident in comments from actors and creators of the show as they reflected on the "nonfiction" aspects of their televised universe; in other words, the UN discussions led to explicit reflections that conflated the potential loss of their own technological dependence to others' energy poverty realities.

This collaboration reflects discursive themes in energy promotions that link energy poverty to a human capabilities lens. The collaboration is perhaps the most mainstream example promoting the energy access–Millennium Development Goals (MDGs) link. Setting goals for addressing poverty, education, gender inequality, disease, and so forth, the MDGs did not specify energy (United Nations Development Programme, 2003). Energy policy specialist Bahareh Seyedi explicitly used the collaboration as a platform to further her efforts in naming energy as the MDGs' "missing link" (Seyedi & Takada, 2010). Others have made similar arguments (e.g., Birol, 2007). González-Eguino (2015) calls for integrating energy poverty into any development plans and, like Seyedi, points to the error in leaving energy access out of the MDGs. Guruswamy (2011) goes a step further, outlining how energy poverty can help address each MDG. This approach is similarly taken with *Revolution*: "Seyedi weaved together the MDGs through a narrative of electricity access as a progressive good, speaking of 'parallels' between the show and real life and the 'mistake' of leaving energy out of the original MDGs" (Cozen, 2017, p. 333). Seyedi, UN Foundation CEO Kathy Calvin, and others would explicitly draw attention to the ways the show's content illustrated how power voids negatively influence and are therefore intimately connected to each goal: from disruptions in schooling (Goal 2) to childbirth mortality (Goals 3–5), to a typhus outbreak (Goal 6), all underscored by the problem of (energy) poverty (Goal 1).

The timing of the *Revolution* collaboration was important in terms of the MDGs giving way to a post-2015 agenda. Through the subsequent Sustainable Development Goals (SDGs), efforts foregrounding energy poverty alleviation's primacy succeeded, as evident with Goal 7, "Affordable and Clean Energy." The United Nations explicitly names energy "especially important as it interlinks with other Sustainable Development Goals" (United Nations Sustainable Development Goals, 2019), such as gender equality, disease prevention, and assisting educational opportunities. Similarly, on the UN's Global Goals Facebook page, a short video post featuring the CEO of Sustainable Energy for All includes this caption: "Energy lies at the heart of both the #2030Agenda for Sustainable Development and the Paris Agreement on Climate Change. Advancement in SDG 7 has the potential to spur progress across the other SDGs" (Global Goals for Sustainable Development, 2019).

In sum, energy poverty discourse often emphasizes the premise that energy access is the starting point for addressing overall human development and achieving UN goals. This narrative emphasizes energy access as central to enabling "social and economic development" in terms of "wider human flourishing, and . . . what people can achieve and do" (Day et al., 2016, p. 258). Energy poverty, through this language strategy, negatively impacts not simply flipping a light on but all of the social practices entailed through modern energy infrastructure. Such advocacy for energy-specific initiatives tied to the UN, including the *Revolution* partnership, name energy as the enabler of opportunities: for work, for well-being, for education, for maintaining relationships. These circulating, mainstream discourses, naming energy's promises, reflect a moral dimension to the capabilities framework (Day et al., 2016). Fallow planes, inoperable smartphones, references to being lost without GPS: the universe of *Revolution* constantly reminds viewers of all that technology affords their life. Articulating this world with energy poverty politics places emphasis not only on secondary and basic capabilities but on the centrality of energy to enable these capacities. Through the UN's focus and influence, energy access pivots from a secondary concern for the show—mainly as a plot device—to the grounding of technological promise without alternative means.

Humanitarian energy: nuclear advocates construct a green (or yellow) elephant

Nuclear energy advocates have recently argued their position in overtly environmental terms, specifically promoting nuclear as an energy source to counter climate change and carbon emissions (Parson, 2012). In particular, a small but vocal collection of ecomodernists have promoted a green nuclearism through a series of communication channels—especially the pronuclear documentary, *Pandora's Promise* (Stone, 2013)—and overlapping grassroots organizations, with names such as Environmental Progress and Generation Atomic. One of these groups, cofounded by *Pandora's Promise* director Robert Stone, is Energy for Humanity. Its mythic origin story follows a simple narrative: Cofounder Kirsty Gogan viewed the documentary and, that night, conceived of the organization (Richardson, 2015). This origin story illustrates how the film's primary purpose is to help frame broader nuclear advocacy, a task that has long perplexed proponents (Cozen, 2015; Richardson, 2015). The documentary's guiding question, according to its DVD marketing, also illustrates how organizations like Energy for Humanity frame a shift in perspective regarding nuclear energy: "whether the one technology we fear most could save our planet from a climate catastrophe, while providing the energy needed to lift billions of people in the developing world out of poverty" (Lorber, 2020). This quote, as well as nuclear advocacy more broadly, situates nuclear as an unexpected, hitherto feared savior, answering the paradox of fighting climate chaos while combating the energy poverty that limits human capabilities.

Energy for Humanity places particular emphasis on a common nuclear advocacy theme: supporting nuclear using claims of humanitarianism and energy poverty. This argument situates nuclear energy as the only viable option that can address climate mitigation and energy poverty alleviation. The midpoint and concluding scenes in *Pandora's Promise* first set up this message, in which those interviewed describe increasing electricity demand as inevitable and necessary, lest we condemn the global poor to further poverty, while coal and fossil fuels are no longer tenable (Cozen, 2015). Gogan continues this argument with Energy for Humanity and the public discourse that surrounds it. For instance, moderating a COP-21 panel, Gogan states the organization directs its efforts toward:

> two of the great environmental and humanitarian challenges that we face in this century. How to dramatically cut carbon emissions to avoid catastrophic climate change

> in our own lifetimes and that of our children, and secondly, lifting billions of people out of poverty to achieve the quality of life that we take for granted.
>
> *(as cited in Adams, 2015, para. 10–11)*

Gogan frames the humanitarian theme in moral terms and underscores how capabilities function as a trope of well-being, naming modern energy as the premise for and the promise of a quality life. Energy for Humanity lays out these value premises, juxtaposes the climate-poverty conundrum, and only then names nuclear and its promise as the elephant in the room: that which can no longer be avoided and that which the organization ultimately supports.

Energy for Humanity's "About Us" page literalizes this elephant trope. In highlighting their "vision, dreams, and aspirations," the page implies "the meaning behind their name" (Casañ-Pitrarch, 2015, p. 71) by constructing nuclear-as-savior—in other words, that this "energy for humanity" will enable humans' capabilities, sustainably. The page's text focuses on unstated premises by which nuclear becomes the implied solution sought in "solving climate change and enabling universal access to modern energy services" (Energy for Humanity, 2020, para. 1). Expanded energy use, in these terms, is necessary. The text's logic seeks to make nuclear appear as the primary solution, even if most would rather dismiss it. The elephant-in-the-room trope indirectly calls forth something so apparent it can no longer be ignored. Nuclear energy is the unstated element that appears as the logical conclusion to these premises. The nuclear elephant and its promise—supplying the "energy" and its attendant elements "for humanity"—are therefore constituted via these premises. The moral framework that links energy to capabilities (while combating climate catastrophe) leaves no room for alternatives.

This argument structure is explicitly elaborated in the "About Us" video, *We (Heart) Electricity* (also found at https://weloveelectricty.org). The opening segment details the green nuclear advocate's first premise: Energy enables human capabilities. The video begins with a declaration of love for power supply and its affordances. As Day et al.'s (2016) second element, power supply functions here as a starting point for the articulated associations made, as the video links infrastructure and access to "love" for all electricity provides: "for light, for heat, for cold drinks, for life" (Energy for Humanity, 2020, 0:00–0:16). "Cold drinks" is a secondary capability (storing food) that implies the energy service of refrigeration, while "life" connotes an open-ended implication to all that energy offers. The video therefore opens with the promise or utopian aspirations not of an energy source but of electricity infrastructures and the functionings (or capabilities enacted) of energy-powered social practice (Day et al., 2016).

The video continues with the second premise: Energy poverty is an unacceptable yet widespread global ill. The video links these ills to lacking in capabilities, such as safety (with sinister eyes appearing in the dark), education (depicting a youth reading by moonlight on the floor), and overall "quality of life, especially for women" (Energy for Humanity, 2020, 0:45). Each example implies specific UN development goals. The video also states that washing machines (a domestic energy service) frees study time, linking Day et al.'s (2016) third through fifth elements (with "washing clothes" a listed secondary capability and "being educated" a basic capability) (p. 260). The third premise builds from these instances that link energy access and services to capabilities: Escaping energy poverty, as the video proclaims, is a moral good. Image and text suggest current electric generation increases will and should continue, "improving the lives of millions" (Energy for Humanity, 2020, 0:59).

Having established a central, binding link from electricity supply to services to capabilities, the fourth premise states that all this good comes from untenable sources. That is, "our high energy planet" (Energy for Humanity, 2020, 1:03) exacts a "heavy price" (1:20) by primarily relying on fossil fuels, especially coal. This premise demonizes fossil fuels as not simply the sources of benevolent electricity but also of pollutants, death, and climate change risks that impact everyone but especially the poor. The final premise returns to the moral good of increased demand, while deeming renewables insufficient. As the "About Us" text emphasizes, "Future leaders will need all the tools at their disposal" (Energy for Humanity, 2020, para. 3). Energy for Humanity claims support for everything and anything that will maintain energy aspirations; however, changing the source to meet those aspirations requires a large-scale substitute.

All of these premises construct the nuclear elephant: "There's an elephant in the room" (Energy for Humanity, 2020, 1:46), the video's text proclaims over a yellow elephant-shaped cutout. The image illustrates nuclear as large scale and therefore best suited to replace fossil fuels. The nuclear elephant also emerges out of the threat of climate change: "If we are serious about climate change the world needs nuclear" (2:00). It further emerges out of naming energy poverty alleviation a human rights issue: "Everyone deserves clean energy" (2:04). Emissions cause the climate crisis, which therefore becomes the point of intervention: change the power source—with the video concluding on its urgency—and the promise of energy for all can be realized without concern (Energy for Humanity, 2020).

Ultimately, nuclear advocacy seeks to place nuclear as the central character in energy future narratives and does so by emphasizing climate change, energy poverty, and the capabilities enabled by energy access. Like the *Revolution* collaboration, this pronuclear discourse encourages people both to consider energy vulnerability (Bouzarovski, 2018)—we all live under precarity amid climate chaos—and the role that energy access and provisions play in their own lives and to reflect on the bonds across elements to all energy affords. In this setup, the fuel source is the source of ill yet the least important to the sanctity of the source-supply-services-capabilities linkage. To support and expand "energy for humanity" is to support what we love energy *for*: "light," refrigeration, "life," or all that energy promises (Energy for Humanity, 2020). Repeating and reiterating the development narrative of aspirational energy access amid climate chaos attempts to deflect any fears associated with nuclear; it may be the elephant in the room that people fear and dismiss, but it is there, constituted as the commonsense solution without a viable alternative, if all of humanity is to realize its potential in "expanding substantive freedoms" (Sen, 1999, p. 3).

Supplying imaginative capabilities: the Shell–GravityLight partnership

The company Deciwatt won the 2015 Shell Springboard program to commercialize its invention, GravityLight, a device using gravity to power an LED ("Lighting Up," 2015). Publicizing their support, Shell Global created *Bedtime Stories*, 50 animated children's fables. Shell's promotion of GravityLight emphasizes how accessible light can overcome key energy poverty challenges, communicating "poverty as capability deprivation" (Dowding, 2006, p. 323) and energy access as the solution. In focusing on imaginative capacities, *Bedtime Stories* situates Shell, primarily a fossil fuel company, as a supplier of capabilities and therefore as morally necessary.

A lifted weight can turn a small generator as it slowly descends, and attaching an LED to the generator produces a gravity-powered light; this process repeats when the weight reaches the ground ("Gravity Light," 2013). Deciwatt's commercialization of this scalable device reflects Cross's (2019) assessment of solar lantern companies, whose humanitarian capitalism discourse "assert[s] the promise that for-profit businesses can 'do well by doing good,' establishing the terrain of humanitarian intervention and chronic global poverty as legitimate arenas for entrepreneurship and corporate activity" (p. 48). These discourses, utopian visions influential for guiding actions and assumed market premises, align ethics and economics. Deciwatt identifies their primary market in two areas: humanitarian aid, including disaster relief and preparation, and replacing kerosene lamps ("Lighting Up," 2015). Although suggesting a market value for widespread emergency kits, the usual promise of GravityLight focuses on its scalability in energy-poor markets, particularly as a means to sidestep conventional energy sources and batteries. As one supporter puts it, GravityLight "is completely clean and green" ("Gravity Light," 2013, p. 60) with "the potential to lift people out of poverty" (p. 61) due to its small but significant role in replacing kerosene purchases. In this use, GravityLight bypasses Day et al.'s (2016) first two elements, source and supply, to directly offer a domestic energy service (i.e., lighting).

In promoting this product, an identical YouTube synopsis previews each animated *Bedtime Stories* fable: "Access to reliable light can mean access to greater opportunities, to employment, and to education. In celebration of light and the joy of reading, Shell #makethefuture and GravityLight have collated a set of 50 classic stories from around the world" (Shell, 2018). These fables come from regions such as southern Africa (the Zulu story, "The Crafty Jackal"), Scandinavia (Sweden's "The Lame Dog"), and eastern Asia (Mongolia's "The Miser and the Judge"), along with numerous Aesop's fables (e.g., "The Fox and the Crow"). Each video begins the same: A character from the fable enters a dark screen and pulls a GravityLight chord, which shines a light over a suddenly visible, opened book. The fable's title appears, and then the text reads, "50 Classic Stories/Presented by Shell and GravityLight/Celebrating Access to Sustainable Light" (Shell, 2018, 0:09–0:14). The rest of the video features original animation illustrating the children's story, often narrated by a celebrity.

The primary thematic across this framing aligns light, reading, and imaginative capabilities. The varied lessons of these 50 fables are less important than the collective theme; they communicate a diversity of expression yet a singularity of desire: for opportunities to access and exercise one's imagination. Light manifests these opportunities. This promise underscores the literal weight at play, when Shell asks, "How can a bag of rocks spark a Kenyan child's imagination?" (Shell, 2017b). Circumventing traditional notions of supply and power source, the product offers an alternative for domestic lighting services, and *Bedtime Stories* discursively renders this intervention in terms of capabilities: accessing information, cultivating relationships, and being educated (Day et al., 2016). The opening image, where the book only materializes with the spotlight, identifies light as the provision that makes actions possible. This promise extends beyond nighttime reading, suggesting the broader opportunities light unlocks. Although findings suggest that solar lighting, while increasing children's light use, does not significantly increase their studying (Rom et al., 2017), this educational promise is clearly the implication with such entrepreneurialism. Indeed, the Shell website emphasizes childhood opportunities as a key impetus and includes a Kenyan educator's optimism that added light will "provide the opportunity to do their homework" (Shell, 2020, "The bright light of an energy idea," para. 3). The promise of light is the promise to spark ideas, opportunities, and future paths for the world's children.

Here, instead of oil and electric grids supporting human potential (Cozen, 2018), the premised ground is light itself. Under the heading, "The Bright Light of an Energy Idea," Shell (2020) names the inspiration of these stories as "created to celebrate and to remind us that something as simple as sustainable light can help create opportunities and inspire imaginations around the world" (para. 4). Igniting opportunity and imagination underscores domestic lighting services as the grounds through which capabilities manifest. An option for reading at night is mere synecdoche for the promise of light, bringing "joy" (Shell, 2018) in numerous ways. The weighted bags of a GravityLight take on added weight, literally and figuratively, opening opportunities implied as closed otherwise. One's imagination is sparked by the bedtime stories told, enabled by light. However, within these premises, "energy poverty" connotes those without such capabilities and therefore not fully actualized. The implication suggests the poor do not have access to social opportunities without help from light but also from the missionary efforts involved in its dissemination.

Finally, Shell, like other fossil fuel companies, situate themselves as "energy companies" to shift attention away from the energy source to the corporate system that supplies infrastructure and know-how to power the world (Cozen, 2018). Shell, in other words, names its work as guided by the purpose of general vitality and providing human capabilities (Day et al., 2016). As stated in YouTube synopses for individual *Bedtime Stories* videos, "We believe that oil and gas will remain a vital part" of energy futures (e.g., Shell, 2017a). This focus on current supply is unsurprising for Shell, but the Deciwatt partnership places specific attention on the vitality of energy and light more generally. In context, the implied association is that all of Shell's efforts serve to light imaginations and enable opportunities. The source and power supply are secondary; fossil fuels may remain "vital," but, the reasoning follows, capabilities generation is what all their activity is actually for. Such campaigns offer a means to reinforce the premises on which larger energy access discussions are understood and morally justified. Framed as lacking social opportunities in turn mobilizes market justifications (Cross, 2019) to provide the "decent life" to the energy poor, by whatever means that may suggest (Schneider et al., 2016). Ultimately, while suggesting alternatives to the elements of source-supply-services (Day et al., 2016), the trope of capabilities in *Bedtime Stories* reinforces an overall tenor of energy provisions as the necessary grounds that enable capabilities and human potential, a gift from energy providers to the energy poor.

Conclusion

This chapter outlined some of the guiding "energy poverty" language of energy service providers and their supporters. In all, I argued that the capabilities framework (Day et al., 2016) functions not merely as a policy and research guide for interventions and conceptualizations but also as a template for assessing the moral frameworks of energy poverty discourses. Examining these communicated narratives is important in that, "Questions about energy are intensely ethical as they encourage, if not demand, reflection on how we feel we ought to live" (High & Smith, 2019, p. 2). In particular, while Day et al.'s (2016) definition emphasizes "alternative means," energy poverty discourse as a whole commonly naturalizes the associations across pertinent elements (source, supply, service, secondary capability, basic capability), interlinking fuel/power with the moral authority of actualizing human needs. The UN–*Revolution* collaboration, which both names energy as development's missing link and encourages audiences to consider one's own life in terms of energy, dissuades imagining or promoting feasible alternatives. Nuclear advocacy intervenes at the fuel source by encouraging reflection on one's secondary capabilities and how energy intensive they are, suggesting that only with consuming more (non–fossil fuel sourced) energy may the global poor satisfy basic needs. *Bedtime Stories* "remind us" (Shell, 2020, "The bright light of an energy idea," para. 4) to imagine imagination as sparked by energy services, here without traditional energy sources but within the context of an oil and gas provider who, by association, aims to offer everyone expanded choice and capabilities through various (energy) means (primarily fossil fuels).

Such circulating public communication functions to "remind" cultural commitments: "Little is more important than naming, marking, and reminding . . . it defines the work of culture" (Schudson, 1986, p. xxi). Naming energy poverty focuses attention on social benefits, a linear association in which energy serves capabilities. Energy poverty defines a problem and so suggests a solution: more energy. As a trope, the capabilities approach illustrates the allure of energy provisions as the only viable solution for human development. However, an alternative starting premise, emphasizing *energetic* capabilities, can mark different cultural commitments, toward the democratic efforts necessary to constitute new ways of thinking, being, and doing.

While "energy poverty" discourse, as examined here, promises energy access through external energy provisions, the process of capabilities expansion best emanates from the energetic participation of emergent publics. The constitution of publics is central to social constructivist approaches to energy research (Chilvers & Longhurst, 2016; Chilvers & Pallett, 2018) and to communication research (Brouwer & Asen, 2010; Warner, 2002), and within energy democracy frameworks, publics are conceived as prosumers, or participants in energy practice (Angel, 2016; Feldpausch-Parker et al., 2019). In particular, direct involvement in energy policy and practice shifts one's commonplace relationship to energy. Cultivating the processes by which dynamic publics can effectively emerge might also redirect understandings of the energy–capability relationship.

Aligning a ritualistic approach to communication and to energy can help reimagine the constitution of such energetic publics. Carey (1992) contrasts communication's ritualistic functions, of crafting cultural norms, to a transmission model where, like electric grids, its function is to transfer information from source to source. A ritual perspective to communication and energy emphasizes the ways by which culture is constituted in and through our engagements with energy practice. The transmission of energy, from Point A to Point B, is a different communicative relationship than the ritualized practices of energy democracy, or community energy (Dusyk, 2017). Energy democracy can name a different sense of one's self and one's relations. What are the habits, practices, and ways of seeing and being reinforced through different energy cultures? For instance, de Onís (2018), analyzing energy's colonial dynamics in Puerto Rico, explores the "shared energies" generated out of collective action, communication, and exerted resistance against power relations (p. 552). "Alternative means" to energy practice, in this way, are not limited to the supply and consumption of energy (Day et al., 2016, p. 260). The constitution of different ways to communicate about our interpersonal and ecological relationships and the "shaping or shifting [of] expectations, customs and practices" (p. 262) can manifest capabilities through and beyond energy use. Contrasting energy poverty, the promise of energy democracy—here a focus on the energetic processes and mechanisms of democratization—does not start with a question of lack that energy access satisfies but with the "energies" of collectives, cultivating the social opportunities for self-determination (de Onís, 2018; Sen, 1999).

References

Adams, R. (2015, December 28). Caldeira, Emanuel, Wigley and Hansen statement at COP21, December 4, 2015. *Atomic Insights*. https://atomicinsights.com/14838-2/

Angel, J. (2016, May). *Towards energy democracy: Discussions and outcomes from an international workshop*. The Transnational Institute.

Birol, F. (2007). Energy economics: A place for energy poverty on the agenda? *The Energy Journal, 28*(3), 1–6. https://www.jstor.org/stable/41323106

Boni, A., & Walker, M. (Eds.). (2013). *Human development and capabilities: Re-imagining the university of the twenty-first century*. Routledge.

Bouzarovski, S. (2018). *Energy poverty: (Dis)assembling Europe's infrastructural divide*. Springer International.

Bouzarovski, S., & Petrova, S. (2015). A global perspective on domestic energy deprivation: Overcoming the energy poverty-fuel poverty binary. *Energy Research & Social Science, 10*, 31–40. https://doi.org/10.1016/j.erss.2015.06.007

Brouwer, D. C., & Asen, R. (Eds.). (2010). *Public modalities: Rhetoric, culture, media, and the shape of public life*. University of Alabama Press.

Carey, J. W. (1992). *Communication as culture: Essays on media and society*. Routledge.

Casañ-Pitrarch, R. (2015). The genre "about us": A case study of banks' corporate webpages. *International Journal of Language Studies*, *9*(2), 69–96.

Chilvers, J., & Longhurst, N. (2016). Participation in transition(s): Reconceiving public engagements in energy transitions as co-produced, emergent and diverse. *Journal of Environmental Policy & Planning*, *18*(5), 585–607. https://doi.org/10.1080/1523908X.2015.1110483

Chilvers, J., & Pallett, H. (2018). Energy democracies and publics in the making: A relational agenda for research and practice. *Frontiers in Communication*, *3*(14), 1–16. https://doi.org/10.3389/fcomm.2018.00014

Cleveland, C. J., & Morris, C. (Eds.). (2015). Energy poverty. In *Dictionary of energy* (2nd ed., p. 200). Elsevier Science & Technology. https://doi.org/10.1016/C2009-0-64490-1

Cozen, B. (2015). *Mediating energy: Rhetoric and the future of energy resources* [Doctoral dissertation, University of Utah]. University of Utah J. Willard Marriott Library Database. https://collections.lib.utah.edu/ark:/87278/s6m93hz3

Cozen, B. (2017). Facting fiction: *Revolution*, the United Nations, and cultural politics of electricity. *Critical Studies in Media Communication*, *34*(4), 329–343. https://doi.org/10.1080/15295036.2017.1325510

Cozen, B. (2018). Stabilizing energies: Intersections between energy promotion texts and rhetorical theory. In B. McGreavy, J. Wells, G. F. McHendry, & S. Senda-Cook (Eds.), *Tracing rhetoric and material life: Ecological approaches* (pp. 348–379). Palgrave Macmillan.

Cross, J. (2019). The solar good: Energy ethics in poor markets. *Journal of the Royal Anthropological Institute*, *25*(S1), 47–66. https://doi.org/10.1111/1467-9655.13014

Day, R., Walker, G., & Simcock, N. (2016). Conceptualising energy use and energy poverty using a capabilities framework. *Energy Policy*, *93*, 255–264. https://doi.org/10.1016/j.enpol.2016.03.019

DeLaure, M. (2011). Environmental comedy: *No impact man* and the performance of green identity. *Environmental Communication*, *5*(4), 447–466. https://doi.org/10.1080/17524032.2011.614952

de Onís, C. M. (2018). Fueling and delinking from energy coloniality in Puerto Rico. *Journal of Applied Communication Research*, *46*(5), 535–560. https://doi.org/10.1080/00909882.2018.1529418

Dowding, K. (2006). Can capabilities reconcile freedom and equality? *The Journal of Political Philosophy*, *14*(3), 323–336. https://doi.org/10.1111/j.1467-9760.2006.00264.x

Dusyk, N. (2017). Community energy: Diverse, dynamic, political. In B. D. Solomon & K. E. Calvert (Eds.), *Handbook on the geographies of energy* (pp. 502–514). Edward Elgar. https://doi.org/10.4337/9781785365621

Energy for Humanity. (2020, May 29). *About us*. http://energyforhumanity.org/en/about-us/

Feldpausch-Parker, A. M., Endres, D., & Peterson, T. R. (2019). Editorial: A research agenda for energy democracy. *Frontiers in Communication*, *4*(53). https://doi.org/10.3389/fcomm.2019.00053

Gabbert, L., & Schein, J. (Directors). (2009). *No impact man* [Film]. Eden Wurmfeld Films.

Global Goals for Sustainable Development [Facebook Page]. (2019, May 24). *Energy lies at the heart of both the #2030Agenda for sustainable development and the Paris agreement on climate change. Advancement* [Video attached] [Status update]. Facebook. www.facebook.com/globalgoalsUN/posts/energy-lies-at-the-heart-of-both-the-2030agenda-for-sustainable-development-and-/10156776771046026/

González-Eguino, M. (2015). Energy poverty: An overview. *Renewable and Sustainable Energy Reviews*, *47*, 377–385. https://doi.org/10.1016/j.rser.2015.03.013

Goodwin, K. J. (2010). Reconstructing automobility: The making and breaking of modern transportation. *Global Environmental Politics*, *10*(4), 60–78. https://www.muse.jhu.edu/article/404383

Gordon, C. (2018). *Troubling "access": Rhetorical cartographies of food (in)justice and gentrification* (Publication No. 10928215) [Doctoral dissertation, University of Colorado]. ProQuest. https://scholar.colorado.edu/comm_gradetds/78

Gravity light. (2013, December). *Appropriate Technology*, *40*(4), 60–61. www.appropriate-technology.com

Guruswamy, L. (2011). Energy poverty. *Annual Review of Environment and Resources*, *36*, 139–161. https://doi.org/10.1146/annurev-environ-040610-090118

High, M. M., & Smith, J. M. (2019). Introduction: The ethical constitution of energy dilemmas. *Journal of the Royal Anthropological Institute*, *25*(S1), 9–28. https://doi.org/10.1111/1467-9655.13012

Hunt, K. P. (2018). The SNAP Challenge: Communicating food security capabilities through anti-hunger advocacy. *Journal of Agriculture, Food Systems, and Community Development*, *8*(2), 87–92. https://doi.org/10.5304/jafscd.2018.082.007

Li, K., Lloyd, B., Liang, X. J., & Wei, Y. M. (2014). Energy poor or fuel poor: What are the differences? *Energy Policy*, *68*, 476–481. https://doi.org/10.1016/j.enpol.2013.11.012

Lighting up developing countries. (2015, April). *Professional Engineering*, *5*.

Lorber, K. (2020, May 28). *Pandora's promise (DVD)*. www.kinolorber.com/product/pandoras-promise-dvd_1
Nussbaum, M. C. (2000). *Women and human development: The capability approach*. Cambridge University Press.
Nussbaum, M. C. (2011). *Creating capabilities: The human development approach*. Harvard University Press.
Oosterlaken, I., & van den Hoven, J. (Eds.). (2012). *The capability approach, technology and design*. Springer. https://doi.org/10.1007/978-94-007-3879-9
Parson, S. (2012). "Climate first"? The ethical and political implications of pronuclear policy in addressing climate change. *Ethics, Policy and Environment, 15*(1), 51–56. https://doi.org/10.1080/21550085.2012.672686
Richardson, J. H. (2015, December 4). The battle to save the planet may come down to nuclear advocates vs. environmentalists. *Esquire*. www.esquire.com/news-politics/news/a40195/paris-climate-conference-nuclear-power/
Rom, A., Günther, I., & Harrison, K. (2017). *The economic impact of solar lighting: Results from a randomised field experiment in rural Kenya*. Summary Report. ETH Zürich. https://ethz.ch/content/dam/ethz/special-interest/gess/nadel-dam/documents/research/Solar%20Lighting/17.02.24_ETH%20report%20on%20economic%20impact%20of%20solar_summary_FINAL.pdf
Schneider, J., Schwarze, S., Bsumek, P. K., & Peeples, J. (2016). *Under pressure: Coal industry rhetoric and neoliberalism*. Palgrave Macmillan.
Schudson, M. (1986). Preface to the paperback edition. In *Advertising, the uneasy persuasion: Its dubious impact on American society* (pp. xiii–xxiv). Basic Books.
Sen, A. (1985). *Commodities and capabilities*. North-Holland.
Sen, A. (1999). *Development as freedom*. Anchor Press.
Sen, A. (2005). Human rights and capabilities. *Journal of Human Development, 6*(2), 151–166. https://doi.org/10.1080/14649880500120491
Seyedi, B., & Takada, M. (2010). Energy for the poor: The missing link for achieving the MDGs. *Spanda Journal, 1*(1), 17–22. www.spanda.org/publications.html
Shell. (2017a, November 21). *The crafty jackal | Shell #makethefuture* [Video]. YouTube. www.youtube.com/watch?v=J7SUX_kzeRw
Shell. (2017b, December 8). *How can a bag of rocks spark a Kenyan child's imagination? | Shell#makethefuture* [Video]. YouTube. www.youtube.com/watch?v=f4VSTPBTX58
Shell. (2018, January 12). *Shell bedtime stories* [Video]. YouTube. www.youtube.com/playlist?list=PLEPIVJVCFQH3o2W8ye9bo7ZXbAPC_Kwk
Shell. (2020, May 28). *How stories can light up lives*. www.shell.us/make-the-future/cleaner-energy-for-homes/how-stories-can-light-up-lives.html
Shove, E., & Walker, G. (2014). What is energy for? Social practice and energy demand. *Theory, Culture & Society, 31*(5), 41–58. https://doi.org/10.1177/0263276414536746
Simcock, N., & Petrova, S. (2017). Energy poverty and vulnerability: A geographic perspective. In B. D. Solomon & K. E. Calvert (Eds.), *Handbook on the geographies of energy* (pp. 425–437). Edward Elgar. https://doi.org/10.4337/9781785365621
Stone, R. (2013). *Pandora's promise* [Film]. Robert Stone Productions.
United Nations Development Programme (UNDP). (2003). *Human development report 2003: Millennium development goals: A compact among nations to end human poverty*. Oxford University Press.
United Nations Sustainable Development Goals. (2019, June 1). *Ensure access to affordable, reliable, sustainable and modern energy*. UN. www.un.org/sustainabledevelopment/energy/
Warner, M. (2002). Publics and counterpublics (abbreviated version). *Quarterly Journal of Speech, 88*(4), 413–425. https://doi.org/10.1080/00335630209384388

11

A BRIEF EXCURSION INTO THE MANY SCALES AND VOICES OF RENEWABLE ENERGY COLONIALISM

Susana Batel

Introduction

Governments, namely in the Global North, are fostering the deployment of large-scale renewable energy generation and associated infrastructures—such as high-voltage power lines—to mitigate climate change (Battaglini et al., 2012; Renewables Directive, 2009). However, the deployment of these infrastructures often raises opposition, mainly from local communities living nearby. Social sciences' research aiming to understand this opposition has been steadily increasing over recent decades and is lately paying more attention to how issues of injustice, inequalities, and exclusion might explain opposition (Batel & Devine-Wright, 2017; Rudolph & Kierkegaard, 2019; Walker, 2009). Concepts and frameworks such as those of environmental and energy justice, energy democracy, and others have been used to further theorize, examine, and address those issues (Baka, 2017; Feldpausch-Parker et al., 2019; Heffron & McCauley, 2017; Walker, 2009). However, that research has so far largely overlooked how opposition to energy infrastructures often stems from, reveals, and may also promote renewable energy colonialist practices (Barry et al., 2008; Batel & Devine-Wright, 2017). The main aim of this chapter is to argue for and illustrate the significance of examining renewable energy colonialism for furthering energy democracies.

Renewable energy colonialism is here defined as materializing, in both symbolic and instrumental ways, sociohistorical, economic, and political power relations as related to the use of renewable energy and the deployment of related infrastructures and practices. In the renewable energy domain, these relations may be between different sociogeographical groups or different scales regarding the relations, ownership, and appropriation with/of space and resources. In this sense, renewable energy colonialism is a new form of colonialism (or neocolonialism) related to renewable energy and the environment. The concept draws from postcolonial and decolonial critiques, given that it also questions modern and hegemonic modes of knowledge production typical of the Global North and recognizes the value of other forms of knowledge (Bhambra, 2014; Escobar, 2018; Mignolo, 2012; Santos, 2015).

In recent decades, debates have already started on what has been called carbon colonialism (Bachram, 2004; Paterson & Stripple, 2010). For example, one of the main measures taken by several countries to comply with the targets set by the Kyoto Protocol and other binding

DOI: 10.4324/9780429402302-13

directives that followed has been carbon trading, namely the offsetting of carbon emissions by deploying energy infrastructures and other carbon-heavy activities in the Global South. This means that carbon credits can be gained, directives complied with, and current levels of consumption maintained in and by the Global North at the expense of the Global South. Carbon colonialism becomes even more blatant if we consider that carbon trading is actually institutionalized by the previously mentioned treaties, that it only opens space for very modest carbon reductions, and that public participation in decision making regarding the deployment of energy and other associated infrastructures in the Global South is in many cases almost nonexistent (see Liverman, 2009; see also Finley-Brook & Thomas, 2010; Paterson & Stripple, 2010). Other authors have also examined related concepts and structures such as those of settler and resource colonialism (Parson & Ray, 2018; Preston, 2013). Resource colonialism has been defined as the "theft and appropriation of land belonging to indigenous people in order to access natural resources" (Parson & Ray, 2018, p. 69), whereas settler colonialism has been defined as:

> "an inclusive, land-centered project that coordinates a comprehensive range of agencies, from the metropolitan center to the frontier encampment, with a view to eliminate Indigenous societies." The operation of this complex social formation is not dependent on "the presence or absence of formal state institutions or functionaries" and thus can manifest itself through mercantilism, or neoliberal private–public partnerships.
>
> *(Wolfe, 2006, as cited in Preston, 2013, p. 44)*

However, these concepts might not fully capture the paradigmatic contradictions of neoliberal capitalism and the way they translate into the relation between energy and the environment, as the concept of renewable energy colonialism does. Further, they fail to systematically identify not only what is renewable energy colonialism but also where and how it is or can be practiced.

This chapter discusses the importance of uncovering and examining renewable energy colonialism—and related practices—in its psychosocial, cultural, institutional, and political dimensions. I first better define renewable energy colonialism in relation to associated concepts and practices. Then I provide examples of how it materializes at different scales and is enacted by different actors. Finally, I discuss potential ways to contest and overcome renewable energy colonialism and to create more democratic relations among energy, societies, and ecological systems.

What is energy colonialism?

Colonialism has been defined academically as the world-historical process by which the modern world was created, based on the imposition and domination of the West's way of thinking and rationalities into the rest of the world, both epistemically and materially—through the incorporation of territories—and in a way that this domination and related physical and symbolic violence appeared naturalized and justified (Bhambra, 2018) and socioculturally as "control by one power over a dependent area or people" and "a policy advocating or based on such control" (Merriam-Webster, n.d.).

The concept of renewable energy colonialism focuses the concept of colonialism by highlighting how it can be practiced regarding the settling of energy-related infrastructures, specifically so-called renewable ones, and the incorporation of territories for the extraction of related resources. Renewable energy colonialism, therefore, exerts power over a—politically, economically, and/or socially—dependent area and/or people in the service of producing renewable energy. The relevance of renewable energy colonialism as a concept lies in highlighting that renewable energy can also be colonial within current supposedly postcolonial neoliberal capitalist societies. This concept, by recuperating ideas embedded in coloniality,[1] highlights how renewable energy colonialism goes beyond other concepts such as those of environmental and energy justice and settler and resource colonialism. Renewable energy colonialism considers the importance not only of a specific spatial-geographical dimension (e.g., the material ownership of a territory) but also of sociohistorical and sociopsychological ones (e.g., the lived ownership of a given territory). As Enns and Bersaglio (2019) highlight, colonialism includes racial and intergroup dimensions,[2] as well as associated material and symbolic dimensions that reflect one group (the colonizer) feeling superior to another group living in a separate spatial area (the colonized) and not acknowledging their sovereignty, as is well illustrated by the *terra nullius* idea (Barry, 2007; Batel & Devine-Wright, 2017). This focus allowed by the concept of renewable energy colonialism enables expansion on the reflection and discussion of the scales and intergroup/territorial relations at which (renewable energy) colonialism has been so far considered. In so being, the concept of renewable energy colonialism contributes to and materializes a critical approach to the social studies of energy and the environment (Batel, 2020a) and clearly differs from and adds to concepts of environmental justice and energy justice.

Environmental justice is normally defined as referring to "the rights of all people to benefit from a healthy environment, to be treated fairly in environmental decision-making and to be meaningfully involved in environmental decision-making" (Feldpausch-Parker et al., 2019, p. 3). It has always been associated with grassroots movements, and so it is inherently both a movement and an area of academic thought and research. Energy justice has also been appropriated and discussed not only at an academic level but also within planning and policy-making circles and by practitioners in the energy sector (e.g., NEA, 2015). Energy justice has translated the concept of environmental justice into the energy field by focusing mainly on issues of recognition, procedural justice, and distributive justice, materialized concretely through energy affordability, availability, due process, good governance, sustainability, intra- and intergenerational equity, and responsibility (Sovacool & Dworkin, 2015). However, its use has mainly aimed at promoting energy justice within current socioeconomic and political systems. This is well illustrated by Heffron and McCauley's (2017) statement that "policy formulation in the energy sector is dominated by economists and industry where economic costing is the prime tool for decision making. To some degree environmental and climate justice as concepts have been naïve in their approach" (p. 664). The renewable energy colonialism concept shares with that of environmental justice not a "naivete in their approach" but instead the aim of uncovering, questioning, and transforming existent power relations and associated energy, economic, and political systems (Batel, 2020a), therefore contributing to fostering energy democracies. However, and in relation to environmental justice, it does this by focusing on how the deployment of renewable energy generation and associated infrastructures reflects, reproduces, and creates new colonial practices embedded in the sociohistorical, economic, and political relations between different social groups within and between nation-states (Batel & Devine-Wright, 2017; Rudolph & Kirkegaard, 2019). It is in this sense that renewable energy colonialism both includes but also goes beyond environmental and energy injustice issues (Endres, 2009; de Onís, 2018; Cozen et al., 2018).

As such, renewable energy colonialism is one lens of analysis through which to look at energy democracies (or their lack thereof) given that it focuses on and addresses issues of participation, justice, and power (Feldpausch-Parker et al., 2019) in specific ways. In so doing, it contributes to the critical turn in social sciences' research on energy and the environment (Batel, 2020a), as it problematizes how renewable energies are already "being fossilized in the sense of becoming akin to the political economy of fossil fuels" and are therefore not "intrinsically democratic and egalitarian" (Raman, 2013, p. 178).

As claimed by Labussière and Nadaï (2018), "energy resources are proposed and not given by nature, entities become energy resources" (p. 51). This process of becoming resources implies that the energy infrastructures that are deployed for harnessing and extracting energy from its milieu embed broader dynamics of political power, identities, and belonging, by imposing energy production and systems as dominant land uses and by defining whom, when, and how they are so (Bridge et al., 2018; Labussière & Nadaï, 2018). In this process, energy becomes commodified (Castree, 2003; Labussière & Nadaï, 2018) and part and parcel of the current larger neoliberal capitalist system or Capitalocene (Haraway, 2015)—including renewable energy generation and associated infrastructures. This raises questions: How are renewable energy generation and related infrastructures practiced, and what is their materiality, including how renewable or fossilized they are? (Raman, 2013). This means that if the deployment of renewable energy and associated infrastructures just perpetuates business as usual within neoliberal capitalist systems—including being focused mainly on economic profit, neglecting democratic processes, and reproducing power inequalities—then low-carbon energy resources can become no different from conventional resource development (Bridge et al., 2018), and their renewability and sustainability become only performative (Nadaï & Labussière, 2017).

Renewable energy colonialism can displace socially, temporally, and geographically to colonized groups many of the social and environmental costs of the increased energy abundance in the Global North. As such, renewable energy colonialism can be seen as one paradigmatic materialization of neocolonialism as the practice of using capitalism, globalization, and cultural imperialism to influence a developing country, region, or area in place of direct military control or indirect political control, based on renewables and under the guise of environmental sustainability. In that sense and contrarily to resource and settler colonialism (Endres, 2009; Frantál, 2017; Mavhunga & Trischler, 2014; Parson & Ray, 2018; Preston, 2013), renewable energy colonialism materializes as a new set of practices and relations that build upon previous resource and settler colonial relations and practices to foster neoliberal environmental sustainability agendas. From the point of view of the colonized, and similarly to other forms of colonialism, renewable energy colonialism still often materializes in what scholars have designated as landgrabbing (Lyons & Westoby, 2014), place stigmatization (Rudolph & Kirkegaard, 2019), energy dispossession (Baka, 2017), and the slow violence (Nixon, 2011) produced and suffered throughout time by material and symbolic domination and inequality (Batel & Devine-Wright, 2017; Batel & Pataco, 2019) underlining any form of colonialism and witnessed by the incremental destruction of landscapes, cultures, and ways of living that we will illustrate and discuss in this chapter.

In this vein, discussing and analyzing renewable energy colonialism and its consequences are part of a critical theorizing of extractive practices (Cozen et al., 2018) or, as others have put it, of a critical approach to the research agenda on the social acceptance of renewable energy and associated infrastructures (Batel, 2020a). This approach is mindful of the impacts of what is said, how, to whom, and by whom regarding both the need for the deployment of renewable energy infrastructures and the location selected for their deployment. Renewable energy colonialism also attends to how renewable energies are deployed in the relations between social groups and assessing how democratic those relations are. As such, renewable energy colonialism is as much a set of material relations of dominance and inequality that build upon specific sociohistorical, economic, and geographical relations, as it is "significantly a rhetorical phenomenon that employs particular discursive strategies for enabling the continuation of colonialism, deliberate exclusion of indigenous voices from decision-making" (Endres, 2009, p. 40), and the perpetuation of neoliberal capitalist resource extractivism (see also de Onís, 2018). This is particularly clear in practices and relations that further colonialist practices in the name of renewable energy and in the associated rhetoric of promoting environmental sustainability.

This makes it important, then, to also discuss which rhetorical devices and discursive strategies and resources have already been identified as enacting and reproducing colonialism specifically regarding energy resources and associated practices. A relevant and vastly illustrated rhetorical strategy is what Endres (2009) has called naming practices. This is demonstrated by Barry et al. (2008) as the rhetoric of "us" and "them" used not only by community members to talk about the "invasion" of large-scale wind farms in their communities but also by developers to talk about community members and by members in some communities to talk about members of other communities—such as residents in the UK promoting the deployment of renewable energy generation infrastructures in the North of Africa instead of in the UK (Batel & Devine-Wright, 2017). Similarly, renewable energy colonialism uses discursive devices to ascertain the *terra nullius* regarding foreign, distant potential places for the deployment of renewable energy infrastructures (Barry, 2007). Edwards (2011) clearly illustrates how this representation of a place is achieved by presenting it as "remote"—as in, "away from where we live"—"empty," and "sparsely populated" (such as the North of Africa in the previous example), which thus legitimates its colonization by energy infrastructures. Also, de Onís (2018) highlights how tropical places are often portrayed as places that deplete energy, motivation, and "productivity," which in turn enables justification and legitimatization of discourses that ascertain the inferiority of local inhabitants and their inability to "manage" the surrounding landscapes appropriately, including more decentralized ways of (renewable) energy harnessing and extraction. Some of these rhetorical strategies and discursive devices used for practicing forms of colonialism will be further illustrated in the examples discussed next of renewable energy colonialism enacted at different scales and through different voices.

Renewable energy colonialism at different scales

Societies' expectations that sustainable energy transitions, including the push toward renewable energy and associated infrastructures, would explicitly avoid environmentally and socially unjust practices has not been realized (see Swyngedouw, 2010; also Batel & Devine-Wright, 2017). In turn, renewable energy colonialisms have become clear not only transnationally—mainly between the Global North and the Global South—but also internationally and intranationally. In fact, the concept of renewable energy colonialism helps to unpack and problematize colonial relations and practices outside of those between so-called developed and developing countries, by examining and diagnosing renewable energy colonialism within regions/countries of the same nation-state (e.g., between England and Wales in the UK, between US states and Puerto Rico) and between different imagined sociogeographical configurations and relations within the same nation-states (e.g., between urban and rural areas).

Transnational scale

Examples of renewable energy colonialism at the transnational scale are those that involve renewable energy–related colonial relations and practices between more than one country from across the world—normally in the Global North—with regard to another country somewhere in the world—normally in the Global South. At this transnational global scale, previously discussed examples of energy colonialism have come in the guise of carbon colonialism (Bachram, 2004; Paterson & Stripple, 2010).

A contemporary example that illustrates renewable energy colonialism at a transnational, global level is the Desertec project. The renewable energy colonialist underpinnings of this particular project have been discussed elsewhere (e.g., Batel & Devine-Wright, 2017; Ragheb, 2010), but this project is basically presented by the Desertec Foundation[3] as a collaboration between/within EUMENA—Europe, the Middle East, and the North of Africa—that "involves making use of the abundant unused solar energy in the deserts and wind on their seashores to promote global energy security and help protect our climate" (Ragheb, 2010, p. 1), or, in other words, to mainly make use of the solar and wind energy potentially generated in the North of Africa and Middle East and export it to Europe where energy consumption is higher.

Howe and colleagues (Howe, 2014; Howe & Boyer, 2015) analyze a similar example that concerns the deployment of wind farms by foreign companies in the Isthmus of Tehuantepec in Oaxaca, Mexico, such as the Mareñas wind park in San Dionisio, financed by an Australian consortium, Mitsubishi, and a Dutch pension fund. The authors demonstrate how these supposedly sustainable energy projects—that highlight a distinction between the sources of energy being renewable/sustainable and how projects are deployed being renewable/sustainable—have disenfranchised local populations and limited their local autonomy to determine how land, air, and water are to be used and managed. In this case, wind power became a site of political contestation regarding the good of the planet versus the good of the people, with local populations in Oaxaca opposed to these projects, seeing these activities as yet another mode of colonial extraction or *la nueva conquista* of green neoliberalism.

International scale

The international scale of renewable energy colonialism is here defined as the renewable energy–related colonial practices and relations between two nation-states, either formally defined as such or perceived as such by the involved groups. This is different from the transnational scale because the latter aims to refer to cases where the main colonialist spatial-geographical axis lies in Global North–Global South relations which, as such, go beyond the relations between any two single nation-states. Considering this international scale also aims and allows to problematize the fact that some renewable energy colonialist practices and relations exist precisely because they are already built into past and current sovereignty struggles, such as those within the UK between Wales, Scotland and Northern Ireland, and England. So whereas institutionally and legally speaking, these would need to be considered intranational relations, considering them instead as international ones fulfills a first analytical potential of the notion of renewable energy colonialism, which is to highlight these symbolic and often also material existent conflicts that also shape renewable energy agendas and transitions (e.g., Ellis et al., 2013).

A clear illustration is provided by de Onís (2018) regarding the carbon-intensive energy colonization of Puerto Rico ("the oldest colony in the world," Monge, 1999) by the United States, and it exposes how renewable energy coloniality is often disguised as trying to advance sustainable development when in fact it is hindering the decentralization and de facto renewability of energy systems, at environmental and social levels, in order to maintain specific power relations and maintain energy privilege by the United States. de Onís (2018) highlights then how "energy colonialism, as an extractivist system, as a discourse, marks certain places and peoples as disposable by importing and exporting logics and materials to dominate various energy forms, from humans to hydrocarbons" (p. 1)—and, as argued here, to renewable energy sources as well, in the guise of renewable energy colonialism.

Another example of the international scale of renewable energy colonialism is given by Batel and Devine-Wright (2017). The authors showed that one important dimension shaping responses of some local communities in Mid Wales (UK) to the deployment of new wind farms and associated high-voltage power lines in the region was the fact that those infrastructures would be giving away electricity produced in Wales to England, after a past (and present, Ellis et al., 2013) history of England exploring, "raiding and milking" Wales. The authors also demonstrate that these local communities easily presented countries in the North of Africa as a good solution to avoid having to deploy wind farms in the UK to generate renewable energy—as those countries are "deserted," "poor," and "filled with sun," naming strategies that, as we have discussed, are very powerful to rhetorically legitimate the deployment of renewable energy infrastructures there and not here.

Lennon and Scott (2015) also examine the Irish–UK intergovernmental agreement to export renewable energy from Ireland to assist the UK in meeting its renewable energy targets. They reveal how some of the storylines developed by local communities in Ireland to oppose this agreement and the deployment of related infrastructures in Ireland became entangled with Ireland's colonial past (in relation to the UK), with reference to "a return of Landlordism" and an allusion to issues of exploitation and dispossession between "them" and "us" that appeals to "fight for the land the same way our forefathers fought" (p. 99).

Intranational scale

Renewable energy colonialism also happens intranationally, within the same (constitutional and symbolic) country—and this might be in fact the scale of renewable energy colonialism that has been more frequently illustrated by social sciences' scholars so far, albeit under different names/conceptualizations. Renewable energy colonialist practices happening in the relation between urban and rural areas in many countries (albeit not all) around the world are a case in point. Taking into account that in many countries, such as in the UK, Denmark, and Portugal, energy systems are still centralized, renewable energy generation and associated infrastructures—such as high-voltage power lines—are usually large-scale and deployed in areas with more natural resources available and that are less populated, which normally equates with rural areas. However, most electricity consumption is in metropolises and other urban centers (Graham & Marvin, 1995), therefore displacing the production of electricity and its costs from its consumption and benefits. The relation between urban and rural areas, then, also often reflects the logics behind developed versus developing areas materialized in conceiving rural areas as *terra nullius*. In turn, that echoes, reproduces, and creates anew energy colonial relations—such as in rural areas with coal mines (e.g., Frantál, 2017)—but now in the guise of renewable energy sustainability. These long standing "resource peripheries" (Murphy & Smith, 2013) often now reflect and suffer renewable energy colonialism.

As an illustrative example, Rudolph and Kierkegaard (2019) have showed how wind farms' developers in Denmark tried to legitimate the deployment of these infrastructures in a specific region of rural Denmark—often called in Denmark "Rotten Banana" or "Outskirts-Denmark" and so already stigmatized—by attempting to rhetorically frame the rural local communities living there as "poor," "in decline," and "in need of development," therefore justifying the purchase and demolition of properties in these marginalized rural areas. The authors then discuss how place stigmatization can not only be the product of hazardous infrastructures and facilities but also be presented as a reason for the siting of stigmatizing facilities, such as wind farms can be, therefore contributing to geographies of exclusion (Massey & Jess, 1995).

A final example of what can also be considered renewable energy colonialism at the intranational scale is the relation between some governments and indigenous communities in their own countries when renewable energy and associated objects are being fostered. This type of relation is also still often based in long standing sociogeographically, economically, and historically embedded relations of resources and land dispossession and exploitation, as still currently happening in several countries around the world, such as Canada and Brazil (Bingham, 2010; Parson & Ray, 2018; Willow, 2016).

Energy colonialism through different voices

As already suggested by the examples discussed in the previous section, renewable energy colonialism is enacted and coconstructed through different voices: those of governments, developers, the media, and in the everyday discourses of communities and publics. All of these voices are very powerful in enacting and reproducing various forms and practices of renewable energy colonialism at local, regional, national, and global levels and, in the same vein and as discussed in the next section, also can have the power to contest those practices.

Research on nuclear colonialism (Edwards, 2011; Endres, 2009) has widely illustrated how governments often rhetorically portray specific places as other, as "far away," "empty," or "sparsely populated," to legitimate both nuclear test sites and sites for nuclear waste disposal. As illustrated in some of the examples given in the previous section, this type of rhetoric has also shaped the deployment of renewable energy and associated technologies and associated sustainability agendas. This is also clear when governments embrace policies to foster centralized renewable energy generation and associated infrastructures and declare some places (mostly rural areas and/or regions in the Global South and populated by indigenous communities) as more adequate than others for the deployment of such infrastructures, often by relying on wind potentials and other apparent objective criteria that disguise how these decisions and associated discourses are based in the politics of land and associated power relations and place representations (Batel, 2020b; Howe & Boyer, 2015). As Howe and Boyer (2015) put it, wind power becomes more than an energetic source, it becomes a "device through which corporate and state interests craft their ecological profile and proclaim moral authority in the practice of climatological care" (p. 42).

This is also very similarly encountered in developers' and companies' practices and discourses. Rudolph and Kierkegaard's work (2019) on how wind farm developers in Denmark present rural areas, livelihoods, and communities in the Rotten Banana in order to justify the deployment of wind farms testifies to the power of developers' and companies' discourses in legitimizing some places and communities and not others as appropriate for energy-related development, based on long standing relations of land and resource grabbing and domination that now feed into renewable energy colonial practices. Other very powerful actors in both reflecting and shaping people's ideas and practices over energy and other social issues are the media (see Batel, 2020b for an example) and also researchers, who also often apply the same type of renewable energy colonialist–related rhetoric that other actors do (see Aitken, 2010; Batel, 2020a, for reviews).

These discourses conveyed by governments, corporations, the media, researchers, and other actors shape citizens and communities' representations of energy issues, who will in turn often reproduce those as well. This is well illustrated in the work by Batel and Devine-Wright (2017), who show how renewable energy colonialism is enacted by UK citizens, members of local communities to be affected by the deployment of large-scale renewable energy generation and associated infrastructures, in order to contest the deployment of those infrastructures near the place where they live and argue instead for their deployment in the Global South, fostering many of the ideas embedded in projects such as the Desertec project. In turn, these ideas are circulated in the media and are then appropriated in the everyday practices of publics in order to negotiate renewable energy issues in their locality, region, and country. Also within the "same" community there are often voices engaging differently with these discourses, as shown by Dunlap (2018), who also reflected upon wind energy colonization at the Isthmus of Tehuantepec, Oaxaca, Mexico. Discourses over what is "progress," what is "local," and what is desirable and just for indigenous and other members of local communities are reproduced, contested, and negotiated also by community members themselves as renewable energy infrastructures are imposed from the outside.

Conclusions and discussion: uncovering renewable energy colonialism to promote energy democracy

In this chapter, I have discussed how different forms of renewable energy colonialism, within the recent green energy transitions turn, are enacted and reproduced at different scales and through different voices. Through that, I have also argued for the importance of considering renewable energy colonialism as a distinct practice in and of itself, to be distinguished from related practices such as environmental injustice and energy injustice. Renewable energy colonialism is a practice embedded in sociohistorical, geographical, economic, and political processes of specific power relations between colonizer and colonized, between developed and developing, that is reappraised and reproduced in new ways to perpetuate those power relations within current neoliberal capitalist discourses and agendas such as those of green growth, renewable energy technologies, and international development and cooperation in sustainable energy transitions. It is precisely because these discourses and agendas hide associated land- and resource grabbing and relations of domination, at both the material and the symbolic levels, that the concept of renewable energy colonialism is relevant, as it lends a critical look to any analyses of renewable/green/sustainable discourses and practices.

As McCreary and Milligan (2018) put it and as discussed above, it is crucial to be aware of the "institutionalization of concern." While issues of environmental and energy injustice start to already be part of institutionalized—governmental, corporate—discourses, policies, and regulations, they are so in a neoliberal capitalist way. In this way, these institutionalizations "obscure and, thereby, normalize operative relations of white supremacy and settler colonialism" (McCreary & Milligan, 2018, p. 2). This makes it important to reflect about new concepts that are able to accompany new sociotechnical configurations and changes, while still serving a critical potential in the sense that, more than aiming to be conceptually and analytically useful to diagnose power relations and their effects, they also aim to uncover and tackle inequality and injustice. The concept of renewable energy colonialism and its application at different scales and analyses in different voices has this potential, as it was argued here. Additionally, focusing on this concept and reflecting about its critical and analytical potential also allowed its use as a lens of analysis and an organizing axis for the scattered social sciences' research on energy that has attempted to uncover associated colonialist practices and to discuss them as resource colonialism, settler colonialism, and energy dispossessions regarding—mostly—nonrenewable energy resources.

In this vein, something that becomes evident in the preceding examples and discussions is the need to critically consider how politics—or institutions, norms, rules, and structures—influence and are influenced by the political—i.e., everyday practices and interactions (Escobar, 2018; Mouffe, 2005). Renewable energy colonialism is enacted and reproduced and can be contested precisely in that interplay between politics and the political. This is especially clear in the example discussed in depth by Santos (2015) regarding the Yasuní project in Ecuador. Ecuador is a poor country and its economy depends heavily on oil exports, therefore causing dramatic environmental destruction. The UN has tried to enforce regulations like the Kyoto Protocol on Ecuador. However, Ecuador proposed an alternative to the developmentalist-extractivist capitalist model of development that is dominant nowadays:

> The Ecuadorian state vouches to leave unexploited in the subsoil oil reserves estimated at 850 million barrels in three blocs of the National Amazonian Park of Yasuní, one of the richest biodiversity regions of the planet, on the condition that the more developed countries reimburse Ecuador by half the income Ecuador would surrender as a consequence of this decision.
>
> *(Santos, 2015, p. 30)*

In turn, this money would go to environmentally appropriate investments. As discussed by Santos (2015), this proposal contests practices of carbon colonialism and instead proposes alternatives to Western epistemologies and conceptions of development based on the idea of *sumac kawsay*, good living. This is undoubtedly a strategy toward decoloniality, based on giving voice to and inscribing other epistemologies in narratives of modernity and neoliberal capitalism and thus creating new geopolitics of knowledge (see also Bhambra, 2014; de Onís, 2018; Escobar, 2018; Mignolo, 2012; Spivak, 1999). This echoes as well with the call of Howe (2014) for new forms of conceptualizing biopower and politics that can be more useful in a time of capitalogenic (Haraway, 2015) neoliberal climate change and that also help us to better understand and contest neocolonialisms such as renewable energy colonialism. Equally important to discussing strategies and resources for decoloniality are redefinitions of citizenship and sovereignty—or considering the moral ownership of places and making it prevail over capital and property (Powell, 2015; Zukin, 2009). Grasping and being able to take decisions based on the lived dimension of territorial sovereignty (Powell, 2015) is a crucial pathway to be able to contest and prevent renewable energy colonialism—even if at points this might imply going against the apparent global drive for mitigating climate change (Powell, 2015). In fact, this might be the most important message to take from discussing renewable energy colonialisms—that no one measure or knowledge fits all and also that, consequently, sometimes avoiding or contesting renewable energy colonialist practices might imply allowing and even fostering some carbon-intensive extractive activities, namely by indigenous communities (e.g., Powell, 2015). The importance of situated knowledges (Haraway, 1988) also uncovers another conundrum with renewable energy colonialism, with examining it and giving it voice at an academic level: Are we all entitled to do it? To which extent and in which sociogeographical configurations? Issues such as colonialism in the research process itself (Coddington, 2017) and overresearch (Neal et al., 2016) also have to be considered.

In turn, reflecting on these issues opens various avenues for future research on renewable energy colonialism and related practices. First and following Howe's (2015) call for a reconceptualized biopolitics of the Anthropocene—or Capitalocene (Haraway, 2015)—it will be important in future research to open up the analysis and conceptualization of renewable energy colonialism to also include nature as an actor, a group, a colonized. Most research on environmental injustice and energy justice has focused on the relation between different social groups with power differentials, but we can no longer leave nature out of any analysis that aims to address environmental issues and justice, including renewable energy colonialist practices. For instance, Glassheim (2007) describes how the coal industry in the Czech Republic has created a landscape of "half-eaten mountains, vast pits inhabited by massive earth-devouring machines, agglomerations of belching smokestacks, row upon row of decaying prefabricated apartment buildings" (as cited in Frantál, 2017, p. 202). This colonization of nature by energy infrastructures needs to be taken into account and contested as much as the colonization of local communities and indigenous groups by governments, developers, and other communities (see also Lyons & Westoby, 2014)—the crucial question being how to do that, as previously noted in the importance of thinking about the process of "giving voice" in a critical way.

Second, it will be relevant to look at renewable energy colonialism beyond the deployment of wind farms and solar plants and also to include all the extractivist practices and relations that come with it, such as mining lithium for energy storage, something crucial for the future of renewable energy transitions, and mining rare earth minerals that are already currently used in wind farms and other green transitions–related technologies (Raman, 2013). Finally, it will also be crucial to further examine renewable energy colonized subjectivities and how they intersect with racialized and other structural inequalities (Fanon, 2007; Lertzman, 2015; Mbembe, 2013): Exploring and addressing the subjective experiences of being colonized by renewable energy infrastructures, in the form of loss, grief, melancholia, alienation, and other experiences, are as important as better understanding the collective consequences of renewable energy colonialism, the associated subject positions, and ways to counteract them for creating real energy democracies.

Notes

1 As in "longstanding patterns of power that . . . continue to define culture, labour, intersubjective relations and knowledge production, long after the end of direct colonialism" (see Enns & Bersaglio, 2019, p. 21).
2 This includes as well interspecies relations.
3 This is a limited liability company based in Germany but composed by a consortium of several European companies, such as the Spanish and Italian electric grid companies (Red Elétrica de España and Terna, respectively).

References

Aitken, M. (2010). Why we still don't understand the social aspects of wind power: A critique of key assumptions within the literature. *Energy Policy*, *38*(4), 1834–1841. https://doi.org/10.1016/j.enpol.2009.11.060

Bachram, H. (2004). Climate fraud and carbon colonialism: The new trade in greenhouse gases. *Capitalism Nature Socialism*, *15*(4), 5–20. https://doi.org/10.1080/1045575042000287299

Baka, J. (2017). Making space for energy: Wasteland development, enclosures, and energy dispossessions. *Antipode*, *49*(4), 977–996. https://doi.org/10.1111/anti.12219

Barry, J. (2007). *Environment and social theory*. Routledge.

Barry, J., Ellis, G., & Robinson, C. (2008). Cool rationalities and hot air: A rhetorical approach to understanding debates on renewable energy. *Global Environmental Politics*, *8*(2), 67–98. https://doi.org/10.1162/glep.2008.8.2.67

Batel, S. (2020a). Research on the social acceptance of renewable energy technologies: Past, present and future. *Energy Research & Social Science*. https://doi.org/10.1016/j.erss.2020.101544

Batel, S. (2020b). Re-presenting the rural in the UK press: An exploration of the construction, contestation and negotiation of media discourses on the rural within post-carbon energy transitions. *Energy Policy*, *138*, 111286. https://doi.org/10.1016/j.enpol.2020.111286

Batel, S., & Devine-Wright, P. (2017). Energy colonialism and the role of the global in local responses to new energy infrastructures in the UK: A critical and exploratory empirical analysis. *Antipode*, *49*(1), 3–22. https://doi.org/10.1111/anti.12261

Batel, S., & Pataco, L. (2019). Portuguese media representations of nuclear facilities in Almaraz, Spain: Beyond borders and risk perception/Representaciones en los medios Portugueses de las plantas nucleares en Almaraz, España: Más allá de las fronteras y la percepción del riesgo. *Psyecology*, 1–12. https://doi.org/10.1080/21711976.2019.1644004

Battaglini, A., Komendantova, N., Brtnik, P., & Patt, A. (2012). Perception of barriers for expansion of electricity grids in the European Union. *Energy Policy*, *47*, 254–259. https://doi.org/10.1016/j.enpol.2012.04.065
Bhambra, G. K. (2014). Postcolonial and decolonial dialogues. *Postcolonial Studies*, *17*(2), 115–121. https://doi.org/10.1080/13688790.2014.966414
Bhambra, G. K. (2018). Colonialism and postcolonialism. *Core Concepts in Sociology*, *37*.
Bingham, A. (2010). Discourse of the dammed: A study of the impacts of sustainable development discourse on indigenous peoples in the Brazilian Amazon in the context of the proposed Belo Monte hydroelectric dam. *Polis Journal*, *4*, 1–47.
Bridge, G., Barr, S., Bouzarovski, S., Bradshaw, M., Brown, E., Bulkeley, H., & Walker, G. (2018). *Energy and society: A critical perspective*. Routledge.
Castree, N. (2003). Commodifying what nature? *Progress in Human Geography*, *27*(3), 273–297. https://doi.org/10.1191/0309132503ph428oa
Coddington, K. (2017). Voice under scrutiny: Feminist methods, anticolonial responses, and new methodological tools. *The Professional Geographer*, *69*(2), 314–320. https://doi.org/10.1080/00330124.2016.1208512
Cozen, B., Endres, D., Peterson, T. R., Horton, C., & Barnett, J. T. (2018). Energy communication: Theory and praxis towards a sustainable energy future. *Environmental Communication*, *12*(3), 289–294. https://doi.org/10.1080/17524032.2017.1398176
de Onís, C. M. (2018). Energy colonialism powers the ongoing unnatural disaster in Puerto Rico. *Frontiers in Communication*, *3*, 2. https://doi.org/10.3389/fcomm.2018.00002
Dunlap, A. (2018). The 'solution' is now the 'problem': Wind energy, colonisation and the 'genocide-ecocide nexus' in the Isthmus of Tehuantepec, Oaxaca. *The International Journal of Human Rights*, *22*(4), 550–573. https://doi.org/10.1080/13642987.2017.1397633
Edwards, N. (2011). Nuclear colonialism and the social construction of landscape in Alaska. *Environmental Justice*, *4*(2), 109–114. https://doi.org/10.1089/env.2010.0023
Ellis, G., Cowell, R., Sherry-Brennan, F., Strachan, P., & Toke, D. (2013). Planning, energy and devolution in the UK. *Town Planning Review*, *84*(3), 397–410. https://doi.org/10.3828/tpr.2013.16
Endres, D. (2009). The rhetoric of nuclear colonialism: Rhetorical exclusion of American Indian arguments in the Yucca Mountain nuclear waste siting decision. *Communication and Critical/Cultural Studies*, *6*(1), 39–60. https://doi.org/10.1080/14791420802632103
Enns, C., & Bersaglio, B. (2019). On the coloniality of 'new' mega-infrastructure projects in East Africa. *Antipode*. https://doi.org/10.1111/anti.12582
Escobar, A. (2018). *Designs for the pluriverse: Radical interdependence, autonomy, and the making of worlds*. Duke University Press.
Fanon, F. (2007). *The wretched of the earth*. Grove, Atlantic, Inc.
Feldpausch-Parker, A. M., Endres, D. E., & Peterson, T. (2019). A research agenda for energy democracy. *Frontiers in Communication*, *4*, 53. https://doi.org/10.3389/fcomm.2019.00053
Finley-Brook, M., & Thomas, C. (2010). Treatment of displaced indigenous populations in two large hydro projects in Panama. *Water Alternatives*, *3*(2), 269.
Frantál, B. (2017). Under the curse of coal. *The Routledge Research Companion to Energy Geographies*, *200*.
Graham, S., & Marvin, S. (1995). More than ducts and wires: Post Fordism, cities and utilities networks. In P. Healey, S. Cameron, S. Davoudi, S. Graham, & A. Madanipour (Eds.), *Managing cities: The new urban context* (pp. 169–189). John Wiley.
Haraway, D. (1988). Situated knowledges: The science question in feminism and the privilege of partial perspective. *Feminist Studies*, *14*(3), 575–599. https://doi.org/ 10.2307/3178066
Haraway, D. (2015). Anthropocene, Capitalocene, Plantationocene, Chthulucene: Making kin. *Environmental Humanities*, *6*(1), 159–165. https://doi.org/10.1215/22011919-3615934
Heffron, R. J., & McCauley, D. (2017). The concept of energy justice across the disciplines. *Energy Policy*, *105*, 658–667. https://doi.org/10.1016/j.enpol.2017.03.018
Howe, C. (2014). Anthropocenic ecoauthority: The winds of Oaxaca. *Anthropological Quarterly*, *87*(2), 381–404. www.jstor.org/stable/43652703
Howe, C., & Boyer, D. (2015). Aeolian politics. *Distinktion: Scandinavian Journal of Social Theory*, *16*(1), 31–48. https://doi.org/10.1080/1600910X.2015.1022564
Labussière, O., & Nadaï, A. (Eds.). (2018). *Energy transitions: A socio-technical inquiry*. Springer.
Lennon, M., & Scott, M. (2015). Opportunity or threat: Dissecting tensions in a post-carbon rural transition. *Sociologia Ruralis*, *57*(1), 87–109. https://doi.org/10.1111/soru.12106

Lertzman, R. (2015). *Environmental melancholia: Psychoanalytic dimensions of engagement.* Routledge.
Liverman, D. M. (2009). Conventions of climate change: Constructions of danger and the dispossession of the atmosphere. *Journal of Historical Geography, 35*(2), 279–296. https://doi.org/10.1016/j.jhg.2008.08.008
Lyons, K., & Westoby, P. (2014). Carbon colonialism and the new land grab: Plantation forestry in Uganda and its livelihood impacts. *Journal of Rural Studies, 36,* 13–21. https://doi.org/10.1016/j.jrurstud.2014.06.002
Massey, D. B., & Jess, P. M. (1995). *A place in the world? Places, cultures and globalization.* Open University Press.
Mavhunga, C., C., & Trischler, H. (Eds.). (2014). *Energy (and) colonialism, energy (in)dependence: Africa, Europe, Greenland, North America.* Rachel Carson Center Perspectives.
Mbembe, A. (2013). *Critique de la raison nègre.* La découverte.
McCreary, T., & Milligan, R. (2018). The limits of liberal recognition: Racial capitalism, settler colonialism, and environmental governance in Vancouver and Atlanta. *Antipode.* https://doi.org/10.1111/anti.12465
Merriam-Webster. (n.d.). *Colonialism.* In Merriam-Webster dictionary. https://www.merriam-webster.com/dictionary/colonialism
Mignolo, W. (2012). *Local histories/global designs: Coloniality, subaltern knowledges, and border thinking.* Princeton University Press.
Monge, J. T. (1999). *Puerto Rico: The trials of the oldest colony in the world.* Yale University Press.
Mouffe, C. (2005). *The return of the political* (Vol. 8). Verso.
Murphy, J., & Smith, A. (2013). Understanding transition—periphery dynamics: Renewable energy in the highlands and islands of Scotland. *Environment and Planning A: Economy and Space, 45*(3), 691–709. https://doi.org/10.1068/a45190
Nadaï, A., & Labussière, O. (2017). Landscape commons, following wind power fault lines. The case of Seine-et-Marne (France). *Energy Policy, 109,* 807–816.
National Energy Action. (2015, March 11). *NEA energy justice campaign response to the NIAUR Northern Ireland sustainable energy programme (NISEP) consultation.* www.nea.org.uk/resources/publications-and-resources/nea-energy-justice-campaign-response-to-the-niaurs-northern-ireland-sustainable-energy-programme-consultation/
Neal, S., Mohan, G., Cochrane, A., & Bennett, K. (2016). 'You can't move in Hackney without bumping into an anthropologist': Why certain places attract research attention. *Qualitative Research, 16*(5), 491–507. https://doi.org/10.1177/1468794115596217
Nixon, R. (2011). *Slow violence and the environmentalism of the poor.* Harvard University Press.
Parson, S., & Ray, E. (2018). Sustainable colonization: Tar sands as resource colonialism. *Capitalism Nature Socialism, 29*(3), 68–86. https://doi.org/10.1080/10455752.2016.1268187
Paterson, M., & Stripple, J. (2010). My space: Governing individuals' carbon emissions. *Environment and Planning D: Society and Space, 28*(2), 341–362. https://doi.org/10.1068/d4109
Powell, D. E. (2015). The rainbow is our sovereignty: Rethinking the politics of energy on the Navajo nation. *Journal of Political Ecology, 22*(1), 53–78. https://doi.org/10.2458/v22i1.21078
Preston, J. (2013). Neoliberal settler colonialism, Canada and the tar sands. *Race & Class, 55*(2), 42–59. https://doi.org/10.1177/0306396813497877
Ragheb, M. (2010, May 19). *Sustainable global energy Desertec concept.* http://magdiragheb.com/NPRE%20475%20Wind%20Power%20Systems/Sustainable%20Global%20Energy%20Desertec%20Concept.pdf
Raman, S. (2013). Fossilizing renewable energies. *Science as Culture, 22*(2), 172–180. https://doi.org/10.1080/09505431.2013.786998
Renewables Directive. (2009). Directive 2009/28/EC of the European Parliament and of the Council of 23 April 2009 on the promotion of the use of energy from renewable sources and amending and subsequently repealing directives 2001/77/EC and 2003/30/EC. *Croatian, 11,* 39–85. https://eur-lex.europa.eu/legal-content/EN/ALL/?uri=CELEX:32009L0028
Rudolph, D., & Kirkegaard, J. K. (2019). Making space for wind farms: Practices of territorial stigmatisation in rural Denmark. *Antipode, 51*(2), 642–663. https://doi.org/10.1111/anti.12428
Santos, B. de S. (2015). *Epistemologies of the South: Justice against epistemicide.* Routledge.
Sovacool, B. K., & Dworkin, M. H. (2015). Energy justice: Conceptual insights and practical applications. *Applied Energy, 142,* 435–444. https://doi.org/10.1016/j.apenergy.2015.01.002
Spivak, G. C. (1999). *Can the subaltern speak?* Harvard University Press.

Swyngedouw, E. (2010). Apocalypse forever? *Theory, Culture & Society*, *27*(2–3), 213–232. https://doi.org/10.1177/0263276409358728

Walker, G. (2009). Beyond distribution and proximity: Exploring the multiple spatialities of environmental justice. *Antipode*, *41*(4), 614–636. https://doi.org/10.1111/j.1467-8330.2009.00691.x

Willow, A. J. (2016). Indigenous ExtrACTIVISM in Boreal Canada: Colonial legacies, contemporary struggles and sovereign futures. *Humanities*, *5*(3), 55. https://doi.org/10.3390/h5030055

Zukin, S. (2009). *Naked city: The death and life of authentic urban places*. Oxford University Press.

12
ENERGY DOMINANCE

Jen Schneider and Jennifer Peeples

> And the oil and gas industry, I want to promise you, has no greater friend than President Donald Trump. And as the President said, in his words, our administration will not only seek American energy independence but will seek American energy dominance. He promised to "eliminate the barriers to domestic energy production, like never before."
>
> —*Then Vice President Mike Pence, at the 2019 Oil and Gas Association Annual Meeting, Columbus, Ohio, March 8, 2019*

Introduction

The election of President Donald Trump and his administrative leadership since were novel for a number of reasons. He upended countless democratic norms that have shaped expectations for how presidents are supposed to communicate, both in form—such as his heavy reliance on Twitter as an outlet for official pronouncements—and in content, including his reliance on "simple, impulsive, and uncivil" messages (Ott, 2017). While stagecraft and showmanship have always been essential to politics, Trump merged the world of entertainment and spectacle with the staid world of policy making in ways his predecessors would not have dared. It is hard to imagine Ronald Reagan, for example (also an entertainment celebrity turned politician) holding a Trump-style rally (Hall et al., 2016). Trump also maintained a singular focus on electoral politics, including repeated and single-minded defenses of his 2016 electoral college victory and comparative inattention paid to the specifics of achieving pragmatic and lasting legislative policy victories, despite holding a Republican supermajority for his first two years in office. Finally, there was Trump's heavy reliance on "unofficial" and unusual ways of doing things, including managing foreign policy, which led to his being impeached twice by the House of Representatives.

Such remarkable departures from the behaviors and norms Americans expect from their president make it difficult to assess just how successful Trump was from a policy perspective. Trump's leadership style veered toward the chaotic, with a rapidly revolving cast of cabinet secretaries and other agency leaders moving in and out of power by tweet. Yet his administration managed to remain remarkably focused on using executive power rather than legislative victories to advance a few key policy areas, namely immigration policy, trade policy, and

DOI: 10.4324/9780429402302-14

environmental deregulation. Disputes over immigration and trade policy were ongoing, with many of the administration's orders making their way through court challenges and renegotiations. Time will reveal what the lasting impacts of the administration's efforts in these areas will be.

In the area of environmental deregulation, however, the administration's actions were decisive, significant in scope, and swift (Gibbens, 2019; "Tracking Deregulation," 2020). Environmental deregulation occurred across federal agencies—from the Environmental Protection Agency (EPA) to the United States Department of Agriculture (USDA) to the Department of Interior. Deregulation was also enacted as a series of federal challenges to state environmental rules, with California—often a national leader in implementing environmental regulation—being a special target (Eilperin et al., 2019). The president and his cabinet officials rolled out this deregulatory agenda under the umbrella of "Energy Dominance," a new form of national energy discourse that picked up elements of energy independence and security discourses that have long typified American approaches to energy policy making, particularly leading up to the Obama administration. But it paired those more traditional discourses with aggressive and aggrieved messages about actions taken by the Obama administration specifically, often alleging conspiratorial "wars" against fuels such as coal. The new discourse called for American energy sources to "dominate" challengers (such as renewables), as well as markets abroad, and provided justification for massive, swift environmental deregulation across sectors (Dolata, 2017).

In this chapter, we describe the contours of Energy Dominance under the Trump administration, including its significant policy elements. We argue that Energy Dominance policy and discourse frequently enact and reinforce what Catalina de Onís (2017, 2018) has termed "energy coloniality," which emphasizes the domination and exploitation of certain lands and people so that others may enjoy energy privileges and profits. Drawing on the work of de Onís and others, we argue that energy coloniality relies on the subjugation of particular peoples and lands, predominantly through harmful or unjust extraction, production, and consumption practices. It also asserts that fossil fuels are universal energy solutions, meaning that they are appropriate and necessary across all contexts. In addition, fossil fuels represent a certain kind of American, masculinist, militarist identity that also must be preserved. Finally, energy coloniality relies on the production of energy privilege, a construct that requires some people to be more deserving of affordable and reliable energy sources, while others must bear the brunt of unjust social and environmental practices. Energy privilege is articulated through a suite of practices and discourses that construct some as more deserving than others. We then conclude the chapter by analyzing how energy dominance is articulated by the Trump administration through a variety of rhetorical forms, including rhetorics of nostalgia, grievance, and American exceptionalism.

Although this chapter focuses almost exclusively on the American context and in particular on energy and environmental policy under the Trump administration, it has implications for countries outside of the United States, and in particular those that are dealing with the threats posed by American foreign policy that centers Energy Dominance. The "foreign markets" that must be "dominated" by the United States under Trump's energy policy may certainly experience the effects of energy colonialism. In addition, energy dominance is touted as leverage for influencing foreign policy (Wald, 2019). Finally, environmental deregulation impacts systems we all share, and American isolationism in the area of climate policy will no doubt have lasting and deleterious consequences for environmental and public health worldwide.

Background: what is Energy Dominance?

When the Trump administration came into power in 2017, it rolled out a new branding of American energy policy called Energy Dominance. President Trump made the following remarks at an Unleashing American Energy event in June 2017 in Washington, D.C.:

> Our country is blessed with extraordinary energy abundance, which we didn't know of, even five years ago and certainly ten years ago . . . We have so much more than we ever thought possible. We are really in the driving seat. And you know what? We don't want to let other countries take away our sovereignty and tell us what to do and how to do it. That's not going to happen. [Applause.] With these incredible resources, my administration will seek not only American energy independence that we've been looking for so long, but American energy dominance.
>
> *("Remarks by President Trump," 2017)*

Commentators have noted that Energy Dominance has never seemed to coalesce around a clear set of stated policy goals, beyond a broadly articulated distaste for regulation (Schneider & Peeples, 2018). Rather, Energy Dominance served as a campaign slogan, a diffuse set of discourses that reflected the administration's environmental, economic, and political values, which primarily privileged fossil fuel resources and American exceptionalism, as well as a suite of administrative and policy preferences and actions based on those values. The rhetoric of Energy Dominance also clearly echoes and reinforces rhetorics that support other important policy priorities of the administration, such as immigration restrictions and removing the United States from international trade agreements. In these areas as well, the former president articulated concerns that the United States was being taken advantage of, paying more than its fair share in global agreements, and had been engaged in "losing" deals. We explore how these rhetorics are interwoven later in this chapter.

One of the challenges of studying any rhetoric or discourse under Trump is that, unlike previous presidential administrations, this administration did not have a disciplined messaging apparatus, and policy priorities and plans rarely were laid out in a clear or systematic fashion (Ott, 2017; Pfiffner, 2017). Rather, one must often track a wide variety of administrative actions across agencies, examine haphazard speech acts (often in the form of tweets or on-the-fly press interactions) and rely on sometimes incomplete and unstructured extemporaneous speeches, which may or may not be officially transcribed or logically comprehensible. Exacerbating matters is the significant amount of administrative turnover in cabinet positions, the lack of permanent appointments made in federal agencies, and the administration's reluctance to hold press conferences or release formal statements or transcripts (not just tweets) to the press and public, as previous administrations have done (Drezner, 2019; Lacatus, 2018; Thrower, 2018).

Trying to characterize Energy Dominance, therefore, demands a fair amount of puzzle-solving and piecework. One must infer what Energy Dominance means from actions the administration has taken in the fields of energy and environmental policy and also analyze rare public speeches given by agency spokespeople, such as former EPA Administrator Scott Pruitt or former Secretary of the Interior Ryan Zinke. Commentators trying to figure out the parameters of Energy Dominance have often been thwarted, furthermore, by internal inconsistencies and lack of specific policy detail (Bordoff, 2017; Kederer, 2018).

That said, after having followed the administration's statements about Energy Dominance and studying its environmental and energy policy record for several years, we argue that Energy

Dominance can be characterized by three elements. First and foremost, Energy Dominance is fundamentally a deregulatory project that aims to undo environmental regulations writ large, and in particular those enacted under the Obama administration (Aldy, 2017). Obama-era regulations raised ire among Trump administration agency leaders and spokespeople, who were fond of calling attention to the previous administration's "war on coal," in which that industry was subject to "punitive regulations" and America's wealth was locked behind a "fortress of regulation and red tape" (Schneider & Peeples, 2018). The administration's attack on these regulations was massive and swift. According to *The Washington Post*, in its first year in power alone, the Trump administration canceled or attempted to cancel 70 environmental regulations enacted by the Obama administration (Eilperin & Cameron, 2018); by September 2019, *The New York Times* reported that 85 rules and regulations had been rolled back or slated to be rolled back (Popovich et al., 2019). Columbia Law School's "Climate Deregulation Tracker" lists deregulatory efforts ranging from dismantling Obama-era climate policies to those that specifically aim to boost coal ("Climate Deregulation Tracker," n.d.; see also Harvard Law's " Regulatory Rollback Tracker").

Second, this deregulation was largely aimed at telegraphing to fossil fuel industries—oil, gas, and coal—that "America is open for business" (Layzer, 2014). Energy Dominance is, at its core, oriented toward facilitating the extraction of fossil fuels and removing any barriers in the way of this extraction, even when those barriers might be essential for protecting human and environmental health. Throughout his presidential campaign and his time in office, Trump asserted that his administration would rescue the coal industry, which he claimed was under attack from Obama-era regulations. In his 2019 State of the Union address, he claimed, "We have ended the war on American energy—and we have ended the war on beautiful, clean coal. We are now very proudly an exporter of energy to the world" ("Remarks by President Trump in the State of the Union," 2019). The influence that former coal baron Robert Murray of Murray Energy had on skillfully directing the president's rhetoric on this issue, as well as laying out a sympathetic deregulatory agenda, has been well documented (Friedman, 2018a; Taddonio, 2019).

Trump also "weaponized" agencies such as the Environmental Protection Agency in the service of fossil fuel interests. The EPA clearly signaled a friendliness to fossil fuel industries and a hostility to environmental interests, and as is the case with many other agencies across the executive branch, it hollowed out or suppressed scientific expertise and public oversight, such as by disbanding advisory groups and advisory councils (Friedman, 2018b; Mooney & Eilperin, 2017). In one of its most high-stakes and visible deregulatory moves, the Trump administration signaled early on that it would not defend the Clean Power Plan, an Obama-era EPA plan that aimed to regulate CO_2 as a pollutant and to collaborate with states to reduce CO_2 emissions from coal plants over time. Trump actually signed an executive order "calling on every federal agency to loosen the regulatory reins on fossil fuel industries, the most significant declaration of the administration's intent to retreat from action on climate change" before directing the EPA not to enforce or defend the Clean Power Plan (Lavelle, 2017). There were rollbacks of methane regulations, which affect natural gas production (and hydrofracturing in particular), as well as attacks on fuel efficiency standards. The Trump administration was pointed and vocal in its critiques of California, a state it saw as a political and ideological opponent, which has set stricter regulations on energy consumption and production than those the federal government requires (Purdum, 2019). And renewable forms of energy production, such as solar and wind, rarely received even a mention, unless it was to mock them for their perceived limitations based on erroneous claims (Shannon, 2019).

Third, Energy Dominance is focused on opening up foreign markets for American energy products, while at the same time withdrawing from existing international trade agreements, climate change agreements, or existing diplomatic arrangements abroad. Implied in the name "Energy Dominance" is the notion that the United States can enter foreign energy markets and dominate them economically, without being beholden to any sort of multilateral environmental agreements in which it might have to compromise. In short, it communicates a philosophy that to the winner should go all the spoils and that winners do not negotiate. Compromise, for the Trump administration, appeared to be the same as "losing," and President Trump made clear that he hates "losers" (Elmer & Todd, 2016). As Trump put it in 2017, when he first announced that the United States was planning to withdraw from the Paris Climate Agreement:

> The Paris Climate Accord is simply the latest example of Washington entering into an agreement that disadvantages the United States to the exclusive benefit of other countries. . . . The Paris Agreement handicaps the United States economy in order to win praise from the very foreign capitals and global activists that have long sought to gain wealth at our country's expense. They don't put American first. I do, and I always will.
> *("Statement by President Trump," 2017)*

Analysts have shown that the Paris Agreement did not in fact put America's "reserves under lock and key," "take away the great wealth of our nation," or put the United States at "grave risk of brownouts and blackouts" (see, for example, Duke, 2019). But Trump raised the fear that other countries are "demeaning" and "laughing at" the United States for signing on to a deal he arguesd would make the United States lose standing internationally. Energy Dominance, on the other hand, promises to make sure American fossil fuel production wins, no matter what.

In the following sections, we argue that Energy Dominance, as a set of administrative practices, policies, and rhetorics, reinscribes what Catalina de Onís (2017, 2018) has called "energy coloniality." Massive environmental deregulation significantly limits efforts to achieve social and environmental justice. A singular emphasis on fossil fuels as "winners" undercuts commitments to climate justice, environmental and public health, corporate accountability, and community self-determination. And the call to "dominate" foreign market echoes exploitative and colonialist practices that overprivilege the rights of fossil fuel companies. Energy Dominance also seeks to diminish or suppress voices of protest in communities affected by energy industrialization and to bolster uneven social arrangements wherein some Americans reap significant benefits from fossil fuel extraction while others bear undue burdens, often with little recourse. It also ignores one of the singular challenges of our time: Climate change. The refusal to acknowledge climate change as a significant threat is a remarkable political stance given that the majority of Americans, across the political spectrum, now believe climate change is happening and that the United States needs to act (Gufstafson et al., 2019). As its name implies, Energy Dominance is fundamentally about the exercise of power, and the ability of some groups to subjugate others for profit and the maintenance of existing social and economic arrangements.

Energy coloniality and privilege

There is a small but growing literature that defines and analyzes the concepts "energy coloniality" and "energy privilege." The work of Catalina de Onís (2017, 2018) has been especially constructive here, though we also draw from scholars who have worked on nuclear colonialism and carbon colonialism as well (Bacon, 2019; Edwards, 2011; Endres, 2009). This literature suggests that three main characteristics define energy coloniality:

1 *The subjugation of some people and lands by others in order to extract energy sources, material or otherwise*: This subjugation is aided by or enacted through rhetoric that frames certain peoples and places as backward, undeveloped, or needing outside saviors and practices that ignore or suppress community organizing and voices of protest.
2 *Privileging "universal" energy solutions or sources—especially fossil fuels—while ignoring alternatives*: This often takes the form of seeing or framing communities primarily as "markets" for particular types of commodities, which have been deemed appropriate or necessary by outside interests.
3 *Energy privilege*: This consists of bolstering the "rights" of certain people—especially those in hegemonic or socially dominant groups—to have access to reliable, affordable energy, while concealing, downplaying, or ignoring unjust social and environmental practices, such as when certain groups must disproportionately bear the burdens of environmental pollution or economic exploitation.

As de Onís (2018) has argued, these practices and discourses often articulate and coproduce other economic and social systems and arrangements, such as neoliberalism:

> Logics of domination, extractivism, and conquest persistently colonize ever deeper subterranean environments, while proposing universal energy solutions that trivialize and ignore geographic, cultural, and other differences, as well as local community efforts to imagine and implement their own alternatives. Characterizations of tropical locales as places that deplete energy, motivation, and "productivity" rely on an environmental determinism that assumes the inferiority of inhabitants and their inability to "manage" the surrounding landscape appropriately.
>
> (*p. 5*)

We also highlight de Onís's call to consider "coloniality" rather than "colonialism" because "coloniality" forces us to acknowledge that these are practices and discourses that must be enacted and re-enacted in multiple ways and across contexts; colonialism is not fixed, nor is it "over," a thing of the past. Instead, it is helpful to think of coloniality as a series of performances, enactments, interpellations, and rhetorics that over and over again reinscribe some people as dominant and others as exploitable and/or expendable. Here, we unpack each of the three characteristics just described as they have been articulated in the literature.

First, in its broadest sense, energy coloniality refers to discursive and systemic efforts to subjugate certain peoples and lands in order to cheaply and quickly extract energy sources, especially fossil and nuclear fuels, that will most likely be consumed elsewhere (de Onís, 2018). Voices of protest and grassroots community organizing are frequently silenced or dismissed. Work on energy coloniality shows that this subjugation and suppression can be enacted differently at different scales. Work on nuclear coloniality, for example, reminds us that colonialism can be enacted at the national scale: Scholars have explored how the sovereignty of native and indigenous peoples can be ignored or exploited in order to extract and process nuclear fuels. For instance, "nuclear colonialism" refers primarily to the extraction of uranium in the service of national nuclear programs in the Global North. These programs—often justified using energy independence discourse and narratives of progress—have had long lasting, deleterious impacts on Native Americans and other indigenous groups, as well as on disenfranchised groups in developing countries (Hecht, 2012).

Other forms of environmental justice scholarship remind us that coloniality can be enacted at local and regional scales, such as when certain groups are treated as "low-use populations" and their lands are designated "national sacrifice zones." For example, Danielle Endres (2009) has written about how American Indian perspectives and sovereignty have been systematically excluded through colonialist practices of nuclearism, which have allowed the nuclear industry to profit specifically at the expense of indigenous public health and lands (see also Churchill & LaDuke, 1986; Edwards, 2011). The notion of sacrifice zones has also characterized oil and gas operations; Gwen Ottinger (2013) has written an in-depth study of how fenceline communities in Louisiana's "Cancer Alley" have worked to organize themselves in opposition to oil and gas refineries and their polluting behaviors over decades yet have faced steady opposition from experts in federal agencies who refused to acknowledge their efforts at data gathering.

Finally, coloniality can also happen at a macroscale, such as when those in the Global North reap the benefits of burning inexpensive and broadly available fossil fuels—as well as enjoy the profits from the extraction and distribution of those fuels—yet deny their responsibilities to and for global forms of pollution, such as carbon pollution. Scholars of "carbon colonialism" remind us that the many forms that subjugation takes under energy coloniality are made possible by the willingness of some groups to see others as less deserving of energy privilege and, in many cases, as less than American, less than human, more able to withstand hardship and deprivation (de Onís, 2018).

In addition to the exploitation of certain people and lands, a second characteristic of energy coloniality is that it privileges what might be called "legacy fuel sources"—primarily coal, oil, gas, and nuclear—and insists that these energy sources are universally appropriate regardless of context. In her work on emergency management discourses in Puerto Rico, de Onís (2018) argues that universal energy solutions are often justified by reading "certain places as exploitable, anachronistic, and in need of development and oversight" (p. 15). As is the case with many colonizing discourses under neoliberalism, such arguments often paternalistically frame people in the Global South or in marginalized communities as needing particular interventions over others, and those interventions are almost always decided by outside energy actors who benefit economically from their implementation. De Onís (2018) argues that energy privilege draws on "colonial logics that legitimize top-down, outsider-knows-best interventions to control local governments, places, and peoples" (p. 3). Such logics draw on a long history of such practices, which have been carefully documented by Development Studies scholars (e.g., Sachs, 1992/2010; Ziai, 2017).

The third characteristic of energy coloniality is that, according to de Onís (2018), it is often deeply intertwined with "energy privilege," in which "some individuals and communities thrive at the expense of others" (p. 5). For example, some communities of color have to expend much more labor or resources to secure energy and may also disproportionately bear the burdens of the pollution created by energy extraction and production (de Onís, 2018). Energy coloniality more broadly refers to an "extractivist system and discourse [that] marks certain places and peoples as disposable by importing and exporting logics and materials to dominate various energy forms, ranging from humans to hydrocarbons" (de Onís, 2018, p. 1). It also circulates the belief that some people deserve to have the lights on—and cheaply—while others must bear untenable environmental and economic expense. At the same time, it allows those in the Global North to refuse to commit to global climate treaties unless the Global South agrees to rely on less carbon-intensive forms of energy (Batel & Devine-Wright, 2017, p. 6).

The argument of this chapter is that Trump's Energy Dominance in the United States continued practices of energy coloniality, such as we have seen in the American southwest under nuclear colonialism and in Puerto Rico, as outlined by de Onís. Such practices have a long history. But Energy Dominance is energy coloniality on steroids; its embrace of explicit rhetorics of domination and privilege mark it as unique to our historical moment. In the following sections, we examine how Energy Dominance both relies on neoliberal logics and buttresses a completely illogical approach to energy markets in the United States. Using the rhetoric of realism, Energy Dominance advances nonsensical approaches to bolstering the faltering American coal market, for example, while at the same time spreading absurd falsehoods about renewable energy sources. Second, while more recent energy discourses such as Energy Independence and Energy Security have maintained cover for problematic practices of energy coloniality, Energy Dominance, by contrast, explicitly names and defends social structures of inequality and exploitation and celebrates them. It echoes other types of chauvinist rhetoric that have gained purchase in contemporary American political discourse. And finally, Energy Dominance uses Trump's America First doctrine to link American energy privilege with American exceptionalism in order to underscore widespread sentiments of resentment and entitlement among Trump supporters who argue they have been disenfranchised by environmentalism, globalization, and liberalism.

Discourses of Energy Dominance

What marks Energy Dominance under Trump as novel is the utter refusal to acknowledge the unmistakable headwinds demanding new responses at this particular historical moment: the increasingly visible threats of climate change, the demands of the global community for the United States to meaningfully address its role in carbon emissions, and calls to distribute the control of and responsibility for energy systems more equitably (Aldy, 2017). It also does so explicitly by employing rhetorics of coloniality. The Trump administration's version of energy coloniality emphasized the necessity and logic of fossil fuels, arguing that previous deviations from that energy path have resulted in a loss of privilege for a chosen few and, most notably, that the chosen few should be incensed by the loss. Those who are harmed by the current course of action (fossil fuel mining, processing, shipping, etc. or climate change) are either belittled, silenced, or erased from the discourse.

Privileging "universal" energy solutions or sources

The Trump administration maintained the dominance of fossil fuels by making traditional fuel sources appear to be the only "reasonable" option, berating those who disagree and packaging this traditional energy plan as "revolutionary."

Discourses of common sense

The arguments for Energy Dominance frequently invoke the high ground of common sense, among other realist rhetoric claims.

> Realism names any view that emphasizes the existence and reality of some kind of object about which humans can make true or false statements. Politically, this amounts

> to individuals who claim unmediated access to "reality", producing a discourse that denies its own politics and discursive character.
>
> *(Reyes, 2006, p. 573)*

President Trump's realist rhetoric often took the form of polarized characterizations: real or fake, true or false, common sense or fanciful. His remarks from the "Unleashing American Energy" event ("Remarks by President Trump," 2017) are characteristic of this rhetorical polarization. After explaining that the American "energy crisis" was a "big, beautiful myth," he contended: "The *truth* is that we have near-limitless supplies of energy in our country. Powered by new innovation and technology, we are now on the cusp of a *true* energy revolution" ("Remarks by President Trump," 2017; our italics). Similarly, in Vice President Pence's short address to the Oil and Gas Association Annual Meeting in 2019, he established "the truth" no less than seven times. For example, while condemning the state of New York for their moratorium on natural gas fracking, he states "the truth is, property values fell by more than 20% because of the decisions that were made" ("Remarks by Vice President Pence," 2019). Similarly, in stating his support for replacing the Clean Power Plan, Rep. Doug LaMalfa (R-CA) stated, "The [Trump] EPA's new proposal is a far more *realistic strategy* that will more accurately interpret the bounds of the Clean Air Act and improve efficiency at power plants, all while empowering states and helping America achieve its goal of energy dominance" ("WTAS," 2018; our italics). Expressing his support for President Trump's State of the Union, Rep. Ken Calvert (R-CA) argued, "There's no doubt that we live in a polarized political environment right now, but if you stop and focus on the actual policies the President is outlining you will see that these are generally *common-sense ideas* that have broad support in our country" ("WTAS," 2018).

For realists, according to Reyes (2006), "the fools are the idealists, the believers of myth who have erringly wedded their political views with ethical ideals. Idealists believe so fully in their ideals that they cannot see reality for what it is—they are betrayed by their ideology, which distorts politics and obscures political truths" (p. 575). Former Secretary of the Interior Zinke exhibited this realist perspective when he called out the "foolish" actions of the Obama administration, stating, "Our nation can't run on pixie dust and hope. And the last eight years showed that" (as cited in Hejny, 2018). Commonsense realism, of course, is almost always in support of fossil fuels' continued control of American energy.

As with other energy coloniality rhetorics, maintaining adherence to, if not enlarging dependence on fossil fuels is the primary goal. In a complementary approach, fossil fuel organizations have attempted to show economic and health *benefits* to those regions and individuals most likely to be harmed by continued fossil fuel energy consumption, production, and climate change (see the "Energy Utopia" chapter in Schneider et al., 2016). The adherents to Energy Dominance take a different tack, arguing that no other energy future is possible, and that only fossil fuels are reasonable, certain, or viable. Making the continued use of coal, gas, and oil use appear "inevitable" is one of the strongest rhetorical strategies utilized by the fossil fuel industries, according to Bill McKibben (for a further discussion of this strategy, see Bsumek et al., 2019). Without question Energy Dominance works to dismiss any other conceivable energy path forward by employing a "rational" approach and declaring that anyone who puts forth a competing vision is foolish or naïve.

Reframing traditional fuels as revolutionary

In an interesting twist, fossil fuels were frequently framed by the Trump administration as common sense but at the same time revolutionary. While signing an executive order on

Energy and Infrastructure in Texas, President Trump boasted, "America's energy revolution has produced affordable, reliable energy for consumers along with stable, high-paying jobs for small businesses—all while dropping carbon emissions to their lowest level in 25 years" ("Remarks by President Trump at signing of executive order," 2019). Vice President Pence concurred in his speech to the Oil and Gas Association Annual Meeting, stating, "And the energy revolution President Trump promised is going strong. Today, I'm here to report on that revolution, but mostly to say thank you. Thank you to all of you who are the foot soldiers in that revolution—the great oil and gas producers of the great state of Ohio" ("Remarks by Vice President Pence," 2019). The energy revolution—though paradoxically hearkening back to a mythical past in which fossil fuels were not contested—promised to overthrow the Obama-era environmental regime, thus uniting its opponents in common cause as partisans.

In addition to the energy "revolution," The Trump administration trumpeted "*renewed* American growth" ("Remarks by Vice President Pence," 2019), the "*revival* of the American energy industry and jobs" ("Remarks by President Trump," 2019), and an American energy *renaissance* ("Remarks by Vice President Pence, 2019). "And this is only the beginning," the President said. "The Golden Era of American Energy is now underway—and all of you will be a part of creating this exciting new future" ("President Trump Vows," 2017). In addition, then Secretary of the Interior Zinke (2017) argued, "This administration and the President believe in American energy *dominance*. . . . Our goal is an America that is the strongest energy superpower this world has ever known." The reference to America as a superpower was also propagated by the Heartland Institute, which argued, "In short, the U.S. has once again become an energy superpower, and the old mantra of 'energy independence' has been replaced by the new mantra of 'energy dominance'" (Weinstein, 2017).

Yet the call to revolution—overthrowing the dominant regime—was primarily used to maintain the status quo. Instead of working for energy innovation, President Trump's energy policies are best characterized as a doubling down on traditional fossil fuel production and consumption along with a call to the political right to renew its commitment to neoliberalism (Schneider & Peeples, 2018). Far from a "revolution," Energy Dominance is better characterized as devolution, signaling a rollback of any environmental policy that curtails the energy industry. It nostalgically reinscribed the values of energy coloniality by shoring up even fossil fuels that are no longer economically competitive, such as coal. Promising a return to a "Golden Era of American Energy" or to the era of national "superpowers" harkened back to a time when coal jobs might have been plentiful, environmental regulation irrelevant, and oppositional voices easily subdued. As with other revivalist movements (Stewart et al., 2001), the Energy Dominance discourse pointed to a past, lost paradise that its adherents maintain is worth resurrecting. The revolution must look backward in order to move forward. Only "moving forward" really means propping up the same old powerful players who have set the terms for energy development in the United States for decades.

It is worth noting that this is not a classically conservative argument about unfettering markets but rather a promise to intentionally boost certain fuel sources over others. As with traditional forms of colonial and imperial nostalgia, which are used to cover historical harms and maintain current injustices, energy nostalgia—even when cloaked in the language of revolution—is thus a particularly important piece of the dominance rhetoric. It seeks to rally certain audiences around a shared sense of a golden past lost, thus triggering a sense of grievance, the need for retribution, and calls to restore a sense of greatness.

Privilege and erasure

Energy coloniality requires the bolstering of the rights and/or privileges of certain people while subjugating others. The approach used by the Trump administration was not simply content to strengthen existing privileges for those who benefit from cheap and reliable energy but worked to create a sense of entitlement and grievance used to enliven political support. Turning away from fossil fuels, the administration argued, has diminished what is "owed" to those who have benefitted from high fossil fuel consumption but who perceive that they have suffered under climate change–focused policies and plans. Energy Dominance discourses name enemies and others who are trying to take away certain privileges of identity and nation. At the same time, the administration silenced and/or rhetorically erased the people and places whose subjugation was necessary for those in socially dominant positions to maintain them.

Constructing the aggrieved

Energy Dominance rhetoric rarely explicitly mentions identity politics in terms of race and gender relations. But we argue that the rhetoric of Energy Dominance clearly speaks to the same sense of grievance that nationalist, racist, and misogynist rhetorics are meant to appeal to under Trumpian populism. "[P]opulism has been mobilized in contemporary conservative politics in the United States to blame economic and social woes on government bureaucrats, immigrants, racial others, and environmental regulations while claiming to stand up for working-class and rural people" (Kojola, 2019, p. 373). Fulfilling this function, discourses of Energy Dominance reinforce the web of identities, meanings, and symbols that fossil fuel industries have built up around their products; in particular, American energy has been synonymized with a neoconservative "American identity," one that is primarily working- or middle-class, heterosexual, and white (Bsumek et al., 2014; Scott, 2010). As such, energy politics is identity politics.

Different types of "chauvinisms" often circulate together rhetorically. Casey Schmitt (2019) argues that "ecochauvinism"—the belief that nature is to be dominated or subdued by humans and that articulates an "ideological resonance between Trumpian conservatism and anti-environmentalism"—frequently echoes and props up nationalist chauvinism, male chauvinism, and so on. We see one form of ecochauvinism—an extreme form of energy privilege—articulated by President Trump when he appealed to miners in his championing of Energy Dominance. During remarks at the signing of the Executive Order on Energy and Infrastructure, which was intended to speed up oil and gas pipeline projects ("Remarks by President Trump," 2019), President Trump surrounded himself with workers in hard hats and reflective vests. He opined, "As I've said from day one, American labor will always have a friend in the White House." He continued, "And we're replacing the previous administration's job-crushing Clean Power Plan and putting our miners back to work" ("Remarks by President Trump," 2019). It is significant that President Trump visually and verbally invokes miners. Scholars such as Kojola (2019) and Scott (2010) have argued that mining frequently is used to articulate commitments to traditional gender norms, racial configurations, and economic entanglements. "Mining is often made meaningful through narratives of masculine labor and nostalgia for when white men could depend on the stability of industrial jobs to sustain vibrant communities" (Kojola, 2019, p. 373).

Energy dominance speaks to and for those on the right who have felt wronged by a perceived loss of power, influence, and privilege as a result of demographic and political changes that have taken place during the last few decades, seemingly exacerbated by eight years of a liberal, African American president (King, 2017; Lopez, 2015). Energy Dominance discourse is thus best understood as a manifestation and articulation of these politics of grievance. As we have argued elsewhere, "It [Energy Dominance] functions as a moral call-to-arms for conservatives to come out from the trials under which they had been tested during the Obama administration, drawing on discourses of American exceptionalism, militarism, and gender, race, and class resentment and grievance" (Schneider & Peeples, 2018, p. 85).

This grievance is often articulated as "violence" perpetuated against historically and/or culturally privileged groups of Americans. President Trump railed against the regulatory violence perpetrated by the Obama administration: "We're ending intrusive EPA regulations that kill jobs, hurt family farmers and ranchers, and raise the price of energy so quickly and so substantially" ("Remarks by President Trump," 2017). He repetitively referred to environmental regulations and the Paris Climate Accord as "job killing" and to President Obama's Clean Power plan as "job crushing." And of course, Energy Dominance repeatedly calls our attention to the violence associated with the "war on American energy," and most significantly, the "war on coal" ("Remarks by President Trump," 2019). "The coal industry seethes with symbolism," wrote journalist Jonathan Thompson (2017). "When Trump demonstrates that he 'digs coal' by rolling back regulations, he's banking on rural nostalgia and pushing back against Obama, who for portions of white America became a symbol of urban elitism, progressivism and blackness." Those named as victims of violence in the discourses of Energy Dominance are the working class and rural people, who, Kojola (2019) argues, would be most influenced by populist appeals: miners, steelworkers, patriots, working families, small businesses, family farms, and ranchers. "Environmental protections are presented as hurting working-class people and threatening the masculinity of male workers in rural extractive industries" (Kojola, 2019, p. 373).

The list of "villains," those perpetrating this violence against the working people of America, was not limited to the Obama administration. Before signing a bill in Texas at the International Union of Operating Engineers International Training and Education Center, President Trump said, "Too often, badly needed energy infrastructure is being held back by special interest groups, entrenched bureaucracies and radical activists" (Lam, 2019). Foreign governments were also suspect. Former Secretary of the Interior Ryan Zinke wrote in an opinion piece in *The Boston Globe*, "Thanks to President Trump's policy of 'energy dominance,' America is now on a path to become a leading global energy exporter and free from being held hostage by foreign powers" (Zinke, 2018). While Zinke did not name the specific foreign powers, *Investor's Business Daily* contended, "America's new Energy Dominance also has important national defense and foreign policy implications. Energy production often comes from troublesome regimes from Russia to Saudi Arabia to Venezuela and others from the Middle East to Africa to South America" (Ferrara, 2018). President Trump pledged, "Our country will no longer be vulnerable to foreign regimes that use energy as an economic weapon" ("President Trump Vows," 2017). We see this same set of enemies—bureaucrats, leftist radicals and environmentalists, and our international partners—named repeatedly by Trump and his associates as worthy of attack and even violence. Trump and Zinke have consistently pointed to these groups as "victimizing" Americans and have then positioned themselves as ideal saviors who can retaliate, hitting back twice as hard (Schneider & Peeples, 2018).

As with other forms of energy coloniality, one of the key elements of Energy Dominance is polarization: separating the "good" Americans (white, masculine, working-class, conservative) from their internal and external enemies, who are lesser or suspect. This rhetorical approach is an amalgamation of populist rhetoric, which is used to differentiate "the people" from radicals, special interests, and bureaucrats, mixed with American exceptionalism, which positions the United States as superior and separate from all other nations.

In contrast to Energy Dominance, other traditional fossil fuel strategies frame the victims of energy coloniality as in some ways benefiting from fossil fuels. Rhetorics of energy poverty and energy utopia justify the development of international markets for their products (Schneider, et al., 2016). They promise that local communities will benefit from development, using narratives of progress, minimizing environmental damage and evading accountability for remediation, and conflating energy access with energy affordability. Multinational energy companies—and particular coal companies, reacting to the collapse of American domestic markets—have begun to prime markets abroad for such interventions.

Energy Dominance discourses focus instead on the tensions and divisions *within* the United States. Energy Dominance erases any vestiges of the colonized, those people and places imperative to maintain American energy supremacy. Their contamination from fossil fuel extraction, their status as climate refugees, or even their role in consuming US fuel exports that are necessary to maintain the perception of "dominance" are absent from the discourse. We argue this is even more damaging than reframing fossil fuel victims as "benefitting" from the status quo. At least in that energy colonialized narrative, those harmed by climate change and fossil fuel extraction are made visible.

National exceptionalism

In a speech given to the Heritage Foundation, former Secretary of the Interior Ryan Zinke argued that Energy Dominance is different from "energy independence" because it recognizes that "America is exceptional" (Zinke, 2017). According to the Trump administration, America's exceptionalism does not end with its energy prowess; it is superior in other aspects as well. During his remarks at the Signing of Executive Order on Energy and Infrastructure, President Trump argued that, in addition to American fossil fuel dominance:

> we're strongly protecting the environment. We have to protect the environment. The United States has among the very cleanest air and water developments in the world. And also we have the cleanest air and water, they say, in the world. We are the best. . . . And we also have, right now, the strongest, by far, economy anywhere in the world. We're the envy of the world.
>
> *("Remarks by President Trump," 2019)*

Whereas energy security promised to protect Americans from energy shocks through protectionism, Energy Dominance promises protection through aggressive movement into global markets while refusing to cede any ground through international agreements, such as the Iran nuclear deal or the Paris Climate accords. As previously articulated, America's exceptionality precludes the United States from needing such international agreements—the country has never been better, the argument goes.

Energy Dominance also posits that the United States should be insulated from vulnerability abroad, but that the country should have unfettered access to and dominance of "global markets," without paying the cost of externalities, such as a degraded environment in the United States or climate change. Energy dominance rhetorics stand in sharp contrast to the majority of U.S and world citizens, especially those engaged in energy democracy movements, who understand both climate change and exclusionary nations as having a direct and deleterious effect on their current and future existence. As we can see, the energy coloniality embedded in the

Trump administration's Energy Dominance discourse supplants the victims of traditional fossil fuels and climate change with the aggrieved white, male, American worker, thereby erasing the plight of those unjustly and disproportionately affected by this administration's decisions.

Conclusion

The argument of this chapter is that Energy Dominance, as discourse and practice, reinscribes energy coloniality. Energy Dominance can be defined as the effort to distribute natural resources quickly and cheaply to some people, at great cost to others, along with all of the political and economic relationships that enable coloniality more generally: the silencing of dissent; the erasure of indigenous sovereignty and the explicit silencing of protest; the propping up of energy privilege; the push for rapid and fossil fuel and nuclear development, where protest is silenced or made illegal; and the accumulation of profit among the few. We have examined how each is enacted through Energy Dominance policies. Seen in this light, Energy Dominance can be understood as an evolving feature of Western energy politics and practices.

Energy Dominance—as practiced by the Trump administration—operationalizes these practices and values, using rhetorics of grievance and privilege to justify them. Undergirding Energy Dominance is the belief that most Americans have a right to affordable, reliable energy and that they should not have to pay for or even be aware of the externalities the production of this energy might create. Similarly, certain Americans—in particular, those representing fossil fuel industries—also have a right to extractive practices free from regulation or protest, as well as the freedom to reassert their *right to dominate* those they oppose. They also have a right to "dominate" global markets, coupled with the right to refuse to participate in international climate and trade agreements. Energy Dominance therefore relies heavily on a belief in Americans' "energy privilege" (de Onís, 2018). Energy privilege rhetoric under Energy Dominance circulates with and amplifies other types of grievance politics, which explicitly resist calls to account for unjust and colonialist practices in American society.

References

Aldy, J. E. (2017). Real world headwinds for Trump climate change policy. *Bulletin of the Atomic Scientists*, *73*(6), 376–381.

Bacon, J. M. (2019). Dangerous pipelines, dangerous people: Colonial ecological violence and media framing of threat in Dakota access pipeline conflict. *Environmental Sociology*, *6*(2), 143–153. https://doi.org/10.1080/23251042.2019.1706262

Batel, S., & Devine-Wright, P. (2017). Energy colonialism and the role of the global in local responses to new energy infrastructures in the UK: A critical and exploratory empirical analysis. *Antipode*, *49*(1), 3–22. https://doi.org/10.1111/anti.12261

Bordoff, J. (2017, July 6). The American energy superpower: Why dominance is about more than just production. *Foreign Affairs*. www.foreignaffairs.com/articles/united-states/2017-07-06/american-energy-superpower

Bsumek, P. K., Schneider, J., Schwarze, S., & Peeples, J. (2014). Corporate ventriloquism: Corporate advocacy, the coal industry, and the appropriation of voice. In J. Peeples & S. Depoe (Eds.), *Voice and environmental communication* (pp. 21–43). Palgrave Macmillan.

Bsumek, P. K., Schwarze, S., Peeples, J., & Schneider, J. (2019). Strategic gestures in Bill McKibben's climate change rhetoric. *Frontiers in Communication*. https://doi.org/10.3389/fcomm.2018.00002

Churchill, W., & LaDuke, W. (1986). Native America: The political economy of radioactive colonialism. *Insurgent Sociologist*, *13*(3), 51–78. https://doi.org/10.1177/089692058601300306

Climate Deregulation Tracker. (n.d.). *Sabin center for climate change law*. Columbia Law School and Columbia University Earth Institute. Columbia University. https://climate.law.columbia.edu/climate-deregulation-tracker

de Onís, C. M. (2017). *Energy remix: Decolonial discourses of decarbonization*. Indiana University.
de Onís, C. M. (2018, January 29). Energy colonialism powers the ongoing unnatural disaster in Puerto Rico. *Frontiers in Communication*. www.frontiersin.org/articles/10.3389/fcomm.2018.00002/full
Dolata, P. (2017). The end of the energy security paradigm under Obama—Trump and energy security: Revival of an old concept? *Journal of Military and Strategic Studies, 17*(3). https://jmss.org/article/view/58240/pdf
Drezner, D. W. (2019). Present at the destruction: The Trump administration and the foreign policy bureaucracy. *The Journal of Politics, 81*(2), 723–730.
Duke, R. (2019, May 19). Leaving the Paris agreement is a bad deal for the United States. *Foreign Policy*. https://foreignpolicy.com/2019/05/19/leaving-the-paris-agreement-is-a-bad-deal-for-the-united-states/
Edwards, N. (2011). Nuclear colonialism and the social construction of landscape in Alaska. *Environmental Justice, 4*(2). www.liebertpub.com/doi/full/10.1089/env.2010.0023?casa_token=0zdGXvikgJcAAAAA%3A8B2UWvPqCh3Kl8fcoDPaFV8bhdvO2ZO_dOPvpOF1Ez-aDEpE07J87PzQDaUgAzRpg0GSvzAlt3xJ0TQ
Eilperin, J., & Cameron, D. (2018, January 20). How Trump is rolling back Obama's legacy. *The Washington Post*. www.washingtonpost.com/graphics/politics/trump-rolling-back-obama-rules/?utm_term=.711cb4d225cb
Eilperin, J., Dennis, B., & Dawsey, J. (2019, September 26). EPA tells California it is 'failing to meet its obligations' to protect the environment. *The Washington Post*. www.washingtonpost.com/climate-environment/epa-tells-california-it-is-failing-to-meet-its-obligations-to-stem-water-pollution/2019/09/26/b3ffca1e-dfac-11e9-8dc8-498eabc129a0_story.html
Elmer, G., & Todd, P. (2016). Don't be a loser: Or how Trump turned the Republican primaries into an episode of *The Apprentice*. *Television and New Media, 17*(7), 660–662.
Endres, D. (2009). The rhetoric of nuclear colonialism: Rhetorical exclusion of American Indian arguments in the Yucca Mountain nuclear waste siting decision. *Communication and Critical/Cultural Studies, 6*(1), 39–60. https://doi.org/10.1080/14791420802632103
Ferrara, P. J. (2018, December 3). Trump's policies bring not just energy independence, but energy dominance. *Investor's Business Daily*. www.investors.com/politics/commentary/energy-dominance-trump-policies/
Friedman, L. (2018a, January 9). How a coal baron's wish list became President Trump's to-do list. *The New York Times*. www.nytimes.com/2018/01/09/climate/coal-murray-trump-memo.html
Friedman, L. (2018b, October 11). EPA to disband a key scientific review panel on air pollution. *The New York Times*. www.nytimes.com/2018/10/11/climate/epa-disbands-pollution-science-panel.html
Gibbens, S. (2019, February 1). 15 ways the Trump administration has changed environmental policies. *National Geographic*. www.nationalgeographic.com/environment/2019/02/15-ways-trump-administration-impacted-environment/
Gufstafson, A., Leiserowitz, A., & Maibach, E. (2019, February 12). Americans are increasingly 'alarmed' about global warming. *Yale Program on Climate Change Communication*. https://climatecommunication.yale.edu/publications/americans-are-increasingly-alarmed-about-global-warming/
Hall, K., Goldstein, D. M., & Ingram, M. B. (2016, Autumn). The hands of Donald Trump: Entertainment, gesture, spectacle. *HAU: Journal of Ethnographic Theory, 6*(2), 71–100.
Hecht, G. (2012). *Being nuclear: Africans and the global uranium trade*. The MIT Press.
Hejny, J. (2018). The Trump administration and environmental policy: Reagan redux? *Journal of Environmental Studies and Sciences, 8*(2), 197–211. https://doi.org/10.1007/s13412-018-0470-0
Kederer, K. (2018, July 23). What is energy dominance? The Trump administration off the leash and unleashing. *N+1 Magazine*. https://nplusonemag.com/online-only/online-only/what-is-energy-dominance/
King, M. (2017). Aggrieved whiteness: White identity politics and modern American racial formation. *Abolition Journal, 1*(1). https://abolitionjournal.org/aggrieved-whiteness-white-identity-politics-and-modern-american-racial-formation/
Kojola, E. (2019). Bringing back the mines and a way of life: Populism and the politics of extraction. *Annals of the American Association of Geographers, 109*(2), 371–381.
Lacatus, C. (2018, January 12). For Donald Trump, campaigning by Twitter limited the public's access to his policy positions and strategies. *American Politics and Policy Blog*. http://blogs.lse.ac.uk/usappblog
Lam, K. (2019). Trump brushes aside environmental concerns, signs 2 executive orders to speed up gas, oil projects. *USAToday*. www.usatoday.com/story/news/nation/2019/04/10/president-trump-orders-speed-oil-gas-pipeline-projects/3431466002/

Lavelle, M. (2017, March 18). Trump's executive order: More fossil fuels, regardless of climate change. *Inside Climate News*. https://insideclimatenews.org/news/28032017/trump-executive-order-climate-change-paris-climate-agreement-clean-power-plan-pruitt
Layzer, J. A. (2014). *Open for business: Conservatives' opposition to environmental regulation*. The MIT Press.
Lopez, I. H. (2015). *Dog whistle politics: How coded racial appeals reinvented racism and wrecked the middle class*. Oxford University Press.
Mooney, C., & Eilperin, J. (2017, June 20). EPA just gave notice to dozens of scientific advisory board members that their time is up. *The Washington Post*. www.washingtonpost.com/news/energy-environment/wp/2017/06/20/trump-administration-to-decline-to-renew-dozens-of-scientists-for-key-epa-advisory-board/
Ott, B. L. (2017). The age of Twitter: Donald J. Trump and the politics of debasement. *Critical Studies in Media Communication*, *34*(1), 59–68.
Ottinger, G. (2013). *Refining expertise: How responsible engineers subvert environmental justice challenges*. New York University Press.
Pfiffner, J. P. (2017, September). The unusual presidency of Donald Trump. *Political Insight*, *8*(2), 9–11.
Popovich, N., Albeck-Ripka, L., & Pierre-Louis, K. (2019, December 21). 95 environmental rules being rolled back under Trump. *The New York Times*. www.nytimes.com/interactive/2019/climate/trump-environment-rollbacks.html
President Trump Vows to Usher in Golden Era of American Energy Dominance. (2017, June 30). www.whitehouse.gov/articles/president-trump-vows-usher-golden-era-american-energy-dominance/
Purdum, T. S. (2019, September 28). Trump's attacks on California are shortsighted. *The Atlantic*. www.theatlantic.com/politics/archive/2019/09/attacking-california-easy-trump-bait/598915/
Regulatory Rollback Tracker. (2020). *Environmental & energy law program*. Harvard University. https://eelp.law.harvard.edu/regulatory-rollback-tracker/
Remarks by President Trump at Signing of Executive Order on Energy and Infrastructure. (2019, April 10). www.whitehouse.gov/briefings-statements/remarks-president-trump-signing-executive-order-energy-infrastructure-crosby-tx/
Remarks by President Trump at the Unleashing American Energy Event. (2017, June 29). www.whitehouse.gov/briefings-statements/remarks-president-trump-unleashing-american-energy-event/
Remarks by President Trump in the State of the Union Address. (2019, February 15). www.whitehouse.gov/briefings-statements/remarks-president-trump-state-union-address-2/
Remarks by Vice President Pence at the 2019 Oil and Gas Association Annual Meeting. (2019, March 8). www.whitehouse.gov/briefings-statements/remarks-vice-president-pence-2019-oil-gas-association-annual-meeting-columbus-oh/
Reyes, G. M. (2006). The swift boat veterans for truth, the politics of realism, and the manipulation of Vietnam remembrance in the 2004 presidential election. *Rhetoric and Public Affairs*, *9*(4), 571–600.
Sachs, W. (1992/2020). *The development dictionary: A guide to knowledge as power*. Zed Books.
Schmitt, C. (2019). *Ecochauvinism: An ideological resonance between Trumpian conservatism and anti-environmentalism* [Conference Paper]. Presented at the Western States Communication Association Convention.
Schneider, J., & Peeples, J. (2018, February 6). The energy covenant: Energy dominance and the rhetoric of the aggrieved. *Frontiers in Communication*. www.frontiersin.org/articles/10.3389/fcomm.2018.00005/full
Schneider, J., Schwarze, S., Bsumek, P., & Peeples, J. (2016). *Under pressure: Coal industry rhetoric and neoliberalism*. Palgrave Macmillan.
Scott, R. (2010). *Removing mountains: Extracting nature and identity in the Appalachian coalfields*. University of Minnesota Press.
Shannon, J. (2019, March 29). 'I know a lot about wind,' Trump says: A government FAQ proves he doesn't. *USA Today*. www.usatoday.com/story/news/politics/2019/03/28/trump-criticism-wind-power-dismissed-department-energy-faq/3307863002/
Statement by President Trump on the Paris Climate Accord. (2017, June 1). www.whitehouse.gov/briefings-statements/statement-president-trump-paris-climate-accord/
Stewart, C., Smith, C. A., & Denton, Jr., R. E. (2001). *Persuasion and social movements*. Waveland Press, Inc.
Taddonio, P. (2019, October 29). Bob Murray helped shape Trump's energy policies. Now, his coal company is facing bankruptcy. *PBS Frontline*. www.pbs.org/wgbh/frontline/article/bob-murray-helped-shape-trumps-energy-policies-now-his-coal-company-is-facing-bankruptcy/
Thompson, J. (2017, September 21). Coal. Guns. Freedom? *High Country News*. www.hcn.org/articles/coal-how-the-trump-administration-has-seized-mythologies-around-coal

Thrower, S. (2018). Policy disruption through regulatory delay in the Trump administration. *Presidential Studies Quarterly*, *48*(3), 517–536.
Tracking deregulation in the Trump Era. (2020, January 31). *Brookings Institute*. www.brookings.edu/interactives/tracking-deregulation-in-the-trump-era/
Wald, E. R. (2019, December 30). Energy dominance is leverage for US foreign relations. *The Hill*. https://thehill.com/opinion/energy-environment/475854-energy-dominance-is-leverage-for-us-foreign-relations
Weinstein, B. (2017). Federal lease sales critical to maintaining America's energy dominance. *Heartland Institute*. www.heartland.org/news-opinion/news/federal-lease-sales-critical-to-maintaining-americas-energy-dominance
WTAS: Support for the Trump Administration's Proposal to Replace the Costly and Overreaching Clean Power Plan. (2018, August 22). www.whitehouse.gov/briefings-statements/wtas-support-trump-administrations-proposal-replace-costly-overreaching-clean-power-plan/
Ziai, A. (2017). Post-development 25 years after the development dictionary. *Third World Quarterly*, *31*(12), 2547–2558.
Zinke, R. K. (2017, September 29). *A vision for American energy dominance*. www.heritage.org/event/vision-american-energy-dominance
Zinke, R. K. (2018, April 16). American energy dominance means mass wind. *The Boston Globe*. https://www2.bostonglobe.com/opinion/2018/04/15/american-energy-dominance-means-massachusetts-wind/xJgMsiIbTAZRYVX9CuWdNM/story.html

13
PART II RESPONSE

Danielle Endres and Stephanie L. Gomez

The chapters in this part of the handbook bring up a number of important ideas and themes that help us to better understand discourses and keywords related to energy democracy, as well as the various ways that energy democracy has been framed. In this brief response, we will begin by noting some themes that emerged across the chapters, particularly as related to the justice, participation, power, and technology framework introduced in Chapter 1. Then we will lay out some future directions for ED research and practice that relate to discourse.

Themes

Perhaps most notably, each chapter engages with notions of power: not only how power influences the development of energy democracy, but also how it influences the ways that publics understand and participate in it. For instance, Demski and Becker (Chapter 9) review research about public attitudes toward and engagement with energy security to reveal that people with different levels of knowledge and expertise regarding energy issues define energy security in very different ways. Moreover, this suggests that energy security and, relatedly, energy democracy require greater levels of transparency and communication, along with a flattening of hierarchical power structures. Cozen (Chapter 10) also engages with issues of power, particularly regarding who has the power to define the role that energy should play in everyday life. Cozen critically assesses the assumption, upheld through a variety of media texts, that energy is the key to human flourishing, which suggests that energy poverty is an exigent problem that needs to be solved for the good of humanity through ensuring future access to status quo levels of energy. Cozen draws attention to the ways that these discourses about energy become reified through unequal power relations, with energy companies having the power to define the importance of energy—and hence the importance of rectifying energy poverty. Batel (Chapter 11) also focuses on questioning taken-for-granted assumptions, particularly regarding renewable energy. Indeed, rather than supporting the belief that renewable energy is inherently democratic, Batel explores the ways that renewable energy systems can enact energy colonialism. Looking at a variety of scalar dimensions, Batel uncovers the ways that large, centralized institutions, such as governmental agencies and the media, have enacted power over communities, imposing renewable energy colonialism against the objections of community members. Laying bare these power relations asks us to question the ways that energy democracy is employed; even if on the surface

DOI: 10.4324/9780429402302-15

these processes appear democratic, it is important to analyze the power relations that undergird energy systems of all kinds.

Schneider and Peeples (Chapter 12) also question power relations, particularly within the context of American exceptionalism. With a specific focus on the former Trump administration, Schneider and Peeples carefully analyze the ways that US energy policy has become an agent of domination with a win/lose orientation toward energy systems. Through discourses of energy domination, Schneider and Peeples reveal that energy production and consumption under Trump was blatantly opposed to ideals of energy democracy and justice.

The process of achieving environmental and energy justice, as these chapters underscore, is only possible through an interrogation of the power structures and relations at play in the production and consumption of energy. Yet these chapters all argue that justice is something for which we should strive. The keywords of "energy security" (Demski & Becker, Chapter 9), "energy poverty" (Cozen, Chapter 10), "energy colonialism" (Batel, Chapter 11), and "energy domination" (Schneider & Peeples, Chapter 12) all suggest problematics within energy democracy discourses that require untangling for environmental and energy justice to be achieved. In the case of Demski and Becker, energy security can contribute to a type of energy justice, with a focus on transparency and clear communication with publics. Cozen encourages us to move away from a frame of energy poverty—lack of energy as deficit—in order to work toward more just understandings of energy distribution. Batel argues for the need to question taken-for-granted practices of energy colonialism in order to achieve justice within energy systems and practices, and Schneider and Peeples similarly ask us to question rhetorics of energy dominance and American exceptionalism in the hopes of achieving both environmental and energy justice.

Ultimately, key to these chapters is the notion that words matter and that they have power—the power to define and frame energy discourses and the ways that we understand energy democracy and justice. How we frame energy and energy democracy has far-reaching implications, on a number of scales, for communities, publics, climate change, and the environment itself. And because words matter and have power, each chapter demands that we question the keywords and discourses that have historically and contemporarily been used to frame energy and energy democracy. It is only through rigorous questioning of taken-for-granted assumptions that we can achieve greater energy democracy and justice.

Future directions

The chapters in this section tell us that renewable energy technologies alone are not a panacea. As such, a future direction for further study of the discourses of energy democracy is to analyze the particular connections between renewable energies and discourses of energy colonialism, security, poverty, and dominance. While Schneider and Peeples articulate energy dominance as a form of energy colonialism, Batel introduces renewable energy colonialism as a way to focus on how renewable energies can also entrench colonial structures and systems transnationally, internationally, and intranationally, providing an important reminder that renewable energies are not guaranteed to be implemented in democratic, anticolonial, and just ways. In a different way, Cozen's analysis highlights how energy poverty lenses—even when linked with solar energy campaigns, for example—do not challenge a growth model of energy consumption, resulting in moral legitimation of fuel/power as a precondition for human achievement and the "good life." Renewables risk being implemented within a system that is wedded to an unsustainable growth model. Demski and Becker analyze the intersections between energy security—which is often conceived on a national level—and locally controlled renewable energy projects, noting that renewables can be incapable of providing the sort of security needed for a national system.

These chapters, then, suggest that an important future direction is to continue to interrogate the perhaps too easy connections made between renewable energies and democracy. Although energy democracy is often explicitly connected with renewable energy technologies, we cannot take for granted that these technologies are inherently more democratic; they still need to be implemented within broader discourses and systems of power than can hinder democracy, justice, and equity. Energy democracy, then, must include processes for framing and reframing discourse and actions toward more just, participatory, and power-responsive approaches to renewable energies.

A second future direction is simply to expand our analysis of the keywords that matter in energy transition. While the chapters in this section honed in on energy security, poverty, colonialism, and dominance, other terms and phrases come up frequently in deliberation about energy transition, including energy justice, just transition, energy commons, energy sovereignty, energy independence, and energy resilience. Each of these could be similarly unpacked via a keywords investigation, contributing to a better understanding of how each of these discourses frame energy transition in different ways with different assumptions and futures.

In closing, no terminology or phrase employed in the debates and discussion over energy transition should go unevaluated. As these chapters suggest, these keywords have histories, evolving meanings, and polysemous interpretations. Viewing them as discourses opens analysis of the ideologies, values, and assumptions about society that underlie these terms, as Cozen deftly highlights by showing how prevailing definitions of energy poverty are informed by assumptions about growth in energy demand. As we noted in the introduction to this part of the handbook (Chapter 8), in addition to critical assessment of the framings, histories, and assumptions that accompany the keyword phrases covered in these chapters, we would be remiss if we did not also take advantage of the inventional possibilities of discourse, framing, and terminologies. How might we compose new terminologies that challenge problematic assumptions and posit new ways of understanding the relations between energy and society?

PART III

Grassroots and critical modes of action

14

GRASSROOTS AND CRITICAL MODES OF ACTION

Introduction

Tarla Rai Peterson

Part III of the handbook focuses on grassroots and critical modes of action for enacting energy democracy by investigating diverse suites of options for fostering energy democracy in particular places. These modes of action suggest ways that such options contribute to democratically oriented compositions for energy transition and reveal how they can enable achievement of previously unrealized possibilities. Each chapter focuses on different ways to compose energy democracy via grassroots action. The strategies of action explored in these chapters make use of a variety of discursive patterns, including some of the keywords highlighted in the previous part (see Chapters 9–12). They also demonstrate opportunities and potential pitfalls of applying the participatory principles discussed in the upcoming part (Part IV; see Chapters 21–24).

In Chapter 15, Becker compares the opportunities and challenges of municipal and cooperative approaches to Germany's *Energiewende*, or transition to a low-carbon, environmentally, and economically sound energy system that is affordable for consumers. While Germany has attained international acclaim both for its concentrated rollout of renewable energy generation and for remunicipalizing its energy system, the systemic reorganization is likely more central to environmental democracy. Remunicipalization of energy infrastructures has meant that local states have acquired energy facilities that had drifted into corporate ownership in response to neoliberal pressures for privatization. Becker argues that a simultaneous increase in the number of member-based energy cooperatives has been at least as important as this return to state ownership and that the two models reflect persistent democratic debates between more anarchist proponents of social self-organization and those promoting strong engagement of the state in forming viable alternatives to capitalist dominion over public services. Becker's fieldwork indicates that while the cooperatives provided more opportunities for public participation among a select body of members, previously marginalized residents remained apprehensive about becoming involved. Although participants in the cooperative were able to use resistive power to transform particular hierarchical dimensions, the public utility provided an institutional anchor needed for scaling up democratic transformation of the energy system.

In Chapter 16, MacArthur and Tarhan provide both temporal and spatial context for Becker's study of contemporary strategies for strengthening environmental democracy in Germany. After reviewing the emergence of electricity cooperatives as part of an international movement for economic democracy, they describe opportunities and challenges facing contemporary cooperatives. What they term the "first wave" of energy cooperatives began emerging in the

DOI: 10.4324/9780429402302-17

1890s, as self-starting efforts to bring electricity to remote rural regions of Europe. A second set of electricity cooperatives, also focused on rural electrification, began developing in the United States during the 1930s. Unlike earlier European efforts, they relied on supportive state policies for both initiation and maintenance. Finally, contemporary cooperatives focus on developing new renewable energy generation from wind, solar, and hydropower sources, often with strong state support at crucial points. Germany's *Energiewende* makes it a hub for these cooperatives, including the particular entity analyzed in Chapter 15. Although in some countries, such as Canada and the United States, electricity cooperatives continue their traditional focus on rural electrical distribution, in others, such as Germany and Denmark, their main roles are to develop localized renewable energy generation and provide legal support for energy democracy. They conclude that, although energy cooperatives hold significant potential for enhancing energy democracy, realization of that potential is limited without appropriate policy support, active participation across diverse memberships, and explicit goals directed toward advancing social justice.

In Chapter 17, Stephens and Allen argue that democratic energy system transformation will be impossible without integrating a feminist perspective into energy system discourses. While recognizing that feminisms are as multifaceted as are democracies, they note that principles embraced by the energy democracy social movement are complementary to feminist, antiracist, and inclusive strategies that would promote empowerment of both individuals and communities. Using bell hooks's definition of feminism, they launch an exploration of possibilities opened by expanding the profile of feminist leadership in energy transformation. With patriarchal systems of power and privilege remaining especially dominant in the energy sector, integrating feminist principles into efforts toward energy democracy may be an especially valuable strategy at this point in time. Stephens and Allen argue for conscious placement of women in leadership positions and, perhaps more crucially, for feminist explorations of energy system realities that can reveal the prevalence of patriarchal values, masculinized authoritarian leadership and the dominance of an associated technocratic approach to innovation of energy systems to participants for whom such thoughts are unlikely to occur without prompting. This chapter reimagines energy transformation as a process that requires carefully crafted strategies that align feminism with energy democracy to address oppressive gender dynamics and accelerate a transition to renewable energy systems.

In Chapter 18, Burke builds from Ostrom's work on cooperative self-governance to challenge the entire neoliberal notion of private ownership and proposes *energy commoning* as a grassroots strategy for revitalizing local communities and their economies. He describes energy commoning as a radical reinterpretation of energy, particularly electricity produced from solar and wind technologies, as a resource commonly available for all living beings and thus existing, both temporally and spatially, beyond systems of private property. Burke argues that, without contesting the commodification of energy sources, the expansion of renewable energy simply extends preexisting patterns of exclusion, which eliminates the democratic principle of justice from the energy transition. He reviews strategies used by emerging institutions of energy commons for reaching beyond market and state control, while also grappling with the significant barriers such institutions face. Burke explores the highly diverse strategies these institutions use to collaboratively strengthen horizontal social relationships and then uses principles drawn from Ostrom's work to compose a set of design principles for energy commoning. One of the most important features of this chapter is its persistently reflexive focus on the action of commoning rather than the achievement of a commons. This allows Burke to emphasize ongoing strategies for resisting the imposition of state- and market-based control over these collaborative approaches to energy.

These chapters highlight examples where citizens of nominally democratic regimes prove astonishingly adaptive to different temporal and spatial contexts and even become change agents within those contexts. They demonstrate Cox's (2010) argument that grassroots campaigns can produce system-level change by positioning themselves at vulnerable junctures within political power structures and then leveraging their position by intervening in ways that redirect the momentum of already powerful forces. The modes of action explored here suggest that, while composing energy democracy remains a challenging and complex endeavor, when framed strategically, grassroots action has the potential to enable democratic energy transitions.

Reference

Cox, J. R. (2010). Beyond frames: Recovering the strategic in climate communication. *Environmental Communication*, *4*(1), 122–133. https://doi.org/10.1080/17524030903516555

15

THE STATE OR THE CITIZENS FOR ENERGY DEMOCRACY?

Municipal and cooperative models in the German energy transition

Sören Becker

Introduction

Energy democracy has evolved into a slogan that unites various and sometimes even conflicting visions for a reconfiguration of technologies, management, and ownership relations of energy systems. Spanning Keynesian approaches to regional transformation to attempts for a renationalization or liquidation of major energy corporations and replacing them with decentralized structures, the notion of energy democracy is inspiring for a wide range of political activities, mainly in the United States and Europe (Angel, 2016).

In an early attempt to provide conceptual orientation through this wide field, Matthias Naumann and I (Becker & Naumann, 2017, p. 4) have proposed two interlinked dimensions to assess what energy democracy might mean: First, we have understood democratization as a political call to open up energy systems to participation. These claims seek to create friction within an energy sector hitherto defined by a historically grown amalgam of state and corporate power, seeking to replace the hegemonic notion of energy as a commodity based on access to resources and ensuring accumulation (Huber, 2009; Mitchell, 2009) by perspectives of even access, public control, and reduced greenhouse emissions. Second, our main distinction from other normative conceptions like "just transitions" and "energy justice" is that approaches to energy democracy often put forward provisions for institutionalizing democratic principles in lasting organizations. In other words, energy democracy proposes an often concrete set of suggestions for how bottom-up participation and ownership can be condensed into organizational forms: Generally these provisions involve either state or a certain kind of community ownership.

Looking at empirical developments in Germany, this chapter takes these conceptual considerations as a vantage point for comparing energy cooperatives and those models of local public ownership that were discussed and rolled out during the so-called "wave of remunicipalization" between 2005 and 2013 (Bauer, 2012). It shows how similar visions and goals for a transition of energy systems may be materialized in different forms of organization and ownership. This chapter will give an overview on the boom of both energy cooperatives and new public utilities in Germany and will discuss important contextual conditions for these developments. Based on empirical work in Berlin and Hamburg—two cities that have witnessed conflicts about the reintroduction of state ownership and that display a variety of different initiatives for

DOI: 10.4324/9780429402302-18

a more democratic energy system—two models are introduced: the cooperative and the new public utility. I then discuss the different organizational principles and challenges for realizing energy democracy in each form.

New forms of ownership in the German energy transition

From 2005 onward, Germany has experienced two major trends that challenged the concentrated structure of energy markets: The increasing numbers of both energy cooperatives and remunicipalizations, meaning the (re)introduction of state ownership. These developments were linked to a technological transition to renewable energy sources that were interpreted as a decentralization of energy provision, both in organizational and spatial terms (Gailing & Röhring, 2016; Klagge & Brocke, 2012). The transition toward renewable energy opened up a new field of energy technology that was occupied by new actors as owners and new spatial dynamics in the context of Germany's energy transition.

Out of these different collective forms of ownership, energy cooperatives gained most attention in research and public discourse, due to an unprecedented dynamic in the formation of energy cooperatives (Klagge & Meister, 2018; Yildiz et al., 2015). A cooperative is defined as a legal form where a group of members share ownership in the entity and its assets, in this case energy plants or grids. A core principle is that each member has one vote in the General Assembly, regardless of how many shares that member owns. Between 2005 and 2015, more than 800 energy cooperatives were founded with an accumulated number of about 165,000 members, producing an average of 223 members per cooperative (DGRV, 2016). A large part of these (86% in 2015) were active in the generation of electricity, mainly through newly installed wind and solar capacities. Their financing model is based on selling electricity generated by cooperatively owned plants according to the feed-in tariff scheme of the Renewable Energy Act. About one out of five cooperatives were running district heating networks, often in rural areas, providing members with heat through mostly insulated networks. Here the cooperative owns not only the capacities for heat generation but also the adjacent heat distribution grids. When it comes to electricity, grid ownership is less relevant for cooperatives, as only 1% is active in running or seeking to run an electricity grid on a local or regional basis. With the stepwise replacement of the feed-in-tariff system by an auction model prompting the allocating of bigger installations after 2014, the dynamic of cooperative foundation has dramatically decelerated (Müller & Holstenkamp, 2015). Most importantly, these cooperatives were discussed as expressions of an alternative economy in the energy sector, as they pursue the interest of their members rather than profit (Becker et al., 2017; Klagge & Meister, 2018).

Remunicipalization refers to a slightly different but no less dynamic phenomenon. Remunicipalization, roughly, means the (re)introduction of local state ownership in energy infrastructures and/or utilities. Remunicipalization describes an international phenomenon that covers a wide variety of sectors, including waste, water, public transport, and other municipal services (Hall et al., 2013). A recent global survey based on information gathered by researchers and trade union activists does account for 835 cases of remunicipalization worldwide, out of which 284 occurred in the German energy sector only (Kishimoto & Petitjean, 2017). Hence, the dimension of remunicipalizations here is without equivalent in other countries or sectors (Cumbers & Becker 2018). On the national level, the phenomenon is of relevance too: Including new entries in commercial registers, Lormes (2016) accounted for 122 newly founded local utilities from 2005 to mid-2014 (p. 334). These make up a considerate part of the 845 local public utilities listed in the official statistics of the year 2011 (Lormes, 2016). Of these, 90%

took place in municipalities with fewer than 50,000 inhabitants, with around 65% occurring in municipalities with 10,001 to 20,000 inhabitants. These numbers also include cases of intercommunal cooperation, which are more likely to occur in smaller communities (Lormes, 2016, p. 334). Other studies relying on different methods of data gathering confirm these numbers and the trend in the size of the municipalities that remunicipalized at least parts of their system (Berlo & Wagner, 2013). Additionally, their data provide insights into the spatial distribution of remunicipalization cases that is characterized by different clusters, the region around Hamburg forming one of them.

The variance among these numbers points not only to methodological challenges of determining the exact number of remunicipalizations but also to the different empirical variants of remunicipalization. Hence, we can identify five ideal types of remunicipalization that may overlap in reality (adjusted from Bauer, 2012):

1. *Remunicipalization by foundation*: A new legal entity (a company under private or public law) can be founded by a municipality. This type is represented in the numerical accounts on business registry entries previously cited.
2. *Remunicipalization by reintegration*: Shares, company units, or entire companies, once privatized, could be bought back into a municipal holding.
3. *Remunicipalization of responsibilities*: Operational services or business activities of a public utility could be extended to another sector. An example is the situation in which an existing water utility would become active in the energy sector. Comments here stress the potential for efficiency gains and synergies when it comes to central units such as costumer service.
4. *Remunicipalization of the supply chain*: Utility activity could be extended along the energy supply chain (i.e., when a grid operator enters into selling energy). In this case, European legal requirements of unbundling energy services and grid operations need to be considered, often leading to different organizational units for the different steps.
5. *Remunicipalization by territorial extension*: The territory of a local utility could be expanded, for example when a municipal grid operator starts to operate grids in suburbs or parts of an inner-city grid that were not integrated into their system before.

Remunicipalizations, on the one hand, are interesting as they redefine the distribution between public and private responsibilities. While objectives of remunicipalization span a wide range of arguments including local benefits, employment, and restoring local control over the direction and quality of energy provision, some instances were also based on prospects for accelerating a local energy transition. Energy cooperatives, on the other hand, present an ideal type of distributed ownership, making communities of citizens co-owners of energy infrastructure and direct profiteers of the transition to renewable energy. However, neither the rise of energy cooperatives nor the dynamics in the field of remunicipalization went without amenable conditions (re) forming the context of German energy policy.

Enabling factors for new forms of energy ownership in Germany

Decentralized ownership structures, energy cooperatives, and cases of remunicipalization in this case have hit a sector that experienced widespread privatization after the liberalization of European energy markets. And indeed, in the late 1990s and early 2000s, many cities and municipalities sold shares or entire utilities to private bidders, resulting in a remarkable process of economic concentration (Bontrup & Marquardt, 2011). The so-called Big Four—E.on, RWE, EnBW, and Vattenfall—were rising: integrated energy corporations growing from previous public and private companies, seemingly dividing the country into interest spheres through subsidiaries, each regionally controlling a large part of energy plants as well as transmission grids. However, just after the new structures of the energy sector had crystallized, various cracks started to show again. Although the widespread occurrence of energy cooperatives and remunicipalizations might come as a surprise to some, several factors enabled remunicipalization and the formation of energy cooperatives.

The energy transition as a discursive and material opening

Originally advocated by a few pioneers, the German energy transition (Energiewende) gained considerable momentum with the reform projects of the Red-Green coalition federal government that came to power in 1998 (Gailing & Röhring, 2016). Connected to the promise to phase out nuclear energy completely, the transition to renewable energies turned into one of the major policy discourses of the new millennium. Notably, the introduction of a feed-in tariff system through the Renewable Energy Act (EEG) two years later rendered renewable energy a worthwhile investment with secure payoff rates. This resulted in the massive buildup of citizen- or farmer-owned wind, solar, and biomass facilities—with some of them organized as cooperatives. In a study for the year 2012, it was found that about half of all renewable energy capacities in Germany were owned by either citizens or farming businesses at the peak of this development (Trend: Research & Leuphana Universität Lüneburg, 2013). Beyond new actors and financial mechanisms, this shift in energy implied a double opening in discourses around energy: First, the private-is-best orthodoxy prominent in privatization and liberalization processes was questioned; second, the normative catalogue of energy policy was diversified to include the new aims of sustainability and climate protection. In other words, the Energiewende underlines that energy provision has become more than a technological and economic issue.

Traditions of local utilities

There is a strong tradition of local utilities providing services in Germany, not only in energy but also in other sectors such as water or transport (Bönker et al., 2016), with multiple sectors integrated into one organization, the so-called *Stadtwerk* (city utility). Although there have been a number of shifts reflecting the changing dominant political economic models over time—from private buildup to a stronger role for the state after World War II, to market-oriented reforms and privatization from the 1980s on—municipalities continued to play an important role in service provision (sometimes also in public–private partnerships) (Wollmann, 2015). While privatizations were often legitimized by questioning the economic efficiency of this model, the notion of a *Stadtwerk* as a strongly embedded local utility remained an important policy option to many.

Disappointment with the performance of private operators

For a long time, the Big Four energy corporations were criticized for failing to respond to demands for renewable energy. While the development of renewable energy had to a large part materialized through decentralized and small-scale projects, renewables did not play a major role in the Big Four companies' business strategies. Instead, these were locked into a situation in which most of the expertise and strategic outlook within the companies was bound to retaining the profitability of their fossil and nuclear plant assets (Kungl, 2015). Additionally, municipal governments felt they had lost control over energy provision in the sense of having little influence over both service quality and available tools for the energy transition. Likewise, only in very few cases did private operators prove to be more efficient than previous municipal ones; instead, prices often rose (Matecki & Schulten, 2013). For the municipalities, in turn, relatively stable revenues from selling energy and running the grid were lacking, which foreclosed the possibility for cross-financing more costly services like swimming pools or public transport as in the past. Lastly, the continuous inertia of energy corporations to deliver a shift to renewable energy has motivated some citizens and municipalities to take the energy transition into their hands.

Termination of concession contracts as a window of opportunity

A key occasional factor that enabled the remunicipalization trend in Germany was the expiration of numerous concession contracts. These contracts set the legal conditions for using streets and other public space for cabling and pipelines—the very material underpinning for running an energy grid in a city. Concession contracts were normally signed for 20 years, and most had to be renewed in the first decade of the 2000s (see Berlo & Wagner, 2013, p. 63ff.). While most contracts were renewed or only partly renegotiated and then renewed, in those places where remunicipalization occurred, the expiration of the concession put the topic of local energy futures on the agenda. And, indeed, in most cases this question has paved the way to remunicipalization, either through giving the concessions to a new (public) enterprise or by buying the private company holding concessions and transferring ownership this way.

Low interest rates on communal and private credits

A further enabling factor was the availability of cheap money for municipal and private investments after the 2009 financial crisis (Libbe, 2013). The European Central Bank's low interest policies also affected the market for communal credits, on which the interest rates are even lower than for private credit. As a constriction, the opportunity to finance remunicipalizations was mainly open to those municipalities that were financially flexible and not under the limits of the strict austerity regime that hit many cities with high debt ratios. Similarly, only those households or farmers who had some financial leeway and security for being eligible for larger credits could invest in renewable energy.

In a broader sense, the ownership transformations in the German energy sector rest on a convergence of traditions of local (public) service provision with the current dynamics of the Energiewende, combined with short-term enabling conditions such as terminating concessions and available credits. In reality, whether these context conditions lead to an actual reorganization of local energy systems depends on political decisions and often even result in political conflicts on the ground.

Conflicts about urban energy in Berlin and Hamburg

The point of departure for the case studies in Berlin and Hamburg is similar. Both cities look back on a long tradition of local energy supply: The Berliner Städtische Electricitätswerke Aktiengesellschaft (BEWAG) and Hamburgische Electricitätswerke (HEW) date back to the late nineteenth century, although they were not publicly owned throughout the entire period.

In the case of Berlin, the local enterprise even survived the period of the city's division and was reunited after the fall of the Berlin Wall (Moss, 2014). Both utilities had their main field of interest in the electricity sector; both were among the largest German energy utilities in the 1990s. However, both were sold step-by-step to the Swedish state-owned company Vattenfall around the year 2000. Just a few years later, however, energy provision in Hamburg and Berlin appeared in the spotlight again, as both cities were the site of intense conflict about remunicipalization. Important for this chapter, both cities saw a competition between the two models of public and cooperative ownership for the urban energy grid, depicting two options for how to organize energy democracy in one city.

Berlin: a roundtable for remunicipalization and an energy grid cooperative

Debates about the future of Berlin's energy supply have attracted widespread attention since two initiatives, the social movement coalition Berliner Energietisch (Berlin Energy Roundtable) and the cooperative BürgerEnergie Berlin (Citizens Energy Berlin), proposed different organizational models as an alternative to the existing private sector–dominated model of energy provision (Angel, 2016; Becker et al., 2015; Blanchet, 2015; Paul, 2018).

The Berlin Energy Roundtable is a grassroots social movement that campaigned for a new participatory form of energy utility. In 2010, even before the onset of a public debate over awarding the concession for the city's electricity grid, a small group of activists had published a concept paper on the formation of a new municipal utility, outlining that it should both produce electricity from renewable energy sources and provide participatory structures (Attac Berlin et al., 2011). This concept laid the foundation for a movement that was coordinated by the Berlin Energy Roundtable, a grassroots coalition to campaign for the remunicipalization of the electricity grid. It involved around 40 different actors such as large environmental organizations, small NGOs, leftist activist groups, and antigentrification initiatives, as well as some professionals from the field of renewable energy (Roundtable Activist, personal communication, August 28, 2013).

As it became clear that the then city government—first run by the Social Democrats and the Left Party in coalition and from 2011 by the Social Democrats and the Christian Democrats—would not support specific steps toward remunicipalization, the Roundtable coalition decided to organize a referendum to pressurize local politics (Roundtable Professional, personal communication, August 27, 2013). As a text to vote on in the referendum, an elaborate law was drafted in a consensus-oriented discussion process within the Roundtable. The draft law laid out different stipulations for a so-called Citizens Utility (Bürgerstadtwerk) that should be democratic, ecologically oriented, and socially just. The draft foresaw public meetings, an Advisory Board with directly elected citizen representatives, and an obligation to make core documents accessible to the public (Berliner Energietisch, 2012; see later in the chapter). Additionally, the entity should serve as a means for increasing the share of renewable energy used in the city and take measures against energy poverty (Berliner Energietisch, 2012). In the actual election on November 3, 2013, however, the referendum failed, as only 24.1% of the required 25% of the city's electorate approved the draft law. It was, nevertheless, an impressive display of opposition to the incumbent utility. While some of the key figures in the campaign successfully ran for the city's parliament in the 2016 elections, the roundtable today exists only as a loosely bound meeting of activists.

Parallel to the calls for a participatory public utility as articulated by the Energy Roundtable, the cooperative Citizen Energy Berlin (BürgerEnergie Berlin, 2019) proposed a second organizational model for citizen involvement that is based on their members collectively owning shares of the city's electricity grid. Their central strategic goal was hence to gather enough capital to become a partner in a cooperation model in which a municipal (or private) utility holds a majority share of the grid operator and the cooperative controls up to 49% (BürgerEnergie, personal communication, September 2, 2013). Applying together with a local utility from Southern Germany, Citizen Energy Berlin was one of the bidders in the concession process. Launched in 2011, its membership counted more than 2,000 in 2015 and registered around 1,000 members in early 2020. It held an overall capital shareholding of about €11 million by the end of 2018 (BürgerEnergieBerlin, 2019). Members are recruited from people genuinely interested in energy issues, spanning various social groups from students to pensioners (BürgerEnergieBerlin, personal communication, September 2, 2013). To become a member, people simply needed to buy shares at €100 each; five shares were recommended. Basically, every-one could buy into the cooperative; ownership was not restricted to Berlin residents.

As with the Roundtable's plans, the objectives of the cooperative cover ecological, democratic, and social issues. Ecologically they aim to strengthen Berlin's potential to achieve an energy transition by preserving 10% of profits for reinvestment in solar and cogeneration plants. Their attempts to advance democracy are seen to be embodied in the principles of "one member/one vote," plus provisions for minority rights in cooperation with the utility, ensuring an aggregate citizen's say on strategic decisions (BürgerEnergieBerlin, personal communication, September 2, 2013). In social terms, the cooperative argues in favor of cost reduction through energy-saving and efficiency measures. The cooperative's core strategy for attracting new members is to appeal to people both as an ecological project and as an investment opportunity (Becker et al., 2015). At the time of writing, the cooperative remained active in lobbying the local government to be incorporated into a future local utility as a minority shareholder, started retailing electricity produced by another cooperative, launched a campaign for Berlin's exit from coal-fired plants, and is seeking new business opportunities around decentralized solar and heat technologies in multifamily houses.

Hamburg: effective remunicipalization and reclaiming the grid

Though attracting less scholarly attention, Hamburg also is an interesting case for new forms of citizen involvement in local energy politics. As with Berlin, the city of Hamburg witnessed a referendum on the remunicipalization of its energy grids in 2013. In contrast to Berlin, the referendum succeeded by a narrow majority of 50.9% of the overall votes. Additionally, the referendum was preceded by the foundation of the public energy enterprise Hamburg Energie in 2009, which became a nationwide model for a state-owned green energy supplier.

Hamburg Energie is a public utility that was founded with a clear ecological orientation and later complemented with participatory provisions to include the voices of citizens and business clients. Against the background of a history of energy-related conflicts in the city, it was the plan to erect a new coal-fired plant of 1.6 MW that proved the most divisive policy issue in the early 2000s. During this debate, the Green Party—previously a strong opponent to the plant—came to form a coalition government with the Christian Democrats. While a previous approval of the plant by city's administration could not be reversed, the Green Party was now given a free hand to establish a public utility for the buildup of renewable energy provision for the city. The utility was founded as an autonomous subsidiary of the local waterworks that had remained publicly owned. Importantly, Hamburg Energie was given a clear mission statement as recorded in the city's commercial register, including commitments to the "provision of energy for the general public and public institutions," the sale of "climate-friendly electricity (non-nuclear and coal-free)," and a requirement that the enterprise would "plan, erect, and run municipal infrastructures." At the same time, it was devised to be a "lean and effective" company that can survive as an actor in the competitive energy market, while at the same promoting renewable energy in the region (Consultant, personal communication, April 20, 2015).

Once established, Hamburg Energie proved highly effective for increasing the share of renewable energy. More than 13 MW in wind power were installed by the end 2015, and a 10-MW solar energy program involving citizens and local businesses as coinvestors was completed. In 2019, about 150,000 consumers were counted; and about 45% of the electricity was generated by the company's renewable energy sources (Hamburg Wasser & Hamburg Energie, 2020). In this sense, Hamburg Energie stands as a case of top-down remunicipalization that has proven to be a very successful instrument for promoting a transition to renewable energy.

Additionally, a Consumer Advisory Board (Kundenbeirat) was installed in 2012. Although this body only has a consultative function, a Hamburg Energie employee emphasized in an e-mail that the company would grant "a transparent look into managerial decisions" and claims that "critical opinions of its members are dealt with in a constructive manner" (personal communication, April 8, 2015). Hamburg Energie, therefore, stands out in two ways from most of the other newly founded local utilities in Germany. First, it was a pioneer in terms of its clear orientation toward the generation of renewable energy in close cooperation with local citizens and business. Secondly, installing a participatory body was another novelty—and a response to pressure exerted by the referendum campaign.

As in Berlin, there was a referendum campaign in Hamburg for remunicipalizing the energy grid. With Hamburg Energie active in the fields of energy generation and sale, the approaching expiry date of the concession with Vattenfall marked the advent of a new political conflict on the city's grids. Here, also, a group of social movement activists and representatives of environmental and social organizations gathered as early as 2010 to consider how to remunicipalize the energy networks. Later they formed the campaigning coalition Our Hamburg—Our Network (OHON), highlighting the local character of possible remunicipalization.

There are four main differences between the developments in Hamburg and those in Berlin. First, the Hamburg campaign not only aimed at the remunicipalization of the electricity grid but also included district heating and gas networks. Second, the Hamburg coalition encompassed both grassroots-level social movements and larger organizations such as the German member of Friends of the Earth (BUND), the Consumer Advice Centre (Verbraucherzentrale, a consumer rights organization by public order), and the charity organization of the regional Protestant-Lutheran Church (Diakonie). Smaller climate action groups and NGOs such as Robin Wood and Stop Moorburg Pipelines (Moorburgtrasse Stoppen) played a supporting role in the coalition, as did the network Enterprises against Nuclear Power. Third, the text subject to the Hamburg referendum consisted of only two sentences, demanding the transition of the city's grids into public ownership and a "socially just, climate compatible and democratically controlled energy provision from renewable sources" as a "mandatory target." Fourth and most important, the Hamburg referendum was successful with a very narrow majority of 50.9% of the valid votes. Although the Hamburg referendum law was less elaborate than that proposed in Berlin, the OHON campaign in Hamburg showed how a coalition of established associations and social movements was able to successfully mobilize and challenge urban energy governance. Today, remunicipalization is completed and the electricity, gas, and district heating networks are run by the city. Especially the last has proven a key asset in shaping the future of the city's energy provision with a new heating concept peeling off fossil plants issued in 2019 (Behörde Stadtentwicklung und Umwelt, 2019).

Returning to the referendum itself, as a campaign organizer said in an interview, the strategy was to "convince 50% + *x*," involving questions on how "to strike the right tone" to appeal to a majority of voters (Campaign Organizer, personal communication, January 8, 2015). However, the quest for remunicipalization provoked resistance from established actors in local energy politics. First, the city government—then Social Democrat—settled on a partial remunicipalization of 25.1% and an "energy concept" detailing future investment measures with each of the utilities in late 2011. The motivation here was to counter the argument that the local state had no influence over energy provision. In the months leading up to the referendum, public debate became increasingly heated as a countercampaign against full remunicipalization was launched. This was backed by a coalition consisting of the main political parties, business associations, and even the major trade unions. Interestingly, social and democratic aims only played a minor role in debates as the discourse revolved around two main issues: the financial costs of remunicipalization and the question of whether grid ownership is a feasible instrument for fostering a transition to renewable energy. Remunicipalization supporters responded to the two issues with the slogan "because it is worth it."

As another parallel to Berlin, Hamburg also had a cooperative that sought to gain a share of the grid operating utility. Called Energy Network Hamburg (Energienetz Hamburg), it was more successful in terms of gathering capital investment than its Berlin counterpart: According to a press release from September 2013, some 3,000 members have accumulated more than €50 million. The cooperative was one of the bidders in the concession awarding process cooperating with the Dutch utility Alliander; yet in public they appeared as clear supporters of the referendum campaign for remunicipalization. However, as the successful referendum requires full remunicipalization by law, the cooperative developed a new strategy that targets investments in renewable generation capacities instead of grid ownership. In line with this new investment strategy, it completed its first project, a citizen-owned photovoltaic installation on the rooftop of an office building, in the summer of 2015. Further, Energy Network has been active in organizing and participating in stakeholder dialogues on energy issues in the city. In a sense, the Hamburg cooperative has charted out not only a new area of activity but also a way of complementing political efforts around the referendum campaign.

New forms of organizing participatory public services: implications for ownership, participation, and justice

In both cities, Berlin and Hamburg, different kinds of initiatives have pushed forward new ideas for an urban energy future that is defined by visions for a more renewable, socially just, and democratic utility. The political processes in the two cities show how the referendum as an element of public mobilization and popular politics was used to alter given constellations of actors and circumvent established relationships of power between the local state and incumbent energy providers. Hence, involving people as activists and as voters was a precondition for political change in both cities. Beyond that, and concurring with the understanding of energy democracy in the beginning of this chapter, another main contribution of these initiatives was to provide details of alternative organizational forms for enabling participation in the management of infrastructure utilities.

The pathways for ensuring citizen participation differ according to the envisioned models of ownership: BürgerEnergie Berlin and Energienetz Hamburg are cooperatives relying on membership, while the two campaigns for remunicipalization envision utility ownership by the local state, complemented by participatory provisions. In the remainder of this chapter, I will detail the specific provisions and discuss their implications along three dimensions: (1) the implications for ownership; (2) envisioned mechanisms for participation and control; and (3) challenges for social justice (see Table 15.1).

The cooperatives aimed for a collaboration with a grid utility in which they would own a part of the shares of the grid operating utility, investing the capital gathered through membership fees and donations. The impetus behind that was to secure citizen influence as a minority shareholder with special rights negotiated with the other owner parties. Thereby the cooperative concept does not necessarily entail a new role for the state in energy provision and governance, although cooperative members stated in interviews that they would rather join forces with a public than a private utility. Ownership of some shares of the energy grid would be transferred to the collective of the citizen members, although those members tend to represent a small fraction of the entire population. The cooperative model hence implies direct ownership of energy infrastructure, but could also be described as *collectivized private ownership*.

Participation in the cooperative model is defined by the principle of membership. That means that the General Assembly is the most important body of control. Here, fundamental decisions on the strategy of the cooperative are made and internal elections for representatives and management are held, while each member therein has one vote regardless of the number of shares owned (Schröder & Walk, 2014). While others have highlighted that energy cooperatives can be interpreted as self-empowerment for people to choose their energy system (van der Schoor & Scholtens, 2015), it is important to note that with energy cooperatives, owners and users are not always identical (cf. Novy 1985, p. 127). Cooperative members could reside anywhere, not necessarily in the area where they seek to obtain ownership of infrastructure. In practice, cooperatives rely on an active membership to fulfill their promise for participation. In Berlin, a new internal body called "team" was introduced to also allow active engagement of members that are not elected to the cooperative's boards.

Members don't just take decisions; they are also direct beneficiaries of potential revenues resulting from the cooperative's operations. To become a member, a membership charge is necessary (€100 for Energy Network Hamburg; recommended 5 × €100 for Citizen Energy Berlin). This implies two issues for social justice: First, there is a potential social imbalance as the fee granting admission could pose an economic obstacle for poorer households to become

Table 15.1 Comparison of the cooperative model and the participatory utility

	Cooperative model	*Participatory utility*
Implications for ownership	Collectivized private ownership: shares of utility in a cooperation model, ideally with a public provider	Participatory public ownership: full public ownership as single utility or public holding
Envisioned mechanisms of control and participation	Member based: General Assembly for strategic decisions and internal elections	Citizen based: mechanisms of control and information, e.g., through advisory boards, public meetings and events
Challenges for social justice	Imbalance in members through economic entry barriers, resulting in uneven benefits and a potential uneven spatial representation	Challenge of refinancing investment costs, with opportunities for cofinancing public services and social tariffs

a member. While both cooperatives in Berlin and Hamburg claim that they represent members from all social classes, general statistics on membership in energy cooperative show a bias toward people with a higher level of education and a familiarity with energy issues (Rommel et al., 2018). Second, this could produce a spatial imbalance in membership structures when there is a higher representation of neighborhoods with more affluent and educated populations within a city. In order to gain as much capital as possible, none of the cooperatives has restricted membership to citizens of the respective cities, so that interested nonlocals could become members, too. Together these could, hypothetically, result in the under representation of local people from poorer neighborhoods in siting and investment decisions.[1] Summing up, the cooperative model provides for a high degree of internal equality and democracy, while its general scope is, at the same time, limited by potential social imbalances.

Targeting full public ownership of the physical grid and the operating utility, the approach of the participatory utility is wider in scope than the membership-based model of the cooperatives. More so, citizens—in this case thought of as inhabitants of Berlin or Hamburg and therefore users of the energy grid—should have a say on its development. Activists were skeptical of the state as the embodiment of general interest per se. Therefore, beyond engaging citizens directly, the participatory approach is meant to ensure the social and ecological ambitions of the project. As a Berlin roundtable spokesperson argued, "[Y]ou cannot ensure the subscription [of politicians and managers] to the public interest by legal means. Only control can secure it" (Roundtable Professional, personal communication, August 27, 2013). Hence, the entitlements of control here are to ensure by institutional design the utility's openness for citizen and social movement demands in the long term. For this reason, voting templates for the referendum either recorded the principle of democracy (as in Hamburg) or deliberately spelled out concrete mechanisms for information and participation (as in Berlin). Table 15.2 presents the provisions devised by Berlin's Energy Roundtable specifically designed to ensure that citizens maintain control and are guaranteed timely access to information. Their model combines elements of direct representation, public gatherings as well as information duties, laying out a kind of constitution for *participatory public ownership*—though one that still awaits its test in practice.

This model also presents challenges regarding social justice. First, and obviously, both in Berlin and Hamburg there was a strong contestation as to whether the investment of buying energy grids was an effective tool for reducing climate emissions and improving social justice given the significant cost (in Hamburg these amounted up to €1.8 billion, including the earlier purchase

Table 15.2 Provisions of control and information drafted in the ownership model of the Berlin Energy Roundtable

Provision	*Details*
Democratic advisory board	Consists of the Senator for Economy, the Senator for Environment, seven employee representatives, six elected members.
Right of initiative	Any initiative gathering at least 3,000 signatures will be considered by the Advisory Board.
Public assemblies	There should be public assemblies to discuss issues of energy generation. These should be held once a year for the entire city and for each of the 13 boroughs. Recommendations of these assemblies are to be discussed by the Advisory Board within three months.
Ombudsperson	The company appoints an ombudsperson as the core contact point for citizen and costumer queries.

of 25.1% shares). Moreover, some feared the amount would draw funds from other kinds of social infrastructure. In Hamburg, conflicts even occurred around what was to be considered socially just in itself, as in the implementation process some actors sought to reduce social justice to ensuring low prices through maintaining fossil energy. Experience also shows that in many cities the leeway for social tariffs after remunicipalization is considered low when municipalities seek to refinance investments (even though no comparative studies exist on this issue). Lastly, even when participatory provisions are actualized, ensuring effective and fair participation in the long run might pose a challenge. Experience shows that debating energy issues requires a high level of expertise, so that often the seats in participatory councils are taken over by representatives of organizations already informed and literate in complex energy issues. Against this, once established, platforms for participation can be used by social movement actors to campaign for changes toward more social business practice in the future.

Conclusion

This chapter started from the observation that, in Germany in the 2000s, there was a spike of new, not privatized forms of generating or distributing electricity. Nurtured by an overall appetite for more climate-friendly sources of energy provision, a favorable financing scheme for decentralized renewables, and beneficial conditions for communal investments, the country experienced a significant growth in energy cooperatives and the reestablishment of local state ownership. Driven by an understanding that energy democracy entails both new visions for an energy system—often merged in the triad of democratic, socially just, and ecological/renewable—and real-world examples for organizing these, this chapter has engaged with different projects for energy cooperatives and remunicipalization in the cities of Hamburg and Berlin. And indeed, the cases analyzed here—envisioned and actualized—spell out different forms of participation that contest given structures around private infrastructure ownership. Moreover, forms and strategies were developed to democratize utilities and state institutions in a way that would open them up to citizen and social movement demands for equal and sustainable public service provision.

The two forms studied here, energy cooperatives and remunicipalization initiatives, offer viable forms of enabling a local transition to a renewable and participatory energy system, each according to its organizational model. In the cooperative model, the group of owners would be widened to those who became members of the cooperative, depicting the cooperative model as form of collective private ownership. As a main challenge, energy cooperatives need to address the potential social imbalance of membership-based models. As relatively small entities, cooperatives just have capacities for targeting a share of generation facilities or energy grids; therefore, their potential for an overall transformation of energy systems appears smaller. This, in turn, renders it easier for cooperatives to dock with existing actor constellations and patterns of energy provision and governance. Social selectivity aside, cooperatives may have considerable effects on the democratization of an energy system when they accumulate and form an ever growing network of different local initiatives.

Against this, remunicipalization stands for a more systemic approach to reclaiming energy systems but also one potentially provoking more opposition; stirring long lasting and resource-intensive conflicts about the future of urban energy systems, even beyond technologies and issues of ownership. It challenges the politicoeconomic imaginary that service provision is best handled by market forces, accordingly, in all of its various forms, remunicipalization explicitly seeks to reverse previous privatizations. In addition, the cases of remunicipalization discussed in this chapter have put forward far-reaching stipulations around rights of information and channels of citizen participation. In both cities (though not in all German cases), remunicipalization relied on popular mobilization and direct democracy to circumvent the established market power of private energy providers and their long-grown relationships in urban politics. The conflicts observed and the failure of remunicipalization in Berlin both are a testimony for the difficulties of actualizing remunicipalization. Once established, however, participatory utilities might provide robust and long lasting energy democracy via channels specifically designed to enable meaningful participation in local energy politics.

Note

1 Looking at cooperatives in electricity generation through wind or solar installations would result in a different picture. Here, citizens as collectives would indirectly gain a higher share in the overall pool of electricity generated.

References

Angel, J. (2016). *Strategies of energy democracy*. Rosa-Luxemburg Foundation.

Attac Berlin, BürgerBegehren Klimaschutz, and PowerShift. (2011). *Neue Energie für Berlin—Netze in Bürgerhand*. https://power-shift.de/wp-content/uploads/2011/05/Brosch%C3%BCre-NeueEnergie-NetzeInB%C3%BCrgerhandBerlin-2Aufl-7-2011.pdf

Bauer, H. (2012). Zukunftsthema "Rekommunalisierung". *Die Öffentliche Verwaltung*, *65*(9), 329–338.

Becker, S., Beveridge, R., & Naumann, M. (2015). Remunicipalization in German cities: Contesting neoliberalism and reimagining urban governance? *Space and Polity*, *19*(1), 76–90.

Becker, S., Kunze, C., & Vancea, M. (2017). Community energy and social entrepreneurship: Addressing purpose, organisation and embeddedness of renewable energy projects. *Journal of Cleaner Production*, *147*, 25–36.

Becker, S., & Naumann, M. (2017). Energy democracy: Mapping the debate on energy alternatives. *Geography Compass*, *11*(8), e12321.

Behörde Stadtentwicklung und Umwelt Hamburg. (2019). *Wärmewende und Energiepark Hafen*. www.hamburg.de/contentblob/12957152/7374b07373873dce6dd7af51f012383c/data/d-waermewende.pdf

Berliner Energietisch. (2012). *Neue Energie für Berlin: Eckpunkte des Gesetzentwurfs für eine demokratische, ökologische und soziale Energieversorgung*. www.berliner-energietisch.net/images/eckpunktepapier%20ge.pdf

Berlo, K., & Wagner, O. (2013). *Stadtwerkeneugründungen und Rekommunalisierungen. Energieversorgung in kommunaler Verantwortung*. Wuppertal Institute. https://wupperinst.org/uploads/tx_wupperinst/Stadtwerke_Sondierungsstudie.pdf

Blanchet, T. (2015). Struggle over energy transition in Berlin: How do grassroots initiatives affect local energy policy-making? *Energy Policy*, *78*, 246–254.

Bönker, F., Libbe, J., & Wollmann, H. (2016). Remunicipalisation revisited: Long-term trends in the provision of local public services in Germany. In H. Wollmann, I. Koprić, & G. Marcou (Eds.), *Public and social services in Europe: From public and municipal to private sector provision* (pp. 71–85). Palgrave.

Bontrup, H.-J., & Marquardt, R.-M. (2011). *Kritisches Handbuch der deutschen Elektrizitätswirtschaft. Branchenentwicklung, Unternehmensstrategien, Arbeitsbeziehungen* (2nd ed.). Ed Sigma.

BürgerEnergieBerlin. (2019). *Geschäftsbericht 2018*. www.buerger-energie-berlin.de/wp-content/uploads/190604_BEB_GB2018.pdf

Cumbers, A., & Becker, S. (2018). Making sense of remunicipalisation: theoretical reflections on and political possibilities from Germany's Rekommumalisierung process. *Cambridge Journal of Regions, Economy and Society*, 11(3), 503–517.

DGRV—Deutscher Genossenschafts- und Raiffeisenverband. (2016). *Ergebnisse der DGRV-Jahresumfrage*. www.dgrv.de/webde.nsf/272e312c8017e736c1256e31005cedff/5f450be165a66e4dc1257c1d004f7b51/$FILE/Umfrage.pdf

Gailing, L., & Röhring, A. (2016). Germany's Energiewende and the spatial reconfiguration of an energy system. In L. Gailing & T. Moss (Eds.), *Conceptualizing Germany's energy transition. Institutions, materiality, power, space* (pp. 11–20). Palgrave Macmillan.

Hall, D., Lobina, E., & Terhorst, P. (2013). Re-municipalisation in the early twenty-first century: Water in France and energy in Germany. *International Review of Applied Economics, 27*(2), 193–214.
Hamburg Wasser & Hamburg Energie. (2020). *Geschäftsbericht 2019.*
Huber, M. T. (2009). Energizing historical materialism: Fossil fuels, space and the capitalist mode of production. *Geoforum, 40*(1), 105–115.
Kishimoto, S., & Petitjean, O. (Eds.). (2017). *Remunicipalisation in public services: How cities and citizens are turning back privatisation.* Transnational Institute.
Klagge, B., & Brocke, T. (2012). Decentralized electricity generation from renewable sources as a chance for local economic development: A qualitative study of two pioneer regions in Germany. *Energy, Sustainability and Society, 2*(1), 5.
Klagge, B., & Meister, T. (2018). Energy cooperatives in Germany: An example of successful alternative economies? *Local Environment, 23*(7), 697–716.
Kungl, G. (2015). Stewards or sticklers for change? Incumbent energy providers and the politics of the German energy transition. *Energy Research & Social Science, 8*, 13–23.
Libbe, J. (2013). Rekommunalisierung der Energiewirtschaft. Erfahrungen und Entwicklungen in Deutschland. In P. Prenner (Ed.), *Kommunaler Ausverkauf. Von der Krise der Privatisierung* (pp. 37–48). Arbeiterkammer (Workers' Chamber).
Lormes, I. (2016). *Kommunalisierung der Energieversorgung. Eine explorative Untersuchung von Stadtwerke-Neugründungen.* Springer.
Matecki, C., & Schulten, T. (Eds.). (2013). Zwischen Privatisierung und Rekommunalisierung. In *Zurück zur öffentlichen Hand? Chancen und Erfahrungen der Rekommunalisierung* (pp. 8–17). VSA.
Mitchell, T. (2009). Carbon democracy. *Economy and Society, 38*(3), 399–432.
Moss, T. (2014). Sociotechnical change and the politics of urban infrastructure: Managing energy in Berlin between dictatorship and democracy. *Urban Studies, 51*(7), 1432–1448.
Müller, J. R., & Holstenkamp, L. (2015). *Zum Stand von Energiegenossenschaften. Aktualisierter Überblick über Zahlen und Entwicklungen.* Leuphana Universität Lüneburg. www.buendnis-buergerenergie.de/fileadmin/user_upload/downloads/Studien/Studie_Zum_Stand_von_Energiegenossenschaften_in_Deutschland_Leuphana.pdf
Novy, K. (1985). Vorwärts immer—rückwärts nimmer: Historische Anmerkungen zu einem aktuellen Problem. In H. Bierbaum & M. Riege (Eds.), *Die neue Genossenschaftsbewegung. Initiativen in der BRD und Westeuropa* (pp. 124–141). VSA.
Paul, F. (2018). Deep entanglements: History, space and (energy) struggle in the German Energiewende. *Geoforum, 91*, 1–9.
Rommel, J., Radtke, J., von Jorck, G., Mey, F., & Yildiz, Ö. (2018). Community renewable energy at a crossroads: A think piece on degrowth, technology, and the democratization of the German energy system. *Journal of Cleaner Production, 97*(2), 1746–1753.
Schröder, K., & Walk, H. (2014). Opportunities and limits of cooperatives in times of socio-ecological transformation. In M. Freise & T. Hallmann (Eds.), *Modernising democracy? Associations and associating in the 21st century* (pp. 301–314). Springer.
Trend: Research & Leuphana Universität Lüneburg. (2013). *Definition und Marktanalyse von Bürgerenergie in Deutschland.* Bremen, Lüneburg. www.buendnisbuergerenergie.de/fileadmin/user_upload/downloads/Studien/Studie_Definition_und_Marktanalyse_von_Buergerenergie_in_Deutschland_BBEn.pdf
van der Schoor, T., & Scholtens, B. (2015). Power to the people: Local community initiatives and the transition to sustainable energy. *Renewable and Sustainable Energy Reviews, 43*, 666–675.
Wollmann, H. (2015). Erbringung öffentlicher und sozialer Dienstleistungen zwischen Kommunen, Staat, Privatem und Dritten Sektor. In M. Döhler, J. Franzke, & K. Wegrich (Eds.), *Der gut organisierte Staat* (pp. 532–559). Ed Sigma.
Yildiz, Ö., Rommel, J., Debor, S., Holstenkamp, L., Mey, F., Müller, J. R., Radtke, J., & Rognli, J. (2015). Renewable energy cooperatives as gatekeepers or facilitators? Recent developments in Germany and a multidisciplinary research agenda. *Energy Research & Social Science, 6*, 59–73.

16

INSTITUTIONALIZING ENERGY DEMOCRACY

The promises and pitfalls of electricity cooperative development

Julie L. MacArthur and M. Derya Tarhan

Introduction

What institutional forms facilitate energy democracy? Calls for democratization of energy sector participants often are accompanied by support for decentralization of energy activities and by expanded benefits to previously marginalized communities in order to meet both environmental and social transformation goals. However, what does it mean to broaden decision making and control of the energy sector in terms of everyday practice and institutional design? How might energy initiatives be set up to allow for decision making control by affected populations and bring in these new users in a meaningful way?

There are both empirical and theoretical reasons to take the cooperative form seriously in the search for institutions of energy democratization. Cooperatives are part of a long standing international social movement for economic democracy, which entails widespread citizen economic decision-making power and control (Cumbers, 2013; Hahnel, 2013; McMurtry, 2004). They are distinct from for-profit firms based on their one-member/one-vote constitution (rather than allocating power by share of capital investment) (Laidlaw, 1980). Cooperatives also bring with them well developed social movement and social democratic roots and a focus on profit (surplus) recirculation via the principle of "people before profit." Electricity cooperatives in particular are legally constituted cooperative firms owned and governed by their membership, and active in power generation, distribution, retail, and energy efficiency activities. Local members collectively pool assets in order to provide—and sometimes also use—energy services. Cooperatives can act as electricity utilities, providing multiple aspects of the electricity service to the public, or can primarily operate in just one or two areas. The activities of these cooperatives depend significantly on their relationship to the state, which has evolved significantly over three "waves" of electricity cooperative development, outlined in the fourth section.

Energy democracy is conceptualized here broadly as both a process and an outcome of increased citizen participation in and benefit from energy sector activities (Szulecki, 2018; van Veelen & van der Horst, 2018). It is important to acknowledge that, like the cooperative movement more broadly, energy democracy contains distinct and sometimes divergent elements within it. For some, the key definition involves an increased role for private or citizen-based actors as prosumers and decision makers (Feldpausch-Parker et al., 2018), while more radical

 DOI: 10.4324/9780429402302-19

variants with energy democracy circles are calling for a return to collective and public ownership of energy assets at a systemic level through remunicipalization, citizen ownership, and public sector expansion (Burke, 2018). These tensions are explored in this chapter as they relate to electricity cooperative development. Our aim is not to put forward one clear and replicable model for energy democracy proponents but to provide enhanced clarity on the history, prospects, and contours of this particular institutional form.

We will begin the chapter with an examination of the emergence and relevance of electricity cooperatives to energy democracy (ED), followed by a historical and comparative account of electricity cooperative emergence. The final sections then assess the potential contribution of electricity cooperatives to energy democracy as well as the challenges faced in realizing this potential. We highlight the uniqueness of this particular form of "community energy," its diversity, and the complex potential that electricity cooperatives hold in current energy transitions.

Cooperative democracy

A range of institutional features of cooperatives mark them as a distinct organizational form, with both legal and historical connections to economic democracy. According to the International Cooperative Alliance, a cooperative is an "autonomous association of persons united voluntarily to meet their common economic, social, and cultural needs and aspirations through a jointly-owned and democratically-controlled enterprise" (International Cooperative Alliance [ICA], 2019). As an organizational form, cooperatives are democratic institutions insofar as they are member controlled rather than shareholder controlled. Most notably, this takes the legal form of an organization based on the principle of one member/one vote (OMOV), as opposed to control based on capital share. The members of a cooperative, in principle, then direct the organization's operations in their interest, and the OMOV principle prevents control from being concentrated in a few large-scale investors, as it can be in shareholder or privately owned firms. Cooperatives are part of the "third sector" of the economy, neither state owned nor operated solely on a private, for-profit basis. They can earn profits or surpluses for their members, but the core goal is serving member interests through the provision of goods, services, and employment rather than profit accumulation per se. While cooperatives are nonstate market actors (social enterprises), they, like all economic actors, depend significantly on government policy settings (Adeler, 2009; Berka et al., 2020; Creamer et al., 2018).

Worldwide, a vast network of cooperatives has developed in more than 146 countries, with more than a billion members controlling 2.6 million cooperatives. At a country level, cooperative economies comprise over 10% of the gross domestic product in four countries: New Zealand (20%), the Netherlands (18%), France (18%), and Finland (14%) (David Grace & Associates, 2014). These cooperatives are diverse, contested, and adaptable organizations. An important feature of cooperatives is that this member-based relationship is flexible, allowing them to adapt to the needs of distinct member groups. As such, a cooperative's membership can be drawn from producers, retailers, consumers, or a mix of stakeholders in the particular good or service.

The democratic impact of cooperatives can also extend beyond the institutional structure of the organization to a more sociological understanding of their contribution. For example, political scientists like Carole Pateman (1970) and Elinor Ostrom (1990) have highlighted how these organizations act as "schools of democracy," helping citizens to individually and collectively develop their capacities and skills. Moreover, by participating in organizational decision making, cooperative structures can facilitate transformative learning wherein members shift how they make decisions rather than simply accepting decisions made by others (Kohn, 2003; Radel, 2005; Wright, 2010). In this view, cooperative organizations help to democratize everyday life and decision making by engendering a form of democracy that is less reliant on governance by periodic elections and more deeply interwoven into participatory practices and economic control of important institutions in education, health, housing, and energy systems (McMurtry, 2009; Wiersma & Devine-Wright, 2014).

The degree to which specific cooperatives contribute to economic democracy is, of course, contested. Cooperatives can be either radical or conservative, local or international, or inclusive or exclusive of marginalized actors depending on their membership and constitution in a particular political context. However, historically, institutionally, and sociologically, there are strong reasons to see them as distinct and potentially empowering avenues for a move toward energy democracy.

Electricity cooperatives

An electricity cooperative provides energy services to its membership, is legally constituted as a cooperative and part of national and international cooperative networks. Its activities in the sector may vary, from power generation and distribution, to retailing electricity services, and energy education. Electricity cooperatives have emerged in distinct forms worldwide, corresponding with specific national legislation, community needs, and resource availability (Wierling et al., 2018). Rural distribution cooperatives focused their attention on extending electrification to underserved communities, often too geographically dispersed to be profitable for commercial power companies. Power generation cooperatives, on the other hand, focus their efforts on owning or operating the assets that produce power, rather than distributing it: solar arrays, wind turbines, hydroelectric, and, in some cases, fossil fuel plants. These assets may be small, able to power just one or two homes, or they may be very large, able to meet the power needs of thousands of citizens. The deciding factor on project size has more to do with local needs, resources, and available policy support than the cooperative form per se. In order to develop, a jurisdiction needs to provide a number of layers of policy support, from enabling cooperative legislation, through to regulations facilitating access to the power grid, financial, and informational support (Berka et al., 2020; REN21, 2016).

Table 16.1 provides an overview of the main activities that electricity cooperatives undertake. In many cases, they branch into new and complementary activities once they are established. For example, cooperative utilities in the electricity sector sell, distribute, store, and sometimes also generate power for consumers, while new renewable generation co-ops may start with energy education and climate advocacy. As we outline in the fourth section, the activities that have become most widespread depend on community needs, resources, and the national policy setting for both the electricity sector and cooperative firms.

Electricity cooperatives around the world

This section will focus on the historical evolution and current activities of electricity cooperatives on a global scale. While cooperatives can undertake any of the activities outlined in the third section, the most prevalent forms in the sector are (1) rural electricity cooperatives and (2) renewable energy generation cooperatives. Our historical analysis has identified three waves of electricity cooperative development that led to their development.

Table 16.1 Cooperative activities in electricity sector

Electricity co-op activity	*Activity*	*Example*
Generation	Members own or invest in power generation infrastructure, such as a wind turbine, hydropower facility, or gas plant; can be a partnership project with public or private sector actors or fully owned by cooperative.	Middlegrunden Wind Turbine Co-operative
Distribution and storage	Members collectively own the lines (grid) infrastructure that moves power from generation facilities to households and firms. A more recent activity is the provision of power storage services through shared battery systems.	Lake Country Power (Rural Electric Co-op)
Retail	Co-op members do not own energy assets but use the cooperative to purchase their power services. Retail electricity co-ops may also be set up by workers to sell energy products (e.g., solar panels or insulation).	Spark Energy (Power retailer)
Education	Educational electricity cooperatives can be set up to assist members with energy education and energy audits, often with a climate focus.	REscoop Plus (Enercoop and Ecopower) efficiency toolbox

First wave: self-starting rural utility cooperatives (1890s–1930s)

Geographically remote and sparsely populated, rural regions across the world have been and continue to be underserved by corporate and public electricity providers. Initial investments in rural electricity provision, such as connection of the rural community to the transmission line or the building of a local grid from scratch, are often considered by private and public actors to be too costly and most likely unprofitable due to the small size of potential consumers (Haanyika, 2006; International Labour Office [ILO], 2013; Reiche et al., 2000). It should therefore not come as a surprise that the first cooperative enterprise known to enter the electricity sector did so in a remote village in the Italian Alps in 1894. The Società cooperativa per l'illuminazione elettrica, is a still active electric distribution cooperative that generates electricity from waterpower and provides it to its members in the small rural community of Chiavenna (Mori, 2012, p. 18). The 40 years following its inception witnessed the emergence of similar cooperatives across Europe, especially in Italy, France, Germany, Austria, and Spain (Mori, 2012). Meanwhile, in Argentina, the rural electric cooperative model emerged in 1926 out of a dissatisfaction with rural electricity pricing policies (ILO, 2013, p. 8). Today, these cooperatives are responsible for 10% of the national energy production and serve 58% of Argentina's rural population (ILO, 2013).

These "first-wave" electricity cooperatives are characterized by serving remote rural areas, being entirely local and of small scale, and are predominantly reliant on hydroelectric power (International Labour Office [ILO], 2013; Mori, 2012). They were established often with no public subsidies and/or a well developed national electricity grid in these countries at the time (International Labour Office [ILO], 2013; Yadoo & Cruikshank, 2010).

Second wave: expansion through rural electrification programs (1930s–1980s)

It was in the 1930s, however, that the cooperative rural utility model began to be scaled up and utilized as a major tool of rural electrification and economic development through partnerships between public agencies and cooperatives. These "second-wave" electricity cooperatives are mostly larger scale compared to their first-wave predecessors and also purchase and/or generate electricity from fossil fuel sources (Johnson & Lewis, 2017; Yadoo & Cruikshank, 2010). Rural electricity cooperatives (RECs) established across the United States, beginning in 1936 with a public partnership of massive scale, offer a prime example of this wave. As part of President Roosevelt's New Deal, the Rural Electricity Administration (REA) was founded in the United States with the aim of kick-starting RECs across the country, helping them build electric lines and provide electricity to rural Americans on a not-for-profit basis (Johnson & Lewis, 2017). RECs were supported through subsidized credit and loan guarantees by the REA. Consequently, between 1936 and 1950 and in spite of the economic downturn due to World War II, RECs tripled the number of rural residents connected to the grid and increased the amount of grid distribution lines more than fivefold (NRECA, as cited in Yadoo & Cruikshank, 2010, p. 2943). As of early 2019, 930 RECs are active in 47 states, which serve approximately 42 million people, roughly amounting to 12% of the US population (NRECA, 2019). Electricity distribution cooperatives also emerged in the Canadian provinces of Quebec and Alberta during the 1950s, similarly driven by policy strategies facilitating decentralized and user-owned power sector development (MacArthur, 2016).

The success of this second-wave electricity cooperative in the United States inspired other jurisdictions to implement the same model for rural electrification. During the 1960s and 1970s, there was a significant surge, especially in the Global South, in establishing public agencies to support the proliferation of rural utility cooperatives. The National Electric Administration in the Philippines in 1969 (Sihag et al., 2004) and the Rural Electrification Board in Bangladesh in 1977 (Barkat et al., 2002; Rahman et al., 2013) with the specific aim of supporting the establishment and proliferation of rural electric cooperatives are prime examples of this form of partnership. Consequently, as of early 2019, most rural households in the Philippines receive their electricity from cooperatives (International Labour Office [ILO], 2013, p. 8), whereas in Bangladesh, over 70 cooperatives provide electricity to 47,650 rural villages (Yadoo & Cruikshank, 2010, p. 2943). Other testimonies to the success of public–co-op partnerships are present in Bolivia, where electricity cooperatives serve over 1 million rural residents, and in Costa Rica, where rural utility cooperatives have been active since the 1960s and have played a major role in the almost universal access to electricity in the country (International Labour Office [ILO], 2013, p. 8).

Furthermore, the public–co-op partnership model continued to be relevant post-1980s, as rural electrification programs have been successfully implemented in Nepal (Yadoo & Cruikshank, 2010), Tanzania (Ilskog et al., 2005), and South Sudan (International Labour Office [ILO], 2017), to name a few. Finally, despite benefiting greatly from public support and at times international funding agencies, these second-wave electricity cooperatives actually required fewer subsidies compared to investor-owned or municipality-owned utilities (Yadoo & Cruikshank, 2010).

Third wave: renewable energy cooperatives (1980s–present)

Up to the 1980s, the predominant form of energy cooperative was the rural electric utility cooperative, providing electricity to underserved rural areas around the globe. With technological advances in renewable energy technologies and a growing environmental movement demanding a transition away from fossil fuels and nuclear energy, the renewable energy cooperative (RE co-op) model emerged during the 1980s. This form of cooperative generates electricity (and sometimes heat) exclusively from renewable sources and often sells the generated electricity to the grid as part of a government procurement program. These co-ops emerged as a result of bottom-up social movement and community pressures and targeted state policy for renewables. For instance, the emergence of "third-wave" co-ops can be linked to the introduction of feed-in-tariff (FIT)[1] programs in Denmark and Germany that also provided specific considerations for cooperative entities (Lipp, 2007). In Germany, besides introducing a FIT program, the *Erneuerbare-Energien-Geset* (*EEG*, or the *Renewable Energy Sources Act*) also facilitated access to financing for RE co-ops through subsidized loans from the state-owned German development bank KfW, which are usually granted by regional development banks or savings banks (Yildiz, 2014).

Today, the renewable energy cooperative movement has spread throughout the continent, albeit with the majority of RE co-op activity still taking place in Germany and Denmark. The European Federation of Renewable Energy Cooperatives (REScoop.eu), the umbrella association for RE co-ops in Europe, reports just over 3,000 RE co-ops across Europe as of 2014, with around 800 of these located in Germany and 650 in Denmark (Wierling et al., 2018). Overall, as we will elaborate further, this third-wave electricity cooperative model remains highly reliant on supportive legislation, and their activities are almost entirely limited to the Global North.

Cooperatives and energy democracy

We now identify various ways electricity cooperatives have and continue to contribute to furthering energy democracy.

Widening/improving access to electricity

As previously demonstrated, rural utility cooperatives played a significant role in the electrification of rural areas that have been underserved by public and/or corporate utilities. Cooperatives, as organizations that balance the economic bottom line with the needs of its members, sprung up in remote rural areas where market efficiency and economic viability formed a barrier to electrification (Haanyika, 2006). Bringing electricity to rural areas meant, among many other things, a significant increase in the quality of life and further economic opportunities for rural residents (Ilskog et al., 2005; Paredes & Loveridge, 2018; Yadoo & Cruikshank, 2010). Overall, electricity cooperatives have proven that they are "more inclined to contribute to a sector that will improve local living conditions even if profit margins remain minimal" (Yadoo & Cruikshank, 2010, p. 2946).

A critical component of energy democracy is making the benefits of electricity available to all—and especially to those who need it most urgently. Especially first- and second-wave cooperatives[2] have widened electricity access for geographically remote and economically marginalized communities, thereby advancing energy democracy. With 1.1 billion people still without access to electricity and the rising threat of energy poverty on a global scale (International Energy Agency [IEA], 2017), there is still much more room for electricity cooperatives to expand and democratize access to this critical good.

Democratizing ownership and control

Besides widening and improving access to electricity, electricity cooperatives help establish collective and community-based ownership in one of the world's most undemocratic sectors.

Through collective ownership, communities are also able to exercise collective control over a key resource for their quality of life and socioeconomic conditions. Decisions within a cooperative are made by community members themselves, maximizing their ability to apply their knowledge about the community's needs and put forth solutions that are locally appropriate. For instance, in India, based on a knowledge of local needs, numerous rural utility cooperatives have linked electricity provision to water pumping for increasing agricultural productivity (International Labour Office [ILO], 2013, p. 11). In short, electricity cooperatives provide community members with the ability to take control of decisions that affect their current and future community lives—within and beyond the electricity sector.

A key outcome for collective ownership and control has been the revitalization of local economies where electricity cooperatives are in operation. Furthermore, collective ownership over a resource that is used locally by members themselves reduces the need for intermediary organizations and thereby eliminates associated administrative and transmission/distribution costs. As a result, rural utility cooperatives have also been able to provide better electricity prices to their members (International Labour Office [ILO], 2013). Thereby, they provide not only direct economic benefits to their members but also indirect economic benefits to the entire community. For this reason, Mori (2012) classifies rural utility cooperatives as "community cooperatives," or cooperatives that "provide services of general interest to a whole community and directly affect the community's welfare through the instruments of open membership and non-member patronage" (p. 7).

Meanwhile, the collective ownership model of RE cooperatives makes otherwise costly investments in renewable energy generation more affordable and accessible to a broader population (Huybrechts & Mertens, 2014; Seyfang & Haxeltine, 2012). RE co-ops, as community-based organizations, also are much more likely to hire and/or contract locally and to help keep financial resources and/or generated surplus within a community (Lantz & Tegen, 2009). Finally, many RE co-ops reinvest their surplus funds back into communities where they operate, supporting other community-based organizations and other local causes (Huybrechts & Mertens, 2014; Tarhan, 2015).

In short, electricity cooperatives advance energy democracy by directly democratizing ownership and control.

Accelerating the energy transition

It is easy to argue that electricity cooperatives (especially RE co-ops) contribute to a transition toward a cleaner electricity system simply by increasingly the uptake of renewable sources. Upon a closer look, the collective and democratic nature of cooperatives strengthens their ability to accelerate this transition in two additional ways.

First, involvement in or exposure to cooperative ownership increases the public acceptance of renewable energy projects (Bauwens & Devine-Wright, 2018; Musall & Kuik, 2011; Viardot, 2013; Warren & McFadyen, 2010). In contrast to community benefit agreements and community consultations implemented by investor-owned projects, the collective nature of cooperatives enables a far more equitable distribution of benefits and provides community members with a direct say in democratic decision-making processes.[3] Consequently, communities where RE projects are owned by an RE co-op respond more positively toward prospects of additional RE projects in their community compared to communities that are home to corporate-owned projects (Musall & Kuik, 2011; Warren & McFadyen, 2010).

Second, involvement in a RE co-op transforms end users of electricity to "prosumers," or individuals who are involved both in the generation and consumption of electricity, resulting in a shift in individual consumption and/or conservation behaviors (Devine-Wright, 2012; Ellsworth-Krebs & Reid, 2016). A study by Bauwens and Eyre (2017) documents that RE cooperative members, as now prosumers, have become more likely to engage in conservation and energy efficiency activities compared to end users of investor-owned or public utilities. Furthermore, many rural utility and renewable energy cooperatives also engage in member education around renewable energy, conservation, and energy efficiency, which further intensifies their potential in accelerating the transition toward a cleaner energy system (Bauwens & Eyre, 2017; Huybrechts & Mertens, 2014; Johnson & Lewis, 2017). Specifically, in the case of rural utility cooperatives, their local, democratic, and transparent management structure[4] enabled their members to learn about how their electricity is generated, distributed, priced, and used (Heras-Saizarbitoria et al., 2018; Ilskog et al., 2005; Yadoo & Cruikshank, 2010). Consequently, electricity ceases to be a technical mystery that can only be understood by experts and becomes part of everyday life.

Although electricity cooperatives clearly advance energy democracy in principle and in specific settings, they also face barriers and limitations that restrict their impact for further democratizing and decarbonizing the electricity system.

Barriers and limitations to electricity cooperatives' impact

We identified three key areas that limit electricity cooperatives' impact and potential in furthering energy democracy: (1) existence and quality of policy support; (2) democratic member engagement (or lack thereof); and (3) limitations of cooperatives in advancing social justice.

Existence and quality of policy support

As has been the case from the second wave onward, supportive public policy providing financial, technical, and organizational support has been critical for electricity co-op development. For rural utility cooperatives, the initial process of setting up the cooperative and securing electricity supply forms a barrier for many communities, especially those in remote rural areas and/or lacking access to financial and technical resources. Despite requiring lower levels of subsidies compared to investor-owned or municipal utilities, rural utility cooperatives still require adequate technical, management, and financial support in establishing and successfully running community-owned and controlled electricity systems (Ilskog et al., 2005; Yadoo & Cruikshank, 2010). In jurisdictions with rural utility cooperatives set up in the second wave, financial hardship and increasing technological complexity have also put pressure on cooperatives to sell community-owned power lines to investor-owned power companies (MacArthur, 2016).

On the other hand, public policy support for renewable energy cooperatives is of fundamental importance, as most cooperatives of this type sell their electricity/heat to the wider grid and hence are in direct competition against corporate entities with immediate access to human and financial resources. Meanwhile, many RE co-ops are volunteer driven and require time for building their membership and raising capital. FIT programs with specific considerations for RE co-ops and other community-owned RE projects have been crucial in their development and proliferation, bypassing competition against corporate entities in electricity procurement tenders and/or a limited capacity available in the grid (Huybrechts & Mertens, 2014; Tarhan, 2015). In short, regulatory frameworks remain a key factor in supporting or hindering RE co-op sectors on a global scale (Hoicka & MacArthur, 2018; Mignon & Rüdinger, 2016).

That being said, there has been an overwhelming transition away from FITs with specific considerations for co-ops across the spectrum: Frontrunning jurisdictions in RE co-op development, such as in Sweden, Germany, France, Denmark, Spain, and Ontario, have moved away from FITs toward more market-based support schemes and technology-neutral tenders/auctions (Capellán-Pérez et al., 2018; McMurtry & Tarhan, 2019; Wierling et al., 2018). Although the expansion of cooperative ownership in the electricity system since the 1980s has been enabled by the expansion of neoliberal governance, off-loading the provision of traditionally public goods onto the private sector, the recent wave of neoliberalism stressing cuts on public spending has eliminated publicly funded support for electricity cooperatives. The result has been a substantial decrease in the founding of new RE co-ops in all these jurisdictions and the inability of many small and medium-sized RE co-ops to survive (Wierling et al., 2018). RE co-ops are attempting to overcome the lack of supportive policies through institutional solutions such as developing new innovative financing tools (Dilger et al., 2017; Herbes et al., 2017) and organizing themselves in associations that share knowledge and expertise and that create political leverage to advocate for the (re)instatement of supportive schemes (Capellán-Pérez et al., 2018).

In addition, RE co-ops' struggle for existence in precarious regulatory environments resulted in many of them having to partner with corporate entities to get projects off the ground (Lipp et al., 2016; MacArthur, 2016). Along with such partnerships came a diminished level of autonomy and control for RE co-ops, limiting their democratizing impact within and beyond the organization. Furthermore, by collaborating with corporate entities that prioritize profit making, cooperatives were also found to place a higher value on the economic bottom line compared to bottom-up cooperatives (Holstenkamp & Kahla, 2016).

Democratic member engagement

A key outcome of a lack of supportive policy schemes is awareness around the cooperative model in the financial and banking sectors, resulting in difficulties accessing financing for RE co-ops. Without supportive policies, RE co-ops are seldom recognized as "viable" borrowers due to their relatively small size and lack of awareness and trust around the cooperative business model (Lipp et al., 2016).[5] Lack of financing, especially in the development phase, results in an inability to purchase technical, financial, and legal expertise and therefore creates an over-reliance on volunteers and a high level of expertise in the board of directors who are willing and able to undertake such tasks (Lipp et al., 2016). The outcome is twofold: First, RE co-ops who have access to such expertise in their board often experience burnout in a quickly shifting political landscape and complicated regulatory processes (Brummer, 2018a; Centgraf, 2018; Huybrechts & Mertens, 2014). Second, cooperatives are unable (or in some cases, unwilling) to focus on building democratic engagement from their member base due to all efforts going into raising funds and legal/technical issues. Consequently, low levels of democratic engagement have been reported for renewable energy cooperatives in Europe (Brummer, 2018b; Radtke, 2014; Yildiz et al., 2015).

Furthermore, many members approach RE co-ops as socially, politically, and ecologically minded investments rather than collectively managed organizations (Holstenkamp & Kahla, 2016). Consequently, the key relationship between the co-op and its members is reduced to individual investments on the part of members and payments of dividends on the part of the co-op. This lack of involvement provides a barrier in building individual and collective capacity in advancing energy democracy.

Rural utility cooperatives also often suffer from low levels of member engagement participation. In certain cases, this lack of engagement is linked to the cooperatives becoming too big in size or being co-opted by local power brokers and other political interests (Yadoo & Cruikshank, 2010). For rural electricity cooperatives in the United States, low levels of democratic engagement have been found to be closely linked to a "separation of ownership and control" (Jeter et al., 2018). Several cooperative boards used their power to reduce transparency and accountability toward their membership base by changing by-laws to ban proxy voting, charging members higher-than-required prices, hiring relatives, and allowing self-dealing transactions by officers and directors (Jeter et al., 2018). Over time, such practices contributed to an increased level of member apathy, who perceived the cooperative as no different from an investor-owned or municipal utility (Farrell et al., 2016; Johnson & Lewis, 2017). Perhaps even more importantly for the cause of energy democracy, these practices also derived from and contributed to the systemic exclusion of Black and other racialized communities and women in the management of cooperatives (Farrell, 2016; Johnson & Lewis, 2017). This critical point will be further evaluated in the next subsection of this chapter.

Overall, an inability or unwillingness to cultivate democracy within the organization is a major challenge in realizing almost all the associated benefits previously identified in this chapter. Furthermore, we contend that democratic governance, engagement, accountability, and transparency are not traits inherent to cooperative enterprises but instead must be deliberately and actively furthered by their members.

Limitations of cooperatives in advancing social justice

Democratization, as part of the energy democracy project or as a general concept, cannot be conceptualized without a social justice component. The growing body of literature on energy justice, closely linked with the field of energy democracy, focuses on the interconnectedness of energy systems and social justice and calls for the prioritization of most marginalized groups in accessing and benefiting from electricity generation and provision (Healy & Barry, 2017; Jenkins et al., 2016). That being said, several electricity cooperatives themselves suffer from a lack of equality and justice within their organizations. For instance, Black communities have historically been excluded from participating and being elected to the boards of numerous rural electricity cooperatives in the US South (Farrell, 2016; Johnson & Lewis, 2017). Although many initiatives are under way to restore racial justice and democracy in these organizations (Johnson & Lewis, 2017), the reflection of a history of institutionalized racism in the social, political, and legal areas can clearly be seen in the makeup of these cooperatives' management (The Rural Power Project, 2016).

On the other hand, as noted earlier, renewable energy cooperatives are capital-intensive undertakings that, even with a certain level of policy support, require significant financial and technical resources. Consequently, the overwhelming majority of RE co-ops have been developed in middle-class communities in the Global North (Radtke, 2014; Tarhan, 2015; Yildiz et al., 2015). Furthermore, studies found that these co-ops' membership base predominantly consists of affluent men, with low levels of participation from low-income, racialized, gendered communities and individuals (Radtke, 2014; Rommel et al., 2018; Yildiz et al., 2015). Without specific policies that prioritize such communities, the uptake of RE co-ops in these communities, which can greatly benefit from them, appears to be difficult. Recent literature called on RE co-ops to the keep minimum investment requirements low (Becker & Kunze, 2014; Tarhan, 2015; Wierling et al., 2018), but marginalized communities are often geographically or socially separated from circles where RE co-ops are being developed. Therefore, a deliberate effort from the side of community activists, policy makers, and the cooperative sector to prioritize marginalized communities in RE co-op development is of essential importance. This is also a reminder that despite being collectively owned and democratically governed, cooperatives still constitute a private form of ownership that, without deliberate efforts, can become exclusionary.

That being said, the number of electricity cooperatives that combine their efforts for the democratization of ownership and control with the furthering of social justice is also on the rise. For instance, the New York City Community Energy Cooperative, in collaboration with the local environmental and social justice advocacy group *Uprose*, is developing a solar project that will primarily benefit low-income and predominantly racialized community members in "a historically underinvested neighborhood" (Fast Company, 2018). Such projects provide a blueprint for electricity cooperatives, policy makers, and community groups, especially in the current neoliberal climate, by demonstrating that addressing institutionalized racism, colonialism, and poverty are critical components in advancing energy democracy through electricity cooperative development.

Conclusion

Electricity cooperatives are diverse organizations, with a democratic member structure, social movement roots, and a long history of activity in the energy sector. With the rise of renewable energy cooperatives in the past three decades, they also occupy a promising space for enhancing the participation of citizens in low-carbon energy sector transitions. Electricity cooperatives provide a wide range of electricity services, concentrating in rural distribution and retail utilities as well as power generation. They are now beginning to meet challenges of electrified transport, energy storage and building public support for new climate and energy initiatives. Other promising benefits for the realization of energy democracy include helping to improve citizens' understanding of energy sector challenges through participation, building policy advocacy and awareness, and distributing the economic benefits of energy services more broadly through society.

However, there are also limitations to the contribution of cooperatives to democratic energy transitions. Despite the commitment of many in the cooperative sector to "people before profit," significant issues remain with respect to how participatory, inclusive, or radical individual cooperatives are. In the energy sector in particular, the daunting task of navigating a sector with high capital costs as well as policy and regulatory complexity can mean that already well resourced citizens are poised to benefit most. The historical development of electricity cooperatives also highlights the key role of strong and sustained government policy support, even though they are sometimes framed as "bottom-up" initiatives. Furthermore, there is nothing necessarily low carbon about the cooperative form.

Ultimately, the contribution of electricity cooperatives to energy democracy depends on the conceptualization of energy democracy one subscribes to. Both organizational form and activities certainly have and continue to contribute to the broadening of citizen participation in the energy sector, as prosumers, investors, and project decision makers. However, more radical conceptualizations of energy democracy see a much more sweeping role for broad public ownership and deep transformation of the sector. Given the continued strength of neoliberal forms of energy transition, as well as active resistance to energy transitions in some jurisdictions, electricity cooperatives hold great potential but are also significantly limited in practice.

Notes

1 A "feed-in tariff" (FIT) is a rate that power producers are guaranteed for a defined period of time (usually 20 years) for every kilowatt hour (kWh) of electricity their contracted project(s) feed(s) into the

grid. The rates provided by FITs are often higher than market rates, thereby incentivizing the generation of electricity from renewable sources.

2 Meanwhile, while third-wave renewable energy cooperatives have also been involved in establishing district heating networks (Bauwens, 2014; Huybrechts & Mertens, 2014; Tarhan, 2015), in many cases engaging both in electricity generation and in distribution; these organizations were predominantly formed in communities already with access to electricity. Hence, their significance for the energy democracy movement will be analyzed in the following subsections.

3 Although democratic control and inclusiveness is a defining feature and principle of cooperative entities, certain cooperatives may still be involved in exclusionary governance practices and inequitable distribution of outcomes (Tarhan, 2015; Walker et al., 2010). This point will be further elaborated on in the following section.

4 As community-based organizations, utility cooperatives have been found to provide better and more transparent services to their members compared to investor-owned utilities (Ilskog et al., 2005). This being said, as will be highlighted in the next section, the level of accountability, transparency, and democratic management varies across cooperatives and cannot be taken as a given simply due to cooperative status.

5 Without adequate policy support and legislative environment, traditional lenders are wary to provide financing to cooperatives, which they see as either too small in size or too risky due to lack of awareness/policy mechanism.

References

Adeler, M. J. (2009). *Enabling policy environments for co-operative development*. Centre for the Study of Cooperatives.

Barkat, A., Ratna, N. N., Majid, M., Maksud, A. K. M., Karim, A., & Islam, S. (2002). *Economic and social impact evaluation study of the rural electrification program in Bangladesh*. Human Development Research Centre. www.hdrc-bd.com/wp-content/uploads/2018/12/6.-Economic-and-Social-Impact-Evaluation-Study-of-Rural-Electrification-Program-in-Bangladesh.pdf

Bauwens, T. (2014). What roles for energy cooperatives in the diffusion of distributed generation technologies? *SSRN*. https://doi.org/10.2139/ssrn.2382596

Bauwens, T., & Devine-Wright, P. (2018). Positive energies? An empirical study of community energy participation and attitudes to renewable energy. *Energy Policy*, *118*. https://doi.org/10.1016/j.enpol.2018.03.062

Bauwens, T., & Eyre, N. (2017). Exploring the links between community-based governance and sustainable energy use: Quantitative evidence from Flanders. *Ecological Economics*, *137*, 163–172. https://doi.org/10.1016/j.ecolecon.2017.03.006

Becker, S., & Kunze, C. (2014). Transcending community energy: Collective and politically motivated projects in renewable energy (CPE) across Europe. *People, Place and Policy*, *8*(3). https://extra.shu.ac.uk/ppp-online/transcending-community-energy-collective-and-politically-motivated-projects-in-renewable-energy-cpe-across-europe/

Berka, A., MacArthur, J. L., & Gonnelli, C. (2020). Explaining inclusivity in energy transitions: Local and community energy in Aotearoa New Zealand. *Environmental Innovation and Sustainable Transitions*, *34*, 165–182. https://doi.org/10.1016/j.eist.2020.01.006

Brummer, V. (2018a). Community energy—benefits and barriers: A comparative literature review of community energy in the UK, Germany and the USA, the benefits it provides for society and the barriers it faces. *Renewable and Sustainable Energy Reviews*, *94*, 187–196. https://doi.org/10.1016/j.rser.2018.06.013

Brummer, V. (2018b). Of expertise, social capital, and democracy: Assessing the organizational governance and decision-making in German renewable energy cooperatives. *Energy Research and Social Science*. https://doi.org/10.1016/j.erss.2017.09.039

Burke, M. J. (2018). Shared yet contested: Energy democracy counter-narratives. *Frontiers in Communication*, *3*, 22. https://doi.org/10.3389/fcomm.2018.00022

Capellán-Pérez, I., Campos-Celador, Á., & Terés-Zubiaga, J. (2018). Renewable energy cooperatives as an instrument towards the energy transition in Spain. *Energy Policy*, *123*, 215–229. https://doi.org/10.1016/j.enpol.2018.08.064

Centgraf, S. (2018). Supporting civic engagement in German energy cooperatives—transdisciplinary research based on the reflection of individual needs. *Energy Research and Social Science*, *44*, 112–121. https://doi.org/10.1016/j.erss.2018.05.003

Creamer, E., Eadson, W., van Veelen, B., Pinker, A., Tingey, M., Braunholtz-Speight, T., Markantoni, M., Foden, M., & Lacey-Barnacle, M. (2018). Community energy: Entanglements of community, state, and private sector. *Geography Compass, e12378*. https://doi.org/10.1111/gec3.12378

Cumbers, A. (2013). *Making space for economic democracy: The Danish wind power revolution*. Zed Books.

David Grace & Associates. (2014). *Measuring the size and scope of the co-operative economy: Results of the 2014 global census on co-operatives*. Retrieved April 20, 2018, from www.u.org/esa/socdev/documents/2014/coopsegm/grace.pdf

Devine-Wright, P. (2012). Energy citizenship: Psychological aspects of evolution in sustainable energy technologies. In J. Murphy (Ed.), *Governing technology for sustainability* (pp. 63–88). https://doi.org/10.4324/9781849771511

Dilger, M. G., Jovanović, T., & Voigt, K. I. (2017). Upcrowding energy co-operatives: Evaluating the potential of crowdfunding for business model innovation of energy co-operatives. *Journal of Environmental Management, 198*, 50–62. https://doi.org/10.1016/j.jenvman.2017.04.025

Ellsworth-Krebs, K., & Reid, L. (2016). Conceptualising energy prosumption: Exploring energy production, consumption and microgeneration in Scotland, UK. *Environment and Planning A: Economy and Space, 48*(10), 1988–2005. https://doi.org/10.1177/0308518X16649182

Farrell, J. (2016). *Being black still a barrier to rural cooperative board membership*. Institute for Local Self-Reliance. https://ilsr.org/being-black-still-a-barrier-to-rural-cooperative-board-membership/

Farrell, J., Grimley, M., & Stumo-Langer, N. (2016). *Re-member-ing the electric cooperative*. https://ilsr.org/report-remembering-the-electric-cooperative/

Fast Company. (2018). *This Brooklyn solar project will be owned by its community*. www.fastcompany.com/90264043/this-co-op-solar-project-will-be-owned-by-the-community-members-it-benefits

Feldpausch-Parker, A., Sprain, L., Peterson, T. R., & Endres, D. (Eds.). (2018). Energy democracy: Research topic. *Frontiers in Communication*. www.frontiersin.org/research-topics/6005/energy-democracy

Haanyika, C. M. (2006). Rural electrification policy and institutional linkages. *Energy Policy, 34*(17), 2977–2993. https://doi.org/10.1016/j.enpol.2005.05.008

Hahnel, R. (2013). *Economic justice and democracy: From competition to cooperation*. Routledge.

Healy, N., & Barry, J. (2017). Politicizing energy justice and energy system transitions: Fossil fuel divestment and a "just transition". *Energy Policy, 108*, 451–459. https://doi.org/10.1016/j.enpol.2017.06.014

Heras-Saizarbitoria, I., Sáez, L., Allur, E., & Morandeira, J. (2018). The emergence of renewable energy cooperatives in Spain: A review. *Renewable and Sustainable Energy Reviews, 94*, 1036–1043. https://doi.org/10.1016/j.rser.2018.06.049

Herbes, C., Brummer, V., Rognli, J., Blazejewski, S., & Gericke, N. (2017). Responding to policy change: New business models for renewable energy cooperatives—barriers perceived by cooperatives' members. *Energy Policy, 109*, 82–95. https://doi.org/10.1016/j.enpol.2017.06.051

Hoicka, C. E., & MacArthur, J. L. (2018). From tip to toes: Mapping community energy models in Canada and New Zealand. *Energy Policy, 121*, 162–174. https://doi.org/10.1016/j.enpol.2018.06.002

Holstenkamp, L., & Kahla, F. (2016). What are community energy companies trying to accomplish? An empirical investigation of investment motives in the German case. *Energy Policy, 97*, 112–122. https://doi.org/10.1016/j.enpol.2016.07.010

Huybrechts, B., & Mertens, S. (2014). The relevance of the cooperative model in the field of renewable energy. *Annals of Public and Cooperative Economics, 85*(2), 193–212.

Ilskog, E., Kjellström, B., Gullberg, M., Katyega, M., & Chambala, W. (2005). Electrification co-operatives bring new light to rural Tanzania. *Energy Policy, 33*(10), 1299–1307. https://doi.org/10.1016/j.enpol.2003.12.006

International Co-operative Alliance (ICA). (2019). *Co-operative identity, values, & principles*. ICA.

International Energy Agency (IEA). (2017). *Energy access outlook 2017: From poverty to prosperity*. www.iea.org/publications/freepublications/publication/WEO2017SpecialReport_EnergyAccessOutlook.pdf

International Labour Office (ILO). (2013). *Providing clean energy and energy access through cooperatives*. Geneva. www.ilo.org/wcmsp5/groups/public/-ed_emp/-emp_ent/documents/publication/wcms_233199.pdf

International Labour Office (ILO). (2017). *Transforming our world: A cooperative 2030—co-operative contributions to SDG 7*. www.ilo.org/wcmsp5/groups/public/-ed_emp/-emp_ent/-coop/documents/publication/wcms_633316.pdf

Jenkins, K., McCauley, D., Heffron, R., Stephan, H., & Rehner, R. (2016). Energy justice: A conceptual review. *Energy Research and Social Science, 11*, 174–182. https://doi.org/10.1016/j.erss.2015.10.004

Jeter, D. C., Thomas, R. S., & Wells, H. (2018). Democracy and dysfunction: Rural electric cooperatives and the surprising persistence of the separation of ownership and control. *Alabama Law Review, 70*.

https://heinonline.org/HOL/Page?handle=hein.journals/bamalr70&id=381&div=14&collection=journals
Johnson, D., & Lewis, A. (2017). Organizing for energy democracy in rural electric cooperatives. In D. Fairchild & A. Weinrub (Eds.), *Energy democracy: Advancing equity in clean energy solutions* (pp. 93–112). Springer. https://doi.org/10.5822/978-1-61091-852-7_6
Kohn, M. (2003). *Radical space: Building the house of the people*. Cornell University Press.
Laidlaw, A. F. (1980). *Co-operatives in the year 2000*. International Co-operative Alliance.
Lantz, E., & Tegen, S. (2009). *Economic development impacts of community wind projects: A review and empirical evaluation preprint*. Proceedings of windpower 2009 conference and exhibition.
Lipp, J. (2007). Lessons for effective renewable electricity policy from Denmark, Germany and the United Kingdom. *Energy Policy*, *35*(11), 5481–5495. https://doi.org/10.1016/j.enpol.2007.05.015
Lipp, J., Tarhan, M. D., & Dixon, A. (2016). *Accelerating renewable energy co-operatives in Canada: A review of experiences and lessons*. http://canada.coop/sites/canada.coop/files/files/documents/en/2016_coop-arecc_report_final_screen.pdf
MacArthur, J. (2016). *Empowering electricity: Co-operatives, sustainability, and power sector reform in Canada*. University of British Columbia Press.
McMurtry, J. J. (2004). Social economy as political practice. *International Journal of Social Economics*, *31*(9), 868–878.
McMurtry, J. J. (2009). *Living economics: Canadian perspectives on the social economy, co-operatives, and community economic development*. Emond Montgomery.
McMurtry, J. J., & Tarhan, M. D. (2019). Consumer (co-)ownership in renewables in Ontario (Canada). In J. Lowitzsch (Ed.), *Energy transition* (pp. 507–531). https://doi.org/10.1007/978-3-319-93518-8_22
Mignon, I., & Rüdinger, A. (2016). The impact of systemic factors on the deployment of cooperative projects within renewable electricity production: An international comparison. *Renewable and Sustainable Energy Reviews*, *65*, 478–488. https://doi.org/10.1016/J.RSER.2016.07.026
Mori, P. A. (2012). *Community and cooperation: The evolution of cooperatives towards new models of citizens' democratic participation in public services provision (14)*. Euricse Working Paper. www.euricse.eu/wp-content/uploads/2015/03/1393492623_n2497.pdf
Musall, F. D., & Kuik, O. (2011). Local acceptance of renewable energy—a case study from southeast Germany. *Energy Policy*, *39*, 3252–3260.
NRECA. (2019). *Our mission*. www.electric.coop/our-mission/americas-electric-cooperatives/
Ostrom, E. (1990). *Governing the commons: The evolution of institutions for collective action*. Cambridge University Press.
Paredes, D., & Loveridge, S. (2018). Rural electric cooperatives and economic development. *Energy Policy*, *117*, 49–57. https://doi.org/10.1016/j.enpol.2018.02.035
Pateman, C. (1970). *Participation and democratic theory*. Cambridge University Press.
Radel, J. (2005). Cooperatives as a community development strategy: Linking theory and practice. *The Journal of Regional Analysis and Policy*, *35*(1), 43–54.
Radtke, J. (2014). A closer look inside collaborative action: Civic engagement and participation in community energy initiatives. *People, Place and Policy Online*, *8*(3), 235–248. https://doi.org/10.3351/ppp.0008.0003.0008
Rahman, M. M., Paatero, J. V, Poudyal, A., & Lahdelma, R. (2013). Driving and hindering factors for rural electrification in developing countries: Lessons from Bangladesh. *Energy Policy*, *61*, 840–851. https://doi.org/10.1016/j.enpol.2013.06.100
Reiche, K., Covarrubias, A., & Martinot, E. (2000). *Expanding electricity access to remote areas: Off-grid rural electrification in developing countries*. World Power 2000. www.martinot.info/Reiche_et_al_WP2000.pdf
REN21. (2016). *Global status report*. Renewable Energy Policy Network. www.ren21.net/
Rommel, J., Radtke, J., von Jorck, G., Mey, F., & Yildiz, Ö. (2018). Community renewable energy at a crossroads: A think piece on degrowth, technology, and the democratization of the German energy system. *Journal of Cleaner Production*, *197*, 1746–1753. https://doi.org/10.1016/j.jclepro.2016.11.114
The Rural Power Project. (2016). *The crisis in rural electric cooperatives in the South*. http://ruralpowerproject.org/wp-content/uploads/2016/02/Rural-Power___Final.pdf
Seyfang, G., & Haxeltine, A. (2012). Growing grassroots innovations: Exploring the role of community-based initiatives in governing sustainable energy transitions. *Environment and Planning C: Government and Policy*, *30*(3), 381–400.

Sihag, A. R., Misra, N., & Sharma, V. (2004). Impact of power sector reform on the poor: Case-studies of South and South-East Asia. *Energy for Sustainable Development*, *8*(4), 54–73. https://doi.org/10.1016/S0973-0826(08)60513-X

Szulecki, K. (2018). Conceptualizing energy democracy. *Environmental Politics*, *27*(1), 21–41. https://doi.org/10.1080/09644016.2017.1387294

Tarhan, M. D. (2015). Renewable energy cooperatives: A review of demonstrated impacts and limitations. *SSRN*, *4*(1), 104–120.

van Veelen, B., & van der Horst, D. (2018). What is energy democracy? Connecting social science energy research and political theory. *Energy Research & Social Science*, *46*, 19–28. https://doi.org/10.1016/j.erss.2018.06.010

Viardot, E. (2013). The role of cooperatives in overcoming the barriers to adoption of renewable energy. *Energy Policy*, *63*, 756–764. https://doi.org/10.1016/j.enpol.2013.08.034

Walker, G., Devine-Wright, P., Hunter, S., High, H., & Evans, B. (2010). Trust and community: Exploring the meanings, contexts and dynamics of community renewable energy. *Energy Policy*, *38*(6), 2655–2663.

Warren, C. R., & McFadyen, M. (2010). Does community ownership affect public attitudes to wind energy? A case study from south-west Scotland. *Land Use Policy*, *27*(2), 204–213.

Wierling, A., Schwanitz, V., Zeiß, J., Bout, C., Candelise, C., Gilcrease, W., & Gregg, J. (2018). Statistical evidence on the role of energy cooperatives for the energy transition in European countries. *Sustainability*, *10*(9), 3339. https://doi.org/10.3390/su10093339

Wiersma, B., & Devine-Wright, P. (2014). Decentralising energy: Comparing the drivers and influencers of projects led by public, private, community and third sector actors. *Contemporary Social Science*, *9*(4), 456–470. https://doi.org/10.1080/21582041.2014.981757

Wright, E. O. (2010). *Envisioning real utopias*. Verso. www.ssc.wisc.edu/~wright/ERU_files/ERU-CHAPTER-7-final.pdf

Yadoo, A., & Cruikshank, H. (2010). The value of cooperatives in rural electrification. *Energy Policy*, *38*(6), 2941–2947.

Yildiz, O. (2014). Financing renewable energy infrastructures via financial citizen participation: The case of Germany. *Renewable Energy*, *68*, 677–685. https://doi.org/10.1016/j.renene.2014.02.038

Yildiz, Ö., Rommel, J., Debor, S., Holstenkamp, L., Mey, F., Müller, J. R., & Rognli, J. (2015). Renewable energy cooperatives as gatekeepers or facilitators? Recent developments in Germany and a multidisciplinary research agenda. *Energy Research and Social Science*, *6*, 59–73. https://doi.org/10.1016/j.erss.2014.12.001

17

A FEMINIST LENS ON ENERGY DEMOCRACY

Redistributing power and resisting oppression through renewable transformation

Jennie C. Stephens and Elizabeth Allen

Why link energy democracy with feminism?

Energy democracy is a growing social movement that prioritizes the transformative social justice potential of moving away from fossil fuels to a renewable-based future (Angel, 2016; Fairchild & Weinrub, 2017; Farrell, 2016; Stephens, 2019; Sweeney, 2012; Szulecki, 2018; van Veelen & van der Horst, 2018). The principles of energy democracy are based on the possibilities for redistributing power more equitably among individuals and communities through renewable transformation. Energy democracy acknowledges how fossil fuel–based energy systems and the associated massive corporate profits of large multinational energy companies have perpetuated inequalities, exacerbated already unequal vulnerabilities, and promoted widespread injustices among and within communities around the world (Burke & Stephens, 2017). By highlighting the negative societal impacts of fossil fuel–based concentration of power and wealth, the principles of energy democracy connect energy system change with an associated transformation toward a more socially just and equal society (Burke & Stephens, 2018).

Feminism is a movement to end sexism, sexist exploitation, and oppression (hooks, 2015). Feminists acknowledge that there are problems with gender equity, and they want to make changes for improvement (Adichie, 2012). Applying a feminist lens to the analysis of any aspect of society involves considering the gender dynamics associated with decision making and positionality (i.e., asking where are the women and the men, how did they get to the positions they are in, and who benefits from this arrangement?) (Enloe, 2017). A feminist lens encourages feminist curiosity, which reveals patriarchal systems—made up of values, beliefs, and relationships—that perpetuate domination, intimidation, and submission (Enloe, 2017). The core hierarchy of patriarchy, the primacy of the male, structures relationships between men and women and also structures almost every other form of power, including head of household over his children, free men over slaves, "natives" over immigrants, rich over poor, racial majority over racial minority, etc. (Schrupp, 2017).

Exploring how society's infrastructure systems, including energy systems, perpetuate oppression and reinforce gender and racial inequities is a valuable way to motivate change that could shift power dynamics (Wilson, 2018). Infrastructure systems are much more than their tangible materiality; infrastructure systems also include formal and informal rules, explicit and implicit

DOI: 10.4324/9780429402302-20

cultural assumptions, policies, institutions, and politics, as well as heterogeneous sets of uses and users (Frischmann, 2013; Parks & Starosielski, 2015; Star, 2015; Star & Ruhleder, 2015). Although energy systems are generally thought about in technological terms, infrastructure systems also organize diverse sets of users, operators, and managers—infrastructure publics—in an interconnected web of relationships (Collier et al., 2016). Given the massive scale of infrastructure transformation associated with the shift from fossil fuel–dominated energy systems to renewable-based energy, more intentional analysis of the gendered relationships, assumptions, institutions, and policies that contribute to innovations in energy systems is critical.

Despite the potential interconnections between feminist normativity and energy democracy, feminist explorations of energy system change have been minimal (Baruah, 2015; Pearl-Martinez & Stephens, 2016; Wilson, 2018). Those calling for more feminist social science research on climate change argue that, without gender analysis, attempts to respond to climate change will be insufficient, unjust, and therefore unsustainable (Denton, 2002; MacGregor, 2009). Several feminist scholars have pointed out that dominant responses to climate change mitigation are stereotypically masculine, focusing on supply-side technological fixes and militaristic "muscle-flexing" (Denton, 2002; Robinson & Odendahl, 2019), while women have focused more on social change and the social dimensions of responding to climate change. (Johnsson-Latham, 2007; MacGregor, 2009). While there is a rapidly growing body of scholarship advancing feminist perspectives on climate change (Robinson, 2018; Robinson & Odendahl, 2019; Terry, 2009), and there is a growing literature on energy systems and energy transitions that applies a women and gender studies lens (Barron, 2014; Willow & Keefer, 2015) (see next section for a review of gender and energy research), we are unaware of an explicitly feminist analysis of the energy democracy movement and energy democracy goals.

This chapter uses a feminist lens to conceptualize both the energy sector and the energy democracy movement. Incorporating a feminist perspective to study energy systems enables analysis of underlying power dynamics and opens possibilities for understanding pathways for energy system transformation that prioritize equity, inclusion, and justice. Applying a feminist lens to the energy democracy movement reveals the power of linking feminism and energy democracy to address oppressive gender dynamics and accelerate a transition to renewable energy systems. This analysis provides a theoretical contribution to our understanding of how fossil fuel–based energy systems are connected to the systemic oppression and marginalization of women and nonwhite people. We begin by briefly reviewing the widely varied research on gender and energy. We then consider specifically ways in which patriarchal values and authoritarian, technocratic leadership have influenced energy system innovation and transformation. Next, we consider points of alignment between feminist principles and energy democracy, including the shared emphasis on resisting, reclaiming, and restructuring current power structures. This analysis of feminism and energy democracy leads us to conclude that these agendas are fundamentally interconnected and aligned. More explicit articulation of these linkages could result in accelerating social change toward a future with greater gender and racial equity based on distributed renewable energy.

Overview of the literature on gender and energy

Male-dominant systems of power and privilege control many sectors of society (Ferrante, 2019; Manne, 2018). New research is revealing multiple ways in which technologies, data, experiments, drug trials, medical interventions, and urban spaces are biased toward men because so much of the world is designed primarily for men (Kern, 2020; Perez, 2019). Energy systems and energy sector decision making has been particularly male dominant, homogeneous, and hierarchical (Pearl-Martinez & Stephens, 2016; Wilson, 2018). As issues of equity, inclusion, exclusion, and justice become more prominent in climate and energy policy (Robinson & Odendahl, 2019), particularly as the proposed Green New Deal advances in United States politics (Aranoff et al., 2019), research connecting gender and energy is increasingly valuable, and a feminist curiosity about power structures associated with energy infrastructure is important.

It is increasingly recognized that gender dynamics influence participation in decision making about responses to climate change (Leduc, 2010; Nagel, 2015). Research has explored the ways in which climate change impacts are differently distributed for men and women globally (Brody et al., 2008; Denton, 2002), how gender identities interface with perceptions of climate change science and policy alternatives (McCright, 2010; Nagel, 2015), and how political and social action to address climate change might reinforce or reduce gender inequality (Arora-Jonsson, 2011; Ergas & York, 2012; Nagel, 2015).

Relationships between women's issues and energy systems have received significant attention in development literature (Baruah, 2015; Farhar, 1998; Osunmuyiwa & Ahlborg, 2019). Historically, international aid for energy development has emphasized large infrastructure projects, which do not always provide improved access to energy for most vulnerable populations (Cecelski, 2000; Farhar, 1998; Winther et al., 2018). In societies lacking access to centralized energy infrastructure, women are typically the primary household energy managers (gathering fuel for cooking, etc.), and thus there are calls for making gender a central consideration in economic development programs focused on renewable energy (Farhar, 1998; Feenstra, 2002; James, 1999). Efforts to integrate gender considerations are often referred to as gender mainstreaming (Alston, 2014; Moser, 2005). Osunmuyiwa and Ahlborg (2019) write that a narrative of "women's poverty" has prompted programs and policies that design electricity systems to address "women's problems" and that focus on improving household energy access. This narrative can reinforce gendered ideas about women's proximity to the home, and recent critiques suggest that electrification may have limited effect on gender equity issues and may be linked to some weakening of the strength of community ties (Listo, 2018; Osunmuyiwa & Ahlborg, 2019; Prebble & Rojas, 2018).

In literature about gender and energy in the developed world, concerns about energy access and vulnerability figure less prominently, and analysis is more likely to focus on gender disparities in energy sector careers (Fraune, 2015; Pearl-Martinez & Stephens, 2016). While the recent boom in oil and gas extraction has led the industry to expand their outreach to women in order to rapidly train and hire a skilled workforce (Neuhauser, 2014), women make up a small minority of leaders and professionals in the energy sector (Pearl-Martinez & Stephens, 2016). Globally, an estimated 22% of the full-time oil and gas workforce is female (IRENA, 2019). Even within the renewable energy industry, gender equity lags behind other sectors of society. According to a recent global survey, an estimated 28% of the STEM-related workforce and 45% of administrative employees in the renewable energy sector are female (IRENA, 2019). In the solar industry in the United States, women made up 21.6% of the workforce in 2014 (Pearl-Martinez & Stephens, 2016). Despite the persistent lack of gender diversity in the energy industry, the benefits of hiring and retaining women in the energy workforce are well documented. Companies that invest in women and include gender-diverse teams have more innovation and revenue growth than their competitors (Pearl-Martinez & Stephens, 2016; Pellegrion et al., 2011; Woolley et al., 2010). Women are more likely to champion sustainability and environmental action in the workplace, and companies with more women on the board of directors are more likely to invest in renewable energy and reduce carbon emissions (Pearl-Martinez & Stephens, 2016). Differences in who holds decision-making power in the energy sector cannot be explained alone by individual preferences and attitudes alone and can be traced back to social conditions including wealth disparities and gender (Fraune, 2015; Kahsar, 2019).

Energy, patriarchy, authoritarian leadership, and technocratic decision making

The energy sector, energy decision making, and energy discourse have been dominated by technocratic perspectives focused largely on the engineering details of energy systems. Given the quantitative absolutism associated with this technocratic lens, energy experts often present themselves as authoritative and capable of providing definitive knowledge to inform energy decision making. Like so many technologically oriented fields, the authoritative attitude has created a culture that underestimates and diminishes the social and political dynamics of energy systems.

This technocratic perspective in energy is coupled with the dominance of scientific and technocratic approaches to considering societal responses to climate change (Robinson, 2018). These scientific and technical approaches to confronting climate change generally assume that humans can control natural systems for their own advantage, which evolves from a technocratic, patriarchal, racist, masculine perspective that dismisses or ignores many aspects of the human experience (Myhre, 2019). This perspective is based on assumptions of control and dominance and therefore oppression. It embodies an arrogance of science and scientists who think that, if only the rest of the world knew what they knew and acted in the rational way that they think people should act, then the world's problems would be solved. From a feminist perspective, this scientific and technocratic arrogance is limiting rather than expanding because many people, perspectives, and impacts are automatically excluded from energy decision making.

An alternative to the narrow, ineffective technocratic approach to responding to climate change, a "people first" perspective on climate change, is advocated for by Mary Robinson, former president of Ireland and UN Special Envoy on Climate Change who is also often associated with the phrase "Climate change is a man-made problem that requires a feminist solution." Robinson (2018) states that "to deal with climate change we must simultaneously address the underlying injustice in our world and work to eradicate poverty, exclusion and inequality." Connecting gender and race, paleoceanographer and thought leader in the climate science communication field, Sarah E. Myhre, has called for the climate movement to look to women of color as the leaders in this space based on the "lens of moral attention" that they bring to their public work (Wang, 2018). Myhre states that "the existing paradigm of climate leadership has been focused on white men's leadership. And their leadership is reductionist, technocratic, top-down, and it does not have a comprehensive moral attention to the people who are in the most danger" (cited in Wang, 2018, para. 10). These calls for broadening the strategy around climate change and energy acknowledge that diversifying who is involved in defining the agenda will change the agenda.

Gender and power dynamics that shape whose voices are heard, whose ideas and expertise are respected in society, play out at multiple scales from the interpersonal level, the organizational level, to the level of national and global decision making. In all parts of society, gender is a defining element of everyday interactions; we enact learned, scripted gender roles that are pervasive and self-perpetuating (Allen, 2011). Male expertise and authority have been often taken as a given, even when the level of confidence exerted by many men is clearly not warranted (Solnit, 2014). As the term "mansplaining" has recently become widely used in popular culture, the gendered nature of confrontational confidence has gained recognition. Solnit (2014) connects this gendered authoritativeness to women's voices and ideas being minimized;

"the out-and-out confrontational confidence of the totally ignorant is, in my experience, gendered. . . . It's the presumption that makes it hard, at times, for any woman in any field; that keeps women from speaking up and from being heard when they dare" (para. 19).

While this masculine authoritative culture is pervasive in many sectors and influences many kinds of public policy; the energy sector and energy policy have been particularly influenced by this tendency. This claim is based on the first author's experiences as a professional woman who has worked on energy technology and energy policy for more than 18 years. The highly technical nature of energy systems and energy policy, coupled with the concentration of wealth and power that has emerged in energy companies around the world, has reinforced gendered assumptions of authority, expertise, and power. Energy is a space in which feminist challenges to authority have been frequently disregarded and marginalized. With growing awareness about how men continue to control dominant narratives in our society (Ferrante, 2019), exposing the gendered nature of energy systems has potential to accelerate change. Identifying and exposing the male-dominated aspects of energy processes and decision making provides an opening for more inclusive engagement and greater diversity of ideas. The resulting broader engagement is essential for more transformative change.

The relationship between feminism and environmental activism has been explored in scholarly, theoretical, and activist literature since the 1960s (Mann, 2011; McCammon et al., 2018; Merchant, 1981; Warren, 1997). In 1974, d'Eaubonne (1974) first coined the term "ecofeminism" as she characterized parallels between the patriarchal suppression of women and the suppression of nature and argued that this suppression results in environmental destruction. Ecofeminist thought has been widely critiqued as oversimplifying and essentializing "women's closeness to nature" and vulnerability (Archambault, 1993; Gaard, 2011). While recognizing the problematic assumptions embedded in some ecofeminist thought, integrating ecological and feminist approaches is vital to analyze linked systems of oppression and to advance global transformation for gender, energy, and climate justice globally (Gaard, 2011). At the level of personal motivations for activism, many women engaged in environmental justice movements invoke racial, ethnic, and religious identities to support their activism, and these intersectional identifies are powerful in promoting women's leadership and activism (McCammon et al., 2018; Pulido, 1997; Taylor, 2000)). Contemporary activists and scholars increasingly recognize explicit connections between the exploitation and degradation of the natural world with the subordination and oppression of women, people of color, and other marginalized people. However, mainstream energy and climate policy has not integrated an intersectional feminist perspective into the discourse.

Revealing gendered assumptions and exploring how patriarchal systems have limited energy transformation and responses to climate change are urgent challenges at this time in human history. Acknowledging that mainstream feminism has been heavily critiqued for being exclusionary and for enforcing white supremacy, the intersectionality of gender and race is important to be integrated. Without a diverse, representative leadership on climate and energy, those individuals and organizations currently profiting from sustaining fossil fuel reliance will continue to exert their power to thwart energy system innovation toward a distributed, heterogeneous mix of renewable-based energy.

Exploring the intersections of energy democracy and feminism

Energy democracy is based on a vision for a just and equitable renewable energy future that links environmental sustainability, community resilience, social justice, racial justice, and economic equity (Burke & Stephens, 2017). The concept of energy democracy draws on energy justice concepts but in contrast has emerged largely from social movements and is less grounded in academic theory and more oriented toward articulating an agenda for system transformation (Szulecki, 2018; van Veelen & van der Horst, 2018). Energy democracy has emerged as a powerful principle around which different climate and energy activist organizations have coalesced. Energy democracy centralizes a vision of active, engaged energy citizenship resulting in a distributed energy system with some level of local control and a local or regionally appropriate mix of different renewable generation. Here it is also important to acknowledge that some challenge the extent to which participation in energy decision making is influenced by social and economic factors, including race, economic status, and home ownership and gender (van Veelen & van der Horst, 2018).

Energy democracy activism is a direct response to the disproportionate influence that large multinational fossil fuel–based energy companies have on energy and climate policy. Corporate interests, rather than the public good, have been increasingly prioritized, resulting in a culture of climate denial (Oreskes, 2015; Supran & Oreskes, 2017) and a systemic minimization and dismissal of the many risks of fossil fuels (Healy et al., 2019). In the United States, the energy democracy movement has gained clarity in its vision and approach as energy democracy is in direct opposition to the Trump administration's policy of Energy Dominance, which is perhaps the antithesis of energy democracy. The Trump administration's principle of Energy Dominance sought to accelerate fossil fuel extraction to strengthen the dominance of the United States over other countries and to strengthen the dominance of the wealthy, powerful individuals and corporations who are profiting over everybody else (Schneider & Peeples, 2018). The blatant misogyny that has characterized Trump's personal behavior and the administration's approach to multiple policy issues was also at play in energy policies that promoted expansion of fossil fuel extraction and infrastructure. The Trump administration unabashedly placed the interests of wealthy, powerful white men above the interests of the public good.

The energy democracy movement has been formulated around three broad goals. The first goal, *resist*, includes trying to slow and eventually cease the extraction of fossil fuels (the "leave it in the ground" campaign), stopping the expansion of fossil fuel energy infrastructure including natural gas pipelines, and phasing out policies that favor fossil fuel energy over renewable and distributed energy. The second goal, *reclaim*, involves the democratization of the energy industry and sources of energy generation. Specifically, this describes individuals and communities reclaiming control over decisions about their energy supply. In many framings of energy democracy goals, the proposed approach to reclaiming control involves returning parts of the energy system that are controlled by private, for-profit corporations to public hands and establishing new publicly owned and managed energy providers. The third energy democracy goal, *restructure*, entails a transformation toward a more democratic, equal, and distributed energy system. Key steps in this process include empowering communities and historically marginalized groups to play a leading role in the energy system and establishing community energy systems (Burke & Stephens, 2017).

In multiple ways, these three energy democracy goals are aligned with the goals of feminism. The goals of the feminist movement can also be characterized in the same three categories of resist, reclaim, and restructure. Feminist principles include resisting the dominant oppressive systems of patriarchy; reclaiming women's voices, agency, and control; and restructuring societal systems, institutions, and cultures to support gender equity and social justice. Both energy democracy and feminism are social movements that challenge the dominant patriarchal systems that have marginalized, excluded, and oppressed some people and communities. Energy democracy and feminism are responding to negative societal implications of authoritarian power, and both movements are fundamentally focused on redistributing power among a greater number of people where once it was concentrated in the hands of a few.

To empirically explore linkages between feminism and energy democracy, we have analyzed two national organizations working toward energy democracy goals that are led by women and guided by feminist principles. *Grid Alternatives* is a solar installation and workforce training nonprofit based in California. *Grid Alternatives* specifically recruits women and people of color into their solar training programs and provides solar power to marginalized communities including workforce training programs that partner with tribal members and programs targeted for people impacted by the criminal justice system. *Mothers Out Front* is a climate and energy activism nonprofit based in Massachusetts. *Mothers Out Front* has distributed operations across the United States and primarily focuses on local issues including protesting fossil fuel infrastructure and lobbying for progressive climate policies. *Mothers Out Front* and *Grid Alternatives* are representative of a relatively small group of nonprofit, nongovernmental organizations working to accelerate the renewable energy transition with a focus on social justice and gender diversity. While neither of these organizations explicitly self-identifies as being part of the energy democracy movement and neither uses the word "feminist" or "feminism" in their materials, their mission statements and organizational objectives are aligned with goals of both feminism and energy democracy, as we will demonstrate in our analysis.

Mothers Out Front is an explicitly nonhierarchical organization. Cofounder Kelsely Wirth states, "We do a lot of relationship building, we share what mothers can do and we share stories about successes that teams are having in other parts of the country. One of the most important pieces of what we do is support development of relationships and among volunteers and staff. The work we do is hard, and what sustains it is the relationships" (Aarons-Mele, 2018). *Mothers Out Front* seeks to engage a broad audience, but it is important to note that many of the organization's members are white middle and upper-class women. Several local chapters are actively working toward achieving greater socioeconomic and racial diversity among its volunteer members, explicitly acknowledging the legacy of white privileged people dominating both the environmental movement and the feminist movement.

Grid Alternatives is focused on the principle that in the current dominant energy system, low-income communities are disproportionately paying for power generation, in terms of environmental and health impacts and in terms of the relative financial cost of energy access. To address inequities in access to renewable energy, *Grid Alternatives* focuses on installing solar electric systems exclusively for low-income families and community organizations. Within the installation process, the organization structures career training opportunities for members of the public who may not otherwise be able to access professional development opportunities in this sector. By installing new renewable energy infrastructure, members of the community come together to gain skills and economic benefits. Installation projects provide an opportunity to get training and experience in the classroom and in the field.

In this next section we review how both *Mothers Out Front* and *Grid Alternatives* are linking feminism with energy democracy by exploring how they are working to resist, reclaim, and restructure.

Resist

Resistance to new fossil fuel infrastructure is a key strategy for energy democracy activists. With recent expansion of natural gas infrastructure and hydraulic fracturing (fracking), protesting natural gas infrastructure resists the dominant agenda of the fossil fuel industry by driving up costs for natural gas companies and making it more difficult to sell and transport fracked gas (Lee & Klump, 2018). Resistance efforts that advocate for ending expansion of new fossil fuel infrastructure serve to raise awareness of climate issues and bring greater scrutiny to fossil fuel industry's activities at a local level. Acts of resistance by *Mothers Out Front* include campaigning against new natural gas pipelines planned for construction near residential zones and raising concern about gas leaks in existing fossil fuel infrastructure. These efforts have centered on empowering local communities to resist fossil fuel infrastructure and protecting communities from the harmful health effects of gas leaks (Mothers Out Front, 2018). The primary motivation of *Mothers Out Front*'s local efforts to thwart the fossil fuel agenda and its harmful effects is to protect communities, which is reflective of the traditionally feminine community caregiving or maternal frame taken by some environmental activists (Willow & Keefer, 2015). As the feminist movement continues to resist the dominant oppressive systems of patriarchy, broader recognition of how fossil fuel–based energy companies are contributing to and benefiting from the oppression can expand the synergistic efforts of activists and policy makers.

Kelsey Wirth, cofounder of Mothers Out Front, explained in a 2018 interview that often the leaders of the campaign against expansion of new gas pipelines are first-time political activists who are motivated to speak out publicly because of the urgency of climate change and human health risks and the strength of social ties within the network (Aarons-Mele, 2018). Wirth notes:

> Mothers vote, and are the primary purchasers in their household. There's nothing gas companies hate more than having a bunch of mothers tagging gas leaks that they're not fixing. Mothers are rate payers, and we're calling out gas companies to fix something that is pretty outrageous.
>
> *(Aarons-Mele, 2018)*

Reclaim

Reclaiming public control of energy system innovation is another key priority for energy democracy. Localizing and democratizing the generation and distribution of energy requires public control of energy infrastructure and different ownership models and financial investment systems that benefit community interests over corporate interests. This is aligned with the feminist principle of reclaiming women's authority and public voice in policy decision making.

One example of reclaiming energy systems to public control can be seen in *Mothers Out Front*'s efforts to bring Community Choice Aggregation (CCA) to the communities where the organization is active. CCA gives communities greater control over electricity purchasing and allows them to choose the source of their electricity supply (Burke & Stephens, 2017). CCA also enables more democratic oversight and participation of residents in the electricity procurement process than typical utility–customer electricity arrangements (Duda et al., 2017). CCA programs can support purchasing a higher-than-required share of renewable energy in the electricity mix. This initiative reclaims citizens' control of the energy sector, including bringing back public decision making about energy sources and purchasing and normalizing a new ownership model of the energy supply with more community control (Burke & Stephens, 2017). These initiatives of *Mothers Out Front* demonstrate alignment between feminists' efforts to reclaim women's voices and women's agency and control and energy/climate activists' efforts in order to take back control from the fossil fuel industry.

Restructure

Restructuring energy systems to enable more distributed and locally appropriate renewable-based energy is another key strategy of the energy democracy movement. The goal of restructuring is to simultaneously redistribute economic and political power as energy technologies become more decentralized and distributed (Stephens, 2019). By restructuring energy systems, energy access and assets are shared more broadly, and community wealth-building is supported. In effect, low-income communities and communities of color have jobs, economic benefits, and decision-making power within energy systems.

An example of how *Grid Alternatives* works toward restructuring energy systems is the organization's career training programs for historically underrepresented and marginalized groups designed to encourage racial and socioeconomic diversity and inclusion in the solar industry. This program focuses on five R's: (1) recruitment to reach members of underserved communities and promote solar job opportunities, (2) real-world experience for people attempting to enter the solar energy workforce, (3) readiness—working with industry leaders to provide relevant job training that meets industry needs, (4) referrals to connect job seekers with industry, and (5) retention—bringing the issue of inclusion to the forefront of the conversation happening in the industry.

Grid Alternatives's cofounder Erica Mackie describes *Grid Alternatives*'s approach to engaging participants as a "community barn raising method with a job training component" (Zipp, 2018). In installing new renewable energy infrastructure, members of the community come together to gain skills and gain economic benefits. Installation projects provide an opportunity to get training and experience in the classroom and in the field. Mackie states, "It's not just about where solar panels go and who benefits from that particular infrastructure, it's about who is going to get those jobs in the future" (Zipp, 2018).

Grid Alternatives's workforce training programs include an initiative for recently incarcerated individuals, a tribal solar energy program, and the Women in Solar Initiative. These programs are meant to create more diversity within the solar industry, to elevate the profile of people from historically marginalized groups working in the industry, and to give representatives of these communities the opportunity to participate in the industry. *Grid Alternatives*'s Women in Solar Initiative was originally started in 2014 with a $1 million donation from Sun Edison (Chrisman, 2014). The campaign is designed to attract more women to the solar industry and to improve retention by making the industry more inclusive to the women already working in it. This includes a series of We Build women-only solar system installations, along with a pledge to train 1,000 women in solar installation and 30 women to be solar installation team leaders (Chrisman, 2014). The campaign also includes a quarterly "We Lead" webcast series revolving around the issues facing women in the solar industry and providing advice for women attempting to enter the solar industry. *Grid Alternatives* also pledged to include at least 20 women in its SolarCorps program in the coming years, ensuring a 50/50 gender ratio (Womens Environmental Network, 2014). Additionally, the campaign launched the We-Give circle of women donors to *Grid Alternatives* in order to form connections among the women involved with the solar industry (Grid Alternatives, 2018). These programs by *Grid Alternatives* link feminist restructuring with energy system restructuring. As the feminist movement continues to strive toward the restructuring of societal systems, institutions, and cultures to support gender equity and social justice, the energy sector is a critically important place to focus, and the efforts of *Grid Alternatives* to connect feminism with energy democracy demonstrate the synergies and intersections.

Conclusions

Applying a feminist lens to energy democracy reveals the power of linking these movements to address oppressive gender dynamics and accelerate a transition to renewable energy systems. Incorporating a feminist perspective to study energy systems enables analysis of underlying power dynamics and opens up possibilities for understanding pathways for energy system transformation that prioritize equity, inclusion, and justice. As energy decision making continues to be driven by powerful political and financial interests rather than the principle of the common good and democratic deliberation and processes, this linking of different social movements focused on social justice seems essential for change. Applying a feminist lens elevates the potential of inclusive transformation, prioritizing investment in marginalized groups including women, people of color, immigrants, and youth.

The energy democracy movement is aligned and consistent with the feminist movement. Both energy democracy and feminism are social movements that challenge the dominant patriarchal systems that allow some members of society to profit from the marginalization, exclusion, and oppression of other members of society. Energy democracy and feminism are both responding to the negative societal implications of oppression, and both movements are focused on resisting, reclaiming, and restructuring to redistribute power more equitably among a greater number of people, in addition to ensuring that both energy democracy and feminism embrace an antiracist and inclusive approach that promotes a vision of empowered individuals and communities.

The energy sector has traditionally been male dominated and technocratic, and the fossil fuel industry is intrinsically linked to the patriarchy. This is reflected in the lack of gender diversity in the industry, the lack of female leadership, and the lack of solutions that consider systems of oppression and that work for greater representation, equity, transparency, and sustainability. Our exploration of the linkages between energy democracy and feminism highlights the critical role of women's leadership in accelerating the transition away from fossil fuels toward a renewable-based future (Allen et al., 2019).

The effectiveness of expanding the explicit intersectionality of different movements was demonstrated by the Global Women's March, which took place January 21, 2017, in reaction to the election of Donald Trump. This march, involving an estimated 7 million marchers globally, was a broad-based call for society to address gender, economic, racial, and environmental injustices. The March represented an amalgamation of progressive resistance causes, including women's reproductive rights, environmental justice, Black Lives Matter, immigrant rights, and religious inclusion. The Women's March and related demonstrations and movements leading up to and following the March represent a pivotal moment in history that demonstrates the increasing importance of women's organizing power and leadership in social change and resistance movements and underscores the power of an integrated approach to political activism (Darrow, 2017; Felsenthal, 2017; Waddell, 2017). The election in 2018 of a record 117 women in Congress is another demonstration of the growing political power of women. The power of elevating the voices of diverse women to resist fossil fuel industry interests can be seen in the leadership of Rep. Alexandria Ocasio-Cortez in the United States Congress in advancing the Green New Deal, a proposed resolution for major national-level investment in renewable energy and energy job creation. By broadening engagement and centralizing women's leadership, organizations and individuals are providing a different kind of collaborative leadership that is redistributing power to promote a transition to more equitable, resilient, and sustainable energy systems.

References

Aarons-Mele, M. (2018, May 22). *Dirty energy is no match for Mothers Out Front* [Audio podcast]. https://podcasts.apple.com/us/podcast/hiding-in-the-bathroom/id1102773197?mt=2

Adichie, C. H. (2012). *We should all be feminists*. Anchor Books.

Allen, B. J. (2011). *Difference matters: Communicating social identity*. Waveland Press.
Allen, E., Lyons, H., & Stephens, J. C. (2019). Women's leadership in renewable energy transitions, energy justice and energy democracy: Redistributing power. *Energy Research & Social Science, 57*.
Alston, M. (2014). Gender mainstreaming and climate change. *Women's Studies International Forum, 47*, 287–294.
Angel, J. (2016). *Strategies of energy democracy*. Rosa-Luxemburg-Stiftung. www.rosalux.eu/publications/strategies-of-energy-democracy-a-report/
Aranoff, K., Battistoni, A., Cohen, D. A., & Riofrancos, T. (2019). *A planet to win: Why we need a green new deal*. Verso.
Archambault, A. (1993). A critique of ecofeminism. *Canadian Woman Studies, 13*(3).
Arora-Jonsson, S. (2011, May). Virtue and vulnerability: Discourses on women, gender and climate change. *Global Environmental Change-Human and Policy Dimensions, 21*(2), 744–751. https://doi.org/10.1016/j.gloenvcha.2011.01.005
Barron, R. (2014, June 30). *The business of bringing more women into solar*. http://solarenergy.net/News/business-bringing-women-in-solar/
Baruah, B. (2015). Opportunities and constraints for women in the renewable energy sector in India. *Women and Environments, 94–95*, 7–10.
Brody, A., Demetriades, J., & Esplen, E. (2008). *Gender and climate change: Mapping the linkages. A scoping study on knowledge and gaps*. https://gsdrc.org/document-library/gender-and-climate-change-mapping-the-linkages-a-scoping-study-on-knowledge-and-gaps/
Burke, M. J., & Stephens, J. C. (2017). Energy democracy: Goals and policy instruments for sociotechnical transitions. *Energy Research & Social Science, 33*, 35–48. https://doi.org/10.1016/j.erss.2017.09.024
Burke, M. J., & Stephens, J. C. (2018). Political power and renewable energy futures: A critical review. *Energy Research & Social Science, 35*, 78–93. https://doi.org/10.1016/j.erss.2017.10.018
Cecelski, E. (2000). *The role of women in sustainable energy development*. National Renewable Energy Laboratory. NREL/SR-550-26889. www.nrel.gov/docs/fy00osti/26889.pdf
Chrisman, K. (2014, December 4). Women apply here: The solar industry is trying to fix its gender issues. *CNBC News*. www.cnbc.com/2014/12/04/women-apply-here-the-solar-industry-is-trying-to-fix-its-gender-issue.html
Collier, S. J., Mizes, J. C., & von Schnitzler, A. (2016). Preface: Public infrastructures/infrastructural publics. *Limn*, 7, 2–7.
d'Eaubonne, F. (1974). *Le féminisme ou la mort*. Pierre Horay.
Darrow, B. (2017, January 23). More than 1 in 100 of all Americans were at women's March events. *Fortune*. http://fortune.com/2017/01/23/womens-march-crowd-estimates/
Denton, F. (2002). Climate change vulnerability, impacts, and adaptation: Why does gender matter. *Gender and Development, 10*(2), 10–20.
Duda, J., Hanna, T., & Burke, M. (2017). *Building community capacity for energy democracy: A deck of strategies*. https://thenextsystem.org/learn/collections/building-community-capacity-energy-democracy-deck-strategies
Enloe, C. (2017). *The big push: Exposing and challenging the persistence of partiarchy*. University of California Press.
Ergas, C., & York, R. (2012). Women's status and carbon dioxide emissions: A quantitative cross-national analysis. *Social Science Research, 41*(4), 965–976.
Fairchild, D., & Weinrub, A. (Eds.). (2017). *Energy democracy: Advancing equity in clean energy solutions*. Island Press.
Farhar, B. C. (1998). Gender and renewable energy: Policy, analysis, and market implications. *Renewable Energy, 15*(1–4), 230–239. http://doi.org/10.1016/S0960-1481(98)00164-5
Farrell, J. (2016, April 6). *Beyond sharing: How communities can take ownership of renewable power (energy democracy initiative)*. https://ilsr.org/report-beyond-sharing/
Feenstra, M. (2002). *Towards a gender aware energy policy*. University of Twente.
Felsenthal, J. (2017). These are the women organizing the women's March on Washington. *Vogue Magazine*. www.vogue.com/article/meet-the-women-of-the-womens-march-on-washington
Ferrante, E. (2019, May 17). A power of our own. *New York Times*. swww.nytimes.com/2019/05/17/opinion/elena-ferrante-on-women-power.html#:~:text=Power%20is%20a%20story%20told,centuries%2C%20men%20have%20colonized%20storytelling.&text=Whether%20it%20was%20at%20school,the%20women%20adopted%20male%20ways
Fraune, C. (2015). Gender matters: Women, renewable energy, and citizen participation in Germany. *Energy Research & Social Science*, 7, 55–65.

Frischmann, B. M. (2013). *Infrastructure: The social value of shared resources*. Oxford University Press.
Gaard, G. (2011). Ecofeminism revisited: Rejecting essentialism and re-placing species in a material feminist environmentalism. *Feminist Formations*, *23*(2), 26–53.
Grid Alternatives. (2018). *Women in solar program*. https://gridalternatives.org/what-we-do/workforce-development/women-in-solar
Healy, N., Stephens, J., & Malin, S. (2019, February 7). Fossil fuels are bad for your health and harmful in many ways besides climate change. *The Conversation*. https://theconversation.com/fossil-fuels-are-bad-for-your-health-and-harmful-in-many-ways-besides-climate-change-107771
hooks, b. (2015). *Feminism is for everyone*. Routledge, Taylor and Francis.
IRENA. (2019). *A new world: The geopolitics of the energy transformation*. www.irena.org/publications/2019/Jan/A-New-World-The-Geopolitics-of-the-Energy-Transformation
James, B. (1999). At the margins of power: Gender and policy in the energy sector. *AGENDA*, *15*, 18–47.
Johnsson-Latham, G. (2007). *A study on gender equality as a prerequisite for sustainable development: What we know about the extent to which women globally live in a more sustainable way than men, leave a smaller ecological footprint and cause less climate change*. Report to the Environment Advisory Council, Ministry of the Environment, Issue.
Kahsar, R. (2019). The psychology of socio-technical systems: Professionalism, power, and personality in the electricity sector. *Energy Research & Social Science*, *53*, 121–125.
Kern, L. (2020). *Feminist city: Claiming space in a man-made world*. Verso.
Leduc, B. (2010). Climate change and gender justice. *Climate and Development*, *2*(4), 390–392. https://doi.org/10.3763/cdev.2010.13054
Lee, M., & Klump, E. (2018, March 8). Pipeline builders, gas drillers fret about protesters. *Energy and Environment News*. www.eenews.net/stories/1060075773
Listo, R. (2018). Gender myths in energy poverty literature: A critical discourse analysis. *Energy Research & Social Science*, *38*, 9–18.
MacGregor, S. (2009). A stranger silence still: The need for feminist social research on climate change. *Sociological Review*, *57*, 124–140.
Mann, S. A. (2011). Pioneers of US ecofeminism and environmental justice. *Feminist Formations*, 1–25.
Manne, K. (2018). *Down girl, the logic of misogyny*. Oxford University Press.
McCammon, H. J., McGrath, A., Hess, D. J., & Moon, M. (2018). *Women, leadership, and the US environmental movement. 100 years of the nineteenth amendment: An appraisal of women's political activism*. Oxford University Press.
McCright, A. M. (2010, September). The effects of gender on climate change knowledge and concern in the American public. *Population and Environment*, *32*(1), 66–87. https://doi.org/10.1007/s11111-010-0113-1
Merchant, C. (1981). Earthcare: Women and the environment. *Environment: Science and Policy for Sustainable Development*, *23*(5), 6–40.
Moser, C. (2005). Has gender mainstreaming failed? *International Feminist Journal of Politics*, 7(4), 576–590. https://doi.org/10.1080/14616740500284573
Mothers Out Front. (2018). *Pipelines*. http://va.mothersoutfront.org/pipelines
Myhre, S. (2019). Why women's climate leadership is vital. *Yes! Magazine*. www.commondreams.org/views/2019/10/01/why-womens-climate-leadership-vital
Nagel, J. (2015). *Gender and climate change: Science, skepticism, security, and policy*. Routledge.
Neuhauser, A. (2014, December 1). Energy's latest target: Women. *US News and World Report*. www.usnews.com/news/special-reports/energy-of-tomorrow/articles/2014/12/01/oil-and-gas-industrys-latest-target-women
Oreskes, N. (2015, October 9). Exxon's climate concealment. *New York Times*. www.nytimes.com/2015/10/10/opinion/exxons-climate-concealment.html?nlid=47634430&src=recpb&_r=0
Osunmuyiwa, O., & Ahlborg, H. (2019). Inclusiveness by design? Reviewing sustainable electricity access and entrepreneurship from a gender perspective. *Energy Research & Social Science*, *53*, 145–158. https://doi.org/10.1016/j.erss.2019.03.010
Parks, L., & Starosielski, N. (Eds.). (2015). *Signal traffic: Critical studies of media infrastructures*. University of Illinois Press.
Pearl-Martinez, R., & Stephens, J. C. (2016). Toward a gender diverse workforce in the renewable energy transition. *Sustainability: Science, Practice and Policy*, *12*(1).
Pellegrion, G., D'Amato, S., & Weisberg, A. (2011). *The gender dividend: Making the business case for investing in women*. https://www2.deloitte.com/content/dam/Deloitte/ru/Documents/public-sector/gender-dividend-en.pdf

Perez, C. C. (2019). *Invisible women: Data bias in a world designed for men*. Harry N. Abrams.
Prebble, M., & Rojas, A. (2018). *Energizing equality: Sub-Saharan Africa's integration of gender equality principles in national energy policies and frameworks*. https://genderandenvironment.org/energizing-equality-sub-saharan-africas-integration-of-gender-equality-principles-in-national-energy-policies-and-frameworks/#:~:text=Sub%2DSaharan%20Africa%20is%20a,its%20policies%20including%20gender%20considerations
Pulido, L. (1997). Community, place, and identity. In N. R. Jones (Ed.), *Thresholds in feminist geography: Difference, methodology, representation* (pp. 11–28). Roman and Littlefield.
Robinson, M. (2018). *Climate justice: Hope, resilience and the fight for a sustainable future*. Bloomsbury Publishing.
Robinson, M., & Odendahl, T. (2019). *Why we must act now for climate and gender justice*. http://news.trust.org/item/20190425202809-2yuhe
Schneider, J., & Peeples, J. (2018). The energy covenant: Energy dominance and the rhetoric of the aggrieved. *Frontiers in Communication*, *3*(5), https://doi.org/10.3389/fcomm.2018.00005
Schrupp, A. (2017). *A brief history of feminism*. MIT Press.
Solnit, R. (2014). *Men explain things to me*. Haymarket Books.
Star, S. L. (2015). The ethnography of infrastructure. In G. C. Bowker, S. Timmermans, A. E. Clarke, & E. Balka (Eds.), *Boundary objects and beyond: Working with Leigh Star* (pp. 473–488). MIT Press.
Star, S. L., & Ruhleder, K. (2015). Steps toward an ecology of infrastructure: Design and access for large information spaces. In G. C. Bowker, S. Timmermans, A. E. Clarke, & E. Balka (Eds.), *Boundary objects and beyond: Working with Leigh Star* (pp. 377–415). MIT Press.
Stephens, J. C. (2019). Energy democracy: Redistributing power to the people through renewable transformation. *Environment: Science and Policy for Sustainable Development*, *61*(2), 4–13. https://doi.org/10.1080/00139157.2019.1564212
Supran, G., & Oreskes, N. (2017). Assessing ExxonMobil's climate change communications (1977–2014). *Environmental Research Letters*, *12*(8), 084019. http://stacks.iop.org/1748-9326/12/i=8/a=084019
Sweeney, S. (2012). *Resist, reclaim, restructure: Unions and the struggle for energy democracy*. http://unionsforenergydemocracy.org/wp-content/uploads/2013/12/Resist-Reclaim-Restructure.pdf
Szulecki, K. (2018). Conceptualizing energy democracy. *Environmental Politics*, *27*(1), 21–41. https://doi.org/10.1080/09644016.2017.1387294
Taylor, D. E. (2000). The rise of the environmental justice paradigm: Injustice framing and the social construction of environmental discourses. *American Behavioral Scientist*, *43*(4), 508–580.
Terry, G. (2009). No climate justice without gender justice: An overview of the issues. *Gender and Development*, *17*(1), 5–18.
van Veelen, B., & van der Horst, D. (2018). What is energy democracy? Connecting social science energy research and political theory. *Energy Research & Social Science*, *46*, 19–28. https://doi.org/10.1016/j.erss.2018.06.010
Waddell, K. (2017, January 23). The exhausting work of tallying America's largest protest. *The Atlantic*. www.theatlantic.com/technology/archive/2017/01/womens-march-protest-count/514166/
Wang, E. (2018, October 8). Climate scientist Dr. Sarah Myhre on the peril and possibilities ahead: 'We need leadership that is not old and white and male'. *Jezebel*. https://theslot.jezebel.com/climate-scientist-dr-sarah-myhre-on-the-peril-and-poss-1829602131
Warren, K. J. (Ed.). (1997). *Ecofeminism: Women, culture, nature*. Indiana University Press.
Willow, A. J., & Keefer, S. (2015). Gendering extraction: Expectations and identities in women's motives for shale energy opposition. *Journal of Research in Gender Studies*, *5*(2), 93.
Wilson, S. (2018). Energy imaginaries: Feminist and decolonial futures. In B. R. Bellamy & J. Diamanti (Eds.), *Materialism and the Critique of Energy*, *31*(2), 5–20.
Winther, T., Ulsrud, K., & Saini, A. (2018). Solar powered electricity access: Implications for women's empowerment in rural Kenya. *Energy Research & Social Science*, *44*, 61–74.
Womens Environmental Network. (2014, July 15). *Erica Mackie and GRID alternatives*. www.wencal.org/spotlights/2014/07/erica-mackie/
Woolley, A. W., Chabris, C. F., Pentland, A., Hashmi, N., & Malone, T. W. (2010). Evidence for a collective intelligence factor in the performance of human groups. *Science*, *330*(6004), 686–688.
Zipp, K. (2018, May 30). Solar power world: Ask a solar vet: Somewhere between social work and engineering, Erica Mackie of grid alternatives [audio podcast]. www.solarpowerworldonline.com/2018/05/ask-a-solar-vet-somewhere-between-social-work-and-engineering-with-erica-mackie-of-grid-alternatives/

18

ENERGY COMMONS AND ALTERNATIVES TO ENCLOSURES OF SUNSHINE AND WIND

Matthew J. Burke

Introduction

Energy democracy opens the possibility for breaking from the conventional social construction of energy-as-commodity, reenvisioning and reconstructing energy systems as energy commons. This chapter aims to support the agenda to democratize energy systems by examining and understanding energy commons and commoning as alternative modes of governance to dominant state-led and especially market-based approaches.

A layered term, the "commons" refers simultaneously to "shared resources, a discourse, a new/old property framework, social processes, an ethic, a set of policies or, in other words, to a paradigm of a pragmatic new societal vision beyond the dominant capitalist system" (Kostakis & Bauwens, 2014, p. 38). In this sense, elements of energy commons include (1) a biophysical resource or process (e.g., sunshine and wind); (2) a community of people having the responsibility to share, manage, sustain, restore, and generally interrelate with this process; (3) the value created through relationships between the biophysical process and the community of people; and (4) the collective rules and property regimes that govern these relationships and collective actions to ensure their ongoing reproduction and defense from enclosure and commodification (Harvey, 2011; Kostakis & Bauwens, 2014; Mattei, 2012; Melville et al., 2017). While histories of the commons extend back to antiquity, recent interest in the commons as an agenda beyond market and state is often traced to the work of Elinor Ostrom (1990) and others on enduring real-world models of collective governance of common-pool resources, and to the autonomous movement of the Zapatistas protesting the dissolution of communal lands in Mexico (Federici, 2012; Linebaugh, 2008). These experiences and modes of scholarship and practice inform the understanding of energy commons as taken up here.

The present moment is pivotal for energy commons and energy democracy. Expectations for the future of solar and wind energy are high, yet these sources currently account for a low proportion of total primary energy conversion. Similarly, while use of renewables has increased, this upward trend holds true for nearly all energy sources, hardly resulting in an energy transition. This failure leaves open the possibility for fundamentally different approaches to energy governance.

 DOI: 10.4324/9780429402302-21

The following section reviews the patterns of flows of Earthly sunlight and wind and explores conventional modes of their enclosure. The third section then provides an overview of the approach of energy commons as a unique paradigm for shifting energy governance. The fourth section addresses practices of energy commons, first by examining specific applications, then by highlighting the barriers that must be confronted in the context of renewable energy transition. Drawing from but not bound to the lessons of traditional commons and common-pool resources[1] (Melville et al., 2017; Ostrom, 2010), the final section suggests a set of design principles for creating and sustaining solar and wind energy commons. Together, this chapter contributes conceptually, theoretically, and practically to alternative discourses, projects, and modes of action of energy democracy as described in this volume.

Grabbing the flows: modes of enclosure

In conventional understandings, energy refers to a wide variety of phenomena or processes including heat, motion, light, electricity, and chemical energies, often defined in terms of a capacity to do work (Heinberg & Fridley, 2016a; Smil, 2009, 2017; Stephenson, 2017). However, "energy" is a notoriously slippery term across disciplines, including the physical sciences, raising several key problems with this understanding (Burke, 2019, p. 18). Broadening the perspective, modern human energy systems include several of countless ways that life on Earth has developed to capture, make use of, and dissipate low-entropy matter-energy. The concept of energy as a universal currency and the standardization of its measurement can minimize important qualitative differences among these diverse phenomena. These differences are increasingly important when understanding the possibilities for renewable energy to substitute conventional fuels.

The amalgamation of diverse processes under a singular and novel abstraction of energy emerged from within a particular capitalist-colonialist historical context, wherein the overriding concern with energy involved its application to certain ends. The problem was to understand how to efficiently convert energy resources into so-called productive work for a developing industrial society, largely to the benefit of those who owned the means of conversion. This conceptualization of energy is associated with the social construction of the industrialized sense of time and labor-as-work (Lohmann & Hildyard, 2014). These understandings have further enabled the enclosure, commodification, and control of these physical phenomena, put to use in the processes of accumulation and dispossession, while increasing societal dependence on high levels of use of finite energy resources (Foxon, 2018; Illich, 2009; Lohmann & Hildyard, 2014). Moreover, this perspective has obstructed and minimized other views of energies, as ecological and social necessities (National Research Council, 1984), bases of social power and power relations (Adams, 1975), or threads of interconnected relationships among living and nonliving phenomena (Frigo, 2017). It is therefore practically and politically relevant to acknowledge that different views on energy carry very different implications for decision making about energy and technologies (Burke, 2019, p. 19).

This same problematic historical and political context frames current understandings of renewables (Burke, 2019, p. 19). The concept of "renewable energy" typically refers to inexhaustible flows of energy, those that the natural environment continuously replenishes on a human timescale. These include primary sources derived from the sun either directly, including thermal and photoelectric energy, or indirectly, such as wind and biomass, or from other natural movements and cycles (Ellabban et al., 2014). As with energy more broadly, a singular definition of renewables minimizes important differences among various sources and their enabling technologies, differences evident when comparing the most mundane bioenergy sources used worldwide with technologically complex, modern systems including wind power, solar photovoltaics, and hydroelectricity (Chatti et al., 2017).

Even their very renewability is debatable and context dependent (Burke, 2019, p. 19). While the flows of these sources may be continuously replenished, the renewability of a host of processes requires consideration in the context of large-scale development designed to meet the demands of industrial societies. These processes, including capture, conversion, storage, movement, use, maintenance, and reproduction, clearly require energy and material inputs at each step. Further, despite the abundance of many of these sources, their uneven distribution in space and time can lead to scarcity, while scarcity can also be created through exclusions that support the profits needed to sustain private investments. Thus, beyond the widely discussed technical challenges of variable energy systems, the idea of renewability is more fundamentally complicated by the ongoing dependence of these technologies on unsustainable systems including fossil fuel infrastructure, industrial production, material extraction and throughput, long-distance supply chains, spatial expansion and enclosures, environmental degradation, continued economic growth and accumulation, and expectations of high levels of financial investment (Foxon, 2018; Georgescu-Roegen, 1984; Heinberg & Fridley, 2016b; Huber & McCarthy, 2017; Raman, 2013).

A troubling inheritance from the fossil fuel era is the primacy of energy as a commodity to be bought and sold on the market. The associated legal and regulatory frameworks allow investors and speculators to buy up locations useful for capturing wind and solar flux and to lay claim to these sources as private property, a form of enclosure of the commons and energy colonialism (Bellamy & Thomas, 2015; Brunette et al., 2013; Martinez, 2017; Vermeylen, 2010). Commodification of renewable energy allows returns and surpluses to concentrate in private hands, while ensuring that development proceeds according to the logics of financialization and profit rather than genuine human and nonhuman needs (Blanchet, 2019). Even as electrification is increasingly viewed as an essential public good, the scope of public policy often remains confined to the facilitation of energy markets. Property rights for energy systems are largely based on ownership of land and access to the grid, which can be regulated to limit rights of access and use by other possible producers (van der Horst & Vermeylen, 2008). This mode of energy development enables green grabbing, or the appropriation of rights of use and control of land and resources, typically by the powerful, for allegedly pro-environmental purposes (Corson & MacDonald, 2012; Fairhead et al., 2012).

In the context of dominant capitalist and colonialist political economies and market societies, the rights of access, withdrawal, management, exclusion, and alienation (rights to sell, lease, or transfer) are thus linked and concentrated. What might be first considered as multitudes of commons are in practice treated primarily as open access regimes. Open access leads to new modes of privatization in terms of rights to wind and sunlight, applying to these renewable flows a poorly adapted legacy of "underground logics" associated with nonrenewable rights of extraction (Hughes, 2017) and expansion of commodity frontiers (Moore, 2000). With increasing conflict over (horizontal, expansive) access to wind and (vertical, localized) access to sunlight across a range of uses and jurisdictions (van der Horst & Vermeylen, 2010; Vermeylen, 2010), the question of who owns the wind and sunlight becomes absurdly more salient. The unfortunate answer increasingly appears to be those with the power to first legitimate their claim (Hughes, 2017). Thus, renewable energy risks reproducing the social and ecological harms of the present energy system (Bellamy & Thomas, 2015). These harms include concentration of wealth and means of production, rising levels of inequality, governance by capital, and exclusion of local communities and a broader public (Byrne & Taminiau, 2016; Feola, 2019; Harvey, 2018; Taylor, 2019). This mode of development can further yield large-scale and poorly sited technologies and grid development, constructed in the pursuit of endless economic growth (Byrne et al., 2009; York & Bell, 2019). Ultimately, this open-access and privatized approach risks more than undermining a genuine energy transition. From an energy democracy perspective, this pattern of renewable energy development also preempts commons-based approaches to using wind and solar fluxes, based on deeper forms of social agency and active citizen and indigenous participation and control (Kunze & Becker, 2015; Platform, 2015).

Recognizing the energy commons

Transitions to renewable energy present an opportunity if not a necessity to view sunlight and wind as energy commons, broadly reinterpreted as shared flows of energy for all living beings, existing prior to and beyond conventional systems of private property. Energy commons allows for diverse and contextually relevant modes of governance, while new material artifacts such as photovoltaic panels can encourage greater participation and energy citizenship and sovereignty (Ryghaug et al., 2018). In a commons-based approach, the value of renewable energy is derived primarily from use rather than exchange (Brunette et al., 2013; Martinez, 2017; Platform, 2015). The commons paradigm sees sharing as the means to ensure resources are adequate for supporting sustainable livelihoods and genuine needs. In other words, energy and energy provision are common goods controlled by the people of a community and oriented toward collective democratic governance, ecological sustainability, and social justice (Becker et al., 2017). There is therefore a need to strongly, democratically, and equitably regulate energy use for the purpose of securing sufficient levels of sustenance for human and nonhuman life, satisfying basic social needs while respecting ecological limits.

Commoning (Linebaugh, 2008) is thus about both reestablishing earlier traditions as well as reinventing and creating new commons-based modes of relations in the context of existing struggles (Federici, 2012). At once assertive and defensive, collective acts of commoning include various processes of reproduction of the commons, using democratic, horizontal, collaborative, participatory, and locally relevant means (Federici, 2012; Platform, 2015). In contrast to decision making organized around market values and economization of the decision arena, the relevant criteria for determining energy pathways first prioritize justice, equity, and sustainability (Byrne & Mun, 2003). More fundamentally, the material and symbolic institutional logics and universalizing ontologies of markets, commodification, and accumulation are kept subordinate to the logics, ontologies, and end goals of diverse modes of the commons (Escobar, 2015; Taylor, 2019), wherein energy systems (constituted of nonhuman and human elements) are viewed as noncommodified means to fulfill social and ecological needs. As with Polanyi's (2001) fictitious commodities of land and labor, renewable energy sources refer to human and nonhuman activities that are inseparable from life itself, neither "produced" in a conventional sense nor produced for sale. From this relational ontology, as Escobar (2015) asserts, "the defense of territory, life and the commons are one and the same" (para. 15).

The core strategy of commoning is one of strengthening social relations based on cooperation, collaboration, and co-responsibility, that is, power-with and power-to rather than power-over (De Angelis, 2012). Key agents of change include networks of diverse and often marginalized communities of resource users, organized through similarly diverse and shared ownership models. Countermovements for energy commons likely require a combination of strategies, including public sector reforms, socially embedded market models, nonmarket and nonstate systems. Commons governance favors decentralized or polycentric modes, bottom-up decision making, self-organization, and self-defined rules of access and rights of use (Bauwens, 2017; Helfrich, 2012b). Public investment and ownership may play a key role, with a stronger shift toward localization (Hammerstein, 2019). Rights of use are tied to individual and collective responsibility to sustainably use resources and protect ecosystem health (Brunette et al., 2013), reversing the invisibility of harms of the existing energy systems, reducing the distance between actions and social and environmental consequences, and thus overcoming patterns of collective separation and irresponsibility (Federici, 2012; Platform, 2015).

To bring about these changes, the transition to renewables must be tied to the project of commoning, of decommodification, of reversal of processes of enclosure and colonization, of restoring relationships among people and of people to the rest of nature, of creating a counter-politics to contemporary valuations and appropriations of nature, based in turn on counter-ontologies of interconnected human–ecological relations (Fairhead et al., 2012). Like energy democracy, energy commoning then implies both legal and political claims as well as the necessary institutional, economic, and material foundations (Linebaugh, 2008).

Claiming and governing the energy commons

Contemporary institutions for collective renewable energy

Institutions for energy commons include diverse, integrated, and collectively organized energy systems that enable a community of users to seek aspirations beyond a renewable energy transition (Acosta et al., 2018; Becker & Kunze, 2014; Brisbois, 2019; Koirala et al., 2016). Several organizational forms and models stand out as most promising.

Renewable energy cooperatives have been widely adopted and viewed as a strong form of energy commons. Cooperatives are embedded, at least potentially, in local communities as well as in the international cooperative movement and its associated principles of solidarity and democratic governance. Through cooperatives, citizens can produce, invest in, and use renewable energy while sharing surplus value (Lambing, 2012; Taylor, 2019). Cooperatives are generally owned and often managed by their members and users rather than investors (Bauwens, 2017). Nonprofit consumer-owned electric cooperatives, including distribution, generation, and transmission cooperatives, provide renewable energy or related services to consumer-members, while worker-owned cooperatives provide employment and financial benefits of ownership to worker-owners and various renewable energy services to customers (Burke & Stephens, 2017). Increasingly, cooperative models orient their work toward serving broader public interest goals rather than seeking to accrue benefits solely to members (Bauwens & Defourny, 2017).

New and existing forms of utilities are fitting for energy commons, including municipally owned utilities and sustainable energy utilities. Municipal utilities are owned and controlled by a city or local government, whereas remunicipalization is the now worldwide process of creating or reestablishing public ownership and governance of energy utilities and infrastructures at the municipal level, potentially in the context of an urban commons (Becker et al., 2017; Blanchet, 2019). A sustainable energy utility is a nonprofit organization or community trust purposed to provide energy services and support commons energy development and management for communities of almost any scale, from local groups to broader regions. Key strategies include helping communities reduce overall energy dependence, supporting renewable generation by and for the communities, redirecting finances and deploying self-financing approaches to achieve conservation goals and build new infrastructure, and instituting cooperative systems of use, monitoring, and enforcement (Byrne et al., 2009; Byrne & Taminiau, 2016). Together these institutions serve as key links between conventional public authorities and emerging energy commoners.

Processes for reclaiming the grid within an energy commons approach are also critical in the context of electrification of energy futures. Community microgrids, minigrids, and democratized grid management support small-scale distributed generation of renewable energy within a defined geography. Here, prosumers, producers, and consumers (i.e., sources and loads) can share and exchange community-based electricity production, allowing local prosumers with equitable grid access to allocate surplus energy where it is needed (Wolsink, 2012). Rules of use are shaped and reshaped by the users themselves based on specific needs and local conditions (Gollwitzer et al., 2018; Melville et al., 2017; Wolsink, 2012). These grid systems can be operated autonomously as controllable units and/or interconnected within networks of microgrids and the larger utility grid (Giotitsas et al., 2015; Grimley & Farrell, 2016; Lambing, 2012; Wolsink, 2012). Such systems could also integrate pooled district heating, energy storage, transport systems, and associated information and communication technologies (Bronin, 2010; Hammerstein, 2019; Roberts et al., 2014; Zhang et al., 2018). Micro- and minigrids offer possibilities for installation in remote areas and may increase reliability through diversification while reducing long-distance transmission (Giotitsas et al., 2015; Peters et al., 2019).

Several other institutional reforms can enable energy commoning. These efforts would include democratization of the technologies themselves, including their manufacture, through models of worker and community ownership, open-source technology transfer, and socially responsible and noncommercial licensing of technologies and energy use data (Hammerstein, 2019). Nonprofit green public service banks and related community development financial institutions can be directed to offer grants, loans, and low-cost credit to community-based renewable energy projects, while improving accountability to local communities for decisions over public financing (Burke & Stephens, 2017). More generally, to advance energy commons, governments at all levels would provide financial and technical support, enable community-based energy planning (Baker, 2017), and ensure at least partial community ownership (Roberts et al., 2014). These actions can be taken to scale, nested within broader networks and institutions, as in energy regions (Späth & Rohracher, 2010). Together, these networks can then challenge overcentralization and private appropriation of wealth and power (Platform, 2015) and integrate resources and users across borders (van der Horst & Vermeylen, 2010).

Indigenous modes of energy governance and experiences with sustaining commons beyond energy provide important practical examples of collective institutions (Martinez, 2017). Tribal Nations and tribally controlled utilities and business entities may prioritize tribal sovereignty, intergenerational social-environmental values, and cultural and spiritual connections to land above purely economic considerations, inspiring a considered and distributed pattern of renewable energy development (MacArthur & Matthewman, 2018; Necefer et al., 2015). Indigenous claims to sunshine and wind demand the agency of indigenous peoples to transform these energy systems on their own land and on their own terms, through recognition of treaty rights and reconstruction of Native Nations (Byrd, 2011). Relatedly, energy commoning also requires land reforms. Reparations, usufruct, periodic redistributions, and the like are needed to ensure that renewable systems continuously resist processes of land concentration through speculation and profit seeking. These reforms aim to reverse ongoing enclosures of common lands while ensuring a greater role for indigenous communities, minorities, labor, and women in the governance of renewable energy commons (Linebaugh, 2008; Platform, 2015). In this way, energy commons can work in support of indigenous energy sovereignty and rights of indigenous peoples to self-determination in order to prevent and redress the removal from and dispossession of lands and territories (UNDRIP, 2007).

At the level of constitutional rules, renewable energy commoning requires establishing these sources and systems as human rights and basic collective needs, while recognizing energy as a relationship to nonhuman nature and interdependent dynamic living systems (Martinez, 2017). Social charters and diverse cultural customs for renewable commons can provide the necessary legal and cultural basis for establishing and defending energy commons. These charters can be operationalized through the formalization of constitutional rules and commons trusts, for example (Linebaugh, 2008; Quilligan, 2012). Without shifts in these fundamental enabling conditions, it is difficult to imagine how commons-based approaches can be recreated and preserved as diverse and durable modes of energy governance, given the rampant patterns of enclosure worldwide.

Obstacles to energy commoning in theory and practice

Broadly, there are challenges that are internal and external to commoning. Experience indicates a strength in overcoming endogenous challenges over time, yet significant vulnerabilities to exogenous factors exist (Dell'Angelo et al., 2017; Wall, 2017). Empirically, it appears that the actual tragedy is less about the misuse or failure of the commons and more about the regularity and brutality of enclosures (Wall, 2017).

Any process of governing the energy commons must nevertheless address widely acknowledged challenges for collective action. These challenges follow from the diversity and unevenness of energy sources and technologies (van der Horst & Vermeylen, 2008), the limited experience and engagement of citizens and nonexperts with energy systems (Bauwens, 2017), and the difficulty of broadening and sustaining involvement (Hoffman & High-Pippert, 2005). As placed-based systems, the way energy is accessed and used will depend upon where the users live. However, the need to accelerate an energy transition away from fossil fuels can too often drive greater centralization (Bollier, 2016). Additionally, energy commoning must confront large capital and up-front investment costs and insufficient financial and technical supports (Gui & MacGill, 2018; Hammerstein, 2019). Self-organization and diverse modes of governance take time to develop and test (Acosta et al., 2018; Lambing, 2012; Melville et al., 2017; Peters et al., 2019). This situation points to a need for experimentation and learning among commoners and to a sharper focus on improving the practices of commoning among advocates of renewables.

Beyond individual sites of commoning, there remains the difficult work of building intersectional coalitions, social movements, and political agendas for energy commons at local, regional, national, and transnational levels across diverse interests and worldviews (Escobar, 2015; Federici, 2012; Kostakis & Bauwens, 2014). This obstacle relates to the unavoidable scale problem, meaning first that what works locally may not work globally, and second that larger scales and nested hierarchies of energy commons are required, raising the political challenge of how to bring together many proliferating and necessarily diverse commons within a cohesive whole (Federici, 2012; Harvey, 2011; Platform, 2015).

Even then, commoning as a broadened set of practices must struggle to take hold in a difficult and often hostile environment. In terms of lock-ins and path dependencies, there exist patterns of centralized and hierarchical governance, regulatory and financial regimes, fossil fuel dependence, and interdependence of established firms and governments (Bauwens, 2017; Brisbois, 2019; Koirala et al., 2016). Regarding the contested role of governments and the dependence of many community-based approaches on the state (Creamer et al., 2018), there is frequently a need for government action and public support for collective energy sharing and the energy commons. For example, the public sector may be pushed to favor cooperatives over for-profit developers and investor-owned firms, while building capacity for participatory processes, learning and adaptation, and cooperative self-governance across scales (Burke & Stephens, 2017; Ostrom, 2010; Taylor, 2019). This shift may require a partner state to empower commoning processes, as public–commons partnerships, to build counterpower and legitimation within the commons and advance the commons not as a "third way" but rather as a means to challenge the alliance between state and private interests (Bollier, 2016; De Angelis, 2012; Kostakis & Bauwens, 2014; Mattei, 2012).

Regarding these incumbent interests, commoners face a host of powerful actors including fossil fuel and nuclear industries, electric utilities, and associated complexes of militarism and industrialism. These interests together hold enormous political influence through regulatory capture, corruption, lobbying, framing of public discourse, dominant decision-making positions, and a merger of state and private interests, all operating within a political economic context that forcibly asserts systems of property rights, large-scale land acquisitions, and inequities of wealth and distribution of capital (Brisbois, 2019; Dell'Angelo et al., 2017; Hammerstein, 2019; Mattei, 2012; Moss et al., 2015). Deeper still are the underlying logics of the commodity form, the orientation toward homogenizing market-based social relations, and the associated capitalistic norms of individualism, competition, profit motive, and economic growth (Bellamy & Thomas, 2015; Feola, 2019; Taylor, 2019). These logics manifest through the development of large-scale and centralized private and public energy systems (Byrne et al., 2009), profit-maximizing commercial business models (Brisbois, 2019), appeals to community power that serve to advance private interests (Taylor, 2019), and accelerated enclosures through green grabbing in response to environmental crises (Corson & MacDonald, 2012).

Given the specific and lasting experiences of enclosures of indigenous peoples, the tensions between energy commons and energy decolonization must also be taken seriously. The very notion of commoning cannot easily be disentangled from ongoing processes of forced removal and erasure of indigenous peoples, raising the prospect of incompatibilities between settler and indigenous claims to land and energy flows (Byrd, 2011; Dunbar-Ortiz, 2014; Tuck & Yang, 2012). Energy democracy likewise draws from settler understandings of development and democracy, which in turn have involved destructive patterns of colonialism, genocide, and dispossession. Any calls to reconsider claims to resources as existing prior to and outside of conventional institutions, especially those made by voices from within settler society, must necessarily account for and ultimately recognize the priority of rights against seizures of lands that precede settler colonialism. While this chapter seeks a process of commoning that can work in parallel to indigenous programs for decolonization and sovereignty, it remains uncertain whether and how these projects can be linked (see also Whyte et al., 2017; Batel, Chapter 11, this volume).

Many of these challenges may be specific to the current context, yet from the longer perspective of struggles for the commons, these dynamics are as long-standing as acts of enclosure. While there must be a time beyond fossil fuels, there will likely be no end to the need to assert and defend the commons. One key opportunity and potential advantage available to commoners now and going forward is the ability to learn from real-world experience.

Elements for enduring energy commons

Design principles serve as a useful diagnostic for communities of energy commoners: for assessing institutional robustness, for organizing internal evaluation or external research, and for sharing and learning from experiences (Melville et al., 2017). The following principles have a strong empirical basis and have been shown to be applicable to many types of social groups and collective action problems (Whyte et al., 2017; Wilson et al., 2013), yet they are not written in stone, nor do they constitute any blueprint for collective action. Principles for energy commons should therefore be approached as dynamic guidelines, elements, or conditions to be tested, monitored, adapted, and refined (Taylor, 2019). Here the starting point for design principles for energy commoning are drawn from those of Ostrom and others, as summarized in Table 18.1. Additionally, experiences of indigenous peoples underscore the importance of related principles of intergenerational involvement, ongoing cross-cultural learning, balance of power and decision making, and respect for indigenous knowledges and control of knowledge mobilization (Whyte et al., 2017).

Under conditions of collective action, of obvious importance is the development of shared rules or collective-choice arrangements for conversion, distribution, use, and monitoring of renewable energy and electricity among the community of interest, for example, a set of users of a micro- or minigrid (Gollwitzer et al., 2018). A precondition here is that a group of people recognize a shared relationship to an energy resource and a common goal or purpose (Principle 1). Yet direct engagement with the energy sector is beyond the experience and daily life of many people, and their entry into this engagement often comes through conflict. There is therefore a need to support the conditions for initiating collective action, for example, recognizing and establishing the potential co-benefits to communities (Principle 2). The rules that such a group develops must be well adapted to local social, cultural, technological, and ecological conditions (Principle 2) and must be seen as fair and legitimate among the users (Principle 3). Commoners must be able to modify, monitor, and enforce rules, while resolving conflicts of interest and minimizing self-serving behavior rapidly and inexpensively through participation of the users themselves (Principles 3–6) (Lambing, 2012; Wilson, 2015; Wolsink, 2012).

Table 18.1 Design principles for long enduring institutions for the commons (Ostrom, 1990; Wilson et al., 2013)

Design principle, element, or condition of success	*Description*
1 Clearly defined boundaries	Both the individuals who have rights of use to the resource and the resource system itself are well-defined.
2 Congruence with local conditions	Appropriation and provision rules are congruent with local social and environmental conditions, and benefits and costs are proportional.
3 Collective choice arrangements	Most individuals affected by the operational rules can participate in modifying the operational rules.
4 Monitoring	Monitors who are accountable to or consist of the users actively audit resource conditions and user behavior.
5 Graduated sanctions	Users who violate operational rules are assessed graduated sanctions proportionate to the seriousness of the offense by other users and/or those accountable to users.
6 Conflict-resolution mechanisms	Users have access to rapid, fair, low-cost arenas to resolve conflicts among users and between users and officials.
7 Minimal recognition of rights to organize	The rights of users to devise their own institutions are recognized by external governmental authorities.
8 Nested enterprises	Appropriation, provision, monitoring, enforcement, conflict resolution, and governance activities are coordinated across multiple layers of nested enterprises.

This set of conditions operates primarily within the bounds of the community of users and, as argued previously, may offer greater potential for success relative to the remaining, essentially exogenous elements.

Recognition of the rights to organize (Principle 7) is arguably the most challenging historically (Dell'Angelo et al., 2017; Wall, 2017). For indigenous peoples, this principle is often expressed as the right to self-determination or sovereignty, involving recognition of treaty rights and autonomy over tribal lands and/or mutual agreement with nonindigenous settler states regarding shared jurisdiction over energy sources (Dunbar-Ortiz, 2014; Whyte et al., 2017). Conventional public authorities are not only unaccustomed to recognizing such rights, autonomy, and effectiveness of self-governing communities (Lambing, 2012) but often play a central role in undermining these efforts and blocking patterns of commoning through active dispossession and enclosure.

A precondition to rights of commoning is that political empowerment not be closed off by privatization. This precondition for recognition means that property or ownership rights must be grounded in a commons perspective and establish a basis for diverse modes of communal rights, broadly recognized to exist prior to and outside of conventional public or private ownership. In other words, *the original claim to flows of sunshine and wind for renewable energy must not be one of accumulation and dispossession but one of commoning and sharing*. This assertion is not made in a general or universal sense but is rather grounded in relation to specific communities and cultural contexts. Such an assertion may appear unrealistic, yet given the inability to achieve even inadequate goals of transition, coupled with the context of severe social and ecological devastation, the claim that we must either learn to share or die (Gorenflo, 2012) seems soundly based in reality. Ideally, defending the rights of commoning would not simply be minimally recognized but also celebrated and encouraged by established officials, making room for the necessary experimentation and diversity in practice.

Finally, within this currently challenging institutional environment, there is a need for coordinated interorganizational action among commoning organizations and institutions to support the formation of higher-level networks (Principal 8) (Bauwens et al., 2016). While locally focused, a given initiative will involve a variety of actors working at different levels of governance, bringing different skills and expertise to the local question. Additionally, the interconnectedness of societies and ecosystems means that the scope of energy and sustainability transitions are ultimately planetary. In this sense, a community initiative is never strictly a local concern (Gui & MacGill, 2018). This need to work across scales also means members and users have a right and responsibility to engage in the work of broader networks (Taylor, 2019). Such networks are critical for addressing multiple issues of scale through bottom-up collective actions, including coordination across initiatives, not only for withstanding inevitable external pressure but in turn reshaping these exogenous factors and energy regimes.

Conclusions

This chapter posits the energy commons as a necessary mode of energy governance for solar, wind, and other replenishable sources to facilitate a just and enduring energy transition. Presently, these global commons are approached as open access regimes, enabling new modes of enclosure, extraction, and commodification of sunlight and wind, with the consequence of replicating social and environmental devastation of the fossil fuel era while ultimately failing to transition. Energy commoning supports direct and durable modes of energy democracy, reconceptualizing energy as a commons rather than as a commodity, owned and managed by communities deploying systems of rules for energy conversion and use. Specific strategies of commoning now in practice can be taken up more broadly by movements seeking to deeply democratize renewable energy systems. These strategies include organizations such as cooperatives and municipal utilities, micro- and minigrids, reforms of land ownership and constitutional rules, and implementation and experimentation as guided by a diverse set of principles for enduring institutions of the energy commons. This inclusive approach to energy transition must not come by way of ongoing enclosures of indigenous lands, however, and therefore, to achieve its potential, energy commoning must be intimately joined to efforts to recognize and redress indigenous rights and sovereignty. While the struggles to create and defend energy and other commons will continue, this transition presents renewable energy advocates with an opportunity to rethink fundamental questions of access, use, value, and relationship and to put into practice lessons learned from diverse real-world experiences of energy commoning.

Note

1 An additional strand of research, largely following neoclassical economics, centers on the classification of common-pool resources as distinct from public, private, and toll goods, depending on the combination and degree of characteristics of rivalry and excludability (Ostrom, 2010). Others argue that no fundamental attribute or physical characteristic necessarily distinguishes a common good from other types of goods; rather, common goods are socially created. For further discussion on this debate, see Helfrich (2012a).

References

Acosta, C., Ortega, M., Bunsen, T., Koirala, B. P., & Ghorbani, A. (2018). Facilitating energy transition through energy commons: An application of socio-ecological systems framework for integrated community energy systems. *Sustainability, 10*(2). https://doi.org/10.3390/su10020366

Adams, R. N. (1975). *Energy and structure: A theory of social power.* University of Texas Press.

Baker, S. H. (2017). Unlocking the energy commons: Expanding community energy generation. In M. Scanlan (Ed.), *Law and policy for a new economy* (pp. 211–234). https://doi.org/10.4337/9781786434524.00021

Bauwens, T. (2017). Polycentric governance approaches for a low-carbon transition: The roles of community-based energy initiatives in enhancing the resilience of future energy systems. In N. Labanca (Ed.), *Complex systems and social practices in energy transitions: Framing energy sustainability in the time of renewables.* Springer International Publishing.

Bauwens, T., & Defourny, J. (2017). Social capital and mutual versus public benefit: The case of renewable energy cooperatives. *Annals of Public and Cooperative Economics, 88*(2), 203–232. https://doi.org/10.1111/apce.12166

Bauwens, T., Gotchev, B., & Holstenkamp, L. (2016). What drives the development of community energy in Europe? The case of wind power cooperatives. *Energy Research & Social Science, 13*, 136–147. https://doi.org/10.1016/j.erss.2015.12.016

Becker, S., & Kunze, C. (2014). Transcending community energy: Collective and politically motivated projects in renewable energy (CPE) across Europe. *People, Place and Policy Online, 8*(3), 180–191. https://doi.org/10.3351/ppp.0008.0003.0004

Becker, S., Naumann, M., & Moss, T. (2017). Between coproduction and commons: Understanding initiatives to reclaim urban energy provision in Berlin and Hamburg. *Urban Research & Practice, 10*(1), 63–85. https://doi.org/10.1080/17535069.2016.1156735

Bellamy, B. R., & Thomas, D. (2015). The green struggle. *Jacobin.* https://jacobinmag.com/2015/10/fossil-fuels-renewables-capitalism/

Blanchet, C. (2019). Energy commons: The missing link between energy transition and climate justice. In S. Bloemen & T. de Groot (Eds.), *Our commons: Political ideas for a new Europe* (pp. 33–38). www.commonsnetwork.org/ourcommons/

Bollier, D. (2016). *State power and commoning: Transcending a problematic relationship.* Deep Dive Report. Common Strategies Group. http://commonsstrategies.org/state-power-commoning-transcending-problematic-relationship/

Brisbois, M. C. (2019). Powershifts: A framework for assessing the growing impact of decentralized ownership of energy transitions on political decision-making. *Energy Research & Social Science*, *50*, 151–161. https://doi.org/10.1016/j.erss.2018.12.003
Bronin, S. C. (2010). Curbing energy sprawl with microgrids. *Connecticut Law Review*, *43*(2). http://papers.ssrn.com/sol3/papers.cfm?abstract_id=1594515
Brunette, C. L., Byrne, J., & Williams, C. K. (2013). Resolving conflicts between renewable energy and wildlife by promoting a paradigm shift from commodity to commons-based policy. *Journal of International Wildlife Law & Policy*, *16*(4), 375–397. https://doi.org/10.1080/13880292.2013.844000
Burke, M. J. (2019). *Energy democracy and the co-evolution of social and technological systems* [Doctoral thesis, McGill University]. https://escholarship.mcgill.ca/concern/theses/4m90dx59n
Burke, M. J., & Stephens, J. C. (2017). Energy democracy: Goals and policy instruments for sociotechnical transitions. *Energy Research & Social Science*, *33*, 35–48. https://doi.org/10.1016/j.erss.2017.09.024
Byrd, J. A. (2011). *The transit of empire: Indigenous critiques of colonialism*. University of Minnesota Press.
Byrne, J., Martinez, C., & Ruggero, C. (2009). Relocating energy in the social commons: Ideas for a sustainable energy utility. *Bulletin of Science, Technology & Society*, *29*(2), 81–94. https://doi.org/10.1177/0270467609332315
Byrne, J., & Mun, Y. M. (2003). Rethinking reform in the electricity sector: Power liberalisation or energy transformation. *Electricity Reform: Social and Environmental Challenges*, 48–76.
Byrne, J., & Taminiau, J. (2016). A review of sustainable energy utility and energy service utility concepts and applications: Realizing ecological and social sustainability with a community utility. *Wiley Interdisciplinary Reviews: Energy and Environment*, *5*(2), 136–154. https://doi.org/10.1002/wene.171
Chatti, D., Archer, M., Lennon, M., & Dove, M. R. (2017). Exploring the mundane: Towards an ethnographic approach to bioenergy. *Energy Research & Social Science*, *30*, 28–34. https://doi.org/10.1016/j.erss.2017.06.024
Corson, C., & MacDonald, K. I. (2012). Enclosing the global commons: The convention on biological diversity and green grabbing. *The Journal of Peasant Studies*, *39*(2), 263–283. https://doi.org/10.1080/03066150.2012.664138
Creamer, E., Eadson, W., van Veelen, B., Pinker, A., Tingey, M., Braunholtz-Speight, T., Markantoni, M., Foden, M., & Lacey-Barnacle, M. (2018). Community energy: Entanglements of community, state, and private sector. *Geography Compass*, *12*(7), e12378. https://doi.org/10.1111/gec3.12378
De Angelis, M. (2012). Crises, capitalism and cooperation: Does capital need a commons fix? In D. Bollier & S. Helfrich (Eds.), *The wealth of the commons: A world beyond market and state*. http://wealthofthecommons.org/contents
Dell'Angelo, J., D'Odorico, P., Rulli, M. C., & Marchand, P. (2017). The tragedy of the grabbed commons: Coercion and dispossession in the global land rush. *World Development*, *92*, 1–12. https://doi.org/10.1016/j.worlddev.2016.11.005
Dunbar-Ortiz, R. (2014). *An indigenous peoples' history of the United States*. Beacon Press.
Ellabban, O., Abu-Rub, H., & Blaabjerg, F. (2014). Renewable energy resources: Current status, future prospects and their enabling technology. *Renewable and Sustainable Energy Reviews*, *39*, 748–764. https://doi.org/10.1016/j.rser.2014.07.113
Escobar, A. (2015). Commons in the pluriverse. In D. Bollier & S. Helfrich (Eds.), *Patterns of commoning*. http://patternsofcommoning.org/contents/
Fairhead, J., Leach, M., & Scoones, I. (2012). Green grabbing: A new appropriation of nature? *The Journal of Peasant Studies*, *39*(2), 237–261. https://doi.org/10.1080/03066150.2012.671770
Federici, S. (2012). Feminism and the politics of the commons. In D. Bollier & S. Helfrich (Eds.), *The wealth of the commons: A world beyond market and state*. http://wealthofthecommons.org/contents
Feola, G. (2019). Capitalism in sustainability transitions research: Time for a critical turn? *Environmental Innovation and Societal Transitions*, *35*, 241–250. https://doi.org/10.1016/j.eist.2019.02.005
Foxon, T. (2018). *Energy and economic growth: Why we need a new pathway to prosperity*. Routledge.
Frigo, G. (2017). Energy ethics, homogenization, and hegemony: A reflection on the traditional energy paradigm. *Energy Research & Social Science*, *30*(Supplement C), 7–17. https://doi.org/10.1016/j.erss.2017.06.030
Georgescu-Roegen, N. (1984). Feasible recipes versus viable technologies. *Atlantic Economic Journal*, *12*(1), 21–31.
Giotitsas, C., Pazaitis, A., & Kostakis, V. (2015). A peer-to-peer approach to energy production. *Technology in Society*, *42*, 28–38. https://doi.org/10.1016/j.techsoc.2015.02.002

Gollwitzer, L., Ockwell, D., Muok, B., Ely, A., & Ahlborg, H. (2018). Rethinking the sustainability and institutional governance of electricity access and mini-grids: Electricity as a common pool resource. *Energy Research & Social Science, 39*(Supplement C), 152–161. https://doi.org/10.1016/j.erss.2017.10.033

Gorenflo, N. (2012). Share or die—a challenge for our times. In D. Bollier & S. Helfrich (Eds.), *The wealth of the commons: A world beyond market and state.* http://wealthofthecommons.org/contents

Grimley, M., & Farrell, J. (2016). *Mighty microgrids.* Institute for Local Self-Reliance. https://ilsr.org

Gui, E. M., & MacGill, I. (2018). Typology of future clean energy communities: An exploratory structure, opportunities, and challenges. *Energy Research & Social Science, 35*, 94–107. https://doi.org/10.1016/j.erss.2017.10.019

Hammerstein, D. (2019). Commons-based renewable energy in the age of climate collapse. In S. Bloemen & T. de Groot (Eds.), *Our commons: Political ideas for a new Europe* (pp. 25–32). Institute of Network Cultures.

Harvey, D. (2011). The future of the commons. *Radical History Review, 2011*(109), 101–107. https://doi.org/10.1215/01636545-2010-017

Harvey, D. (2018). *The limits to capital.* Verso.

Heinberg, R., & Fridley, D. (Eds.). (2016a). Energy 101. In *Our renewable future: Laying the path for one hundred percent clean energy* (pp. 15–34). Island Press.

Heinberg, R., & Fridley, D. (Eds.). (2016b). Energy supply: How much will we have? How much will we need? In *Our renewable future: Laying the path for one hundred percent clean energy* (pp. 115–130). Island Press.

Helfrich, S. (2012a). Common goods don't simply exist—they are created. In D. Bollier & S. Helfrich (Eds.), *The wealth of the commons: A world beyond market and state.* http://wealthofthecommons.org/contents

Helfrich, S. (2012b). The logic of the commons and the market: A shorthand comparison of their core beliefs. In D. Bollier & S. Helfrich (Eds.), *The wealth of the commons: A world beyond market and state.* http://wealthofthecommons.org/contents

Hoffman, S. M., & High-Pippert, A. (2005). Community energy: A social architecture for an alternative energy future. *Bulletin of Science, Technology & Society, 25*(5), 387–401. https://doi.org/10.1177/0270467605278880

Huber, M. T., & McCarthy, J. (2017). Beyond the subterranean energy regime? Fuel, land use and the production of space. *Transactions of the Institute of British Geographers, 42*(4), 655–668. https://doi.org/10.1111/tran.12182

Hughes, D. M. (2017, October 23). Who owns the wind? *Boston Review: A Political and Literary Forum.* Retrieved March 28, 2019, from http://bostonreview.net/science-nature/david-mcdermott-hughes-who-owns-wind

Illich, I. (2009). The social construction of energy. *New Geographies, 2*, 11–19.

Koirala, B. P., Koliou, E., Friege, J., Hakvoort, R. A., & Herder, P. M. (2016). Energetic communities for community energy: A review of key issues and trends shaping integrated community energy systems. *Renewable and Sustainable Energy Reviews, 56*, 722–744. https://doi.org/10.1016/j.rser.2015.11.080

Kostakis, V., & Bauwens, M. (2014). *Network society and future scenarios for a collaborative economy.* Palgrave Macmillan.

Kunze, C., & Becker, S. (2015). Collective ownership in renewable energy and opportunities for sustainable degrowth. *Sustainability Science, 10*(3), 425–437. https://doi.org/10.1007/s11625-015-0301-0

Lambing, J. (2012). Electricity commons—toward a new industrial society. In D. Bollier & S. Helfrich (Eds.), *The wealth of the commons: A world beyond market and state.* http://wealthofthecommons.org/contents

Linebaugh, P. (2008). *The Magna Carta manifesto: Liberties and commons for all.* http://ebookcentral.proquest.com/lib/mcgill/detail.action?docID=345552

Lohmann, L., & Hildyard, N. (2014). *Energy, work and finance.* www.thecornerhouse.org.uk

MacArthur, J., & Matthewman, S. (2018). Populist resistance and alternative transitions: Indigenous ownership of energy infrastructure in Aotearoa New Zealand. *Energy Research & Social Science, 43*, 16–24. https://doi.org/10.1016/j.erss.2018.05.009

Martinez, C. (2017). From commodification to the commons: Charting the pathway for energy democracy. In D. Fairchild & A. Weinrub (Eds.), *Energy democracy: Advancing equity in clean energy solutions* (pp. 21–36). Island Press.

Mattei, U. (2012). First thoughts for a phenomenology of the commons. In D. Bollier & S. Helfrich (Eds.), *The wealth of the commons: A world beyond market and state*. http://wealthofthecommons.org/contents
Melville, E., Christie, I., Burningham, K., Way, C., & Hampshire, P. (2017). The electric commons: A qualitative study of community accountability. *Energy Policy*, *106*, 12–21. https://doi.org/10.1016/j.enpol.2017.03.035
Moore, J. W. (2000). Sugar and the expansion of the early modern world-economy: Commodity frontiers, ecological transformation, and industrialization. *Review (Fernand Braudel Center)*, *23*(3), 409–433.
Moss, T., Becker, S., & Naumann, M. (2015). Whose energy transition is it, anyway? Organisation and ownership of the Energiewende in villages, cities and regions. *Local Environment*, *20*(12), 1547–1563.
National Research Council. (1984). *Energy use: The human dimension* (P. C. Stern & E. Aronson, Eds.). https://doi.org/10.17226/9259
Necefer, L., Wong-Parodi, G., Jaramillo, P., & Small, M. J. (2015). Energy development and Native Americans: Values and beliefs about energy from the Navajo Nation. *Energy Research & Social Science*, 7, 1–11. https://doi.org/10.1016/j.erss.2015.02.007
Ostrom, E. (1990). *Governing the commons: The evolution of institutions for collective action*. Cambridge University Press.
Ostrom, E. (2010). Beyond markets and states: Polycentric governance of complex economic systems. *American Economic Review*, *100*(3), 641–672. https://doi.org/10.1257/aer.100.3.641
Peters, J., Sievert, M., & Toman, M. A. (2019). Rural electrification through mini-grids: Challenges ahead. *Energy Policy*, *132*, 27–31. https://doi.org/10.1016/j.enpol.2019.05.016
Platform. (2015). Energy beyond neoliberalism. *Soundings*, *59*(59), 96–114. https://doi.org/10.3898/136266215814890521
Polanyi, K. (2001). *The great transformation: The political and economic origins of our time* (2nd Beacon paperback ed.). Beacon Press.
Quilligan, J. B. (2012). Why distinguish common goods from public goods? In D. Bollier & S. Helfrich (Eds.), *The wealth of the commons: A world beyond market and state*. http://wealthofthecommons.org/contents
Raman, S. (2013). Fossilizing renewable energies. *Science as Culture*, *22*(2), 172–180. https://doi.org/10.1080/09505431.2013.786998
Roberts, J., Bodman, F., & Rybski, R. (2014). *Community power: Model legal frameworks for citizen-owned renewable energy*. ClientEarth.
Ryghaug, M., Skjølsvold, T. M., & Heidenreich, S. (2018). Creating energy citizenship through material participation. *Social Studies of Science*, *48*(2), 283–303. https://doi.org/10.1177/0306312718770286
Smil, V. (Ed.). (2009). Energy in our minds: Concepts and measures. In *Energy: A beginner's guide* (pp. 1–25). Oneworld Publications.
Smil, V. (2017). *Energy and civilization: A history*. The MIT Press.
Späth, P., & Rohracher, H. (2010). 'Energy regions': The transformative power of regional discourses on socio-technical futures. *Research Policy*, *39*(4), 449–458. https://doi.org/10.1016/j.respol.2010.01.017
Stephenson, J. (2017). What does energy mean? An interdisciplinary conversation. *Energy Research & Social Science*, *26*, 103–106. https://doi.org/10.1016/j.erss.2017.01.014
Taylor, K. (2019). *Governing the wind energy commons: Renewable energy and community development*. West Virginia University Press.
Tuck, E., & Yang, K. W. (2012). Decolonization is not a metaphor. *Decolonization: Indigeneity, Education & Society*, *1*(1). https://jps.library.utoronto.ca/index.php/des/article/view/18630
UNDRIP. (2007). *United Nations declaration on the rights of indigenous peoples (A/RES/61/295)*. www.un.org/development/desa/indigenouspeoples/declaration-on-the-rights-of-indigenous-peoples.html
van der Horst, D., & Vermeylen, S. (2008). The new energy commons: Exploring the role of property regimes in the development of renewable energy systems. *12th Biennial Conference of the International Association for the Study of Commons*, 14–18. http://hdl.handle.net/10535/952
van der Horst, D., & Vermeylen, S. (2010). Wind theft, spatial planning, and international relations. *Renewable Energy Law and Policy Review*, *1*, 67–76.
Vermeylen, S. (2010). Resource rights and the evolution of renewable energy technologies. *Renewable Energy*, *35*(11), 2399–2405. https://doi.org/10.1016/j.renene.2010.03.017
Wall, D. (2017). *Elinor Ostrom's rules for radicals: Cooperative alternatives beyond markets and states*. Pluto Press.
Whyte, K., Reo, N., McGregor, D., Smith, M. A. (Peggy), Jenkins, J., & Rubio, K. (2017). Seven indigenous principles for successful cooperation in Great Lakes conservation initiatives. In E. Freedman &

M. Neuzil (Eds.), *Biodiversity, conservation, and environmental management in the Great Lakes Basin* (pp. 182–194). Routledge.

Wilson, D. S. (2015). Generalizing the commons. In D. Bollier & S. Helfrich (Eds.), *Patterns of commoning*. http://patternsofcommoning.org/contents/

Wilson, D. S., Ostrom, E., & Cox, M. E. (2013). Generalizing the core design principles for the efficacy of groups. *Journal of Economic Behavior & Organization*, *90*, S21–S32. https://doi.org/10.1016/j.jebo.2012.12.010

Wolsink, M. (2012). The research agenda on social acceptance of distributed generation in smart grids: Renewable as common pool resources. *Renewable and Sustainable Energy Reviews*, *16*(1), 822–835. https://doi.org/10.1016/j.rser.2011.09.006

York, R., & Bell, S. E. (2019). Energy transitions or additions? Why a transition from fossil fuels requires more than the growth of renewable energy. *Energy Research & Social Science*, *51*, 40–43. https://doi.org/10.1016/j.erss.2019.01.008

Zhang, C., Wu, J., Zhou, Y., Cheng, M., & Long, C. (2018). Peer-to-peer energy trading in a microgrid. *Applied Energy*, *220*, 1–12. https://doi.org/10.1016/j.apenergy.2018.03.010

19
PART III RESPONSE

Tarla Rai Peterson

The chapters in this part of the handbook describe modes of action that help us to better understand how people instantiate energy democracy, as well as the various challenges that complicate the responses to those actions. In this brief response, I will begin by noting interconnections that emerged across the chapters, particularly as related to the justice, participation, power, technology (JPPT) framework introduced in Chapter 1. Then, I will lay out some future directions for energy democracy research and practice that relate to the modes of action highlighted in these chapters.

Navigating possible activities to promote energy democracy

Perhaps most notably, each chapter engages much more directly with the confusing and sometimes opposing interactions within the JPPT framework than with any individual dimensions of energy democracy. For instance, Becker's (Chapter 15) comparative studies of state-based and locally initiated energy transitions in Germany differentiates between how the two forms appear to enable different levels of participation that are not necessarily matched by their contributions to justice and power. Although the locally initiated cooperative significantly expanded opportunities for participation in dialogue and discussion, participation venues were filled with people who were veteran community leaders. Despite participants initiating changes in local (social and political) power differentials related to their energy system, he found little evidence that those changes contributed to increased justice: Citizens who had previously seemed peripheral to decision making remained marginalized. On the other hand, residents of the community where energy transformation took the form of remunicipalization experienced little change in availability of participatory options; being incorporated into the municipality's preexisting institutional arrangements provided a convenient link for scaling up changes prompted by local action. The municipalized system also had an inside track with engineers and other technology experts when exploring ways technology could be used to support democratic aims, which granted citizens involved in this project greater access to decision making.

The scale of MacArthur and Tarhan's (Chapter 16) historicization of energy cooperatives provides important contextualization for the cases analyzed by Becker and also helps situate the emphasis on power and justice in Stephens and Allen's (Chapter 17) endorsement of actions to shift the power dynamics of energy leadership away from being an almost exclusively male bastion

DOI: 10.4324/9780429402302-22

and Burke's (Chapter 18) advocacy for energy commoning. Within this context, grassroots actions designed to democratize energy systems materialize as recurring rather than new phenomena. Further, it indicates that no single location or political entity holds a monopoly on such activity. And, finally, modes of action that are highly productive in a remote Italian village during the 1890s may need significant revamping for application to an urban neighborhood in 2021.

Stephens and Allen (Chapter 17) focus directly on actions to subvert what they see as the energy industry's unobtrusive slippage across actions that perpetuate power-over relationships, social injustice, and an assumption that technology should lead decisions. They attack the patriarchal power evidenced across the energy industry, highlighting male privilege as a key element in need of change. While some readers might consider this central argument passé, the authors demonstrate its currency and, more importantly, its urgency. The dangerous alignment of patriarchal privilege with *technological psychosis*, or a societal tendency to see and describe anything in terms that centralize immediate utilitarian values (Burke, 1954), provides an impetus for their bold recommendation to affirmatively seek to place women in leadership positions within the industry. They point out that energy systems are notoriously hierarchically organized and that oft touted attempts to redress injustice tend to be hollow public relations ploys. Further, no industry more fully embodies modern society's technological psychosis than does energy. While Stephens and Allen do not presume that all female leaders automatically behave more democratically than males, they justify their recommendation with leadership research indicating a strong tendency in this direction. Further, their chapter presumes that such obviously change-oriented action offers one means of alleviating the overwhelming acceptance of hierarchical power relations and mitigating the industry's tendency toward monolithic, centralized systems of control.

Matthew J. Burke (Chapter 18) takes the most overtly radical approach to the potentially contentious field of grassroots action leading to energy democracy. While he shares an awareness that "the commons" is variously understood in numerous ways, including as an ethic, a grammar, and a philosophy, he chooses to approach it from an action orientation. For Burke, *energy commoning* describes a set of practices that may (or not) lead to energy democracy, which, in turn, may (or not) transform energy systems in numerous surprising ways. Succumbing to the need for a central object around which all this action swirls, he follows other scholars in describing commons as living (and therefore changeable) systems and then immediately turns to an exploration of how those systems operate. Burke's focus on dynamic action is central for purposes of understanding and practicing energy democracy. As he argues, commoning relies on exploring and nurturing social values that extend beyond privatization, which makes it a particularly difficult ontology to justify, especially as the very possibility raises the hackles of the largest and most powerful human institutions, whether corporate or state. Such an approach obviously has the potential to disconnect long standing hegemonic relations, which includes recognizing and then struggling to overcome injustice. And it requires widespread and frequent participation in conversations about what a particular community wants and needs. Commoners return to technology as a tool for composing sustainable societies, with the emphasis on the plurality of those societies. Lastly, Burke argues that commoning offers repeated opportunities for community members to reprise democracy, as they continually seek to deepen and broaden the options available for democratic practice, all the while consciously avoiding the trap of expecting to reach the perfect solution.

Future directions

The process of democratizing energy transitions, as these chapters emphasize, is not something that occurs automatically along with shifting from fossil fuels to renewable energy sources.

All four chapters suggest challenges within grassroots and critical modes of action that require untangling for energy democracy to flourish or even to emerge. As such, one important direction for further study of grassroots efforts to promote energy democracy is to analyze the particular connections between renewable energies and particular modes of action. Burke's chapter on commoning only begins to explore ways that two sources of renewable energy, sunlight and wind, can be reimagined as common resources available to all living beings. At the same time, Becker's comparative study of remunicipalization versus cooperatives simultaneously offers two models for organizing the management of those resources once they have been transformed into electricity, at the same time it suggests cautions. In what kinds of grassroots activities should people engage to warrant that whatever model they select is equitable and fulfills the needs of even the most vulnerable members of the community? Are some activities more conducive to equity than others? If so, how can activists, managers, and scholars assist and support those activities rather than others?

Another concern is that democratic possibilities of grassroots activities may be minimized if those activists fail to understand and, where possible, make use of current forms of institutionalization. At the very least, successful attempts to change a power dynamic require some understanding of current hegemonic relations, as well as a plan for how to mobilize resistive power against that configuration. For example, we need research extending discoveries such as the problems Becker found with scaling up activities of a small energy cooperative. And we need to identify points of vulnerability where critical modes of action such as commoning have at least a minimal chance of gaining a foothold. Placing women in leadership positions throughout the energy sector may be a good starting point for changing the power dynamic and minimizing the control of societal technological psychosis, but it is only a beginning. Research on how to deepen the feminist critique of energy transitions initiated in Stephens and Allen's chapter will need to explore the myriad ways in which patriarchy shapes the energy industry, as well as important concepts such as producer and consumer.

Lastly, MacArthur and Tarhan's historical interrogation of energy cooperatives suggests an especially fruitful direction for future research. It is one thing to lament the overreliance on technology that seems to characterize modern societies and quite another to propose an alternative perspective that has broad appeal. Deeper and longer study of energy transitions might open currently excluded possibilities for how to match felt needs to available technologies. For example, although digitization may be instrumental in making renewable resources a practical source of electricity at scales needed in the twenty-first century, this is not the first time such resources have been harnessed for human use. What the current energy transition affords is an unprecedented opportunity to connect the production and use of energy with democratic principles. Understanding the myriad ways that previous transitions both promoted and blocked social equity and justice can offer valuable guidance.

Reference

Burke, K. (1954). *Permanence and change* (2nd ed.). University of California Press.

PART IV

Democratic and participatory principles

20

DEMOCRATIC AND PARTICIPATORY PRINCIPLES OF ENERGY DEMOCRACY

Introduction

Tarla Rai Peterson

Part IV of the handbook focuses on democratic and participatory principles of energy democracy research and practice. It introduces principles and constructs that characterize energy democracy in widely varied contexts. Chapters in this section theorize and analyze a broad array of democratic practices in a wide variety of political and technological contexts. The point of this part is not to identify a best way to formulate energy democracy but to explore the diverse ways participants can make meaningful contributions to energy system transformation.

In Chapter 21, Kinsella uses nuclear energy to frame a discussion of democratic deliberation, arguing that the "high-consequence" risks associated with failure, the immeasurable lifespan of nuclear waste, intrinsic links between civilian and military use, and complicated political ecology of the nuclear industry simultaneously restrict and promote democratic participation. He notes, for example, that despite a political ecology that enabled proliferation of nuclear energy and disabled associated public protest in Japan, the catastrophic Fukushima failure of 2011 led to a wide variety of public participation (including but not limited to protest) on multiple continents. While this participation was not necessarily instrumental to changes in all locations, it did achieve both specific aims to shutter projects and influence broader contexts such as the national nuclear program in Germany. Although public participation appears to have been less systematically influential in the United States, it contributed to termination of specific projects, such as the Vermont Yankee reactor. Kinsella turns to France for a demonstration of the complex international relations that insulate nuclear energy from democratic intervention. The political and economic configuration of Électricité de France (EDF) means that both the state and the private sector have strong interests in advancing the use of nuclear energy throughout the globe, which leads to substantial international capital investments in nuclear infrastructure. The placement of these projects enables EDF to more easily transfer risks to host populations and to minimize public engagement in decisions. Perhaps most problematic for energy democracy, this demonstrates how reliance on nuclear power includes continued commitment to a centralized and technocratic energy system that seems antithetical to democracy.

In Chapter 22 Barnes begins from the premise that social sustainability requires comprehensive reinvention of the ways society uses energy and that such reinvention necessitates a systematically invigorated model of public participation. After contextualizing public participation

DOI: 10.4324/9780429402302-24

within trends toward *decarbonization*, *digitalization*, and *decentralization* that implicate—while also transcending—energy, he explores opportunities for deep democratization enabled by these sociotechnical trends. From this perspective, energy publics shift from being identified as passive consumers to becoming active stakeholders. For Barnes, then, the interesting question is not whether energy democracy emerges but *how democratic* future energy systems may become and particularly how publics may agitate for richer participatory forms that deepen democratic governance. As described by Barnes, the composition of democratic energy systems demands treating public participation as a primary site for experimentation to enable institutional responses that, in turn, foster vigorous public debate about the outcomes of those experiments. He notes that, within the contexts of rapid digitalization and decentralization, diverse groups of people have begun actively seeking meaningful participation opportunities, with associated capacities to explore, dissent, and even produce energy system transformations. Excluding them from active participation in composing new energy systems leaves the resulting systems less inclusive, equitable, and environmentally sustainable. The overarching goals for both researchers and advocates, then, become better understanding and then maximizing how participation, along with justice, (social) power, and technology, operates as a constituent part of energy democracy.

In Chapter 23, Scherhaufer attempts to navigate the dangerous shoals between seeking social acceptance and participation. This chapter returns the reader to a more traditional approach to participation grounded in instrumental logics of conflict management, as it examines the role of local and regional participatory processes in composing social acceptance of renewable energy technologies. Scherhaufer explicitly recognizes the tensions between promoting public acceptance and nurturing meaningful opportunities for citizens to participate in decision making. While noting the substantive differences between these efforts, he also claims that the two can and should complement each other. One reason for this complementarity is that disputes form the core of planning and implementation processes for most energy projects, with the felt needs of those most directly affected often facing off against the broad and ephemeral benefits promised to an illusory general public. From this vantage point, public participation fulfills an ethical imperative to enable all interested parties to participate in an open negotiation process between different perspectives and values. Scherhaufer reminds readers that new approaches to participation need not eliminate but often can work in tandem with more traditionally formalized approaches. He suggests that process facilitators can enable this productive expansion of what counts as participation by helping project leaders shift from considering themselves primarily as the purveyors of information to self-identifying as learners whose primary responsibility is to gain a deeper understanding of what matters to the publics with whom they interact.

In Chapter 24, Delina explores the novel possibility of energy democracy in nondemocratic states. Using a case from Thailand, he excavates a set of principles that could provide a guide for shaping participation opportunities in both democratic and nondemocratic states.

In this case, residents of an isolated forest village created and nurtured spaces to practice energy democracy via communal deliberation. With neither state intervention nor support from external civic institutions, they negotiated the contradictions and tensions associated with providing their community with access to electricity for basic needs. This experiment in public participation demonstrates how localized energy democracy can emerge and even thrive despite the limitations of existing within a larger nondemocratic state. Delina offers the principles displayed in this case as a framework that could transfer to other nondemocratic locations. The energy democracy that emerged and reemerged in this community was coproduced in tandem with other sociotechnical orderings, with participants prioritizing mutual learning as a means of understanding the deep cultural dynamics in play rather than prioritizing any particular deliverables or predetermined end point. All members participated voluntarily in arranging the always temporary forms of engagement that emerged from an interdependent collective rather than from an amalgamation of independent individuals. And, finally, rather than shy away from the concept of insider/outsider, they accepted exclusion as a fundamental characteristic of any social organization.

The goal of this part of the handbook is not to celebrate public participation so much as to tease out perspectives and components that seem to accompany processes that enable more diverse contributions and yield richer democratic contexts. Although public participation is essential to healthy democratic functioning, there is no consensus regarding the single best type of participation. These chapters, however, identify principles that can be used to guide project proponents, citizen advocates, process facilitators, and others. Although we recognize that the fear of being mired in endless public discussions haunts project developers and managers, we share the normative commitment demonstrated in these chapters to privilege a full unfolding of participatory principles. Although that unfolding may or may not align with prior expectations of those who initiated the process, it will encourage a more democratic energy transition.

21

SPLITTING (OVER) THE ATOM

Nuclear energy and democratic conflict

William J. Kinsella

Nuclear energy emerged as a topic of public concern in the context of a violent global conflict, with the destruction of Hiroshima and Nagasaki at the end of the Second World War. This potent energy source has been a matter of political conflict ever since, with both its military and civilian uses demanding democratic deliberation and decision making. Advocates of civilian nuclear power argue that it is essential for a growing world population and an energy-intensive, consumption-driven global economy, while critics stress its safety, public health, and ecological hazards. Additional debates surround the relationships between civilian nuclear power and nuclear weapons proliferation. Appraising the dilemmas of nuclear energy, historian Alice Kimball Smith (1971) described it as simultaneously "a peril and a hope." That duality persists in nuclear discourses today, expressed in policy debates, political controversies, and divergent technical opinions.

Radioactive by-products of human uses of nuclear energy provide key markers of the condition now named the Anthropocene, "the current epoch in which humans and our societies have become a global geophysical force" (Stefan et al., 2007, p. 614). Meanwhile, another feature of the Anthropocene, global climate change, calls for urgent action and offers nuclear advocates a new argument for their technology. Framing nuclear power as the best low-carbon alternative to fossil fuels, its proponents suggest a path to addressing climate change without forsaking commitments to ever expanding consumption and energy use. Those commitments, in turn, are embedded in a broader political, economic, and institutional system that resists efforts to move toward less centralized and more ecologically sound energy sources.

This chapter examines these and other aspects of nuclear energy as questions of *energy democracy*. Stephens et al. (2018) describe energy democracy as "an emergent social movement focused on advancing renewable energy transitions by resisting the dominant energy agenda while reclaiming and democratically restructuring energy regimes" (p. 2). Within that framework, they argue that "[r]esisting the legacy centralized fossil and nuclear dominated energy systems is key to the energy democracy movement" (p. 2). Feldpausch-Parker et al. (2019) regard energy democracy as both a research agenda and a critical-analytical perspective that can inform such pragmatic goals while illuminating broader relationships between energy technologies and society.

At least four aspects of nuclear energy make it a matter of democratic concern. First, as demonstrated periodically by dramatic "focusing events" (Birkland, 1997), such as those at

DOI: 10.4324/9780429402302-25

Three Mile Island (1979), Chernobyl (1986), and Fukushima Daiichi (2011), the construction and operation of nuclear power plants have always entailed risks of high-consequence failures with the potential for deadly releases of radioactivity. Accordingly, *nuclear power plant siting and safety* has long been an issue demanding democratic attention (see, for example, Aldrich, 2008; Venables et al., 2012).

Second, the so-called nuclear fuel cycle (IAEA, 2011; Wilson, 1996) involves a chain of hazardous waste production that begins with uranium mining and milling, continues through a series of industrial processing steps, and results in the accumulation of highly potent radioactive materials posing multimillennial dangers. The stakeholders constituted by those hazards are geographically, racially, culturally, and socioeconomically diverse and extend many generations into an indefinitely distant future. Thus complex questions of social, intergenerational, and ecological justice, democratic participation, and institutional power surround issues of *nuclear waste production, management, storage, transportation, and disposal* (Calyx & Jessup, 2019; Clarke, 2010; Endres, 2009a, 2009b, 2012, 2013; Fan, 2006a, 2006b; Freudenburg & Davidson, 2007; Ishiyama, 2003; Kinsella, 2016, 2020; Peeples et al., 2008; van Gerven, 2014).

Third, commercial/civilian uses of nuclear power are linked intrinsically with the potential for *nuclear weapons proliferation*, although the directness of those links is often a matter of technical disagreement and policy debate. Historically, antinuclear protests have focused on both nuclear weapons and civilian nuclear power, often emphasizing the links between them (Findlay, 2010; Kinsella et al., 2015).

Fourth, looking to nuclear energy as a response to climate change or issues of energy security requires recognizing its broader *economic and political-economic aspects*. Nuclear energy may entail problematic opportunity costs if chosen in place of alternative, renewable energy options and is now competing with those options at the very time when they are becoming more technologically viable (see Suna & Resch, 2016). Commitments to nuclear power may also distract from the consideration of changes to the fundamental inequities that characterize the present energy-intensive, consumption-driven global economy and its reliance on centralized energy sources. Thus what might be called a "nuclear energy alibi" may provide a rationale for parties resistant to energy democratization (cf. Anshelm & Hultman, 2015, pp. 54–55).[1] Meanwhile, carbon-based energy sources, particularly natural gas made available by using hydraulic fracturing and horizontal drilling techniques, are posing their own competitive threats to nuclear power.

The following sections of this chapter explore these democratic tensions further. I begin by briefly examining selected concepts related to energy democracy. I then identify some particular features of nuclear power that pose challenges to basic principles of energy democracy. Informed by these considerations, I then focus on current conflicts between nuclear power and other energy sources and between business-as-usual perspectives and others that envision a more transformative future. I then consider some of the social, ethnic, racial, economic, intergenerational, and ecological implications of nuclear power. I close with suggestions for informing theory and practice in the growing field of energy democracy.

Concepts of energy democracy

Feldpausch-Parker et al. (2019) conceptualize energy democracy as both a political practice and a research agenda, identifying three interdependent analytical dimensions: justice, participation, and power. Questions of *justice* involve, in part, outcomes of energy politics, such as equitable or inequitable distribution of risks and benefits, and the empowerment or disempowerment of affected or potentially affected parties. Questions of *participation* involve forms of public engagement and decision making and their inclusiveness, equality or inequality of access, and influence on energy outcomes. Questions of *power* involve structural, relational, and institutional conditions that enable and constrain energy deliberations and decisions at levels ranging from local to global. While these three dimensions clearly are intertwined, identifying them as distinct spaces of inquiry provides a useful analytical starting point. This perspective reframes these questions in ways that differ from the established environmental justice, energy justice, and climate justice paradigms (see Jenkins, 2018) by considering justice as one of multiple elements that are fundamentally problematic for energy democracy. Energy democracy as foregrounded here is an ethical imperative in itself, involving reconstructions of communicative processes, political conditions and institutions, social and cultural practices, and power relations.

As Chilvers and Pallett (2018) observe, static and generalized concepts of "democracy" or of a monolithic democratic "public" fail to capture the evolving and variegated manifestations of energy democracy in practice. Applying this observation to the nuclear energy arena highlights the need for understanding justice, participation, and power in specific historical, geographic, cultural, and political contexts where this technology is considered, implemented, challenged, or otherwise consequential. For example, populations in the Global South and indigenous communities worldwide are affected by continued resource colonialism in the form of uranium mining, while the benefits of nuclear power production accrue to energy consumers in distant, more affluent locations (Banjeree, 2000; Falk et al., 2006; Hecht, 2012; Malin, 2015). Meanwhile, developing countries are pressed to aspire to their own atomic power programs by established nuclear nations seeking to expand their markets for technologies and services, irrespective of discrepancies between this centralized energy source, national and local energy needs, and sometimes incompatible electricity grid infrastructures (Thomas, 2017, 2018; Yi-chong, 2011). In another example, efforts to ensure robust nuclear safety regulation must respond to varied forms of political organization, institutional arrangements, cultures of transparency or opacity, expectations for protection, and possibilities for public influence across diverse national contexts, while at the same time they are expected to follow models developed in the established nuclear countries (see, for example, IAEA, 2018).

Burke (2018) notes that narratives and counternarratives play key roles in democratic energy transitions by framing problems and solutions requiring collective action, articulating societal norms and values, envisioning possible futures, providing historical continuity, and identifying "specific agents and adversaries to change" (p. 24). Accordingly, and as illustrated next, nuclear energy narratives have taken on a variety of forms and have led to a range of outcomes across varied local, national, and historical contexts (for examples, see Chernus, 2002; Edberg & Tarasova, 2016; Felt, 2015; Jasanoff & Kim, 2009; Leadbeater, 2013; Medhurst, 1987; Tarasova, 2018).

Distinctive aspects of nuclear energy

While it is important to recognize the specificity of contexts for nuclear energy conflicts and debates, the technology does present some fundamental characteristics with broad implications for energy democracy. Here I consider three sets of characteristics: (1) dominant technocratic logics of governance; (2) tensions between public antinuclear protest and industry–government promotion; and (3) inherent issues of centralization, scale, financing, risk-shifting, secrecy, and securitization.

Dominant technocratic logics

One basic aspect of nuclear power decision making and governance is its dominant technocratic framework, which seems inevitable in light of its technical foundations, complexity, and institutional history. Governmental and corporate actors typically embrace a logic of rational risk assessment, cost/benefit calculation, and expert authority within an overarching "Promethian" narrative (Dryzek, 2005) of social progress through technological innovation. Gwin (1990) has examined a related "promotional heritage" in nuclear power discourses linking corporate, governmental, and expert communities in a network of support for nuclear power. Within this framework, nuclear power is presented as a commonsense solution to problems of energy scarcity, energy security, and more recently, climate change. Meanwhile, critics emphasize the inherent risks associated with nuclear activities, hazards to public health and safety, ecological threats, disruptive effects on communities, and the dangers of accumulating nuclear wastes before viable disposal solutions have been demonstrated.

Thus two divergent versions of "common sense" are at odds in nuclear power debates (Kinsella, 2015). "Informal logics of risk" (Horlick-Jones, 2005) expressed by ordinary citizens and local communities are often discounted or dismissed by authorities as emotional, irrational, and lacking in scientific grounding. Lack of support for nuclear activities is attributed to a "deficit" of public understanding, thus ignoring essential questions of values and recognition while also excluding relevant local forms of knowledge (Sturgis & Allum, 2004; Wynne, 1991, 1996). For parties thus treated, those technocratic responses appear "contemptuous" (Katz & Miller, 1996) and exclusionary, undermining the legitimacy of decision processes and, by extension, the entire nuclear enterprise (see Kinsella, 2016).

Antinuclear protest and institutional promotion

Antinuclear protest has evolved in parallel with the history of nuclear power. In the United States, a number of significant protests have centered on particular nuclear power plants in advance of, during, or after their construction (Balogh, 1991; Wellock, 1998; Wills, 2006).

Other protests, including ones enacted largely by local communities, have been directed at nuclear waste storage and disposal projects (Gowda & Easterling, 1998, 2000; Easterling & Kunreuther, 1995; Peterson, 2001). In Europe, nuclear waste has been a persistent focus of controversy, prompting a history of intense confrontations between protesters and police. In one example, Blowers (2017) argues that protests at the proposed Gorleben nuclear waste repository in Germany were instrumental in stopping that project and challenging the nation's nuclear power program more broadly.

In Japan, antinuclear protest has long existed in tension with a system of pronuclear alliances among municipal and prefectural governments, the national government, and the nuclear industry. Observers have used the term "nuclear village" (*Genshiryoku Mura*) to describe the hegemony of this close-knit set of actors and their interests (see Kingston, 2012). Aldrich (2008) contrasts "strategies of soft social control and complex incentives" (p. 119), used within that framework in Japan to promote nuclear development and undermine opposition, with the "highly coercive and hard social control tools" he identifies in the case of France (p. 152). A distinctive aspect of Japanese antinuclear protest has been the leadership roles played by nuclear scientists who dissent from the positions typically identified with their profession. According to Avenell (2016):

> The antinuclear advocacy and activism of scientific experts arguably helped to slow down nuclear power development, resulting in a less nuclearised Japan than imagined by its advocates. Antinuclear experts and their movements, however, never became players in mainstream nuclear policymaking and had to be content with pressure tactics from the political peripheries.
>
> *(p. 88)*

In the Indian context, Haines (2019) has examined the case of antinuclear activists who have been marginalized despite their professional qualifications. Like their counterparts in Japan, these experts also have had to adopt tactical approaches to working within the prevailing bureaucratic system. As these examples suggest, the dynamics and narratives of nuclear promotion and antinuclear protest vary across national contexts and specific issue areas as well as over time (cf. Flam, 1994; Joppke, 1993). There are certainly overarching global conditions that inform nuclear power discourses and deliberations across national contexts: The 2011 Fukushima disaster, climate change, and the increasing internationalization of the market for nuclear technologies are three examples. However, those conditions are understood and appropriated differently across diverse argumentative settings.

Centralization, scale, financing, risk-shifting, and securitization

A debate dating at least to the work of Louis Mumford (1964) concerns whether some forms of technology are inherently undemocratic. Although scholarship in the field of science and technology studies now emphasizes the mutual coproduction of social and technological systems, nuclear power remains an oft cited example of an inherently undemocratic technology (e.g., Beck, 2011). Nuclear power generation to date has involved a highly centralized model, with typical power plant capacities increasing over time due to economies of scale and technological innovations that favor maximizing production. Complex safety and radiological protection systems contribute to high capital costs, requiring elaborate financing arrangements that often involve consortia of industry and government actors. Government support is essential for the construction of such facilities, whether directly in state-controlled economies or through tax incentives, loan guarantees, and other financial instruments in market economies. Insurance against the financial risks of nuclear disasters presents a particular democratic challenge, discussed in more detail in this chapter, with risks shifted from facility operators to potentially affected communities and the general population.

These properties of scale, complexity, centralization, financial concentration, and risk-shifting all pose challenges for the forms of grassroots engagement associated with energy democracy scenarios. These features are simultaneously obstacles to democratic participation and entry points for democratic intervention (e.g., Kinsella, 2015; Kinsella et al., 2013, 2015). Additional concerns involve issues of secrecy and securitization. The potential military uses of nuclear facilities and materials have always provided warrants for secrecy (Kinsella, 2001, 2005; Taylor et al., 2007). However, these concerns are also invoked more broadly to constrain public debate through acts of rhetorical securitization invoking tropes of threat, urgency, and national interest (Cotton, 2020; Kinsella, 2019; Szulecki, 2020). Likewise, security concerns have always accompanied the operation of complex industrial facilities housing materials that pose potent hazards and can be vulnerable to intrusion or operational interference. Those concerns have only become more acute in the present environment of terrorist threats and cyber-vulnerabilities and are now subject to new forms of rhetorical appropriation. In these contexts, discourses of securitization depoliticize issues of democratic concern, shifting the authority for decision making to closed, elite communities (see Kinsella, 2020).

Nuclear power as a conflicted energy option

In light of these considerations, the uses of nuclear energy for electric power production present a range of democratic conflicts. Following a period of promotional optimism in the early 2000s marked by visions of a "nuclear renaissance," nuclear power now faces challenges that make such a renaissance appear unlikely in any near-term timeframe. The events at Fukushima in 2011 and their continuing aftermath have resurfaced public concerns regarding nuclear reactor safety with varied consequences in differing national contexts. In Japan, as public opposition has strengthened, a condition characterized by Duffield (2016) as "nuclear ambivalence" has emerged, challenging the nation's strong historical reliance on nuclear power (see also Ogawa, 2013). In Europe, Germany and a few other nations have committed to ending their nuclear power programs (Schneider & Froggatt, 2019). In other nations, including the United States, the impact of the Fukushima disaster has been more limited and may be diminishing over time (Hoffman & Durlak, 2018). Divergent forms and levels of grassroots engagement have contributed to these varied outcomes.

Beyond reactor safety concerns, other controversies involve the economics of nuclear plant construction and operations. Globally, nuclear construction projects are running consistently over budget and behind schedule, often dramatically (Schneider & Froggatt, 2019). In the United States, which operates the largest number of nuclear plants and produces the greatest total output of nuclear energy, multiple construction projects have failed and only one (involving two new reactors) remains active. Democratic interventions have contributed to these project terminations, as consumers mobilize to protest electricity rate increases proposed by the industry in efforts to recover escalating costs (e.g., Kinsella et al., 2013). Meanwhile, a number of operating plants have closed. Plant owners typically cite their inability to compete in the energy marketplace, often highlighting the costs of compliance with safety regulations, but grassroots activism has also contributed to these closures. In the case of one plant in the state of Vermont, Stephens et al. (2018) suggest that the "closing of Vermont Yankee, the state's only nuclear power plant in 2014, can be viewed as an example of operationalizing energy democracy" (p. 5). They argue that despite the "powerful resistance of the legacy centralized energy system . . . Vermont experienced widespread citizen activism that contributed to . . . closing Vermont Yankee" (p. 5). The full context for this plant closure is complex, involving elements of market competition, regulatory compliance challenges, and the costs of maintaining an aging facility. However, decades of persistent citizen opposition to this plant, often coordinated by public interest organizations with long lasting commitments to the issue, maintained financial and political pressures that helped lead to the closure decision. Cases such as this one do not stand in isolation but often are understood as bellwethers by both the nuclear industry and its challengers (Kinsella, 2015; Kinsella et al., 2013).

The growing capacity of wind and solar power, and the improvement of battery and other energy storage technologies that help address the diurnal cycle of solar energy availability, pose another threat to nuclear power. Here grassroots activism plays another role by highlighting sustainable energy sources as viable alternatives, strengthening the case for the elimination of nuclear energy. These democratic efforts simultaneously build a climate of support for further technical advancements in sustainable energy, call attention to the possibilities those advancements enable, and create political demand for their implementation.

From a technical and economic perspective, nuclear power may be viewed as one of multiple, competing energy options, each presenting its own benefits and challenges (see Stern et al., 2016). For example, Pacala and Socolow (2004) provided a useful heuristic for considering such energy options in efforts to address climate change, identifying a portfolio of choices for stabilizing global carbon emissions. This model seems scalable across levels and sectors, offering a framework usable by municipalities, subnational governments, or national governments, as well as by industrial, commercial, and institutional actors seeking to reduce their carbon footprints. Some of these carbon-reduction approaches, including nuclear power, are well established and capable of providing substantial amounts of energy with limited (but not zero, as often claimed) greenhouse gas emissions.[2] Others, like wind and solar power, are growing rapidly in technological maturity and economic viability and have advanced dramatically since Pacala and Socolow's 2004 analysis. Other suggested approaches include carbon capture and storage technologies, improved agricultural soils management (e.g., limiting tillage) to reduce greenhouse gas releases, hydrogen fuels (essentially a form of energy storage and transportation rather than an original source of energy), biofuels, and energy conservation and efficiency measures.

The benefits, limitations, risks, and viabilities of all of these approaches are matters of ongoing and often contentious democratic engagement. In those settings, technical analysis intersects with forms of political and institutional power such as lobbying, corporate advocacy, and differential access to decision makers. Meanwhile, atmospheric greenhouse gas levels have increased alarmingly in the years since Pacala and Socolow (2004) proposed their model, while action on alternative energy options has proceeded more slowly than is demanded by the problem. A model of this kind has the potential to serve as a framework for democratic policy debate but operates in an essentially technocratic mode, focusing on criteria such as the relative capacities of various energy options and the economics of their deployment. Debates focused on these criteria are thus vulnerable to interminable dispute with evidence and counterevidence offered by advocates and opponents of various energy options. A more democratized debate would be a broader one about fundamental meanings and values associated with energy use and the related implications of various options for energy system transformation.

For some influential commentators, such as climate scientist James Hansen and *Whole Earth Catalog* editor Stuart Brand, the climate crisis is so urgent and consequential that it overrides all other considerations. These opinion leaders have come to view the expansion of nuclear capacity as an essential tool for combating climate change. However, as Kopytko and Perkins (2011) argue, "Mere absence of greenhouse gas emissions is not sufficient to assess nuclear power as a mitigation for climate change" (p. 318). In light of the long construction times and complex financing arrangements for nuclear plants, it is difficult to see how the needed capacity could be built in time to avoid the immanent catastrophic effects of accumulating greenhouse gases (see Lovins, 2009). The expanded capacity would greatly amplify the already intractable problem of nuclear waste. Accelerated construction efforts and premature deployment of new reactor designs could compromise nuclear plant safety. Further, as Kopytko and Perkins (2011) observe, nuclear plants are vulnerable to safety challenges and operational interruptions as the climate warms, sea levels rise, and extreme weather and flooding events become more common, affecting plant cooling systems and other critical functions.[3] Resources committed to nuclear expansion may displace resources and policy decisions that could otherwise support the development of alternative, renewable sources. Finally, nuclear expansion may undermine a sense of necessity for pursuing alternative energy directions including conservation and efficiency, sanctioning a business-as-usual approach to ever increasing energy consumption.

Expanding nuclear power as a response to climate change involves a commitment to a centralized, megaproject-based model of energy production and a technocratic approach to governance, whether in the context of a neoliberal market economy or an authoritarian command economy. Nuclear advocates appeal to the prospect of a future generation of smaller reactors operating in a more distributed arrangement, but those designs remain speculative and would still entail many of the same political-economic arrangements (Sovacool & Ramana, 2015). During the time period that matters for effective climate action, which is shrinking rapidly as the effects of a warming planet manifest, large, centralized, and costly nuclear projects remain the only option for this energy source.

Implications for present and future nuclear power stakeholders

Decisions about the utilization of nuclear energy have implications across a broad range of stakeholder communities. As previously considered and discussed further here, those implications are particularly salient for stakeholders affected by the siting (or potential siting) of nuclear waste facilities. Communities that host nuclear power plants are also exposed to nuclear waste risks, as highly radioactive used fuel is routinely stored on-site, along with the risks of reactor-related accidents.[4] Nuclear waste transportation, essential to any permanent disposal solution, poses risks for communities far beyond the locations where plants and waste facilities are sited.

These forms of risk have health and safety dimensions as well as economic dimensions. An analysis prepared by two nongovernmental organizations, International Physicians for the Prevention of Nuclear War and Physicians for Social Responsibility (2016), estimates that "just under 10,000 excess cases of cancer are to be expected in Japan in the coming decades" (p. 5) as a result of the Fukushima disaster, using conservative data and assumptions. The report cautions, as well, that "[i]f independent data and more modern risk factors are used, estimates of the rise in cancer incidence are significantly higher at around 66,000 additional cancer cases, approximately half of which would be fatal" (p. 5). The report recognizes that some parties may interpret those numbers as small relative to Japan's overall population size, the multidecade timeframe for these illnesses, and the population's baseline cancer risks. Indeed, nuclear advocates often respond to such numbers by comparing them to health and safety risks associated with other forms of energy production, such as coal, and to the hazards posed by climate change. Here, however, a technocratic logic prevails, enumerating and comparing risks in abstract terms that ignore the people, families, and communities affected. Such comparisons also disregard different alternatives to nuclear power, such as renewable sources, energy efficiency improvements, and conservation.

As for the economic impacts of Fukushima, a March 2019 estimate by the Japanese government projected a cost of ¥22 trillion (approximately US$200 billion) to address the damages, while an independent organization estimated a cost of ¥35–81 trillion (US$315–728 billion) (Komori, 2019). Those figures do not include the broader disruptive effects of the disaster on the Japanese economy and society, including intangible effects associated with displacement and community disruption. Political and legal disputes are likely to continue well into the future regarding questions of responsibility, compensation, and financing for the Fukushima remediation efforts.

As Nadesan (2013) has observed, the Fukushima disaster illustrates a massive shifting of health, ecological, and economic risk from government and industry to affected individuals, families, and communities. That disaster, of course, represents an unusual and severe scenario, but it is not the worst-case scenario for possible nuclear failures. Under only minimally different conditions, for example, the city of Tokyo could have been directly exposed to radiation releases from the plant (Kan, 2017). Worldwide, many other nuclear plants are located in proximity to major cities, making the prospects real for an even greater disaster than was seen at Fukushima.

If risk is understood not as manifest harm but as the potential for harm and the burdens that accompany that potential, then nuclear risks are routinely transferred from institutions to populations. As one example of economic risk, US nuclear plants are required to contribute annually to an insurance arrangement that provides approximately $13 billion for addressing the consequences of potential major nuclear disasters (Gilinsky, 2020; NAIC, 2019). As the Fukushima figures indicate, this is a small fraction of the possible costs of such an event. The remainder of the costs would presumably be borne by the national treasury and ultimately by US taxpayers (although the mechanism for that process has not been clarified) and/or by uncompensated or undercompensated victims. In terms of justice, understood as one component of energy democracy, consequences are evident regarding the potential health, ecological, and economic implications for stakeholders. Additional issues of process and power involve how such an insurance apparatus, which legitimates and underwrites nuclear activities while containing costs to the industry, was developed and is perpetuated, and how alternative arrangements might be developed democratically.

Nuclear waste, temporality, and energy democracy

Nuclear waste issues extend these kinds of risk problems not only across space but also across time. Nuclear waste comprises many types of material of varying persistence and potency, but the most infamous issues involve the highly radioactive by-products of civilian and military reactor operations. Some radioisotopes contained in used reactor fuel persist for tens and even hundreds of millennia. The multimillennial lifetimes of nuclear wastes raise baffling questions regarding what Okrent (1999) describes as "*intergenerational* equity and its clash with *intragenerational* equity" (italics added; see also Okrent & Pidgeon, 2000; Shrader-Frechette, 2000). Disposing of nuclear waste may address health, safety, and ecological threats to the present generation, as well as problems of practicality and legitimacy for the nuclear power enterprise. However, if containment of those materials cannot be guaranteed for periods exceeding the present age of written language and human history, then future generations are placed at risk.

Irrespective of future choices regarding society's commitments to nuclear energy, a dangerous inventory of waste already exists, created before a solution to the disposal problem has been achieved. The controversial proposed US repository site at Yucca Mountain in Nevada, which remains an uncertain prospect, even if completed would not have the capacity to accept all of the nation's present high-level wastes. Nevertheless, industry advocates and their political allies do not view this situation as an impediment to continued and accelerated nuclear power expansion.

As Burke (2018) has argued, narratives are central to processes of imagining, implementing, and transforming societal energy relationships (see also Jasanoff & Kim, 2009). In the case of nuclear wastes, the production of those materials may be understood as the outcome of multiple narratives, including narratives of technological progress, ever expanding energy consumption, rational risk assessment, and subordination of the interests of affected communities to a purportedly common interest defined by elites without meaningful participation or consent by those communities (Kinsella, 2020). In the domain of intergenerational equity, this situation is compounded further by our lack of knowledge regarding the conditions and interests of temporally distant stakeholders. Arguably, the present decision makers are not qualified to serve as representatives for future parties whose needs they cannot imagine. In these ways, commitments to nuclear energy pose fundamental challenges to concepts of justice, participation, and power within an energy democracy framework.

Nuclear political economy and energy democracy

The political economy of nuclear power provides another area for scrutiny within an energy democracy framework. As centralized megaprojects, nuclear power plants have always relied on various forms of government subsidy, varying in approach by national context. For example, in France, the nation with the greatest commitment to nuclear power as a share of its total energy generation, nuclear plants are operated by the public–private hybrid company Électricité de France (EDF). According to the World Nuclear Association (2018), the French government owns approximately 85% of the company. EDF maintains significant capital investments in nuclear infrastructures not only in France but in many European countries and beyond, and it is an active exporter of nuclear technologies and services. Through elaborate connections such as these, nuclear power has become embedded in the French political and economic system (see Hecht, 1998), linking France to a global network of commitments and activities.

One EDF project particularly illustrates the system of international nuclear entanglements, while also providing a case study of domestic nuclear subsidies in the United Kingdom. There, EDF is the primary party constructing two new reactors at the Hinkley Point nuclear power plant, with additional participation by the state-owned China General Nuclear Power Group (CGN). Like other new reactor construction projects worldwide, the Hinkley Point project has been chronically behind schedule and over budget. The project has been politically controversial both in the United Kingdom and in France, where its costs have posed a high-level risk to the financial stability of EDF. As a condition of the construction agreement, the British government has guaranteed a future sales price for electricity from the plant that is approximately 50% higher than the projected cost of offshore wind power (Harrabin, 2017), subsidizing the project at consumers' expense. While the projected construction cost for the two reactors was estimated at £20 billion in 2017, the cost to consumers who would purchase its electricity over a projected 60-year lifetime is forecasted to be approximately £50 billion above the market value of electricity. This discrepancy is likely to increase further as wind energy and other renewable sources develop greater capacity and effectiveness. Thus British electricity consumers have been locked into a dubious and controversial economic arrangement promoted by the nuclear industry and its political allies as a necessary response to climate change.

Jenkins et al. (2017) have examined the Hinkley Point project from a perspective framed in terms of environmental justice, but their analysis also illuminates issues of participation and institutional-political power. In this case, various parties' attributions of responsibility for just decision making have been related to their assessments of transparency (or lack of transparency) and trust (or lack of trust) in the decision process. Broadly similar patterns have characterized nuclear construction projects in many national contexts (see Kinsella et al., 2013; Schneider & Froggatt, 2019).

Conclusion: informing energy democracy theory and practice

Society's present uses of nuclear energy strongly exemplify the problems identified by the energy democracy perspective. Nuclear power is arguably the most centralized form of electricity production, relying on large-scale facilities that demand high degrees of technical, bureaucratic, and political control. As Mumford (1964) and others have suggested, such technological systems may exhibit inherently authoritarian characteristics. As a matter of energy justice, the risks posed by nuclear power are imposed not only on its presumed beneficiaries, the consumers of that energy, but also on parties far distant from the point of consumption, from uranium miners and communities affected by uranium extraction to future generations exposed to the risks of nuclear waste. As a matter of participation and power, decision making regarding nuclear energy is closely managed by an elite community of technical experts, industry actors, and their political allies. In the high-cost and high-stakes realm of nuclear politics, lobbying and the cultivation of political patronage are central activities affecting policy choices, regulatory standards and practices, and nuclear energy governance more broadly. In light of the breadth and depth of these challenges, it is difficult to formulate simple responses that might enhance the theory and practice of energy democracy; nevertheless, I close this chapter with a few broad suggestions.

First, as illustrated by some of the examples considered here, it is important to recognize that a strong tradition of nuclear critique and activism has always existed and continues to thrive, encompassing grassroots activists, allied technical experts, and political actors concerned with the problems of nuclear energy. Many nuclear projects have in fact been challenged, modified, and in some cases stopped or closed due to this network of opposition. Energy democracy proponents can look to this history for insights regarding how to strengthen and grow the critique of nuclear energy and forms of related political action. Organizations that provide independent public expertise—that is, technical knowledge that empowers grassroots critique and action—can play important roles in this regard (Kinsella, 2004, 2015).

Second, a broader program of energy literacy, with concepts of energy democracy as one component, can help overcome the alienation that many citizens experience regarding nuclear energy issues. To many people, nuclear issues often appear esoteric and mysterious, leading to a deference to technical and political authority (Kinsella, 2005). Of course, accurate and detailed technical understanding is essential to nuclear decision making. Nevertheless, many of the issues at stake involve fundamental questions of value and societal choice that extend beyond the technical domain. Greater energy literacy, understood as a broad understanding of the existing energy options as well as of the policy process involved in their selection, can help citizens engage more confidently with these crucial choices.

Third, proponents of energy democracy can look for ways to transcend the technocratic framework that prevails in energy decision making. Here again, scientific knowledge, systematic appraisals of risk, and other technical elements are crucial components of energy decision making. However, these elements alone are not sufficient. As Walter Fisher (1989) has argued, purely technocratic debates can easily devolve into duels between competing experts and analyses, relegating those affected to the roles of spectators. Providing space for the expression of values and interests in vernacular forms of discourse—ranging from ordinary language utilized at public hearings to performative and artistic statements—can help escape that technocratic trap (Kinsella, 2015; Kinsella et al., 2013).

Fourth, even with the technocratic framework, a more complete accounting of the costs and benefits of energy options, including nuclear power, is needed. Many of the costs of nuclear energy, from the effects of uranium mining, to government subsidies for nuclear power plant construction, to the shifting of risks of nuclear disasters to the general population, to the legacy of nuclear waste faced by future generations, are treated as economic externalities within the technocratic framework. As Verbruggen et al. (2014) have argued, it is necessary to fully evaluate the "external costs, benefits, risks and irreversible impacts inherent to the lifespan of a nuclear fission cycle. End users should also pay the full cost; if not, the bill will be footed by others, now or in the future" (p. 18).

Finally, energy democracy advocates can look for ways to monitor and ensure principles of transparency, openness, and independence in nuclear policy making and regulation. Unequal access to policy makers, mediated by institutional structures of lobbying and corporate advocacy, constrains the possibilities for inclusive democratic engagement regarding energy choices. The phenomena known as regulatory "capture" (co-optation of regulatory authorities by regulated industries; see Laffont & Tirole, 1991) and "recreancy" (broader failures of regulatory oversight; see Freudenburg, 1993) are problematic across the energy domain and perhaps especially difficult in the case of nuclear energy due to its technical and bureaucratic opacity.

With areas such as these in mind, proponents of energy democracy can continue to address the social, economic, and political dimensions of nuclear technologies. Conflicts are inevitable regarding justice, participation, power, and how best to address society's relationships with energy, amplified in the context of climate change. Those conflicts are key spaces in which social and ecological futures are created. In those spaces, visions and practices inspired by concepts of energy democracy can lead to more ethical and sustainable outcomes.

Notes

1 Here my use of the term "alibi" is informed by its use by Spivak (1999) in other contexts.
2 Although nuclear power advocates often claim that the technology is free of greenhouse gas emissions, it is not. Uranium mining and milling, nuclear fuel fabrication, nuclear plant construction, transportation of fuel and waste materials, and other related activities all produce a greenhouse gas footprint. Thus "low carbon" may be a more accurate term than "carbon free" to describe nuclear power (see Sovacool, 2008).
3 Nuclear power plants require substantial and reliable supplies of water for reactor cooling, and for that reason are often sited in locations that have become vulnerable to the effects of climate change.
4 Technical, industry, and government discourses commonly use the term "spent nuclear fuel" to describe the materials removed from reactors after energy production. That term, however, obscures the fact that used nuclear fuel, which has been irradiated in the reactor core, is far more radioactively dangerous than fresh fuel.

References

Aldrich, D. P. (2008). *Site fights: Divisive facilities and civil society in Japan and the West.* Cornell University Press.

Anshelm, J., & Hultman, M. (2015). *Discourses of global climate change: Apocalyptic framing and political antagonisms.* Oxford University Press.

Avenell, S. (2016). Antinuclear radicals: Scientific experts and antinuclear activism in Japan. *Science, Technology & Society, 21*(1), 88–109. https://doi.org/10.1177/0971721815622742

Balogh, B. (1991). *Chain reaction: Expert debate and public participation in American commercial nuclear power, 1945–1975.* Cambridge University Press.

Banjeree, S. B. (2000). Whose land is it anyway? National interest, indigenous stakeholders, and colonial discourse: The case of the Jabiluka uranium mine. *Organization & Environment, 13*(3), 3–38.

Beck, U. (2011, July 30). Germany is right to opt out of nuclear power. *Dissent.* www.dissentmagazine.org/online_articles/germany-is-right-to-opt-out-of-nuclear-power

Birkland, T. A. (1997). *After disaster: Agenda setting, public policy and focusing events.* Georgetown University Press.

Blowers, A. (2017). *The legacy of nuclear power.* Routledge.

Burke, M. J. (2018). Shared yet contested: Energy democracy counter-narratives. *Frontiers in Communication, 3*(22). https://doi.org/10.3389/fcomm.2018.00022

Calyx, C., & Jessup, B. (2019). Nuclear citizens jury: From local deliberations to transboundary and transgenerational legal dilemmas. *Environmental Communication, 13*(4), 491–504. https://doi.org/10.1080/17524032.2018.1464489

Chernus, I. (2002). *Eisenhower's atoms for peace.* Texas A&M University Press.

Chilvers, J., & Pallett, H. (2018). Energy democracies and publics in the making: A relational agenda for research and practice. *Frontiers in Communication, 3*(14). https://doi.org/10.3389/fcomm.2018.00014

Clarke, T. (2010). Goshute Native American tribe and nuclear waste: Complexities and contradictions of a bounded-constitutive relationship. *Environmental Communication, 4*(4), 387–405. https://doi.org/10.1080/17524032.2010.520724

Cotton, M. (2020). Nuclear power and environmental justice—the case for political equality. In A. Kaijser, M. Lehtonen, J. H. Meyer, & M. D. M. Rubio-Varas (Eds.), *Engaging the atom: The history of nuclear energy and society in Europe from the 1950s to the present.* West Virginia University Press.

Dryzek, J. S. (2005). *The politics of the earth: Environmental discourses* (2nd ed.). Oxford University Press.

Duffield, J. S. (2016). Japanese energy policy after Fukushima Daiichi: Nuclear ambivalence. *Political Science Quarterly, 131*(1), 133–162. https://doi.org/10.1002/polq.12431

Easterling, D., & Kunreuther, H. (1995). *The dilemma of siting a high-level nuclear waste repository.* Springer.

Edberg, K., & Tarasova, E. (2016). Phasing out or phasing in: Framing the role of nuclear power in the Swedish energy transition. *Energy Research & Social Science, 13,* 170–179.

Endres, D. (2009a). Science and public participation: An analysis of public scientific argument in the Yucca Mountain controversy. *Environmental Communication*, *3*(1), 49–75. https://doi.org/10.1080/17524030802704369
Endres, D. (2009b). The rhetoric of nuclear colonialism: Rhetorical exclusion of American Indian arguments in the Yucca Mountain nuclear waste siting decision. *Communication and Critical/Cultural Studies*, *6*, 39–60. https://doi.org/10.1080/14791420802632103
Endres, D. (2012). Sacred land or national sacrifice zone: Competing values in the Yucca Mountain controversy. *Environmental Communication*, *6*(3), 328–345.
Endres, D. (2013). Animist intersubjectivity as argumentation: Western Shoshone and Southern Paiute arguments against a nuclear waste site at Yucca Mountain. *Argumentation*, *27*(2), 183–200.
Falk, J., Green, J., & Mudd, G. (2006). Australia, uranium and nuclear power. *International Journal of Environmental Studies*, *63*(6), 845–857.
Fan, M. F. (2006a). Environmental justice and nuclear waste conflicts in Taiwan. *Environmental Politics*, *15*(3), 417–434. https://doi.org/10.1080/09644010600627683
Fan, M. F. (2006b). Nuclear waste facilities on tribal land: The Yami's struggles for environmental justice. *Local Environment*, *11*(4), 433–444. https://doi.org/10.1080/13549830600785589
Feldpausch-Parker, A. M., Sprain, L., Endres, D., & Peterson, T. R. (Eds.). (2019). *Energy democracy: A research agenda*. Frontiers Media. https://doi.org/10.3389/978-2-88963-197-1
Felt, U. (2015). Keeping technologies out. In S. Jasanoff & S. Kim (Eds.), *Dreamscapes of modernity* (pp. 103–125). University of Chicago Press.
Findlay, T. (2010). *Nuclear energy and global governance: Ensuring safety, security, and non- proliferation*. Routledge.
Fisher, W. R. (1989). *Human communication as narration: Toward a philosophy of reason, value, and action*. University of South Carolina Press.
Flam, H. (Ed.). (1994). *States and antinuclear movements*. Edinburgh University Press.
Freudenburg, W. R. (1993). Risk and recreancy: Weber, the division of labor, and the rationality of risk perceptions. *Social Forces*, *71*(4), 909–932.
Freudenburg, W. R., & Davidson, D. J. (2007). Nuclear families and nuclear risks: The effects of gender, geography, and progeny on attitudes toward a nuclear waste facility. *Rural Sociology*, *72*(2), 215–243.
Gilinsky, V. (2020, February 26). The US government insurance scheme for nuclear power accidents no longer makes sense. *Bulletin of the Atomic Scientists*. https://thebulletin.org/2020/02/the-us-government-insurance-scheme-for-nuclear-power-plant-accidents-no-longer-makes-sense/
Gowda, M. V. R., & Easterling, D. (1998). Nuclear waste and Native America: The MRS siting exercise. *Risk: Health, Safety & Environment*, *9*, 229–258.
Gowda, M. V. R., & Easterling, D. (2000). Voluntary siting and equity: The MRS facility experience in Native America. *Risk Analysis*, *20*(6), 917–929.
Gwin, L. (1990). *Speak no evil: The promotional heritage of nuclear risk communication*. Praeger.
Haines, M. B. (2019). Contested credibility economies of nuclear power in India. *Social Studies of Science*, *49*(1), 29–51.
Harrabin, R. (2017, September 11). Offshore wind power cheaper than new nuclear. *BBC News*. www.bbc.com/news/business-41220948
Hecht, G. (1998). *The radiance of France: Nuclear power and national identity after World War II*. MIT Press.
Hecht, G. (2012). *Being nuclear: Africans and the global uranium trade*. MIT Press.
Hoffman, S. G., & Durlak, P. (2018). The shelf life of a disaster: Post-Fukushima policy change in the United States and Germany. *Sociological Forum*, *33*(2), 378–402. https://doi.org/10.1111/socf.12419
Horlick-Jones, T. (2005). Informal logics of risk: Contingency and modes of practical reasoning. *Journal of Risk Research*, *8*(3), 253–272.
International Atomic Energy Agency (IAEA). (2011). *The nuclear fuel cycle* (Document 11-2522). IAEA.
International Atomic Energy Agency (IAEA). (2018). *Functions and processes of the regulatory body for safety*. IAEA Safety Standards Series No. GSG-13. IAEA.
International Physicians for the Prevention of Nuclear War (IPPNW) and Physicians for Social Responsibility (PSR). (2016). *5 years living with Fukushima: Summary of the health effects of the nuclear catastrophe*. IPPNW, PSR.
Ishiyama, N. (2003). Environmental justice and American Indian tribal sovereignty: Case study of a land-use conflict in Skull Valley, Utah. *Antipode*, *35*(1), 119–139.
Jasanoff, S., & Kim, S. H. (2009). Containing the atom: Sociotechnical imaginaries and nuclear power in the United States and South Korea. *Minerva*, *47*, 119–146. https://doi.org/10.1007/s11024-009-9124-4

Jenkins, K. (2018). Setting energy justice apart from the crowd: Lessons from environmental and climate justice. *Energy Research & Social Science, 39*, 117–121. https://doi.org/10.1016/j.erss.2017.11.015
Jenkins, K., McCauley, D., & Warren, C. R. (2017). Attributing responsibility for energy justice: A case study of the Hinkley Point nuclear complex. *Energy Policy, 108*, 836–843. https://doi.org/10.1016/j.enpol.2017.05.049
Joppke, C. (1993). *Mobilizing against nuclear energy: A comparison of Germany and the United States*. University of California Press.
Kan, N. (2017). *My nuclear nightmare: Leading Japan through the Fukushima disaster to a nuclear-free future* (J. S. Irish, Trans.). Cornell University Press.
Katz, S. B., & Miller, C. R. (1996). The low-level radioactive waste-siting controversy in North Carolina. In C. G. Herndl & S. C. Brown (Eds.), *Green culture: Environmental rhetoric in contemporary America* (pp. 111–139). University of Wisconsin Press.
Kingston, J. (2012). Japan's nuclear village. *Asia-Pacific Journal, 10*(37), 1–29.
Kinsella, W. J. (2001). Nuclear boundaries: Material and discursive containment at the Hanford nuclear reservation. *Science as Culture, 10*(2), 163–194. https://doi.org/10.1080/09505430120052284
Kinsella, W. J. (2004). Public expertise: A foundation for citizen participation in energy and environmental decisions. In S. P. Depoe, J. W. Delicath, & M. A. Elsenbeer (Eds.), *Communication and public participation in environmental decision making* (pp. 83–95). SUNY Press.
Kinsella, W. J. (2005). One hundred years of nuclear discourse: Four master themes and their implications for environmental communication. *Environmental Communication Yearbook, 2*, 49–72.
Kinsella, W. J. (2015). Rearticulating nuclear power: Energy activism and contested common sense. *Environmental Communication, 9*(3), 346–366. https://doi.org/10.1080/17524032.2014.978348
Kinsella, W. J. (2016). A question of confidence: Nuclear waste and public trust in the United States after Fukushima. In R. Hindmarsh & R. Priestly (Eds.), *The Fukushima effect: A new geopolitical terrain* (pp. 223–246). Routledge.
Kinsella, W. J. (2019). Ecological communication and security. In B. C. Taylor & H. Bean (Eds.), *The handbook of communication and security* (pp. 54–70). Routledge. https://doi.org/10.4324/9781351180962
Kinsella, W. J. (2020). Extracting uranium's futures: Nuclear wastes, toxic temporalities, and uncertain decisions. *The Extractive Industries and Society*. https://doi.org/10.1016/j.exis.2020.01.003
Kinsella, W. J., Andreas, D., & Endres, D. (2015). Communicating nuclear power: A programmatic review. In E. Cohen (Ed.), *Communication yearbook 39* (pp. 277–309). Routledge.
Kinsella, W. J., Kelly, A. R., & Kittle-Autry, M. (2013). Risk, regulation, and rhetorical boundaries: Claims and challenges surrounding a purported nuclear renaissance. *Communication Monographs, 80*(3), 278–301. https://doi.org/10.1080/03637751.2013.788253
Komori, A. (2019, March 10). Think tank puts cost to address nuke disaster up to 81 trillion yen. *Asahi Shimbun*. www.asahi.com/ajw/articles/AJ201903100044.html
Kopytko, N., & Perkins, J. (2011). Climate change, nuclear power, and the adaptation-mitigation dilemma. *Energy Policy, 39*, 318–333. https://doi.org/10.1016/j.enpol.2010.09.046
Laffont, J. J., & Tirole, J. (1991). The politics of government decision-making: A theory of regulatory capture. *Quarterly Journal of Economics, 106*(4), 1089–1127.
Leadbeater, M. (2013). *Peace, power and politics: How New Zealand became nuclear free*. Otago University Press.
Lovins, A. B. (2009). *Four nuclear myths*. Rocky Mountain Institute.
Malin, S. A. (2015). *The price of nuclear power: Uranium communities and environmental justice*. Rutgers University Press.
Medhurst, M. J. (1987). Eisenhower's "atoms for peace" speech: A case study in the strategic use of language. *Communication Monographs, 54*, 204–220. https://doi.org/10.1080/03637758709390226
Mumford, L. (1964). Authoritarian and democratic technics. *Technology and Culture, 5*(1), 1–8.
Nadesan, M. H. (2013). *Fukushima and the privatization of risk*. Palgrave Macmillan.
National Association of Insurance Commissioners (NAIC). (2019). *Nuclear liability insurance*. www.naic.org/cipr_topics/topic_nuclear_liability_insurance.htm
Ogawa, A. (2013). Young precariat at the forefront: Anti-nuclear rallies in post-Fukushima Japan. *Inter-Asia Cultural Studies, 14*(2), 317–326. https://doi.org/10.1080/14649373.2013.769760
Okrent, D. (1999). On intergenerational equity and its clash with intragenerational equity and on the need for policies to guide the regulation of disposal of wastes and other activities posing very long-term risks. *Risk Analysis, 19*(5), 877–901.
Okrent, D., & Pidgeon, N. (2000). Introduction: Dilemmas in intergenerational versus intragenerational equity and risk policy. *Risk Analysis, 20*(6), 759–762.

Pacala, S., & Socolow, R. (2004). Stabilization wedges: Solving the climate problem for the next 50 years with current technologies. *Science, 305*(5686), 968–972.
Peeples, J. A., Krannich, R. S., & Weiss, J. (2008). Arguments for what no one wants: The narratives of waste storage proponents. *Environmental Communication, 2*(1), 40–58.
Peterson, T. V. (2001). *Linked arms: A rural community resists nuclear waste.* SUNY Press.
Schneider, M., & Froggatt, A. (2019). *World nuclear industry status report 2019.* Mycle Schneider Consulting. www.worldnuclearreport.org/-World-Nuclear-Industry-Status-Report-2019-.html
Shrader-Frechette, K. (2000). Duties to future generations: Proxy consent, intra-and intergenerational equity: The case of nuclear waste. *Risk Analysis, 20*(6), 771–778.
Smith, A. K. (1971). *A peril and a hope: The scientists' movement in America, 1945–47.* MIT Press.
Sovacool, B. K. (2008). Valuing the greenhouse gas emissions from nuclear power: A critical survey. *Energy Policy, 36,* 2940–2953. https://doi.org/10.1016/j.enpol.2008.04.017
Sovacool, B. K., & Ramana, M. V. (2015). Back to the future: Small modular reactors, nuclear fantasies, and symbolic convergence. *Science, Technology, & Human Values, 40*(1), 96–125.
Spivak, G. C. (1999). *A critique of postcolonial reason: Toward a history of the vanishing present.* Harvard University Press.
Stefan, W., Crutzen, P. J., & McNeill, J. R. (2007). The anthropocene: Are humans now overwhelming the great forces of nature? *Ambio, 36*(8), 614–621.
Stephens, J. C., Burke, M. J., Gibian, B., Jordi, E., & Watts, R. (2018). Operationalizing energy democracy: Challenges and opportunities in Vermont's renewable energy transformation. *Frontiers in Communication, 3*(43). doi.org/10.3389/fcomm.2018.00043
Stern, P. C., Sovacool, B. K., & Dietz, T. (2016). Towards a science of climate and energy choices. *Nature Climate Change, 6,* 547–555. https://doi.org/10.1038/NCLIMATE3027
Sturgis, P., & Allum, N. (2004). Science in society: Reevaluating the deficit model of public attitudes. *Public Understanding of Science, 13,* 55–74.
Suna, D., & Resch, G. (2016). Is nuclear economical in comparison to renewables? *Energy Policy, 98,* 199–209. https://doi.org/10.1016/j.enpol.2016.08.023
Szulecki, K. (2020). Securitization and state encroachment on the energy sector: Politics of exception in Poland's energy governance. *Energy Policy, 136,* 111066. https://doi.org/10.1016/j.enpol.2019.111066
Tarasova, E. (2018). (Non-) alternative energy transitions: Examining neoliberal rationality in official nuclear energy discourses of Russia and Poland. *Energy Research & Social Science, 41,* 128–135.
Taylor, B. C., Kinsella, W. J., Depoe, S. P., & Metzler, M. S. (Eds.). (2007). *Nuclear legacies: Communication, controversy, and the US nuclear weapons complex.* Lexington.
Thomas, S. (2017). China's nuclear export drive: Trojan Horse or Marshall Plan? *Energy Policy, 101,* 683–691. doi.org/10.1016/j.enpol.2016.09.038
Thomas, S. (2018). Russia's nuclear export programme. *Energy Policy, 121,* 236–247. https://doi.org/10.1016/j.enpol.2018.06.036
van Gerven, J. P. (2014). "It is laced with faults": American Indians, public participation and the politics of siting a high-level nuclear waste repository. *Societies without Borders, 9*(2), 161–187. https://scholarlycommons.law.case.edu/swb/vol9/iss2/4
Venables, D., Pidgeon, N. F., Parkhill, K. A., Henwood, K. L., & Simmons, P. (2012). Living with nuclear power: Sense of place, proximity, and risk perceptions in local host communities. *Journal of Environmental Psychology, 32*(4), 371–383.
Verbruggen, A., Laes, E., & Lemmens, S. (2014). Assessment of the actual sustainability of nuclear fission power. *Renewable and Sustainable Energy Reviews, 32,* 16–28.
Wellock, T. R. (Ed.). (1998). *Critical masses: Opposition to nuclear power in California, 1958- 1978.* University of Wisconsin Press.
Wills, J. (2006). *Conservation fallout: Nuclear protest at Diablo Canyon.* University of Nevada Press.
Wilson, P. D. (1996). *The nuclear fuel cycle: From ore to wastes.* Oxford University Press.
World Nuclear Association. (2018). *Nuclear power in France.* www.world-nuclear.org/information-library/country-profiles/countries-a-f/france.aspx
Wynne, B. (1991). Knowledges in context. *Science, Technology, & Human Values, 16*(1), 111–121.
Wynne, B. (1996). May the sheep safely graze? A reflexive view of the expert-lay knowledge divide. In S. Lash, B. Szerszynski, & B. Wynne (Eds.), *Risk, environment and modernity: Towards a new ecology* (pp. 44–83). Sage.
Yi-chong, X. (Ed.). (2011). *Nuclear energy development in Asia: Problems and prospects.* Palgrave Macmillan.

22

PUBLIC PARTICIPATION AND ENERGY SYSTEM TRANSFORMATIONS

Jake Barnes

Introduction

Making energy systems more sustainable will require large-scale changes to the way society produces, distributes, and consumes energy (IPCC, 2018). Such large-scale changes are increasingly recognized as requiring the active participation of citizens, contributing to decision making at a variety of scales (Chilvers & Kearnes, 2020; EU, 2019; Stirling, 2014). This represents a significant departure from the past, wherein citizens were largely viewed as passive energy consumers and public participation typically amounted to periodic consultations on singular issues or topics.

Today, public participation in energy systems is widely recognized as being much more diverse than this (Chilvers & Kearnes, 2016; Denecke et al., 2016; Feldpausch-Parker et al., 2019). Citizens are now regularly engaging though consultations, opinion polls, and behavior change programs. Protests against planning applications, as well as activism and lobbying, continue unabated. Across Europe and elsewhere, communities have come together to install, manage, and sometimes benefit from renewable energy installations (Bauwens et al., 2016; Seyfang et al., 2014; Tosun et al., 2019). Regional social movements have formed with the intention of influencing local energy policy and, in some cases, aspiring to own and manage urban energy infrastructure (Becker et al., 2017; Blanchet, 2015; Hess, 2018). "City labs," "living labs," and urban experiments, through which diverse local actors including municipalities, citizens, and businesses come together to experiment in new ways of using and distributing energy services, are increasingly advocated and deployed (Bulkeley et al., 2011; Nevens et al., 2013). Visioning exercises ask citizens to explore and shape possible energy futures while "transition arenas" are encouraged as a tool to facilitate and act on collective visions (Frantzeskaki et al., 2012). Meanwhile, "hacker" or "maker" spaces are appearing in cities around the world and facilitating citizens (de)constructing new and old energy technologies (Smith et al., 2017). As a result of these diverse participatory forms, there have been calls to "remake" or "reclaim" participation (Chilvers & Kearnes, 2016; Denecke et al., 2016).

As public participation practices have evolved, so too has our understanding of it. For a long time, energy participation was depicted in instrumental and deliberative terms, as necessary for legitimizing the development of energy infrastructures. The idea entailed a variety of processes designed to inform, consult, or involve publics in decisions that affect them (Rowe & Frewer,

DOI: 10.4324/9780429402302-26

2000). Since the turn of the century, this traditional conceptualization has been challenged, resulting in the successive broadening of the term and research area. This broadening initially followed a relational turn and has emphasized the performative, material, and situated nature of participation. Relational perspectives have consequently opened up for analysis a wider diversity of "participatory collectives" than under traditional approaches. More recently, a further broadening of the term and research area can be identified, through which systemic conceptualizations have been advocated (Chilvers & Kearnes, 2020; Chilvers & Longhurst, 2016; Chilvers & Pallett, 2018). These emerging systemic perspectives aim to situate energy participation within the energy systems of which they are part and call for a greater awareness and understanding of systems of participation. In this chapter, I seek to further develop this emerging systemic perspective on energy participation by drawing on research that studies sociotechnical systems and their transformation (Smith et al., 2005; Stirling, 2014).

Research under the broad umbrella of sustainability transitions advocates a focus on the diverse yet pervasive systems that provide basic societal functions such as shelter, transportation, or energy provision (Köhler et al., 2019). For the most part, such research has not foregrounded citizens as significant actors in transformation processes. Nonetheless, this burgeoning area of work has amassed a rich conceptual armory for understanding and seeking to intervene in systemic change processes. In this chapter, I argue this work aids emerging thinking on systems of participation by directing attention away from the study of individual, discrete participatory events and how they can be perfected; it reframes the relationship between participation, governance, and energy systems by conceptualizing participation and governance as coconstituent parts of sociotechnical energy systems; it expands the agency of diverse publics from being solely deliberative to incorporating a far wider range of performative and potentially transformative capacities; and it challenges the relationship between participation and change. Collectively, this points to a variety of future research avenues and the study of regional participation as a possible point of departure.

To this end, I first explain why contemporary energy participation has become increasingly diverse through recourse to three societal megatrends. On the basis of these trends, I unpack why public participation in the evolution of energy systems is increasingly necessary, if not a vital component in deciding their future qualities, design, and realization. I then review how the study of energy participation has been successively broadened since the turn of the century, before outlining how insights from the study of sociotechnical systems and their transformation could further advise thinking on systems of participation. This leads to a variety of open questions about how diverse participatory collectives interact, potentially influencing governance processes and coshaping energy systems over time. I conclude with a synopsis of the arguments made and suggestions for future research.

In this chapter, I consequently focus on public participation as an important research topic and essential component in the study of energy democracy (Feldpausch-Parker et al., 2019; Szulecki, 2018; van Veelen & van der Horst, 2018). By attending to energy participation directly (c.f. examining it through the lens of energy democracy, research, and practice), I review the evolution of research on energy participation and set out future directions.

Decarbonization, digitalization, decentralization, and democratization

To understand how contemporary energy participation has evolved, we must also understand broader changes within global energy systems and society as a whole. Three "megatrends," thought to be influencing societal development, provide a useful entry point (Di Silvestre et al., 2018).

The first of these trends, **decarbonization**, is a political commitment in response to the urgent need to curtail and ultimately halt societal reliance on carbon-emitting energy systems. Global decarbonization targets were set for the first time at COP21 in Paris with the aim of keeping global temperatures well below 2°C by the end of the century and with the ambition to limit global temperature increases to 1.5°C. Different regions and countries, including the EU and China, have subsequently pledged to reduce carbon emissions and set out strategies to achieve this. This political commitment has been led by national governments but with subnational actors, particularly cities, playing a more prominent role than previously, and has fundamental implications for contemporary energy systems.

The deployment of renewable energy generation technologies is probably the most striking change to have manifested in response to this commitment (Mitchell, 2016). Supportive policy contexts and rapid reductions in the price of renewables over the last two decades mean that their continued deployment appears nearly all but inevitable. Small renewable energy generation technologies such as solar photovoltaics (PV), have facilitated new forms of material participation as citizens start producing as well as consuming energy (Inderberg et al., 2018; Parag & Sovacool, 2016). As the penetration of renewable energy technologies increases, the challenge of incorporating variable power sources alongside the growth of demand response technologies, such as batteries, at local and household scales are expected to open up further avenues through which citizens can materially participate in future energy systems (Brown et al., 2019; Parag & Sovacool, 2016). Meanwhile, the deployment of household solar photovoltaic (PV) has been linked to the development of a more climate-informed and active citizenry in Germany and the United States (Graziano, 2019).

Through a second trend, **digitalization**, the world is experiencing a fifth "great surge in development" (Perez, 2002). The Information and Communication Technology (ICT) revolution that began in the 1970s is resulting in structural changes across all sectors of the economy (production, distribution, communication, and consumption). It is also resulting in profound, qualitative changes to society. The energy system is far from immune. The deployment of "smart" meters has been a European continental imperative since at least 2010 and is opening up new avenues for citizens to engage with energy systems (Darby, 2010; Marres, 2011). Increasingly intelligent or "smart" energy systems are thought essential in balancing variable power generated by renewable sources of electricity, for decoupling power generation and use and managing multiple forms of energy storage, for facilitating the development of increasingly complicated time-of-use- and location-based energy tariffs, and for managing increasingly diverse, localized, and dynamic regional energy systems. As the energy system becomes more "active," opportunities for increased material participation by consumers and citizens are multiplying (Renström, 2019).

Digitalization, as increased technological connectivity, has implications for how consumers engage with energy systems (Hansen et al., 2020). The growth of smart home appliances and associated communication infrastructures is facilitating new business models (Brown et al., 2019; Burger & Luke, 2017). Digital-enabled energy trading via peer-to-peer platforms are emerging, with the promise of households trading locally produced electricity, individually, through community organizations, or through aggregators who make decisions and act on their behalf (Hardy, 2018). Digitalization is changing the way consumers engage with energy networks, suppliers, and markets. Digitalization is also creating new avenues for publics to contribute to decision making. Government consultations and petitions increasingly employ digital methods while internet-based social movements have also proliferated in the last decade. Meanwhile, as Certomà et al. (2015) argue, the pervasiveness of digital infrastructures within cities is opening up new urban governance processes based on crowdsourcing. Digitalization is subsequently creating new participatory spaces for citizens to engage with energy systems and offering new means to influence decision making.

The third trend, **decentralization**, strongly connects to the previous two. Decentralization concerns the distribution of power, decision-making authority, security, and control as well as energy hardware and systems of production and consumption. Contemporary communication systems and the internet have facilitated a return to more decentralized markets and social relationships. In energy systems, decentralization manifests in the move toward managing supply and demand over increasingly short spatial scales, as innovation increasingly moves to the "grid's edge," (Sioshansi, 2017) if not "beyond the meter," and as power ebbs away from large incumbent energy utilities to a greater number of smaller players. In large part, this is a direct consequence of the rise of modular, scalable renewable energy generation technologies. It is also a somewhat self-reinforcing process, as the challenges associated with incorporating large amounts of renewably sourced electricity encourages and requires citizens (and others) to be more flexible in their local use of energy. Decentralization is consequently resulting in new possibilities for material and economic participation by owning and managing energy system artifacts (through renewable energy cooperatives, for example) and infrastructures (through municipally owned distribution networks), and it is taking power away from the previously dominant, large energy suppliers. Nonetheless, and as work by Judson et al. (2020) shows, decentralization is a malleable term encompassing a multiplicity of locally distinct, energy pathways. In short, decentralization further expands opportunities for citizens to get more actively involved in the design, governance, and use of energy.

When viewed collectively, these trends suggest new roles for citizens in energy systems, as prosumers (actors who both produce and consume energy) (Parag & Sovacool, 2016), for example. They imply changes to the rules that guide participation and signal new rights and responsibilities of energy citizens (Devine-Wright, 2007). To Judson et al. (2020), Thompson and Bazilian (2014), and Soutar (2018), this points to a fourth underlying trend: **democratization**. More emergent and controversial than the previous three trends, democratization is identified by these authors in the changing roles and associated responsibilities of citizens as central stakeholders in energy systems. That is to say, they take a pragmatic view of system transformation and observe how, by moving from a time when citizens performed largely passive roles, the variety of new opportunities for participation in energy systems is thereby making them more democratic. This leads to a crucial point: If energy systems are becoming more democratic by virtue of the increased opportunity afforded to citizens in increasingly smart localized energy systems, then what is at stake can be more appropriately framed as a question of how democratic energy systems might become. Viewed in this light, democratization is an observable trend and social movements advocating energy democracy are agitators of richer participatory forms. The depth of energy democratization, however, is uncertain and open to debate. As Soutar (2018) points out, "deeper" shades of energy democracy imply not only new opportunities to engage but also the capacity to contest and shape "the rules of the game." Implicated in this is a significant step change, as to date, consumers have had little say in the rules shaping how or when they might participate. The emergence of community and peer-to-peer energy trading enables early engagement, but for deeper, more transformational and participatory systems to emerge, greater awareness of and debate over future citizens' roles and responsibilities in energy systems are required.

Deeper forms of energy democracy can be conceived through at least two entry points. On the one hand, diverse bottom-up, community-orientated energy trading schemes, such as the infamous Brooklyn microgrid, create spaces through which future rules and responsibilities of actively prosuming citizens can be negotiated in practice. On the other hand, traditional state-led democratic processes, such as voting and consultations, present avenues for energy citizens in shaping future energy policy. Either way, efforts to cultivate more effective "deliberative systems" (Parkinson & Mansbridge, 2012) will require a combination of effective, strong, responsive, and responsible governance institutions, the fostering of public debate and the willingness and capacity of institutions to shape energy system developments in more prosocial ways (Soutar, 2018; Szulecki, 2018). Although national governments will remain crucial actors in the process, the emergence of city and regional energy policy-making bodies is increasingly viewed as important experimental sites for new participatory decision making to emerge (Certomà et al., 2015; Hodson et al., 2017; Hodson & Marvin, 2009). This leads us back to this chapter's core topic, namely public participation, governance, and energy system transformation, all pivoting around a single, largely overlooked issue.

Despite movement to more decarbonized, digitalized, and decentralized energy systems there remains no clear, uniform, or single postcarbon destination. A variety of technically feasible and economically viable futures have the potential to meet carbon targets (Sovacool & Watts, 2009; Stirling, 2009). In the power sector, options include the fitting of carbon capture and storage technologies to fossil-fueled power stations; the use of large, centralized or small, modular nuclear reactors; the creation of centralized, continent-sized renewable energy infrastructures; or decentralized, "smart" renewable energy grids. Potential heat decarbonization pathways include the increased use of local heat networks or sustainably sourced biomass, the electrification of heat provision, or the development of gas networks delivering hydrogen or other "green gases." Each of these pathways present orientating technical options. Not all can be realized at the same time and in the same place, while costs are kept low. This implicates a variety of technical pathways as possible, even before discussing social, cultural, and economic permutations within each (Walker & Cass, 2007). In short, societies face real social, political, and technical choices about the direction in which energy systems develop (Stirling, 2009; Thombs, 2019). Choices that, to date, seem largely unrecognized and undebated.

This has fundamental implications for our understanding of public participation and the governance of systems change. Since the postwar period, the development of energy systems was largely uncontested (Chilvers & Kearnes, 2020): National infrastructures were created from scratch or through regional electricity networks being linked together; large, centralized and predominantly fossil-fueled power stations were built, and power flowed in one direction, from generation plants through the "grid" to businesses and households. The system was managed either by the state or by private interests. Although alternative development pathways clearly existed (Lovins, 1976) and there were fierce societal debates over the development of nuclear power, energy systems evolved in a remarkably settled and depoliticized way. Today, because diverse economically viable and socially desirable decarbonization pathways exist—each entailing winners and losers—future transition pathways are contested, fiercely, if not openly. Public participation thus conceived is an inherently political process. No longer can citizens be expected to feed into the objective, relatively uncontested management of "the energy system." Political decisions need to be made about the qualities of future energy systems. Choices, involving value-laden judgments that stretch far beyond the objective appraisal of technical options alone, have to be made (Stirling, 2007). Consequently, traditional views of "the public" feeding into the objective management of "the system" no longer appear adequate. Instead, energy transformations fundamentally question the relationship between public participation, governance, and energy systems evolution.

Recognizing the multiplicity of possible decarbonization pathways and the value-laden political process required to make progress, it necessarily follows that a broad societal consensus about the direction and extent of change is required. Costs, after all, will ultimately fall on citizens (paying through taxes, their energy bills, and the products they buy). Fundamental changes to individual and collective energy practices (such as shifting from individual to collective mobility) will require public willingness to adapt. Moreover, energy infrastructures—from pylons and wind farms to district heating and smart meters—permeate communities and their landscapes, and the potential for citizens resisting energy developments being imposed on them is high (Devine-Wright, 2013). Unless citizens can engage in meaningful ways, through shallower or deep forms of energy democracy (Soutar, 2018) with associated capacities to contest, explore, and construct future pathways, energy system transformations are likely to be less inclusive, equitable, and environmentally sustainable.

Diverse governance actors recognize this and are exploring varied means to engage citizens as active participants. Broad, large-scale attempts to promote citizens' deliberation over climate and energy policy have been commissioned in the UK (e.g., Pidgeon et al., 2014). Contemporary calls from protesters, activists, and think tanks for "Climate Citizens Assemblies" (Hennig, 2020)—in which randomly selected citizens review the evidence on and deliberate over a particular issue of national importance—presents another example of possible means to engage "the public" and contribute to decision making. Nonetheless, these developments largely sit within traditional understandings of the role of public participation in energy governance. "The challenge" is viewed as a "problem of extension": how to enroll publics into the decision-making process (Collins & Evans, 2002). At the same time and largely ignored is how diverse participatory collectives, like Transition Towns (Seyfang & Haxeltine, 2012) or community-based energy movements (Blanchet, 2015), have emerged at diffuse scales and locales to explore energy system futures largely detached from traditional governance forums. These examples further question traditional conceptualizations of energy participation and support calls for "reclaiming" or "remaking participation" (Chilvers & Kearnes, 2016; Denecke et al., 2016).

Decarbonization, digitalization, and decentralization consequently help explain how and why contemporary participation in energy systems is more diverse than has historically been the case. Collectively, they point to the increasing democratization of energy, while opening up questions of how participatory energy systems might become. They position public participation as an increasingly vital element in exploring and deciding upon the future qualities, design, and realization of energy systems. In short, the multiplicity of potential decarbonized futures, all with different social, cultural, technological, and economic foundations, requires citizens to engage with and contribute to the development of energy systems. These trends also challenge traditional conceptualizations of public participation as being primarily instrumental and deliberative. To further demonstrate this, I now turn to briefly review how understandings of energy participation have successively broadened since the turn of the century. In outlining three successive "waves," I build on the work of Chilvers and Kearnes (2016).

Mainstream, relational, and emerging systemic conceptions of energy participation

Mainstream perspectives

Mainstream, or traditional, perspectives on energy participation are closely linked to historical energy system developments, including their centralized design, one-way power flows, and limited role for citizens as passive consumers (Chilvers & Kearnes, 2016). With a limited role in energy systems, public participation subsequently concerned the periodic consultation of citizens on particular energy issues, such as the siting of new energy infrastructure (Cowell & Devine-Wright, 2018). Public consultations, surveys, and focus groups all fit within this framing, designed to elicit information from particular segments of society so that citizens' views can be fed into larger decision-making processes. At the core of these mainstream conceptions of public participation is a focus on deliberation: Participation is conceived as a purely deliberative process in which citizens contribute to decision making, the overall objective being to manage system developments through the objective appraisal of sociotechnical options and rational deliberation. The outputs of public participation are subsequently viewed as one piece of the decision-making jigsaw, placed alongside other considerations. As a result, each form of participation typically entails singular events on particular issues or topics and is typically led by experts with participants carefully selected to represent a particular area or a cross section of society.

To these formal, top-down forms of participation can also be added protests and some forms of social movements where the aim is to vocalize an opinion or challenge a decision. Meanwhile, more contemporary approaches for engaging citizens such as "transition management" approaches (that seek to steer actors through a set of coherent interventions) (e.g., Rotmans et al., 2001; Rotmans & Kemp, 2008) or behavior change campaigns, like the Japanese government's "Coolbiz" campaign (Tanabe et al., 2013), can also be grouped with more historically dominant forms. Although there is less focus on participation being a deliberative process, they all share a common set of assumptions, including fixed ideas about what it means to participate, how participation occurs, who is to be involved, and what is to be explored (Chilvers et al., 2018). They all assume there is an external public "out there," ready to be known or consulted. Furthermore, publics are perceived as being separable from governance processes and energy systems. Framed in this way, public participation becomes the technical application of pregiven methods, that can be perfected, scaled up, or rolled out through objective evaluation of "best practice" criteria in order to better understand, incorporate, or placate(!) what a predefined public think. Public participation is subsequently viewed as one input into the objective management (governance) of "the system."

This understanding of public participation is highly influential in policy making and practice. Yet, as centralized energy systems collapse and as diverse, distributed, participatory, and postcarbon energy systems emerge, this framing has been increasingly challenged. It is now widely recognized that "the public" cannot be conceived as a single homogeneous mass but is rather constituted by an ever emerging variety of publics depending on the time, space, and issue at hand (Pesch, 2019). This framing has also been criticized for closing down deliberation over potential energy futures, for ignoring the potential of publics to create new energy futures, and for underplaying the role of publics to challenge and critique current policy and practice (Pidgeon et al., 2014; Stirling, 2014). At the same time, mainstream approaches and the philosophies underpinning them have been criticized for neglecting diverse and emergent forms through which citizens are actively getting involved in and contributing to energy system change (Smith & Stirling, 2016, Chilvers & Longhurst, 2015; Cowell & Devine-Wright, 2018).

Relational perspectives

Largely analytical and influenced by advances in the study of science and technology studies, relational approaches have challenged traditional conceptualizations of public participation by emphasizing how individuals never participate alone or in isolation: Participatory "engagements" are actively constructed and shaped by—as well as shaping—their wider cultural, material, social, and institutional settings. This sets relational approaches apart from mainstream understanding because participation is no longer viewed as fixed or predefined. Instead, participation is understood as being "performative," shaped by the participants and the elements involved in and constructing participation. Participation is subsequently viewed as emergent rather than fixed or pregiven (Chilvers & Pallett, 2018).

A variety of different relational approaches have subsequently been developed. These include object-orientated approaches that emphasize "*material participation*" (Marres, 2012) and that challenge the idea of public participation as being centered on discursive or deliberative processes. While such material participation can occur through everyday mundane objects (Ryghaug et al., 2018), the approach also gives rise to a broader understanding about how some objects appear to have distinctive or special characteristics that in particular situations can play a role in the enactment of public participation. A related approach has explored how particular participatory approaches or "technologies of participation"—like focus groups, deliberative workshops—give rise to particular ideas or framings of the "public" and what effects this has (Lezaun & Soneryd, 2007).

By questioning the "what" (i.e., issue spaces), the "who" (i.e., participants/publics), and the "how" (i.e., through what means) of participation, relational approaches open up for analysis a far wider diversity of "participatory collectives" than mainstream or traditional approaches. Relational approaches subsequently challenge what it means to participate and how participation occurs, as well as exploring questions of who is and is not included. This has clear implications for our understanding of politics and power—who gets to shape societal developments. It asks which participatory collectives have a voice in governance processes, which are viewed as legitimate and which are not. This relational approach also speaks and responds to the diverse material ways through which publics actively participate in contemporary energy system developments, through, for example, smart technology trials or eco-open home days, where members of the public open their doors and invite local residents to explore and learn about their practical experiences with energy retrofits. Nonetheless, the extent to which relational approaches engage with and have implications for the governance of energy systems change is less clear. Although increasingly influential, this relational view tends to critically examine individual instances of participation. They remain largely analytical and leave unanswered how diverse participatory collectives influence one another and the energy system of which they are a part.

Emerging systemic perspectives

To answer these questions, a third, emerging perspective on participation seeks to take a more systemic, whole-systems view. Less cohesive than the previous two frames, emerging systemic approaches to participation have largely emerged as an extension of science, technology, and society studies (STS)–informed relational and coproductionist approaches.

In some instances, these relational STS approaches have also moved to take more systemic views on participation. One strand of this work looks across collective participatory approaches, typically at the national level, and explores the interactions and relations between the "constitutional contexts" of participation and collective participatory entanglements. Emphasis is placed on the interplay of political cultures, histories, and institutions that might explain particular "socio-technical imaginaries" or shared visions of the future (Jasanoff & Kim, 2009). An alternative, object-inspired approach seeks to map "issue" or "controversy" spaces to inquire into the different forms of participation taking place and of different publics being mobilized, as well as the links between them (Marres, 2015).

Both of these perspectives have subsequently been molded into an "ecologies of participation" framework (Chilvers et al., 2018; Chilvers & Longhurst, 2016) that creates a means to compare qualitatively different manifestations of participation on the basis of differing models (how), subjects (who), and objects (what) of participation. This framework has been applied in a systematic study of energy participation in the United Kingdom (between 2010 and 2015). The results (Chilvers et al., 2018) map diverse instances of participation according to how dominant, divergent, or emergent they are and has resulted in an alternative means to visualize energy participation as being multiple, diverse, and continually emerging. While the value and insights generated from this approach are still being explored, much of the framework's resulting strength appears to lie in demonstrating the linkages between diverse participatory collectives. This research is also one of the first studies that actively seeks to engage with contemporary "challenge" of participation as a problem of "relevance." That is, the research calls for the "burden" of participation to no longer be placed on "publics to participate"—as might be conceived under traditional perspectives where the challenge of participation is conceived as a problem of "extension"; rather, the question is one of inadequate institutional reflexivity, requiring rectification through greater "institutional listening" (Chilvers et al., 2018; Pallett et al., 2019).

This work represents a significant advance. Accounting for and responding to the multiple ways in which citizens are already engaged in energy system transformation is important. But how to interpret the existence of diverse participatory practices and their relevance to policy questions requires further attention. A further weakness of this approach is the extent to which it perpetuates an analytical separation of "participation" and "governance" as two separate processes and areas of study. In order to advance a systemic understanding of how publics contribute to energy systems transformations, this divide needs breaking down, and a greater understanding of how participatory collectives may already be contributing to energy governance and system change is required. Implicated in this is a greater appreciation of the diverse ways in which participatory collectives contribute to "doing" system change. One means to advance these emerging systemic perspectives on public participation, as well as to enhance an understanding of participation and governance as being mutually coconstituent in broader energy systems specifically, is to incorporate insights from the study of sociotechnical systems and their transformation (Leach et al., 2007; Smith et al., 2005, 2010).

Advancing systemic perspectives of participation

Over the last two decades, a new area of research under the umbrella of "sustainability transitions" has reframed the challenge of sustainability by calling for a focus on meso level regimes as a unit for analysis and as the target for policy and action. Regimes are thought to comprise sets of interlinking social and technical elements and are defined by their delivery of societal systems of provision, such as food, mobility, shelter, and energy (Geels, 2002). Transitions to a more sustainable future are thought to occur through reconfiguring unsustainable systems of provision, overcoming lock-in and path dependency in the process. Taking this systemic view, initial attention focused on bridging micro and macro forms of change to understand broad patterns and dynamics of transformation as well as governance processes. This has led to an appreciation of sociotechnical pathways (Geels & Schot, 2007; Geels et al., 2016; Leach et al., 2007) and the agency and roles of actors driving change (Fischer & Newig, 2016). This research approach has subsequently developed a growing conceptual armory for understanding and seeking to intervene in sociotechnical systems and their governance. However, for the most part, such research has not foregrounded "publics" as significant actors in transformation processes. This is curious considering how the perspective shifts focus away from the greening of industry or changing individual behaviors to the systems that fulfill societal (public?) needs and functions (Meadowcroft, 2009). Nonetheless, this broad area of research offers a variety of insights that might advance systemic perspectives on public participation.

First, a focus on sociotechnical regimes implicates publics as important component parts of broader systems. It suggests publics cannot be analytically separated from governance processes or broader energy systems easily. This framing challenges traditional conceptualizations of public participation as an activity separable from governance. Smith and Stirling (2007) have elaborated on this difference by articulating two competing and idealized governance perspectives. The first, termed "governance on the outside," speaks to mainstream perspectives of public participation as discrete from formal decision making. Decision making is, in turn, viewed as entirely separated from the energy systems that it seeks to influence, while the energy system is conceived as a discrete, knowable object. This form of governance, they suggest, is managerial in style and seeks objective knowledge of the energy system and public views, which can be fed into the objective management of "the system." In contrast, "governance on the inside" is founded on (1) recognition that there are multiple ways of knowing energy systems and (2) the inseparability of governance participants (such as publics) from broader sociotechnical systems. Because each position or participant view of energy systems is inherently contestable, this subsequently implies that making decisions on the future shape and direction of energy systems changes is fundamentally a political process (c.f. an exercise in objective appraisal). In outlining these two contrasting positions, Smith and Stirling (2007) argue that the objectification of governance processes and energy systems is not necessarily unhelpful but that what is required is greater reflexivity in governance processes and greater recognition of the politics involved.

This line of thinking has resulted in calls for culturing supportive environments where multiple forms of participation can interconnect and flourish: Public participation and democratic struggle are viewed as the principle, most effective means of fostering radical emancipatory transformations (Stirling, 2014). This conceptualization can also be used to advance emerging systemic perspectives of participation because it no longer views participation, governance, and energy systems as discrete objects or areas. In other words, it transcends arguments of participation being a problem of "extension" (requiring better methods of engagement) or of "relevance" (requiring greater "institutional listening" by governance actors). Whilst both are important and worthy of increased attention, by conceiving participation and governance as coconstituting elements of broader sociotechnical systems, the approach suggests that future work should give greater attention to what diverse publics are actually doing, how they are seeking to influence the development of energy systems, and how they get involved in shaping governance processes.

This leads to a second area where insights from the study of sociotechnical systems and their transformation could add value. To date, significant attention has been dedicated to the roles of different actors in developing and advocating more sustainable solutions. One such branch has focused on the role of "grassroots innovations," variously described as diverse, collective, and participatory networks of activists and organizations generating novel bottom-up solutions to sustainable development (Seyfang & Smith, 2007). This lens has been applied to the study of community currencies, ecovillages (Boyer, 2015), the transition town movement (Feola & Nunes, 2013; Seyfang & Haxeltine, 2012), and community energy projects (Hargreaves et al., 2013), among others, and has clear crossovers to the diverse forms of public, primarily civil society–led participation in contemporary energy systems. This work has subsequently pointed to multiple "public good" outcomes of grassroots innovations, including the development of alternative, increasingly sustainable alternatives, the creation of critical knowledge (about power, institutions, and systems) and the ability to challenge prevailing discourses and contest possible futures (Seyfang & Smith, 2007; Smith, 2012; Smith & Stirling, 2016). More broadly, publics and social movements are thought to play a variety of central roles in different stages of transitions, including shaping societal values and worldviews, legitimating new technologies and transition pathways, while lobbying for and against particular sociotechnical systems (Schot et al., 2016; Smith, 2012). This work is subsequently useful to emerging systemic understandings of public participation because it ascribes broader forms of agency to diverse participatory collectives (publics) and thereby extends the performative characterization of participation as conceived under relational approaches. It consequently shifts attention away from participation being a purely deliberative process and places greater emphasis and explanatory power of publics to contribute to the "doing" of change.

Third, sustainability transitions research recognizes how multiple, diverse transformation pathways are possible, emerging from existing or new path dependencies, the alignment of resources, actors, and technologies (Geels & Schot, 2007; Geels et al., 2016). Diverse actors, including publics, are thus conceived as influencing how transformations unfold through the decisions and actions they take. Studies that zoom in on actors subsequently show how they can collectively contribute to the "steering" or "directionality" of progress (e.g., Bird & Barnes, 2014). Yet to date few studies have specifically addressed the role of diverse publics in guiding change. Exceptions to this include studies focused on how particular social movements have influenced regional or urban decision making (e.g., Blanchet, 2015; Hess, 2019). This work offers a point of departure from which wider and more diverse forms of public participation in energy systems could be investigated: A focus on emerging pathways creates space for investigating how diverse participatory collectives influence one another and the energy system of which they are a part.

To this, we can also add a fourth area of research inquiring into how diverse regional and urban energy system transformations are emerging, shaped by their particular contexts, including histories, resources, actor constellations, and governance processes (e.g., Bulkeley et al., 2011; Dudley et al., 2019; Ehnert et al., 2018; Hodson et al., 2017; Torrens et al., 2018). This work further underlines the multiplicity of potential energy system decarbonization pathways and positions regional energy systems as a potentially fruitful site for studying systems of participation.

Regional systems of participation

Collectively these insights suggest a potentially fruitful line of inquiry, studying regional systems of participation. Developing systemic conceptualizations will not be easy. Delineating system boundaries will be an important but challenging step. In practice, boundaries are perhaps best drawn for conceptual and practical reasons, as has been the case in much of the existing work to date. On top of this, prior studies by Chilvers et al. (2018) and by Jasanoff and Kim (2009) indicate the considerable work involved in mapping and tracing public participation at national scales. For these reasons, the study of urban or regional areas offers an interesting point of departure because they appear large enough to capture a diversity of participatory processes (as demonstrated in Barnes, 2019), while being sufficiently contained to allow for the tracing of interactions and relations between participatory events. They also appear to contain a diversity of competing transformation pathways (Torrens, Johnstone et al., 2018; Torrens, Schot et al., 2018).

A variety of avenues offer interesting starting points. One avenue includes exploring how diverse participatory collectives coproduce one another over time, that is, how diverse instances of participation emerge, interact, overflow, and connect. From this arises questions of how different "patterns" of interrelating participatory collectives emerge over time and evolve and how such patterns and energy systems coproduce each other over time. A second avenue might explore how regional participatory practices interact (or not) with diverse regional governance processes. One means to approach this is to follow participatory collectives as they engage diverse governance arenas (as in Barnes et al. [2018], or Hess [2018]). Such an approach may facilitate knowledge of and for the "transformative capacities" of publics (Wolfram, 2016). Equally, participation could be investigated from the vantage of contemporary governance arenas and processes in a search for the ways in which they listen and respond (or not) to what diverse participatory collectives are saying and doing. In some respects, these entry points continue the unhelpful analytical separation of participation and governance. But this need not be a retrograde step if the central challenge is kept in focus: viewing participation and governance as two sides of the same coin, mutually influencing and coproducing each other and being coconstitutive parts of broader energy systems.

Conclusions

This chapter examined public participation as a research topic and important component of energy democracy. It started by recognizing how energy participation has grown increasingly diverse in recent decades and argued that previously stable expectations of what public participation is and does are no longer adequate.

This is perhaps no surprise given the rapid evolution of contemporary energy systems. On the one hand, decarbonization, digitalization, and decentralization are opening up new possibilities for citizens to participate in energy systems. Indeed, many new forms of participation can be viewed as experimental responses to these trends and, in turn, further drivers of them. On the other hand, encapsulated within the idea of transformation is how citizens and society as a whole face fundamental decisions about the direction and extent of change. For while decarbonization, digitalization, and decentralization provide a direction of travel, there remains no single identifiable destination. A multiplicity of postcarbon, local, and smart energy systems remain possible and deciding which paths to take means navigating diverse political choices. This implicates public participation as a central, if not vital, component in the transformation of energy systems. It puts greater emphasis on participation as a diverse and evolving process, and it potentially sets out a challenge to energy democracy advocates: If energy systems are becoming more participatory, how can deeper forms of democratization be fostered in which citizens actively shape "the rules of the game"?

The chapter also suggested a concomitant evolution of the ways we conceive of public participation. Traditional conceptualizations have been challenged and the research area successively broadened: first, by scholars emphasizing the material, situated, and performative nature of contemporary participatory collectives and, second, by scholars advocating systemic conceptualizations of energy participation. Nonetheless, work advancing systemic approaches remains at an early stage of development and have tended to focus in on particular forms of participation and their circulation rather than taking a more systemic view of how diverse participatory collectives interact and influence equally diverse forms of decision making. To advance thinking on systems of participation and to recognize how publics cannot be easily separated from governance processes, insights from the study of sociotechnical systems and their transformation appear helpful. In the meantime, continued movement toward more decarbonized, digitalized, and decentralized energy systems are creating openings for increased citizen participation and for ultimately serving as a catalyst for deeper forms of energy democracy within energy system transformations.

Finally, the chapter argued for and outlined an emerging research agenda based on the study of regional systems of participation. This agenda argues for the meta-analysis of energy participation (c.f. the study of individual, discrete instances of participation). It asks how diverse participatory collectives interact, building capacities to engage in governance processes and thereby steer energy system developments, of which they are a part. Only through further theoretical and empirical work will we know if this suggestion is indeed fruitful. Either way, further theoretical and methodological advances seem required, given the scale of change currently unfolding in societies and energy systems in particular.

References

Barnes, J. (2019). *Public participation in a West of England energy transition: Key patterns and trends*. www.renewableenergy.ox.ac.uk/public-participation-in-a-west-of-england-energy-transition/

Barnes, J., Durrant, R., Kern, F., & MacKerron, G. (2018). The institutionalisation of sustainable practices in cities: How sustainability initiatives shape local selection environments. *Environmental Innovation and Societal Transitions*. https://doi.org/10.1016/j.eist.2018.04.003

Bauwens, T., Gotchev, B., & Holstenkamp, L. (2016). What drives the development of community energy in Europe? The case of wind power cooperatives. *Energy Research and Social Science*, *13*, 136–147. https://doi.org/10.1016/j.erss.2015.12.016

Becker, S., Naumann, M., & Moss, T. (2017). Between coproduction and commons: Understanding initiatives to reclaim urban energy provision in Berlin and Hamburg. *Urban Research and Practice*, *10*(1), 63–85. https://doi.org/10.1080/17535069.2016.1156735

Bird, C., & Barnes, J. (2014). Scaling up community activism: The role of intermediaries in collective approaches to community energy. *People, Place and Policy Online*, *8*(3), 208–221. https://doi.org/10.3351/ppp.0008.0003.0006

Blanchet, T. (2015). Struggle over energy transition in Berlin: How do grassroots initiatives affect local energy policy-making? *Energy Policy*, *78*, 246–254. https://doi.org/10.1016/j.enpol.2014.11.001

Boyer, R. H. W. (2015). Grassroots innovation for urban sustainability: Comparing the diffusion pathways of three ecovillage projects. *Environment and Planning A*, *47*(2), 320–337. https://doi.org/10.1068/a140250p

Brown, D., Hall, S., & Davis, M. E. (2019). Prosumers in the post subsidy era: An exploration of new prosumer business models in the UK. *Energy Policy*, *135*. https://doi.org/10.1016/j.enpol.2019.110984

Bulkeley, H., Castan Broto, V., Hodson, M., & Marvin, S. (Eds.). (2011). *Cities and low carbon transitions*. Routledge.

Burger, S. P., & Luke, M. (2017). Business models for distributed energy resources: A review and empirical analysis. *Energy Policy*. https://doi.org/10.1016/j.enpol.2017.07.007

Certomà, C., Corsini, F., & Rizzi, F. (2015). Crowdsourcing urban sustainability: Data, people and technologies in participatory governance. *Futures*, *74*, 93–106. https://doi.org/10.1016/j.futures.2014.11.006

Chilvers, J., & Kearnes, M. (Eds.). (2016). *Remaking participation*. Routledge.

Chilvers, J., & Kearnes, M. (2020). Remaking participation in science and democracy. *Science Technology and Human Values*, *45*(3), 347–380. https://doi.org/10.1177/0162243919850885

Chilvers, J., & Longhurst, N. (2016). Participation in transition(s): Reconceiving public engagements in energy transitions as co-produced, emergent and diverse. *Journal of Environmental Policy & Planning*, *18*(5), 585–607. https://doi.org/10.1080/1523908X.2015.1110483

Chilvers, J., & Pallett, H. (2018). Energy democracies and publics in the making: A relational agenda for research and practice. *Frontiers in Communication*, *3*, 1–16. https://doi.org/10.3389/fcomm.2018.00014

Chilvers, J., Pallett, H., & Hargreaves, T. (2018, February). Ecologies of participation in socio-technical change: The case of energy system transitions. *Energy Research & Social Science*, *42*, 199–210. https://doi.org/10.1016/j.erss.2018.03.020

Collins, H. M., & Evans, R. (2002). The third wave of science studies: Studies of expertise and experience. *Social Studies of Science*, *32*(2), 235–296. https://doi.org/10.1177/0306312702032002003

Cowell, R., & Devine-Wright, P. (2018). A 'delivery-democracy dilemma'? Mapping and explaining policy change for public engagement with energy infrastructure. *Journal of Environmental Policy and Planning*, 1–19. https://doi.org/10.1080/1523908X.2018.1443005

Darby, S. (2010). Smart metering: What potential for householder engagement? *Building Research and Information*, *38*(5), 442–457. https://doi.org/10.1080/09613218.2010.492660

Denecke, M., Ganzert, A., Otto, I., & Stock, R. (Eds.). (2016). *Reclaiming participation*. Transcript Verlag.

Devine-Wright, P. (2007). Energy citizenship: Psychological aspects of evolution in sustainable energy technologies. In J. Murphy (Ed.), *Governing technology for sustainability* (pp. 63–86). Earthscan. https://doi.org/10.4324/9781849771511

Devine-Wright, P. (2013). Explaining "NIMBY" objections to a power line: The role of personal, place attachment and project-related factors. *Environment and Behavior*, *45*(6), 761–781. https://doi.org/10.1177/0013916512440435

Di Silvestre, M. L., Favuzza, S., Riva Sanseverino, E., & Zizzo, G. (2018). How decarbonization, digitalization and decentralization are changing key power infrastructures. *Renewable and Sustainable Energy Reviews*, *93*, 483–498. https://doi.org/10.1016/j.rser.2018.05.068

Dudley, G., Banister, D., & Schwanen, T. (2019). The dynamics of public participation in new technology transitions: The case of dockless bicycle hire in manchester. *Built Environment*, *45*(1), 93–111. https://doi.org/10.2148/benv.45.1.93

Ehnert, F., Frantzeskaki, N., Barnes, J. J., Borgström, S., Gorissen, L., Kern, F., Strenchock, L., & Egermann, M. (2018). The acceleration of urban sustainability transitions: A comparison of Brighton, Budapest, Dresden, Genk, and Stockholm. *Sustainability*, *10*(3), 612. https://doi.org/10.3390/su10030612

EU. (2019). *Clean energy for all Europeans*. https://ec.europa.eu/energy/en/topics/energy-strategy-and-energy-union/clean-energy-all-europeans

Feldpausch-Parker, A. M., Endres, D., & Peterson, T. R. (2019). Editorial: A research agenda for energy democracy. *Frontiers in Communication*, *4*, 1–8. https://doi.org/10.3389/fcomm.2019.00053

Feola, G., & Nunes, R. J. (2013). *Failure and success of transition initiatives: A study of the international replication of the transition movement*. Walker Institute for Climate System Research. https://core.ac.uk/download/pdf/42147521.pdf

Fischer, L. B., & Newig, J. (2016). Importance of actors and agency in sustainability transitions: A systematic exploration of the literature. *Sustainability*, *8*(5), 476. https://doi.org/10.3390/su8050476

Frantzeskaki, N., Loorbach, D., & Meadowcroft, J. (2012). Governing societal transitions to sustainability. *International Journal of Sustainable Development*, *15*(1–2), 19. https://doi.org/10.1504/IJSD.2012.044032

Geels, F. W. (2002). Technological transitions as evolutionary reconfiguration processes: A multi-level perspective and a case-study. *Research Policy*, *31*(8–9), 1257–1274. https://doi.org/10.1016/S0048-7333(02)00062-8

Geels, F. W., Kern, F., Fuchs, G., Hinderer, N., Kungl, G., Mylan, J., Neukirch, M., & Wassermann, S. (2016). The enactment of socio-technical transition pathways: A reformulated typology and a comparative multi-level analysis of the German and UK low-carbon electricity transitions (1990–2014). *Research Policy*, *45*(4), 896–913. https://doi.org/10.1016/j.respol.2016.01.015

Geels, F. W., & Schot, J. (2007). Typology of sociotechnical transition pathways. *Research Policy*, *36*(3), 399–417. https://doi.org/10.1016/j.respol.2007.01.003

Graziano, M. (2019). Participation is the fuel. *Nature Energy*, *4*, 1012–1013. https://doi.org/10.1038/s41560-019-0511-2

Hansen, P., Morrison, G. M., Zaman, A., & Liu, X. (2020). Smart technology needs smarter management: Disentangling the dynamics of digitalism in the governance of shared solar energy in Australia. *Energy Research and Social Science*, *60*, 101322. https://doi.org/10.1016/j.erss.2019.101322

Hardy, J. (2018). Consumer participation in future smart low-carbon energy systems. *Institution for Public Policy Research*. www.jstor.org/stable/resrep21885.7?seq=1#metadata_info_tab_contents

Hargreaves, T., Hielscher, S., Seyfang, G., & Smith, A. (2013). Grassroots innovations in community energy: The role of intermediaries in niche development. *Global Environmental Change*, *23*(5), 868–880. https://doi.org/10.1016/j.gloenvcha.2013.02.008

Hennig, B. (2020). The case for citizens' assemblies. *Policy Network*. https://policynetwork.org/?opinion=the-case-for-citizens-assemblies

Hess, D. J. (2018). Energy democracy and social movements: A multi-coalition perspective on the politics of sustainability transitions. *Energy Research and Social Science*, *40*, 177–189. https://doi.org/10.1016/j.erss.2018.01.003

Hess, D. J. (2019). Coalitions, framing, and the politics of energy transitions: Local democracy and community choice in California. *Energy Research and Social Science*, *50*, 38–50. https://doi.org/10.1016/j.erss.2018.11.013

Hodson, M., Geels, F. W., & McMeekin, A. (2017). Reconfiguring urban sustainability transitions, analysing multiplicity. *Sustainability (Switzerland)*, *9*(2). https://doi.org/10.3390/su9020299

Hodson, M., & Marvin, S. (2009). Cities mediating technological transitions: Understanding visions, intermediation and consequences. *Technology Analysis & Strategic Management*, *21*(4), 515–534. https://doi.org/10.1080/09537320902819213

Inderberg, T. H. J., Tews, K., & Turner, B. (2018). Is there a prosumer pathway? Exploring household solar energy development in Germany, Norway, and the United Kingdom. *Energy Research and Social Science*. https://doi.org/10.1016/j.erss.2018.04.006

IPCC. (2018). Summary for policy makers. In *Global warming of 1.5°C: An IPCC special report on the impacts of global warming of 1.5°C above pre-industrial levels and related global greenhouse gas emission pathways, in the context of strengthening the global response to the threat of climate change*. IPCC.

Jasanoff, S., & Kim, S. H. (2009). Containing the atom: Sociotechnical imaginaries and nuclear power in the United States and South Korea. *Minerva*, *47*(2), 119–146. https://doi.org/10.1007/s11024-009-9124-4

Judson, E., Fitch-Roy, O., Pownall, T., Bray, R., Poulter, H., Soutar, I., Lowes, R., Connor, P. M., Britton, J., Woodman, B., & Mitchell, C. (2020). The centre cannot (always) hold: Examining pathways towards energy system de-centralisation. *Renewable and Sustainable Energy Reviews*, *118*. https://doi.org/10.1016/j.rser.2019.109499

Köhler, J., Geels, F. W., Kern, F., Markard, J., Wieczorek, A., Alkemade, F., Avelino, F., Bergek, A., Boons, F., Fünfschilling, L., Hess, D., Holtz, G., Hyysalo, S., Jenkins, K., Kivimaa, P., Martiskainen, M., McMeekin, A., Mühlemeier, M. S., Nykvist, B., . . . Wells, P. (2019). An agenda for sustainability transitions research: State of the art and future directions. *Environmental Innovation and Societal Transitions*, *December 2018*, 1–32. https://doi.org/10.1016/j.eist.2019.01.004

Leach, M., Scoones, I., & Stirling, A. (2007). *Pathways to sustainability: An overview of the STEPS centre approach*. https://steps-centre.org/publication/pathways-to-sustainability-an-overview-of-the-steps-centre-approach/

Lezaun, J., & Soneryd, L. (2007). Consulting citizens: Technologies of elicitation and the mobility of publics. *Public Understanding of Science*, *16*(3), 279–297. https://doi.org/10.1177/0963662507079371

Lovins, A. B. (1976). Energy strategy: The road not taken? *Foreign Affairs*, 65–96.

Marres, N. (2011). The costs of public involvement: Everyday devices of carbon accounting and the materialization of participation. *Economy and Society*, *40*(4), 510–533. https://doi.org/10.1080/03085147.2011.602294

Marres, N. (2012). *Material participation: Technology, the environment and everyday publics*. Palgrave Macmillan.

Marres, N. (2015). Why map issues? On controversy analysis as a digital method. *Science Technology and Human Values*, *40*(5), 655–686. https://doi.org/10.1177/0162243915574602

Meadowcroft, J. (2009). What about the politics? Sustainable development, transitions management, and long term energy transition. *Policy Sciences*, *42*(4), 323–340. https://doi.org/10.1007/S11077-009-9097-Z

Mitchell, C. (2016). Momentum is increasing towards a flexible electricity system based on renewables. *Nature Energy*, *1*(2), 15030. https://doi.org/10.1038/nenergy.2015.30

Nevens, F., Frantzeskaki, N., Gorissen, L., & Loorbach, D. (2013). Urban transition labs: Co-creating transformative action for sustainable cities. *Journal of Cleaner Production*, *50*, 111–122. https://doi.org/10.1016/j.jclepro.2012.12.001

Pallett, H., Chilvers, J., & Hargreaves, T. (2019). Mapping participation: A systematic analysis of diverse public participation in the UK energy system. *Environment and Planning E: Nature and Space*. https://doi.org/10.1177/2514848619845595

Parag, Y., & Sovacool, B. K. (2016). Electricity market design for the prosumer era. *Nature Energy*, *1*(4), 16032. https://doi.org/10.1038/nenergy.2016.32

Parkinson, J., & Mansbridge, J. (2012). *Deliberative systems: Deliberative democracy at the large scale*. Cambridge University Press.

Perez, C. (2002). *Technological revolutions and financial capital: The dynamics of bubbles and golden ages*. Edward Elgar.

Pesch, U. (2019). Elusive publics in energy projects: The politics of localness and energy democracy. *Energy Research and Social Science*, *56*. https://doi.org/10.1016/j.erss.2019.101225

Pidgeon, N., Demski, C., Butler, C., Parkhill, K., & Spence, A. (2014). Creating a national citizen engagement process for energy policy. *Proceedings of the National Academy of Sciences*, *111*(Supplement 4), 13606–13613. https://doi.org/10.1073/pnas.1317512111

Renström, S. (2019). Supporting diverse roles for people in smart energy systems. *Energy Research and Social Science*. https://doi.org/10.1016/j.erss.2019.02.018

Rotmans, J., & Kemp, R. (2008). Detour ahead: A response to Shove and Walker about the perilous road of transition management. *Environment and Planning A*, *40*(4), 1006–1012. https://doi.org/10.1068/a4004let

Rotmans, J., Kemp, R., & van asselt, M. (2001, September). More evolution than revolution: Transition management in public policy. *Foresight, 3*, 1–5.

Rowe, G., & Frewer, L. J. (2000). Public participation methods. *Science, Technology & Human Values, 25*(1), 3–29. https://doi.org/10.1177/016224390002500101

Ryghaug, M., Skjølsvold, T. M., & Heidenreich, S. (2018). Creating energy citizenship through material participation. *Social Studies of Science, 48*(2), 283–303. https://doi.org/10.1177/0306312718770286

Schot, J., Kanger, L., & Verbong, G. (2016). The roles of users in shaping transitions to new energy systems. *Nature Energy, 1*(5), 16054. https://doi.org/10.1038/nenergy.2016.54

Seyfang, G., & Haxeltine, A. (2012). Growing grassroots innovations: Exploring the role of community-based initiatives in governing sustainable energy transitions. *Environment and Planning C: Government and Policy, 30*(3), 381–400. https://doi.org/10.1068/c10222

Seyfang, G., Hielscher, S., Hargreaves, T., Martiskainen, M., & Smith, A. (2014). A grassroots sustainable energy niche? Reflections on community energy in the UK. *Environmental Innovation and Societal Transitions, 13*, 21–44. https://doi.org/10.1016/j.eist.2014.04.004

Seyfang, G., & Smith, A. (2007). Grassroots innovations for sustainable development : Towards a new research and policy agenda. *Environmental Politics, 16*(4), 584–603. https://doi.org/10.1080/09644010701419121

Sioshansi, F. P. (Ed.). (2017). *Innovation and disruption at the grid's edge*. Academic Press. https://doi.org/10.1016/B978-0-12-811758-3.00001-2

Smith, A. (2012). Civil society in sustainable energy transitions. In G. Verbong & D. Loorbach (Eds.), *Governing the energy transition: Reality, illusion or necessity?* (pp. 180–202). Routledge.

Smith, A., Rogge, K., Nightingale, P., Martin, B., & Bloom, M. (2017). Innovation, democracy and makerspaces editorial assistance. *Science Policy Research Unit*. https://papers.ssrn.com/sol3/papers.cfm?abstract_id=2986245

Smith, A., & Stirling, A. (2007). Moving outside or inside? Objectification and reflexivity in the governance of socio-technical systems. *Journal of Environmental Policy & Planning, 9*(3–4), 351–373. https://doi.org/10.1080/15239080701622873

Smith, A., & Stirling, A. (2016). Grassroots innovation & innovation democracy. *STEPS Centre*. https://steps-centre.org/wp-content/uploads/Grassroots-innovation-and-innovation-democracy.pdf

Smith, A., Stirling, A., & Berkhout, F. (2005). The governance of sustainable socio-technical transitions. *Research Policy, 34*(10), 1491–1510. https://doi.org/10.1016/j.respol.2005.07.005

Smith, A., Voß, J., & Grin, J. (2010). Innovation studies and sustainability transitions: The allure of the multi-level perspective and its challenges. *Research Policy, 39*, 435–448. https://doi.org/10.1016/j.respol.2010.01.023

Soutar, I. (2018). *We need to talk about (deep) energy democracy*. Exter Energy Policy Group Blog.

Sovacool, B. K., & Watts, C. (2009). Going completely renewable: Is it possible (let alone desirable)? *Electricity Journal, 22*(4), 95–111. https://doi.org/10.1016/j.tej.2009.03.011

Stirling, A. (2007). "Opening up" and "closing down": Power, participation, and pluralism in the social appraisal of technology. *Science, Technology & Human Values, 33*(2), 262–294. https://doi.org/10.1177/0162243907311265

Stirling, A. (2009). *Direction, distribution and diversity! Pluralising progress in innovation, sustainability and development*. Steps Working Paper No. 32. https://steps-centre.org/publication/direction-distribution-and-diversity-pluralising-progress-in-innovation-sustainability-and-development/

Stirling, A. (2014). Emancipating transformations: From controlling 'the transition' to culturing plural radical progress. *Politics of Green Transformations, 64*.

Szulecki, K. (2018). Conceptualizing energy democracy. *Environmental Politics, 27*(1), 21–41. https://doi.org/10.1080/09644016.2017.1387294

Tanabe, S. I., Iwahashi, Y., Tsushima, S., & Nishihara, N. (2013). Thermal comfort and productivity in offices under mandatory electricity savings after the Great East Japan earthquake. *Architectural Science Review, 56*(1), 4–13. https://doi.org/10.1080/00038628.2012.744296

Thombs, R. P. (2019). When democracy meets energy transitions: A typology of social power and energy system scale. *Energy Research & Social Science, 52*, 159–168. https://doi.org/10.1016/j.erss.2019.02.020

Thompson, G., & Bazilian, M. (2014). Democratization, energy poverty, and the pursuit of symmetry. *Global Policy, 5*(1), 127–131. https://doi.org/10.1111/1758-5899.12103

Torrens, J., Johnstone, P., & Schot, J. (2018). Unpacking the formation of favourable environments for urban experimentation: The case of the bristol energy scene. *Sustainability (Switzerland), 10*(3). https://doi.org/10.3390/su10030879

Torrens, J., Schot, J., Raven, R., & Johnstone, P. (2018, June). Seedbeds, harbours, and battlegrounds: On the origins of favourable environments for urban experimentation with sustainability. *Environmental Innovation and Societal Transitions*, 1–22. https://doi.org/10.1016/j.eist.2018.11.003

Tosun, J., Zöckler, L., & Rilling, B. (2019). What drives the participation of renewable energy cooperatives in European energy governance? *Politics and Governance*, 7(1), 45. https://doi.org/10.17645/pag.v7i1.1782

van Veelen, B., & van der Horst, D. (2018). What is energy democracy? Connecting social science energy research and political theory. *Energy Research and Social Science*, *46*(June), 19–28. https://doi.org/10.1016/j.erss.2018.06.010

Walker, G., & Cass, N. (2007). Carbon reduction, 'the public' and renewable energy: Engaging with socio-technical configurations. *Area*, *39*(4), 458–469. https://doi.org/10.1111/j.1475-4762.2007.00772.x

Wolfram, M. (2016). Conceptualizing urban transformative capacity: A framework for research and policy. *Cities*, *51*, 121–130. https://doi.org/10.1016/j.cities.2015.11.011

23

THE COMPLEX RELATIONS BETWEEN JUSTICE AND PARTICIPATION IN COLLABORATIVE PLANNING PROCESSES FOR A RENEWABLE ENERGY TRANSITION

Patrick Scherhaufer

Fostering energy democracy or the democratization of the energy transition is a twisted and complex endeavor. One of the big challenges of the energy transition is that the shift toward renewable energy technologies (RETs) depends on hundreds of thousands of local decisions. The new energy system will definitely be more decentralized than the one we are familiar with, and therefore many people will be involved, will be directly affected by negative impacts, and may expect a say in decision-making processes. As "participation is the *elixir of life* for democracy" (van Deth, 2014, p. 350), this could be a great opportunity for better forms of citizen's engagement and deliberation (Feldpausch-Parker et al., 2019). At the same time, it opens room for conflicts or even manipulation and tokenism (Arnstein, 1969). The worst-case scenario from an ecologically aware perspective is that the energy transition is delayed or even fails due to competing actors, actor coalitions, and interests or irreconcilable norms, values, and beliefs.

Hence, there is an extraordinary need for better conflict management and participation in planning and decision-making processes. The central challenge from an RET perspective is to coordinate two substantively different goals: (1) to provide people with meaningful opportunities to be part of deliberation and decision-making processes without prejudging the results and (2) to persuade people to accept RETs. Although these objectives are not mutually exclusive, a tension between the demands of democratic participation and its role in "overcoming feelings of powerlessness and alienation, and contributing to the legitimacy of the political system" (Fiorino, 1990, p. 229) and raising acceptance as "a positive and desirable outcome of planning projects" (Busse & Siebert, 2018, p. 235) becomes apparent.

This chapter will discuss the challenging role of participation in collaborative planning and how different parties may be involved in fair and just decision-making processes. It focuses on the prospects and obstacles of democratizing the energy transition from a local to regional perspective. The chapter starts looking at the challenges of the energy transition toward large-scale RETs from two different angles. One is represented by the literature on energy democracy and

DOI: 10.4324/9780429402302-27

its attempts at making decisions within the transition as democratic as possible, with the other represented by the literature on acceptance aiming at fostering public support. This link is of utmost importance because, although people generally support transitioning to RETs, they may oppose a specific renewable energy project or feel dissatisfied with their lack of control over the energy sector in general. Approaching the transition to RETs from an energy democracy perspective helps scholars and decision makers better understand what matters to the publics with whom they interact and may suggest more productive ways to guide those interactions. Then the chapter discusses different functions of participation based on the division among normative, substantive, and instrumental rationales of participation and the chance to realize them in a participatory setting. At the end, tangible recommendations on a procedural and distributional level will be provided. Overall, the chapter explores ways to integrate the goals of public or social acceptance into a democratic and just energy transition.

Challenges of the energy transition

The contemporary practice of citizen participation in local or urban planning can be traced to the late 1960s in policy fields like city planning and urban development and was introduced by giving procedural rights to an affected public (Fagence, 1977; Godschalk & Mills, 1966). However, from these early days on the exchange of information with an affected neighborhood or the deliberation among stakeholders was not a crucial part of the planning structure for critical infrastructures like power plants. Since the 1970s and until today, nuclear, fossil, and large hydropower stations often face public protests, but this did not significantly change the hierarchical and centralized planning culture in the field. The envisaged transition of the energy system toward the deployment of RETs changed the situation dramatically. With the mass deployment of RETs, far more individual plants must be planned, sited, and built in areas with different place identities, cultures, and societies. In addition, numerous technoeconomic studies have demonstrated that renewable energy resources can supply the majority of a country's or region's electricity demand (Cochran et al., 2014; Jacobson & Delucchi, 2011). The vision is that human society would use energy responsibly and obtain its energy demands 100% from decentralized renewable energy sources in the near future. Thus this approach to an energy transition has already become a global narrative due to pressure from climate change, the loss of biodiversity, and the pursuits of energy security and independence. Unfortunately, there is no unified approach for tackling the complex problems of transition (Rotmans, 2001; Shove & Walker, 2007).

A very promising indicator in handling the energy transition lies in public opinion, which, in many countries, is rather accommodating regarding RETs (Anderson et al., 2017; Stokes & Warshaw, 2017). However, that cannot hide the fact that single projects are often faced with local protests. The baseline is that conflicts and disputes form the core of any planning and implementation process, with the pros and cons of those directly affected facing the broad but rather unspecific costs and benefits of the general public. RET infrastructure projects such as photovoltaic, hydropower, and wind turbines can suffer from acceptance problems at different levels. On a community level, people, for example, are burdened by noise, visual impacts on landscapes, decline in property values, reflections, dazzling effects, or more traffic. Disagreements associated with these burdens are often subsumed under the term "NIMBY" (Not in My Back-Yard). Other potential conflict areas are competitions with regard to different types of land use or the noncompliance with environmental and social standards (Frantál et al., 2018; Poggi et al., 2018). For example, photovoltaic units, if not installed on roofs, consume additional land and are always in potential conflict with nature protection laws and other forms of land use like food production and recreation. Scholars who work with the concept of place identity and place attachment have demonstrated the relevance of beliefs and attitudes on energy infrastructures and have verified their arguments with surveys or small *n* studies (Devine-Wright et al., 2017). In addition, right wing populist parties question the influence of human activities on global climate change and thus dispute the need for an energy transition at all (Fraune & Knodt, 2018). Whatever the proximate causes of their discontent, publics are organizing protest groups and nongovernmental organizations (NGOs) that delay and sometimes stop projects (Wolsink, 2010). These different social conflicts can result in skyrocketing costs in regard to time and money invested in infrastructure projects, never ending lawsuits, and delays but also in a growing dissatisfaction of local communities about the planning and decision-making processes and a questioning of the energy transition in general.

Most of the preceding research focuses on different conceptualizations of acceptance (for an overview, see Busse & Siebert, 2018) and are eager to share methods for understanding public concerns as a precursor to increasing public acceptance (Devine-Wright, 2011; Scherhaufer et al., 2017; Wüstenhagen et al., 2007). The continuum here ranges from comprehensive information activities, through consultation with stakeholders, to roundtables with active citizens and local plebiscites. Fournis and Fortin (2017) introduced an important difference between the term acceptance and acceptability. Acceptance is often labeled as something to be managed and aiming at a positive outcome, which could be expressed by passive approval or active support. Whereas "acceptability" serves as a descriptive term and "complex scientific concept" (Busse & Siebert, 2018, p. 243) that includes nonacceptance, opposition, and rejection. Acceptability research attempts to explore the reasons people oppose an infrastructure project, such as a fear of losing property values, and posits that such reasons should be respected.

However, another literature branch shows that egoism is not the only factor that leads to nonacceptance and that self-interest is only one of many reasons for the emergence of environmental conflicts (Müller, 2012; Syme, 2012; Walker et al., 2011). Although the concentration on supposed NIMBYism has a strong influence in how politicians, commentators, or special interest groups frame the issue of opposition to infrastructure projects, it neglects the diversity of local objections (from social inequalities to environmental protection reasons) and the need of early, frequent, and meaningful participation (Devine-Wright, 2009, 2011; Feldpausch-Parker et al., 2019; Few et al., 2007; Wolsink, 2006). This research is much more inspired by notions of democracy, such as justice, participation, and power (Feldpausch-Parker et al., 2019) than by the theoretical or empirical impacts of acceptance and acceptability. These authors stress that from a democratic perspective these conflicts must be discussed and negotiated in well functioning procedures and institutions (Dahrendorf, 1959). From this angle, the implementation of politically legitimate modes of conflict management are at least as important as discovering and addressing the proclaimed causes of conflicts.

The concept of deliberation is a good starting point in this regard. Deliberation is about using public discourse to weigh arguments for or against something and coming to a more reasoned and informed decision (Dryzek, 2000; Fishkin, 1991; Habermas, 1994). However, the concept of deliberation focuses on the style of decision making but without providing the "institutional infrastructure for implementing it" (Fiorino, 1989, p. 508). This is where participatory democracy steps in. It considers the principle of self-determination of individuals as a central point of reference. Representatives of this political conviction understand democracy as a process of learning and enlightenment. The focus is on political and social equality and the learning of citizen virtues in social institutions. Prominent examples of more bottom-up-oriented processes of political engagement are all forms of local to regional initiatives of codetermination and self-regulation (e.g., Local Agenda 21, Citizen Juries and Councils). Hence, the process of participation has a value and meaning independent from the output of decision making. Citizens are engaged in institutions, where they can express their own ideas and preferences, and learn essential citizen's competences (Fiorino, 1990).

Likewise, a growing research agenda (Chilvers & Longhurst, 2016; Chilvers & Pallett, 2018; Cowell et al., 2011; Feldpausch-Parker et al., 2019; Phadke, 2013) in the field of energy transition attempts to integrate knowledge about acceptability, participation, and democracy. This research characterizes public participation as one of the core elements of energy democracy and explores democratic ideals like empowerment, fairness, and an understanding of democracy as a learning process. Normative commitments of this agenda include a belief that "the decisions that shape our lives should be established jointly" (Kunze & Becker, 2014, p. 8) and that energy democracy scholarship should "contribute to making energy transitions and decisions as democratic as possible" (Feldpausch-Parker et al., 2019, p. 1). From this perspective, public participation is much more nuanced, both in its conceptualization and in its practice, than means to facilitating public acceptance.

Despite some epistemological and praxiological disputes (e.g., about the role of power and resistance in decision making), most of these authors agree that collaborative planning approaches serve as one of the missing links in the relationship between the research agendas just discussed. It is a modus operandi somehow between the bottom-up- and the top-down-oriented methods of (public) participation, and it critically depends on (1) the authority of local to regional decision makers to act and (2) the involvement of interested and affected parties in the negotiation process. The following sections discuss the strengths and weaknesses of this exchange in energy transitions toward RETs with a focus on the roles of participation and justice. The analyses shows that this is not an easy task but maybe one of the most challenging issues in changing energy systems because it makes a strong contribution "to both democratizing energy and addressing the existential climate crises" (Feldpausch-Parker et al., 2019, p. 1).

Participatory planning and rationales of public participation

One already well established mode of conflict management and regulation in democracies, which recognizes decision making "at multiple scales as well as experimentation and learning from experience with diverse policies" (Ostrom, 2010, p. 550) is collaborative or participatory planning. "Participatory planning" is an umbrella term that subsumes activities to discuss and adopt planning strategies and options collectively. The participants are invited by local to regional authorities (administration, policy makers, planners). The idea behind the approach is that a relatively small group of interested and/or affected people can contribute to better knowledge production and ensure democratic decisions. Participation serves as a tool to guarantee shared influence and to transform power in politics.

The literature about participatory planning is rich and manifold. A quick search in Google Scholar in April 2020 revealed 1,180 articles with "participatory planning" in the title, with 109 published since 2018. Articles were linked to urban development, natural resource management, disasters, and risk perception, but only eight were linked to the keyword "energy." There were 3,100 entries with "energy planning," but only 12 when the word "participatory" was added as a search parameter. This a strong indicator that the discussion about participatory planning in regard to the energy transition is in its starting blocks and that there is strong need for clarifying the role and impact of participatory planning in making the energy transition more democratic.

The framing of public participation as one of the crucial elements of energy democracy is conceptualized by Fiorino's division between normative, substantive, and instrumental rationales of participation (Fiorino, 1989, 1990; Scherhaufer, 2013; Stirling, 2006). The normative ideal is characterized by people being able to codetermine their own living environment. At the level of the instrumental rationale, the focus is on the quality of the process of participation. The substantive rationale focuses on the integration of different values and norms that can be used to improve the quality of knowledge or output (Scherhaufer, 2013). This perception of participation falls in line with the legitimacy concept by Fritz Scharpf (1999) and Zürn, (2000), who distinguish between an input-oriented authenticity (government by the people) and an output-oriented effectiveness (government for the people), while focusing at the same time on how to organize a credible decision-making process (throughput).

To be more concrete, public participation in any form of collaborative decision making and planning is dependent on the following four criteria (Fiorino, 1990, pp. 229–230). The first two criteria mostly refer to the normative rational of participation, the third criterion to the instrumental rationale, and the fourth to the substantive argument. However, according to Fiorino (1990), these criteria should be handled "as a continuum, and the assessment of each mechanism as a judgement about its capacity to fulfill the criteria" (p. 229).

1. The availability of opportunities for nonexperts (citizens) to get involved in decision making.
2. The extent to which citizens can take part in decision-making processes.
3. The existence of framework conditions and structures for direct exchange over a period of time.
4. The possibility for citizens to acquire expertise and to negotiate with representatives of the administration or technical experts on an equal basis.

The following assessment is not so much about encouraging participatory planning in general (Forester, 1999) or about power politics (Monno & Khakee, 2012) or learning in participatory planning (Meléndez & Parker, 2019), as it is about the democratic aspects of collaboration in the energy transition.

Assessing drawbacks and lessons learnt

In this section, Fiorino's four criteria will be used to assess the potential advantages and drawbacks of participatory planning processes within the energy transition. It emphasizes the questions of who participates, when, on what level, and with what intensity in a decision-making process.

Criterion 1: opportunities for participation

In conventional planning exercises of RETs, laypeople or citizens were excluded from the negotiations. In contrast, participatory planning offers a forum of cocreation, where in the ideal case active and empowered citizens are able to raise their ideas, beliefs, and attitudes in a trustful atmosphere. Among the participants, no hierarchy is presumed or expected. The aim is to bring experts, interested stakeholders, and laypeople together to discuss their views and preferences and to conclude on the most preferred or more appropriate renewable energy projects. Participatory planning exercises can at best prevent conflicts or at least put diverse policies on the table and contribute to a constructive dialogue.

Using participatory planning is clearly a process imposed, structured, and managed by the authorities (policy makers, public administration, planners) and not by the users (citizens, laypeople, stakeholders). The problem is that securing broad-based public participation includes "practical and conceptual difficulties"—in particular on the question of who is a legitimate participant or not (Few et al., 2007, p. 48f.). It is clear that increasing participation is always limited to a selected group of participants. However, there is strong evidence that in many participatory processes only an elite in terms of income, education, or status (e.g., students, retirees) is willing and able to contribute actively (Fiorino, 1989, p. 527). Hence, methods of stakeholder selection (e.g., stakeholder mapping) are needed to gain a better level of representation and to reduce social exclusion. The participatory planning process should be in principle open for everyone but reluctant to involve anyone interested in the negotiations. On a conceptual level, this refers to long lasting debates on elitist versus egalitarian or deliberative versus participatory forms of democracy (Fishkin, 1991); in practical terms, it reveals the necessity to address the voices of the citizens not participating in the process. This should be ensured by comprehensive measures of documentation (e.g., minutes, brochures, audio and video recording, live streams) and public relations (e.g., press release, press conferences, public events, websites).

Criterion 2: levels of codetermination

Participation implies "a degree of active involvement in taking decisions" (Few et al., 2007, p. 49). It demands a setting where participants not only can express their opinions and preferences but also have a voice in the decision-making process. Unfortunately, participatory planning often takes place solely at the levels of information (e.g., public hearings, leaflets, information brochures) and consultation (e.g., surveys, public opinions) but not on the important and decisive levels of codecision making and self-determination. According to the seminal work of Arnstein (1969) and her ladder of citizen participation, the first two levels pave the way for tokenism, while forms of delegated power, partnership, or even citizen control are basic pillars and methods of any serious participatory effort. Top-down or one-way information meets the minimum requirements of exchange, but speaking truly about participation starts with the principle of dialogue or two-way communication (Rowe & Frewer, 2000, p. 6).

Opportunities of codetermination in participatory planning are important, but they are not sufficient when it comes to processes of political steering and decision making. Hence, the implementation of any assumptions made within the planning exercise is dependent on a strong political commitment from policy makers and public administration or at least a verbal or written expression to take the results seriously. A participatory planning exercise should take any chance to gain credibility in the shadow of hierarchical decision making.

Criterion 3: structured participation

Most scholars in the field assume that participatory planning is a complex, time-consuming, sometimes cost-intensive endeavor where laypeople, citizens, or stakeholders should be involved as early as possible (e.g., Forester, 1999; Rowe & Frewer, 2000). In this context participatory research points to a "participatory dilemma" or "participation paradox." It describes the experience by which the interest and engagement of citizens, when they are directly affected in an implementation process, often develop. However, direct involvement often comes about only when the strategic phase of the implementation process, in which decisions are made, has already been completed. The result of this relationship is that commitment to discuss a particular program or project in the public domain emerges bottom-up when the ability to strategically influence the process is already limited.

But not only is the time decisive, just as important is the degree of structuring the participation process. Participatory methods are defined "as methods to structure group processes in which non-experts play an active role in order to articulate their knowledge, values and preferences" (van Asselt Marjolein & Rijkens-Klomp, 2002, p. 168). This definition helps to distinguish participatory efforts from lower levels of participation like information and consultation (see Arnstein's ladder of citizen participation).

A very common interactive setting seemed to be participation in a workshop format. However, workshops are not a participatory method by themselves. Various other and better methods are available to engage citizens, the public, or stakeholders in decision-making processes, like, for example, citizen panels, consensus conferences, planning cells, citizen juries, stakeholder dialogues, or role-play simulations (for an overview, see Rowe & Frewer, 2000; van Asselt Marjolein & Rijkens-Klomp, 2002). All of these methods ensure structured and interactive group processes within a participatory planning approach and are adequate tools to elaborate knowledge.

Criterion 4: knowledge integration

In participatory planning, practical knowledge, experiences, and perspectives from different sources and interested parties come together. The integration of knowledge is a gainful task if values, ideas, and norms of citizens improve the output and the quality of knowledge. Ideally speaking, in a participatory planning exercise "the development and usage of fundamental knowledge, applied knowledge and practical knowledge go hand-in-hand and do influence each other directly and interactively" (Rotmans, 2006, p. 41).

The problem is that without input of expert knowledge, participation can lead to inefficient debates, dilettantism, and undesirable consequences. Therefore, participation processes should account for the expertise of participation practitioners and moderators as well as researchers, which cover a broad spectrum of knowledge. These external inputs are essential in a cooperative conflict management, but the interpretation and assessment of this specialist knowledge should always remain the responsibility of the respective forum or group of participants. Otherwise, the chance of discussions and negotiations on an equal level will be very limited.

In addition, credible and interactive visualization methods can help to better communicate planning processes, make information easier to grasp, and thus support the local or regional discussion process. Several studies already pointed out that landscape issues are very important in explaining support or opposition to RETs and in particular to wind farms (Scherhaufer et al., 2017; Wolsink, 2007; Wüstenhagen et al., 2007).

In sum, participatory planning exercises are timely and, as regards resources, restricted processes, highlighting questions and claims about efficiency and effectiveness. Efficiency is an economic principal of a cost/benefit relationship. An assessment could be labeled as efficient if the costs are minimized in regard to the goal(s) and output(s) or if the return of investments is high on the basis of the available resources. Effectiveness corresponds with the achievement of objectives and can be understood as an output matching the predefined focal points. No wonder that the relationship between different parties in the participatory planning process is regarded as something to be "managed." However, conflicts of interest should not only be managed, and the management of conflicts is not synonymous with their solution. In any case, in order to handle conflicts in the energy transition, they must be perceived as such, and the competing objectives of different groups must be accepted. Therefore, raising dissenting opinions or alternative options in participatory planning is allowed or even encouraged through different professional moderation techniques and a necessary prerequisite for the legitimacy of the process. In addition, seeing failure as a success could reduce the pressure within the participatory projects and open up room for creative (noneffective) results that you did not ask for.

A meaningful participatory planning exercise is better geared toward smaller groups and toward using many different participatory methods. The aim is to clearly detach different purposes or levels of participation with different methods of participation, which obviously needs professional management, structured planning, transparency, and plenty of time and resources.

Discussion and recommendations

The previous discussion of participatory planning has attempted to emphasize the claim that participation does not necessarily bring better results, although, at the same time, it highlights the problematic nature of the term "better." The power of group dynamics is often insufficient to resolve conflicts with a multitude of parties and many interrelated issues. If the exchange of interests fails at the local or regional level—e.g., if the advocates cannot convince the opponents (or vice versa) or if no viable compromise is found—the fundamental problem remains that a reconciliation of interests between different "worldviews" with respect to RETs may not be possible (for different worldviews in the wind energy sector, see Scherhaufer et al., 2017). Thus there is a need for strong democratic institutions administering the energy transition, more power to the people in terms of reliable and sufficient opportunities for public participation and supplementary compensation mechanisms.

Justice and fairness—or as Rawls (1971, 2001) labeled it, "justice as fairness"—are interpreted as core principles of energy democracy. The perception of the degree of justice is an important motive for individual action (or inaction) and helps to evaluate and understand decisions of others (Montada & Kals, 2000). If individuals or communities experience the decision-making process as fair and just, which means that democratic and participatory standards are appropriately applied or that transparency and openness are guaranteed, the chance of gaining people's satisfaction at the end increases. It is important to note that this political philosophy focuses on the quality of decision-making processes and not on convincing residents, citizens, or stakeholders to accept something that runs counter to their preferences and beliefs. However, the scientific concept of acceptability, which is more open to occasions of nonacceptance and rejection, is compatible with these notions of justice and fairness (Cowell et al., 2011; Fournis & Fortin, 2017; Scherhaufer et al., 2017).

Taking these assumptions into account, the following paragraphs discuss some effects of procedural and distributional justice on different levels of local to regional policy making. They show that requirements of energy democracy and acceptability are linked and can contribute to one another. In addition, each section concludes with a summary of tangible recommendations to enable (or constrain) the energy transition (cf. Tables 23.1 and 23.2). The suggestions are based on a literature review, including my own transdisciplinary projects in the field of renewable energy and social acceptance (Scherhaufer et al., 2017, 2018).

Procedural justice

At the local levels, citizen participation in decision making, trust in the negotiation process, and transparency are increasingly important. Comprehensive, honest, and reliable information is a prerequisite for legitimate public discourses. Citizens want early and frequent information about the expected location, investment costs and profits, the impact on humans and

Table 23.1 Recommendations in a nutshell on a procedural level

Increase the quality of planning processes in terms of democratic governance!
✓ Inform early, frequently, and comprehensively.
✓ Communicate transparently and confidently.
Strengthen participation and openness!
✓ Involve citizens through structured participatory processes and procedures.
✓ Involve local decision makers and citizens in the discussion of the number and location of the renewable energy facilities.
✓ Use appropriate and credible presentation, modeling, and visualization methods that leave enough room for interaction.

Table 23.2 Recommendations in a nutshell on a distributional level

Spread monetary profits locally!
✓ Fair compensation and distribution of revenues to citizens and communities (e.g., fund solutions, intermunicipal compensation mechanisms)
Consider tasks of political coordination!
✓ Diversification of RETs according to technical and economic potentials and linked to regional and national energy strategies
✓ A successful transition of the energy system requires the expansion of renewable energy production while at the same time limiting energy use through efficiency and sufficiency measures.

the environment, and the opportunities for participation and codecision making. To involve interested or affected parties, common learning processes—like activities of participatory planning—should be initiated.

The legitimate interests of nature conservation and environmental and climate protection must not be played off against each other. Locals must be informed about the meaningfulness of renewable energy production and its social, environmental, and economic value. At the same time, the expected effects and negative impacts on humans and nature must be presented transparently and taken into account in the (official) licensing and approval procedures. Appropriate methods of information sharing include articles in local newspapers and consultation processes (e.g., info days, excursions, and home visits). These tasks not only should be left to the involvement of the operating companies but should be carried out by local leaders, civil society, or outsourced to a neutral third party.

Operators or authorities should prepare several scenarios of the design of the RET project(s) (e.g., with a flexible number of installations and different distances to sensitive areas), in order to be able to respond better to the interests of citizens and local decision makers. In addition, making previously invisible planning processes visible is crucial to participatory planning. The different options of RET can be presented and discussed by means of participatory modeling approaches (Höltinger et al., 2016), public participatory geographic information system (GIS) (Brown et al., 2014), and visualizations and virtual reality assessments (Miller et al., 2016; Schipper, 2016). Conducting a referendum is another possibility for strengthening democratic legitimacy but may simply reinforce polarization within a community (Kapeller & Biegelbauer, 2020). Taking an all-or-nothing approach only increases the pressure and leads to unnecessary conflicts. At the same time, operating companies must establish a high degree of operational reliability of the renewable energy systems and installations and intensively communicate these operational realities.

Distributional justice

Compensation payments to the local population or community, for example in the form of leases, are an important aspect of the level of distributional justice (Langer et al., 2016). For economically underdeveloped communities, RET projects can be an important source of income. Nevertheless, concerns about equity between citizens or the local population create conflicts and clashes (Kirwan, 2007; Langer et al., 2016; Walker et al., 2010). If all consumers pay while few benefit, social inequalities could be increased, and energy democracy is endangered. Therefore, an equal distribution of benefits and losses, for example via models of community ownership, lower electricity bills, or tax rebates for those affected (Walker & Baxter, 2017), is fundamental to a fair and just transition.

Questions of human and nature protection are dealt with mainly within the licensing procedure, where residents might have a say, if an environmental impact assessment is necessary depending on the country's regulatory requirements. If adverse effects on protected goods, the wildlife, or individual households are expected, compensation payments or compensatory measures may be required by the authorities. Hence, these questions have a great influence on the realization of a project. However, in opposition to impacts on human, nature protection issues are usually of interest to the population only insofar as projects can be prevented or at least slowed down.

At the level of local government, it is important to understand RETs as local energy projects and to respond to the fears and worries of those concerned. The task of government authorities is to assume a leadership role in project management and planning and to link the possible implementation with other, visible accompanying measures (such as the financing of municipal services or the implementation of energy efficiency and sufficiency measures). At the regional level, RET projects must also be embedded in higher-level energy strategies.

RETs could have a major impact on the perception of the landscape. A possible reduction of the recreation function or the loss of "identity" cannot be compensated economically in many cases. In contrast, RETs are also seen by citizens as a symbol of sustainable energy production. In addition, experts and stakeholders bring habituation effects into play. Growing up in a community characterized by RETs can help participants regard these facilities as an integral part of the landscape and thus not perceived as disturbing infrastructure buildings.

Conclusion

This chapter has explored ways that participatory planning of RET installations can enable energy democracy. To achieve this potential, the normative, instrumental, and substantive rationales for participation have to be considered. If we do not take this seriously, participatory projects in the renewable energy sector run the risk of becoming an instrument of control, exercising power-over, or even manipulation.

Appropriately structured and well planned participatory planning has the potential to incorporate stakeholders and citizens in a trustful atmosphere. In this atmosphere, citizens can have a genuine voice in decision making about RETs and hold politicians more accountable, public authorities gain legitimacy, and operators can trust in the quality of planning and implementation processes.

From the perspective of energy democracy, the normative commitment to an equitable, fair, transparent, and as broad as possible participatory process is sufficient justification itself irrespective of whether it leads to public acceptance or the outcome traditionally sought by project operators. This is a crucial claim because there is no guarantee that just and fair participatory planning processes will convince people and therefore increase the acceptance of RETs. One reason for this is embedded in the normative rationale for participation. People have a wide variety of reasons for supporting and opposing RETs, and these reasons do not necessarily match those pushed forward by scientists, project developers, and policy makers.

Participatory planning is not a cheap and, by some standards, not an efficient process, but the money and time are well invested compared to the scale of the infrastructure projects. If the concept of energy democracy is accepted as fundamental to sustainable energy transitions, it makes sense to complement indirect and informal participation efforts with more direct, active, and sophisticated means of participation that enables both procedural and distributional justice. The manifold polices, measures, and efforts listed in this chapter can be used to increase meaningful participation and strengthen the legitimacy of decisions related to energy system changes. All the better if, along with strengthening political legitimacy, such processes increase public acceptance, promote energy transitions toward RETs, and defeat global climate change. However, we should avoid reducing participatory planning to managerialism and must consider public acceptance as a possible outcome, rather than as the primary goal, of public participation. Only then can collaborative planning have a democratic function and provide great opportunities to increase fairness, procedural and distributional justice, and social learning.

References

Anderson, B., Böhmelt, T., & Ward, H. (2017). Public opinion and environmental policy output: A cross-national analysis of energy policies in Europe. *Environmental Research Letters, 12*(11). http://doi.org/10.1088/1748-9326/aa8f80

Arnstein, S. R. (1969). A ladder of citizen participation. *Journal of the American Institute of Planners, 35*(4), 216–244.

Brown, G., Donovan, S., Pullar, D., Pocewicz, A., Toohey, R., & Ballesteros-Lopez, R. (2014). An empirical evaluation of workshop versus survey PPGIS methods. *Applied Geography, 48*, 42–51. https://doi.org/10.1016/j.apgeog.2014.01.008

Busse, M., & Siebert, R. (2018). Acceptance studies in the field of land use—a critical and systematic review to advance the conceptualization of acceptance and acceptability. *Land Use Policy, 76*, 235–245. https://doi.org/10.1016/j.landusepol.2018.05.016

Chilvers, J., & Longhurst, N. (2016). Participation in transition(s): Reconceiving public engagements in energy transitions as co-produced, emergent and diverse. *Journal of Environmental Policy & Planning, 18*(5), 585–607. https://doi.org/10.1080/1523908X.2015.1110483

Chilvers, J., & Pallett, H. (2018). Energy democracies and publics in the making: A relational agenda for research and practice. *Frontiers in Communication, 3*(14). https://doi.org/10.3389/fcomm.2018.00014

Cochran, J., Mai, T., & Bazilian, M. (2014). Meta-analysis of high penetration renewable energy scenarios. *Renewable and Sustainable Energy Reviews, 29*, 246–253. https://doi.org/10.1016/j.rser.2013.08.089

Cowell, R., Bristow, G., & Munday, M. (2011). Acceptance, acceptability and environmental justice: The role of community benefits in wind energy development. *Journal of Environmental Planning and Management, 54*(4), 539–557. https://doi.org/10.1080/09640568.2010.521047

Dahrendorf, R. (1959). *Class and class conflict in industrial society*. Stanford University Press.

Devine-Wright, P. (2009). Rethinking NIMBYism: The role of place attachment and place identity in explaining place-protective action. *Journal of Community & Applied Social Psychology, 19*(6), 426–441. https://doi.org/10.1002/casp.1004

Devine-Wright, P. (Ed.). (2011). *Renewable energy and the public: From NIMBY to participation*. Earthscan.

Devine-Wright, P., Batel, S., Aas, O., Sovacool, B., Labelle, M. C., & Ruud, A. (2017). A conceptual framework for understanding the social acceptance of energy infrastructure: Insights from energy storage. *Energy Policy*, *107*, 27–31. https://doi.org/10.1016/j.enpol.2017.04.020
Dryzek, J. S. (2000). *Deliberative democracy and beyond*. Oxford University Press.
Fagence, M. (1977). *Citizen participation in planning*. Pergamon Press.
Feldpausch-Parker, A. M., Endres, D., & Peterson, T. R. (2019). Editorial: A research agenda for energy democracy. *Frontiers in Communication*, *4*(53). https://doi.org/10.3389/fcomm.2019.00053
Few, R., Brown, K., & Tompkins, E. L. (2007). Public participation and climate change adaptation: Avoiding the illusion of inclusion. *Climate Policy*, 7(1), 46–59.
Fiorino, D. J. (1989). Environmental risk and democratic process: A critical review. *Columbia Journal of Environmental Law*, *14*(2), 501–547.
Fiorino, D. J. (1990). Citizen participation and environmental risk: A survey of institutional mechanisms. *Science, Technology, and Human Values*, *15*(2), 226–243.
Fishkin, J. S. (1991). *Democracy and deliberation: New directions for democratic reform*. Yale University Press.
Forester, J. (1999). *The deliberative practitioner: Encouraging participatory planning processes*. MIT Press.
Fournis, Y., & Fortin, M. J. (2017). From social 'acceptance' to social 'acceptability' of wind energy projects: Towards a territorial perspective. *Journal of Environmental Planning and Management*, *60*(1), 1–21. https://doi.org/10.1080/09640568.2015.1133406
Frantál, B., van der Horst, D., Martinát, S., Schmitz, S., Teschner, N., Silva, L., Golobic, M., & Roth, Mi. (2018). Spatial targeting, synergies and scale: Exploring the criteria of smart practices for siting renewable energy projects. *Energy Policy*, *120*, 85–93. https://doi.org/10.1016/j.enpol.2018.05.031
Fraune, C., & Knodt, M. (2018). Sustainable energy transformations in an age of populism, post-truth politics, and local resistance. *Energy Research & Social Science*, *43*, 1–7. https://doi.org/10.1016/j.erss.2018.05.029
Godschalk, D. R., & Mills, W. E. (1966). A collaborative approach to planning through urban activities. *Journal of the American Institute of Planners*, *32*(2), 86–95. https://doi.org/10.1080/01944366608979362
Habermas, J. (1994). *The theory of communicative action I: Reason and the rationalization of society*. Beacon Press.
Höltinger, S., Salak, B., Schauppenlehner, T., Scherhaufer, P., & Schmidt, J. (2016). Austria's wind energy potential—A participatory modeling approach to assess socio-political and market acceptance. *Energy Policy*, *98*, 49–61. https://doi.org/10.1016/j.enpol.2016.08.010
Jacobson, M. Z., & Delucchi, M. A. (2011). Providing all global energy with wind, water, and solar power, part I: Technologies, energy resources, quantities and areas of infrastructure, and materials. *Energy Policy*, *39*(3), 1154–1169. https://doi.org/10.1016/j.enpol.2010.11.040
Kapeller, S., & Biegelbauer, P. (2020). How (not) to solve local conflicts around alternative energy production: Six cases of siting decisions of Austrian wind power parks. *Utilities Policy*, *65*, 101062. https://doi.org/10.1016/j.jup.2020.101062
Kirwan, K. L. (2007). *Social-psychological aspects of domestic renewable energy: A study of low-income tenants responses to solar photovoltaics* [unpublished Doctoral thesis, De Montfort University].
Kunze, C., & Becker, S. (2014). *Energy democracy in Europe: A survey and outlook*. Rosa Luxemburg-Foundation.
Langer, K., Decker, T., Roosen, J., & Menrad, K. (2016). A qualitative analysis to understand the acceptance of wind energy in Bavaria. *Renewable and Sustainable Energy Reviews*, *64*, 248–259. https://doi.org/10.1016/j.rser.2016.05.084
Meléndez, J. W., & Parker, B. (2019). Learning in participatory planning processes: Taking advantage of concepts and theories across disciplines. *Planning Theory & Practice*, *20*(1), 137–144. https://doi.org/10.1080/14649357.2018.1558748
Miller, D., Morrice, J., McKeen, M., Donaldson-Selby, G., Wang, C., & Munoz-Rojas, J. (2016). *Use of digital and 3D visualisaton technology in planning for new development*. Report for Scottish Government. www.gov.scot/binaries/content/documents/govscot/publications/factsheet/2017/01/use-of-digital-3d-visualisation-technology-research-report/documents/review-planning-system-research-3d-visualisation-technology-final-report-pdf/review-planning-system-research-3d-visualisation-technology-final-report-pdf/govscot%3Adocument/Review%2Bof%2Bthe%2BPlanning%2BSystem%2B-%2BResearch%2B-%2B3D%2Bvisualisation%2Btechnology%2B-%2BFinal%2BReport.pdf
Monno, V., & Khakee, A. (2012). Tokenism or political activism? Some reflections on participatory planning. *International Planning Studies*, *17*(1), 85–101. https://doi.org/10.1080/13563475.2011.638181

Montada, L., & Kals, E. (2000). Political implications of psychological research on ecological justice and proenvironmental behaviour. *International Journal of Psychology*, *35*(2), 168–176. https://doi.org/10.1080/002075900399466

Müller, M. M. (2012). Justice as a framework for the solution of environmental conflicts. In E. Kals & J. Maes (Eds.), *Justice and conflicts: Theoretical and empirical contributions* (pp. 239–250). Springer.

Ostrom, E. (2010). Polycentric systems for coping with collective action and global environmental change. *Global Environmental Change*, *20*(4), 550–557. https://doi.org/10.1016/j.gloenvcha.2010.07.004

Phadke, R. (2013). Public deliberation and the geographies of wind justice. *Science as Culture*, *22*(2), 247–255. https://doi.org/10.1080/09505431.2013.786997

Poggi, F., Firmino, A., & Amado, M. (2018). Planning renewable energy in rural areas: Impacts on occupation and land use. *Energy*, *155*, 630–640. https://doi.org/10.1016/j.energy.2018.05.009

Rawls, J. (1971). *A theory of justice*. Clarendon Press.

Rawls, J. (2001). *Justice as fairness: A restatement*. Belknap.

Rotmans, J. (2001). More evolution than revolution: Transition management in public policy. *Foresight*, *3*(1), 15–31. https://doi.org/10.1108/14636680110803003

Rotmans, J. (2006). Tools for integrated sustainability assessment: A two-track approach. *The Integrated Assessment Journal*, *6*(4), 35–57.

Rowe, G., & Frewer, L. J. (2000). Public participation methods: A framework for evaluation. *Science, Technology, & Human Values*, *25*(1), 3–29. https://doi.org/10.1177/016224390002500101

Scharpf, F. (1999). *Governing in Europe: Effective or democratic?* Oxford University Press.

Scherhaufer, P. (2013). Bridging the gap between the theory and practices of stakeholder participation in integrated vulnerability assessments of climate change. *Systemic Practice and Action Research*, *27*(5), 449–463. https://doi.org/10.1007/s11213-013-9294-8

Scherhaufer, P., Höltinger, S., Salak, B., Schauppenlehner, T., & Schmidt, J. (2017). Patterns of acceptance and non-acceptance within energy landscapes: A case study on wind energy expansion in Austria. *Energy Policy*, *109*, 863–870. https://doi.org/10.1016/j.enpol.2017.05.057

Scherhaufer, P., Höltinger, S., Salak, B., Schauppenlehner, T., & Schmidt, J. (2018). A participatory integrated assessment of the social acceptance of wind energy. *Energy Research & Social Science*, *45*, 164–172. https://doi.org/10.1016/j.erss.2018.06.022

Schipper, A. (2016). The power of virtual reality for wind energy. *Renewable Energy Focus*, *17*(2), 62–63. https://doi.org/10.1016/j.ref.2016.02.006

Shove, E., & Walker, G. (2007). Caution! Transitions ahead: Politics, practice, and sustainable transition management. *Environment and Planning A: Economy and Space*, *39*(4), 763–770. https://doi.org/10.1068/a39310

Stirling, A. (2006). Analysis, participation and power: Justification and closure in participatory multi-criteria analysis. *Land Use Policy*, *23*(1), 95–107.

Stokes, L. C., & Warshaw, C. (2017). Renewable energy policy design and framing influence public support in the United States. *Nature Energy*, *2*(8), 17107. https://doi.org/10.1038/nenergy.2017.107

Syme, J. G. (2012). Justice and environmental decision making. In E. Kals & J. Maes (Eds.), *Justice and conflicts: Theoretical and empirical contributions* (pp. 283–295). Springer.

van Asselt Marjolein, B. A., & Rijkens-Klomp, N. (2002). A look in the mirror: Reflection on participation in integrated assessment from a methodological perspective. *Global Environmental Change*, *12*(3), 167–184.

van Deth, J. W. (2014). A conceptual map of political participation. *Acta Politica*, *49*(3), 349–367. https://doi.org/10.1057/ap.2014.6

Walker, C., & Baxter, J. (2017). "It's easy to throw rocks at a corporation": Wind energy development and distributive justice in Canada. *Journal of Environmental Policy & Planning*, *19*(6), 754–768. https://doi.org/10.1080/1523908X.2016.1267614

Walker, G., Devine-Wright, P., & Barnett, J. (2011). Symmetries, expectations, dynamics and contexts: A framework for understanding public engagement with renewable energy projects. In P. Devine-Wright (Ed.), *Renewable energy and the public: From NIMBY to participation* (pp. 1–14). Earthscan.

Walker, G., Devine-Wright, P., Hunter, S., High, H., & Evans, B. (2010). Trust and community: Exploring the meanings, contexts and dynamics of community renewable energy. *Energy Policy*, *38*(6), 2655–2663. https://doi.org/10.1016/j.enpol.2009.05.055

Wolsink, M. (2006). Invalid theory impedes our understanding: A critique on the persistence of the language of NIMBY. *Transactions of the Institute of British Geographers*, *31*(1), 85–91. https://doi.org/10.1111/j.1475-5661.2006.00191.x

Wolsink, M. (2007). Wind power implementation: The nature of public attitudes: Equity and fairness instead of 'backyard motives'. *Renewable and Sustainable Energy Reviews*, *11*(6), 1188–1207. https://doi.org/10.1016/j.rser.2005.10.005

Wolsink, M. (2010). Near-shore wind power—protected seascapes, environmentalists' attitudes, and the technocratic planning perspective. *Land Use Policy*, *27*(2), 195–203. https://doi.org/10.1016/j.landusepol.2009.04.004

Wüstenhagen, R., Wolsink, M., & Bürer, M. J. (2007). Social acceptance of renewable energy innovation: An introduction to the concept. *Energy Policy*, *35*(5), 2683–2691. https://doi.org/10.1016/j.enpol.2006.12.001

Zürn, M. (2000). Democratic governance beyond the nation-state: The EU and other international institutions. *European Journal of International Relations*, *6*(2), 183–221. https://doi.org/10.1177/1354066100006002002

24

PARTICIPATION IN NONDEMOCRACIES

Rural Thailand as a site of energy democracy

Laurence L. Delina

Introduction

Sustainable energy transitions, a system evolution from fossil fuel–fired toward sustainable energy systems, are key in meeting the normative agendas of climate action and sustainable development. To date, however, state and nonstate actions to achieve the transition remain slow, and the dynamics of contemporary energy systems remain largely unsustainable and unattuned to climate mitigation (Michaelowa & Michaelowa, 2017; Rogelj et al., 2016). The persistence of environmentally destructive energy systems owes to their being largely and complicatedly entwined with social and political systems in addition to being technological and economic (cf. Jasanoff, 2004), albeit replacement technologies have become increasingly less expensive and more efficient, signaling renewable energy systems' growing competitive advantage. Addressing fossil fuel incumbency challenges—alongside the need to scale and speed up the transition—requires not only continuous technological innovation but also persistent advances in policy making. This means policy makers will need to revise the current regime's almost exclusive dependence on technoeconomic assumptions to a more pluralistic perspective that includes social dimensions. After all, technological systems are coproduced alongside the ordering of societies—a process that is considerably political and hence requires the understanding of social and power dynamics (Jasanoff, 2004). Thus understanding energy(ies) as a sociotechnical system may facilitate a transition that simultaneously moves toward effective climate action, greater sustainability, and more inclusive democracy (Stephens et al., 2008).

In 2015, the global community agreed on two key normative international agendas: to address the climate change challenge and to work toward sustainable development for all. The Paris Agreement and the 2030 Agenda on Sustainable Development provide the key texts for these directions and the impetus for moving toward sustainability. These norm-changing trajectories further signal an important moment in energy transition, opening up new opportunities to increase the democracy of an otherwise largely undemocratic sector: energy generation. The Paris Agreement contains nationally determined contributions to mitigate climate change impacts through reductions in key emitting sectors, of which energy often takes the lion's share (United Nations Framework Convention on Climate Change, 2015). The 2030 Agenda contains 17 interrelated goals, including climate action (Goal No. 13) and energy access and transition (Goal No. 7) (United Nations General Assembly, 2015). Based on these ambitions,

 DOI: 10.4324/9780429402302-28

the normative shift toward sustainability extends to multiple scales—from the international to regional to national to subnational—and to multiple sectors—education, livelihoods, water, land, and energy, among others. Energy transition is pivotal in this shift and manifests across scales and sectors. Thus we are witnessing examples by which energy transition operates at the nation-state (e.g., a 100% renewable energy target in several Pacific island states [International Partnership on Mitigation and MRV, 2015]), within states or provinces (e.g., in the State of California [California Air Resources Board, 2016]), and in agglomerations of city-based initiatives (e.g., Global Covenant of Mayors for Climate and Energy, 2019).

In liberal democracies, where the "voice" of the people supposedly drives state policy, national sustainable energy transition programs are increasingly being reflected in terms of heightened recognition of the role of community energy transition (e.g., Bauwens & Devine-Wright, 2018; van der Schoor & Scholten, 2015). These dynamics involved many bottom-up sociotechnical approaches taken and led largely by community energy project developers, and they required extensive public participation in the generation of distributed, often small-scale, renewable energy systems, which the public also directly consumes or, in case of surplus, exports to the grid. In most arrangements, the same public also owns, manages, and sustains the operations of these systems through institutionalized social systems such as cooperatives and neighborhood associations (e.g., Bauwens et al., 2016; Yildiz et al., 2015). Inarguably, these arrangements—community-owned and organized renewable energy generation systems—are one of the finest manifestations of democratically produced energy transitions. While the literature is already rich about how these arrangements, over time, prosper (and fail) in many wealthy and democratic societies (e.g., Bauwens et al., 2016; Yildiz et al., 2015), we still know little about how community energy manifests and thrives in nondemocratic places or in states where governments may rhythmically move from liberal democracies to dictatorships and vice versa.

Energy democracy can take many forms (Burke, 2018; Chilvers & Pallett, 2018; Feldpausch-Parker et al., 2019). The generation of renewable energy by communities, whereby the public at large holds and owns stakes in its institutional and/or commercial arrangements, is a phenomenon commonly identified as energy democracy. As a social process, energy democracy can be exercised in multiple ways and under multiple labels (Bauwens et al., 2016; Burke, 2018). This variability gives rise to the question of whether and how energy democracy may emerge in nondemocratic places, where citizen participation in political life is sharply curtailed (Delina, 2018c). If so, how might these citizens mobilize democratic energy transitions? What elements of public engagement are present in these conditions such that they constitute meaningful expressions of energy democracy in a nondemocracy?

In this chapter, I respond to these questions and argue that energy democracy can manifest (and eventually persist and thrive) in a nondemocracy. I use a case study from rural Thailand—a state currently under military rule, hence a nondemocracy—to provide evidence of the capacity of Thai citizens to reflexively and deliberatively engage with energy systems. Based on this empirical evidence, I argue that the sociotechnicality of this public engagement further reveals energy democracy as a collective, cultural, consequential, coproduced, coexistent, and critical governance phenomenon.

The evolving interpretations of "energy democracy"

Although energy democracy as a concept remains open for multiple interpretations (Burke, 2018; Chilvers & Pallett, 2018; Delina, 2018a; Feldpausch-Parker et al., 2019), its social agenda is relatively straightforward: It involves processes tantamount to social mobilizations for performing the social, economic, and political projects of energy transitions (Burke, 2018; Burke & Stephens, 2018; Delina, 2018b; Marquardt & Delina, 2019). Beyond electric power, energy democracy creates, sustains, and uses other types and manifestations of "power," specifically social, economic, and political (Stirling, 2014). As a source of social-economic-political power, energy democracy encompasses several public advocacies: from eschewing fossil-based institutions and corporate profits to addressing historical economic and political inequalities by providing avenues for ordinary citizens to chart their own futures (e.g., Bauwens et al., 2016), and, collectively, advance political agendas (Marquardt & Delina, 2019). These practices have become sites of prefigurative social movements, where, as pockets of sociotechnical innovations, they usually blossom within and among neighbors, associations, and communities—and with adequate internal and external support, may scale up to regional, national, and even international energy transitions (Bauwens & Devine-Wright, 2018; Marquardt & Delina, 2019).

Energy democracy exists in a socially-economically-and-politically dynamic continuum, where actors generate, supply, use, and export renewable energy via practices that are highly characterized by hits and misses, stabilities and uncertainties, consensus and dissensus, and ebbs and flows (Delina, 2018a). By looking at energy democracy from this temporal perspective (and beyond its spatial orientation), one can appreciate the durability (or nondurability) of practices that characterize energy democracy. Such moments of change make energy democracy open-ended, provisional, and a continuing exercise of public engagement (Chilvers & Kearnes, 2016; Delina, 2018a; Stirling, 2015). While there is a preponderance of literature on the dynamics of energy democracy as practiced in liberal democracies, particularly in the Global North, we still see little research about the emergence of these dynamics in nondemocracies.

As a sociotechnical practice, energy democracy, in both the temporal and the spatial scales, can be systematically identified. In my own work, I have empirically observed and analyzed how certain public exercises could have constituted energy democracy in a nondemocracy (Delina, 2018a, 2018b, 2018c; Marquardt & Delina, 2019). In this work, I have found that, despite differing locations and regardless of political arrangements, energy democracy can still manifest in the public engagement continuum as long as the practice exhibits certain qualities and elements, which I summarized in Table 24.1.

While this list is not exhaustive and remains provisional, it provides some guidance for analysts seeking to evaluate whether certain energy transition practices are within the remit of energy democracy. In the next section, I discuss how these elements manifest in an actual practice.

Energy democracy in a nondemocracy: an empirical case from Thailand

Background, methods, and study site

The southeast Asian country of Thailand is an upper-middle-income economy with steadily rising emissions and pronounced social and economic inequality (Chiengkul, 2019). On May 22, 2014, a military coup d'état, led by General Prayut Chan-o-cha, successfully removed from power a democratically elected government and established a junta called the National Council for Peace and Order to govern Thailand (Montesano et al., 2018). Prayut eventually became the new prime minister when he won the national election on March 24, 2019, the first since the 2014 coup.

Natural gas-fired generation, with coal and renewable energy (primarily from hydroelectricity imported from Laos), dominate the country's energy landscape. In 2016, the Government

Table 24.1 Qualities of the practice of energy democracy in rural Thailand

Qualities	*Brief description*
Coexistent	Cooperation needs to be willingly agreed to and is more likely when people share a common fate and common challenges.
Collective	Rather than a simple amalgamation of autonomous individuals and their independent needs, energy democracy emerges from the collective nature of public deliberations.
Cultural	Publics are internal, not external to energy democracy; both formal and informal deliberations are processes of cultural embedding and active mediation whereby publics shape their own engagement.
Contingent	Energy democracy deliberations can occur in both formally and informally sanctioned and fostered public spaces.
Coproduced	Energy democracy is a coproduced interweaving of the social, the normative, and the material contingent upon issues, cultures and contexts; hence, its tools, instruments, and mechanisms could never be standardized.
Critical	Exclusions are accepted as part of the practice of democracy.
Consequential	Energy democracy does not promise the delivery of actionable outcomes; closure and consensus are not its end state.
Connected	Energy democracy occurs in diverse, entangled and interrelated interactions to describe, negotiate, and decide a confluence of issues.

Source: Adapted from Delina (2018b)

of Thailand approved its 20-year national energy strategic plan called the Thailand Integrated Energy Blueprint (TIEB) 2017–2036. The TIEB aims to enhance the country's domestic energy security, national development, and regional connectivity and envisages an energy mix characterized by decreasing reliance on imported natural gas and an increasing role for "clean coal." The TIEB also introduces a 5% role for nuclear energy. The TIEB brings under one coherent plan five of the government's independently pursued plans: the Power Development Plan, the Energy Efficiency Plan, the Alternative Energy Development Plan, the Oil Plan, and the Gas Plan. A state-owned and highly centralized grid distributes energy across the Thai population and industries. This distribution system extends all the way to rural communities, except for villages inside national parks where grid extension is forbidden by state regulation.

This study was conducted in one of these national park communities using a "grounded" approach and that included interviews, focus group discussions, participant observations, site transects, and a diary. Between November 2016 and January 2017, five sites were visited, and 31 interviews were conducted. Each interview, where respondents were selected according to their availability, lasted between 40 and 90 minutes. I employed snowball sampling, asking respondents to suggest additional contributors. I conducted five pair interviews, two small groups (one was a group of only women and the other a mixed gender group), and the rest individually. During the site transects and direct observations, attention was paid to renewable energy technologies and systems. I observed a monthly forum, which is the principal data source for this chapter. Interviews were audio recorded, while notes were made during transects and observations. These were then transcribed and coded, with emergent themes relative to the practice of energy democracy analyzed vis-à-vis the extant literature on energy democracy.

In Kaeng Krachan National Park, the Pa Deng community of mostly farmers and semi-skilled workers relies on small-scale farming of crops such as pineapples, maize, vegetables, plums, and jackfruits; rural industries such as cattle raising; and employment in nearby Cha-am municipality, a premium tourist area in the Gulf of Thailand. Located 275 km southwest of Thailand's capital, Bangkok, in the province of Phetchaburi, Pa Deng has more than 100 households situated in ten *moo ban* (villages). Five of these moo bans are within the jurisdiction of a royal dairy production development project; the remaining five are inside the park, meaning that households therein have no electricity grid access. Cooking and lighting fuels for these households were provided by expensive kerosene or by illegally harvested firewood (Interview, 2017a, 2017b, 2017c, 2017d, 2017e, 2017f, 2017g, 2017h). Wealthier households purchased liquified petroleum gas (LPG) tanks for cooking and diesel generation systems for lighting and, in some cases, for irrigation (Interview, 2017a). Thaksin Shinawatra's government (2001–2006) attempted to address this energy access gap through a national solar photovoltaic (PV) program, which distributed home solar PV systems; however, the program failed largely due to the lack of skilled workers to maintain these new technologies (Interview, 2017a, 2017c, 2017e).

In 2006, energy-poor households in Pa Deng created a *moso* (moderation society) network, patterned after King Bhumibol Adulyadej's "sufficiency economy philosophy"—a rural development model underscoring the "true Buddhist way of life" where people would live self-sufficiently without greed or overexploitation of natural resources (Piboolsravut, 2004). Moso networks are common in rural Thailand, usually state organized, and often attached to royal development projects where the monarchy provides income-generating programs such as the previously mentioned dairy production project (United Nations Development Programme, 2007). Over time, moso networks have become increasingly popular ways of transcending the imagination of rural Thais, further elevating positive regard of their monarchs. While the Pa Deng network was not a state-led moso mobilization effort, its members considered the King as their exemplar (Connors, 2011; Delina, 2018a, 2018b, 2018c). With the fieldwork conducted barely a month after the King died, references to him consistently recurred in the interviews (Interview, 2017a, 2017b, 2017c, 2017d, 2017e, 2017f, 2017g, 2017h).

Collectivism and the production of democratic sustainable energy transitions

The Pa Deng moso network's foremost reason for creation was for neighbors to collectively discover options to address their pressing energy challenge: how to replace expensive kerosene or LPG tanks for cooking and diesel for generators (Interview, 2017a, 2017g, 2017h). A biogas digester for cooking, which some of them saw in neighboring Myanmar, offered one of the most promising technologies given the abundance of local feedstock from cow manure, kitchen and food waste, and grass trimmings (Interview, 2017b, 2017c). The network decided to construct a digester as a pilot demonstration project in one of the households; instead of a typical reinforced concrete biogas digester system, the local farmers experimented with cheaper, easy-to-install, and movable plastic (Interview, 2017a, 2017b, 2017c). In parallel, the network also revived their unused solar PV systems, this time providing their members with the necessary skills to maintain these technologies (Interview, 2017g, 2017h).

Ten years after it was created, informants described the Pa Deng moso network as a best practice in Thailand (Interview, 2018). Other moso networks have studied its practices and have attempted to model themselves after it (Interview, 2017a). These social innovations include microloan provision mechanisms (funded by members' monthly contributions and used to support the construction/purchase of energy systems) and in-network trainings (Interview, 2017a, 2017c, 2017d, 2017g). The network also developed and institutionalized monthly deliberative meetings where network members would decide on future development and priorities (Author's diary). Altogether, these formal (e.g., monthly assemblies) and informal (e.g., interactions on roadways and evening gatherings over local wine) practices of non-state-led energy democracy reveal how participants were able to transform their places into sites for constituting energy democracy. The next section highlights how this practice exemplifies the qualities of energy democracy outlined in Table 24.1.

The practice of energy democracy in Pa Deng

The Pa Deng moso network exemplifies how civil society in a nondemocratic state such as Thailand could mobilize, form, and organize people's preferences and mutual interests. The network—a product of people's realization of their *coexistence*—can be regarded as a proactive response to the shortcomings of Bangkok-led traditional politics and as the failure of the Bangkok establishment to provide solutions to ordinary people's basic needs: in this case, household energy services. Some interviewees highlighted how their communal practice essentially reverses the traditional role of the state in implementing a development project, and, instead, "they would take the matter in their own hands" (Interview, 2017a, 2017c, 2017e). *Cooperation* manifested in the Pa Deng network because its members recognized their shared common fate: They had similar problems, and they willingly agreed to communally address their common challenges (Mansbridge, 2015). The Pa Deng moso network demonstrates energy democracy as internally, rather than externally driven.

The network's desire to "join together" was rooted in meanings generated by the members. They described themselves as progressively developing a common desire to live sufficiently, to improve their livelihoods, and to contribute to their benevolent King's ideals of "rural development" (Interview, 2017a, 2017b, 2017d). This *collective* response also recognized their ability to make use of their local resources and capacities (Interview, 2017a). The network's emphasis on energy democracy in mundane practices such as cooking, for example, exemplifies what Keane (2009) calls "monitory democracy." This practice was visible during interviews, when network members recalled their intrinsic motivations to deliberate together, thereby constituting a public sphere, in a space that was neither donor nor state fostered. For them, rather than a simple consolidation of independent needs of citizens, energy democracy is a practice underlining the collective nature of public deliberation (Heller & Rao, 2015, p. 3; Interviews, 2017a, 2017b, 2017c, 2017d, 2017e, 2017f, 2017g, 2017h). Further evidence of this collectivity can be found in the network's communally defined priorities that are deeply aligned with their desire to "produce" sustainable, less expensive, and readily available fuel (Interview, 2017a, 2017b, 2017d). This commonly identified need, regardless of their socioeconomic differences, eventually became the glue that bound the network together.

In the Pa Deng moso network, publics are internal, not external to the processes and practices of energy democracy. Their formal and informal deliberations had become processes of *cultural* embedding and active mediation (Appadurai, 2015). A premium was placed on culture, which drove farmers to participate largely because moso networks are attached to the King, who, in Thai culture, enjoys demigod-like respect and adoration (Connors, 2011; Elinoff, 2014; Fong, 2009). This cultural influence is also related to Thailand's *lèse majesté* laws, which ban criticism of royalty and, by extension, their "royal development projects" (Connors, 2011), which are subject to various contradictory interpretations (Hewison, 2008). Regardless of the obvious constraints posed by this cultural rationale, it positions moso networks to drive energy transition in Thailand (Interview, 2017i). This informant saw great potential for moso networks, which proliferate throughout the Thai countryside, to transform into a "strong force for facilitating and deploying community-oriented renewable energy systems that can later be scaled up" (Interview, 2017i).

The network also resists normative assumptions about energy democracy as a deliberative forum occurring primarily in formally sanctioned and fostered public spaces (Mansbridge, 2015; Marres & Lezaun, 2011). The sociotechnical energy innovations observed in the Pa Deng network relied on neither donor nor state support (Author's diary; Interview, 2017a); rather, they are *contingent* upon traditional expectations for rural Thai behavior. Interviewees frequently referenced "communality" as a behavior that is not only inherent to being Thai but also expected from Thais (Interview, 2017c, 2017e, 2017f, 2017h). This notion of "Thainess" is similar to notions found in other rural communities where residents elected to discuss issues first with their village chiefs—behaviors documented, for instance, in Malawi (Swidler, 2014; Swidler & Watkins, 2015) and India (Sen, 2006). The communal structures for common resource management that predate industrialization (Ostrom, 1990) provide additional examples of this concept of *contingency*.

The Pa Deng moso network also counters the conventional understanding that the tools, instruments, and mechanisms of energy democracy could be standardized or universalized. Rather, interviewees underlined that energy democracy should be understood and appreciated as a *coproduced* interweaving of social, normative, and material contingencies upon the specificities of issues, cultures, and contexts (Appadurai, 2015; Jasanoff, 2004). The coproduction of technological innovation (the shift toward sustainable fuel), alongside social arrangements (the sense of communality through camaraderie and cohesion), that underpins the Pa Deng energy transition reveals that energy democracy in this place differs substantially from other spatial and temporal sites. The uniqueness of this practice, therefore, effectively rejects what Evans (2004) calls "institutional monocropping."

This network also recognizes exclusion as part and parcel of energy democracy itself. As some interviewees mentioned, not everyone in Pa Deng joined the moso network (Interview, 2017a, 2017h). A *critical* appreciation of this fact would run counter to the orthodox understanding of public engagement as one that is heavily focused on methods aimed at guaranteed "inclusion" of all actors (Callon et al., 2009). Although inclusivity is important for effective deliberation (Dryzek, 2009), it is impossible to be completely inclusive within a single sociomaterial collective of participation (Chilvers & Kearnes, 2016). For the Pa Deng network, the virtues of reflexivity and humility are far more important than the methodological focus on ensuring full inclusivity. Such virtues further illustrate the critical appreciation by network members of what should matter to them (cf. Jasanoff, 2011). Reflexivity and humility are evident in interviewees' communal practices, where they would critically assess their own normative biases and commitments when making collective decisions, as observed especially during a deliberative forum (Author's diary). Network members have learned that their preferences could always adapt or change, meaning that their positions remained malleable and open for nudging toward options that could be mutually accommodative (Dryzek, 2005).

The network also steers away from the conventional understanding that energy democracy should always deliver actionable outcomes and that either closure or consensus should always be the end state of deliberation. By contrast, acceptable *consequences* of their deliberations (Author's diary) could range from specifically binding decisions, which legitimize their deliberative processes (Mansbridge, 2015), to agreed-upon stopping points that could serve as the first in a series of steps through which preferences could be eventually translated into material actions (Heller & Rao, 2015). This openness highlights a shift toward understanding energy democracy as occurring in diverse, entangled, and interrelated—in short, *connected*—interactions to describe, negotiate, or decide variety of issues, rather than a single and stationary one (Smith & Stirling, 2016; Stirling, 2015). The *connectedness* of issues that the Pa Deng moso network members expected meant that osmosis across issues is the norm, not the exception. Pa Deng moso network's sociotechnical energy innovation was embedded in a constellation of many interconnected and interdependent interests (Author's diary).

Conclusion

As this chapter has shown, energy democracy may emerge and even thrive in a nondemocratic state. However, the extent to which it can actually make a difference is context dependent. The publics of Pa Deng have shown that the dynamics of legitimacy- and meaning-making could materialize even in the absence of state support. Furthermore, their story reveals that when people collectively reflect on their mutual needs, critically weigh their options, and experiment, they can make and remake energy democracy. The deliberative spaces these publics had produced have become productive channels for serving the needs of the less powerful, while promoting public awareness of the capacity to mobilize power.

The features of energy democracy highlighted by this context differ from those generally noted within democratic regimes. First, energy democracy relies on a willing awareness of *coexistence* among the demos, who embrace a common destiny. Second, energy democracy is *collective*, meaning it is more than the amalgamation of autonomous individuals with independent needs. Third, it is *culturally* defined by publics who are internal to the process. Fourth, energy democracy is *contingent* on both spatial and temporal conditions that remain always shifting. Fifth, the participants *coproduce* their democracy by interweaving these contingencies such that boundaries between the normative and the material are often blurred. Sixth, energy democracy is *critical*, where exclusions are acknowledged, embraced, and regarded as part of the practice of democracy. Seventh, its *consequential* quality focuses more on reflection and learning than on closure. Lastly, energy democracy occurs in *connections* among confluences of issues and challenges, multiplicities of options to address them, and the acceptance of the heterogeneity of the impacts of the decisions made.

These conclusions reveal energy democracy as a set of dynamic, malleable, and reflexive practices that may emerge in both democratic and nondemocratic regimes. This chapter contributes to the ongoing quest to identify qualities of energy democracy that enable scholars to analyze its practices in situ.

Acknowledgment

This chapter is an output of the author's Future of Energy Systems in Developing Countries project at the Frederick S. Pardee Center for the Study of the Longer-Range Future at Boston University.

References

Appadurai, A. (2015). Success and failure in the deliberative economy. In P. Heller & V. Rao (Eds.), *Deliberation and development: Rethinking the role of voice and collective action in unequal societies* (pp. 67–84). The World Bank Group.

Bauwens, T., & Devine-Wright, P. (2018). Positive energies? An empirical study of community energy participation and attitudes to renewable energy. *Energy Policy*, *118*, 612–625. https://doi.org/10.1016/j.enpol.2018.03.062

Bauwens, T., Gotchev, B., & Holstenkamp, L. (2016). What drives the development of community energy in Europe? The case of wind power cooperatives. *Energy Research & Social Science*, *13*, 136–147. https://doi.org/10.1016/j.erss.2015.12.016

Burke, M. (2018). Shared yet contested: Energy democracy counter-narratives. *Frontiers in Communication*, *3*, 22. https://doi.org/10.3389/fcomm.2018.00022

Burke, M., & Stephens, J. (2018). Political power and renewable energy futures: A critical review. *Energy Research & Social Science, 35*, 78–93. https://doi.org/10.1016/j.erss.2017.10.018

California Air Resources Board. (2016). *The governor's climate change pillars: 2030 greenhouse gas reduction goals*. https://ww3.arb.ca.gov/cc/pillars/pillars.htm

Callon, M., Lacoumes, P., & Barthe, Y. (2009). *Acting in an uncertain world: An essay on technical democracy (inside technology)*. The MIT Press.

Chiengkul, P. (2019). Uneven development, inequality and concentration of power: A critique of Thailand 4.0. *Third World Quarterly, 40*(9), 1689–1707. https://doi.org/10.1080/01436597.2019.1612739

Chilvers, J., & Kearnes, M. (2016). *Remaking participation: Science, environment and emergent publics*. Routledge.

Chilvers, J., & Pallett, H. (2018). Energy democracies and publics in the making: A relational agenda for research and practice. *Frontiers in Communication, 3*, 14. https://doi.org/10.3389/fcomm.2018.00014

Connors, M. K. (2011). When the walls come crumbling down: The monarchy and Thai-style democracy. *Journal of Contemporary Asia, 41*(4), 657–673. https://doi.org/10.1080/00472336.2011.610619

Delina, L. L. (2018a). Energy democracy in a continuum: Remaking public engagement on energy transitions in Thailand. *Energy Research & Social Science, 42*, 53–60. https://doi.org/10.1016/j.erss.2018.03.008

Delina, L. L. (2018b). Climate mobilizations and democracy: The promise of scaling community energy transitions in a deliberative system. *Journal of Environmental Policy & Planning, 22*(1), 30–42. https://doi.org/10.1080/1523908X.2018.1525287

Delina, L. L. (2018c). Can energy democracy thrive in a non-democracy? *Frontiers in Environmental Science, 6*, 5. https://doi.org/10.3389/fenvs.2018.00005

Dryzek, J. (2005). Deliberative democracy in divided societies alternatives to agonism and analgesia. *Political Theory, 33*(2), 218–242. https://doi.org/10.1177/0090591704268372

Dryzek, J. (2009). Democratization as deliberative capacity building. *Comparative Political Studies, 42*(11), 1379–1402. https://doi.org/10.1177/0010414009332129

Elinoff, E. (2014). Sufficient citizens: Moderation and the politics of sustainable development in Thailand. *Political and Legal Anthropology Review, 37*(1), 89–108. https://doi.org/10.1111/plar.12052

Evans, P. (2004). Development as institutional change: The pitfalls of monocropping and potentials of deliberation. *Studies in Comparative International Development, 38*, 30–52. https://doi.org/10.1007/BF02686327

Feldpausch-Parker, A. M., Endres, D. E., & Peterson, T. R. (2019). Editorial: A research agenda for energy democracy. *Frontiers in Communication, 4*, 53. https://doi.org/10.3389/fcomm.2019.00053

Fong, J. (2009). Sacred nationalism: The Thai monarchy and primordial nation construction. *Journal of Contemporary Asia, 39*(4), 673–696. https://doi.org/10.1080/00472330903077030

Global Covenant of Mayors for Climate and Energy. (2019). www.globalcovenantofmayors.org/

Heller, P., & Rao, V. (Eds.). (2015). *Deliberation and development: Rethinking the role of voice and collective action in unequal societies*. The World Bank Group.

Hewison, K. (2008). Review of Thailand human development report: Sufficiency economy and human development. *Journal of Contemporary Asia, 38*, 212–219.

International Partnership on Mitigation and MRV. (2015). *100% renewable energy targets in the Pacific Islands*. www.transparency-partnership.net/system/files/migrated_document_files/2012ongoingrenewableenergy_pacificislands.pdf/

Interview. (2017a). Interview with Khun Kosol (name used with permission), adult male, Pa Deng network's leader.

Interview. (2017b). Interview with an adult male, Maize farmer, name withheld.

Interview. (2017c). Interview with an adult female, vegetable farmer, name withheld.

Interview. (2017d). Interview with an adult male, vegetable farmer, name withheld.

Interview. (2017e). Interview with an adult female, housewife, name withheld.

Interview. (2017f). Interview with an adult male, jackfruit farmer, name withheld.

Interview. (2017g). Interview with an adult female, farmer, name withheld.

Interview. (2017h). Interview with an adult male, farmer, name withheld.

Interview. (2017i). Interview with a Thai academic in a Bangkok-based university, name withheld.

Interview. (2018). Interview with a development worker from the Ministry of Energy, name withheld.

Jasanoff, S. (2004). *States of knowledge: The co-production of science and the social order*. Routledge.

Jasanoff, S. (2011). Constitutional moments in governing science and technology. *Science and Engineering Ethics, 17*, 621–638. https://doi.org/10.1007/s11948-011-9302-2

Keane, J. (2009). *The life and death of democracy*. W.W. Norton and Company.

Mansbridge, J. (2015). A minimalist definition of deliberation. In P. Heller & V. Rao (Eds.), *Deliberation and development: Rethinking the role of voice and collective action in unequal societies* (pp. 27–50). The World Bank Group.

Marquardt, J., & Delina, L. (2019). Reimagining energy futures: Contributions from community energy transitions in Thailand and the Philippines. *Energy Research & Social Science*, *49*, 91–102. https://doi.org/10.1016/j.erss.2018.10.028

Marres, N., & Lezaun, J. (2011). Materials and devices of the public: An introduction. *Economy & Society*, *37*(4), 489–509. https://doi.org/10.1080/03085147.2011.602293

Michaelowa, K., & Michaelowa, A. (2017). Transnational climate governance initiatives: Designed for effective climate change mitigation? *International Interactions*, *43*, 129–155. https://doi.org/10.1080/03050629.2017.1256110

Montesano, M. J., Chong, T., & Heng, M. (Eds.). (2018). *After the coup: The national council for peace and order era and the future of Thailand*. ISEAS Publishing.

Ostrom, E. (1990). *Governing the commons: The evolution of institutions for collective action*. Cambridge University Press.

Piboolsravut, P. (2004). Sufficiency economy. *ASEAN Economic Bulletin*, *21*, 127–134.

Rogelj, J., Den Elzen, M., Höhne, N., Fransen, T., Fekete, H., Winkler, H., Schaeffer, R., Sha, F., Riahi, K., & Meinshausen, M. (2016). Paris agreement climate proposals need a boost to keep warming well below 2C. *Nature*, *534*, 631–639. https://doi.org/10.1038/nature18307

Sen, A. (2006). *Argumentative Indian*. Picador.

Smith, A., & Stirling, A. (2016). *Grassroots innovations and democracy*. STEPS Centre.

Stephens, J. C., Wilson, E. J., & Peterson, T. R. (2008). Socio-political evaluation of energy deployment (SPEED): An integrated research framework analyzing energy technology deployment. *Technological Forecasting and Social Change*, *75*, 1224–1246. https://doi.org/10.1016/j.techfore.2007.12.003

Stirling, A. (2014). Transforming power: Social science and the politics of energy choices. *Energy Research & Social Science*, *1*, 83–95. https://doi.org/10.1016/j.erss.2014.02.001

Stirling, A. (2015). From controlling 'the transition' to culturing plural radical progress. In I. Scoones, M. Leach, & P. Newell (Eds.), *The politics of green transformations* (pp. 54–67). Routledge.

Swidler, A. (2014). Cultural sources of institutional resilience: Lessons from chieftaincy in rural Malawi. In P. Hall & M. Lamont (Eds.), *Social resilience in the neo-liberal era* (pp. 319–345). Cambridge University Press.

Swidler, A., & Watkins, S. C. (2015). Practices of deliberation in rural Malawi. In P. Heller & V. Rao (Eds.), *Deliberation and development: Rethinking the role of voice and collective action in unequal societies* (pp. 133–166). The World Bank Group.

United Nations Development Programme. (2007). *Thailand human development report 2007: Sufficiency economy and human development*. UNDP.

United Nations Framework Convention on Climate Change. (2015). Adoption of the Paris Agreement. FCCC/CP/2015/L.9/Rev.1.

United Nations General Assembly. (2015, September 25). *Transforming our world: The 2030 agenda for sustainable development*. Resolution adopted by the General Assembly. A/RES/70/1.

van der Schoor, T., & Scholten, B. (2015). Power to the people: Local community initiatives and the transition to sustainable energy. *Renewable and Sustainable Energy Reviews*, *43*, 666–675. https://doi.org/10.1016/j.rser.2014.10.089

Yildiz, O., Rommel, J., Debor, S., Holstenkamp, L., Mey, F., Muller, J. R., Radtje, J., & Rognli, J. (2015). Renewable energy cooperatives as gatekeepers or facilitators? Recent developments in Germany and a multidisciplinary research agenda. *Energy Research & Social Science*, *6*, 59–73. https://doi.org/10.1016/j.erss.2014.12.001

25
PART IV RESPONSE

Tarla Rai Peterson

The chapters in this part of the handbook identify and investigate principles and constructs characterizing participatory processes that empower citizen efforts to compose and sustain energy democracies. In many ways, they posit potential contexts that influence the viability of grassroots activities such as those described in the previous section (Part III). This response focuses on the themes that span the chapters in this section and then offers suggestions for future directions in research on the importance of democratic and participatory principles to energy democracy.

Interconnecting themes

The most striking theme to cut across the chapters in this section is the fraught nature of the term "participatory democracy." Although the term "democracy," as derived from the Greek *demokratia*, is commonly translated to mean something akin to "rule of the people," the increasing complexity and size of nominally democratic states has effectively precluded direct involvement of the entire polity (Barber, 1984; Dahl, 1956). Although only one of many interconnected factors, the scaling up of democracy has contributed to the ascendency of a governance mode commonly labeled as representative democracy, as a more practical alternative for modern states. In representative, as opposed to participatory democracy, the governed cede decision making to elites, albeit elites that have been selected by the populace. As should be apparent from the justice-participation-power-technology framework developed in the first chapter, our normative perspective follows from an ethical imperative to continually reach toward participatory democracy, or governance that relies primarily on self-rule by its citizens. Part of the normativity demonstrated in Part IV of this handbook is a commitment to strive toward participatory democracy, in whatever ways are allowed in a particular sociopolitical configuration. It follows participatory democracy theorists such as Dewey (1954) in arguing that citizen education can remediate supposed deficits sufficiently to enable widespread participation in governance. It diverges, however, in expanding the concept of citizen education far beyond that envisioned by Dewey. In these chapters, citizen education is more about facilitating a mutual learning across any potentially interested citizens than about one-way information sharing.

Democratic deliberation is a participatory concept that carries through all four chapters in this section. For Barnes (Chapter 22), the challenge is to use energy system transformation to

 DOI: 10.4324/9780429402302-29

discover and push against preconceived limits of energy democracy. From his perspective, the technological advances that have enabled energy's *decarbonization*, *digitalization*, and *decentralization* present opportunities to compose entirely new systems that upend traditional hegemonic political relationships and enable energy justice across diverse groups of citizens. The deep democratization enabled by these sociotechnical trends offers possibilities never before experienced, which fuels his advocacy for experimentation. Barnes does not succumb to naïve expectations that every experiment will be universally successful but instead encourages us to use whatever emerges as the basis for inventing new participatory venues and for reimagining how old venues might function. Barnes's enthusiastic embrace of experimentation grows out of his awareness that the systemic confluence of decarbonization, digitalization, and decentralization has created a rare opportunity for resonance at the internal boundaries of modern society, which opens the possibility for new social configurations to emerge (Luhmann, 1989). The urgency of experimentation comes from awareness among systems theorists that preexisting system configurations always stand ready to defend against new configurations, which means that energy democracy remains an unlikely, if possible, outcome of energy system transitions. A key point for Barnes is that, although energy democracy does not exist at any meaningful scale today because energy systems are ripe for transformation, now is the time to reinvent the entire system.

Kinsella's chapter (Chapter 21) offers public participation in decisions about nuclear energy as a highly productive case for exploring the complex political ecology of energy deliberation that public participation organizers will face. Although public perceptions of nuclear energy are much more negative than public perceptions of any other low-carbon energy source, this difference is one of the greatest strengths of the chapter. The possibility of a new nuclear reactor, or even possible expansion or decommissioning of an already operating facility, is big news that extends beyond activists to whom project managers apply the pejorative term "usual suspects." Organizers of both formal and informal participation venues face participants who are unlikely to accept contrived distinctions between appropriate discussion topics and topics that project proponents designate as falling outside of the purview available for public comment. While other low-carbon energy sources may elicit deeply felt fears when it comes to siting projects, nuclear energy elicits those fears immediately upon the first mention of a project. Whether the levels and types of public responses are grounded in harms associated with past failures, the immeasurable lifespan of nuclear waste, close connections to military use, or all three, project proponents have come to expect it, and organizations such as the International Atomic Energy Agency (n.d.) and the World Association of Nuclear Operators (n.d.) produce numerous materials for persuading a nervous public that nuclear energy is not only safe but highly desirable. Kinsella enumerates recent examples where the nuclear industry has minimized the effects of negative perceptions by erecting economic, legal, and political barriers to public participation; at the same time he explores examples where participants have breached these barriers and have influenced decisions ranging from decommissioning an individual project to changing the course of national policy.

Scherhaufer's chapter (Chapter 23) directly responds to the problem of conflating participation and acceptance. For many energy project promoters, the only justification for devoting resources to public participation is to obtain approval. Nothing else so predictably spans the differences between proponents of coal mining, hydraulic fracturing, solar collection, wind farming, and other approaches to providing energy for modern societies, as the frustration over time and effort spent trying to persuade interested members of the public to support or at least not to oppose a project. This problem, of course, extends beyond renewable energy and beyond energy itself to plague evaluations of public participation in numerous settings. One of

Scherhaufer's most valuable contributions to this section is his reminder to recognize and negotiate the tensions between promoting public acceptance and nurturing meaningful participation opportunities rather than simply abandoning the quest for acceptance. Given our experience as researchers and facilitators, the editors of this handbook can attest that, at the time of this writing, institutional support for public participation in energy system decisions relies, at the very least, on tacit support from both government and corporate sectors and that the primary missions of these sectors rarely include providing robust opportunities for all interested citizens to participate fully in deliberations. For this reason, if for no other, Scherhaufer's attempts to compose approaches that simultaneously develop and sustain meaningful participation opportunities and contribute to public acceptance require serious consideration.

Despite the value of obtaining some level of external support for participatory processes, Delina (Chapter 24) uses an example of energy democracy in Thailand to demonstrate that such processes can not only survive but even thrive in the absence of such support. In a similar way that lessons can be learned from nuclear energy, the very oddity of this case enables Delina to extract a set of design principles that can be constructively applied to a highly diverse set of public participation efforts. These principles are characterized by their process orientation rather than outcome orientation. One lesson to take from these principles is that, more than anything else, public participation offers an opportunity to develop a stronger sense of joint citizenship among all participants, whether project proponents, opponents, or merely observers.

Future directions for research

As these chapters indicate, participatory principles can be applied in a wide variety of ways by those seeking to democratize energy transitions. Kinsella's approach to energy democracy suggests one especially fruitful direction for future research. Risk research has been applied to the (generally negative) public perceptions of nuclear energy to derive simplistic formulations wherein participation is organized to transfer correct (usually positive) information to publics who suffer from information deficits that lead them to perceive risk incorrectly. Although social science has theoretically and empirically debunked the deficit approach, the ideology remains prominent within the framing of participation venues. One possible reason for its persistence is the absence of a compelling alternative, which is something that future scholarship might develop. Kinsella's chapter (Chapter 21), as well as his overall research program (Kinsella et al., 2015), could provide a core for the arduous process of theorizing an alternative explanation for public resistance to energy, as well as to other technologies.

Empirical data gleaned from all four chapters demonstrate both how important it is to appreciate the complex interactions between strategies of action and modes of participation as components of sociotechnical change and how thin our current understanding of those interactions remains. The chapters suggest numerous directions for such research, ranging from in-depth case studies of unlikely and highly localized energy democracy (Chapter 24) to systemic analyses that postulate alternative organizational frameworks for energy transitions (Chapter 23). Despite the differences among these analyses, they could by unified by their focus on discovering what dimensions of these communicative interactions enable sociotechnical change and how they do so.

The dilemmas these chapters identify and explore remind us that, by the time energy democracy became a topic of social scientific research, it had emerged as a social movement, complete with heroes and villains. Social movement studies (i.e., Crick, 2020; Kern, 2020) offer a rich and underused literature for energy democracy scholarship. From the context of this handbook, questions posed by Obregón and Tufte (2017) seem especially productive. They join other social movement scholars in asking how communicative interactions both enhance and impede social change, what should count as evidence of that change, and how theory and practice intersect within social movements. While they join other researchers in framing social movements as loosely organized collectives seeking redress for some social ill, their explicit directive to pursue such studies from an interdisciplinary perspective is particularly apt guidance, given that energy democracy is thoroughly embedded within complex sociotechnical changes. Lastly, these chapters suggest the importance of affirmatively striving to sustain a praxis-based orientation that encourages networking between research and practice (Endres et al., 2008).

References

Barber, B. R. (1984). *Strong democracy: Participatory politics for a new age.* University of California Press.
Crick, N. (2020). *The rhetoric of social movements: Networks, power, and new media.* Routledge.
Dahl, R. (1956). *Preface to democratic theory.* University of Chicago Press.
Dewey, J. (1954). *Democracy and education.* Free Press.
Endres, D., Sprain, L., & Peterson, T. R. (2008). The imperative of praxis-based environmental communication research: Suggestions from the step it up 2007 national research project. *Environmental Communication, 2*(2), 237–245. https://doi.org/10.1080/17524030802141794
International Atomic Energy Agency. (n.d.). *Atoms for peace and development.* www.iaea.org/
Kern, T. (2020). Social movements. In B. Hollstein, R. Greshoff, Y. Schimank, U. Weiß, & A. de Gruyter (Eds.), *Soziologie: Sociology in the German-speaking world* (pp. 399–413). De Gruyter. https://doi.org/10.1515/9783110627275-027
Kinsella, W. J., Andreas, D. C., & Endres, D. (2015). Communicating nuclear power: A programmatic review. In *Communication yearbook* (Vol. 39, pp. 277–309). Routledge. https://doi.org/10.1080/23808985.2015.11679178.
Luhmann, N. (1989). *Ecological communication* (J. Bednarz, Jr., Trans.). University of Chicago Press.
Obregón, R., & Tufte, T. (2017). Communication, social movements, and collective action: Toward a new research agenda in communication for development and social change. *Journal of Communication, 67*(5), 635–645.
World Association of Nuclear Operators. (n.d.). www.wano.info/

PART V

Energy resource tensions

26
ENERGY RESOURCE TENSIONS
Introduction

Andrea M. Feldpausch-Parker

In this part, we address different types of energy resources and technologies and the conflicts that arise from their continued and often contested use in global energy systems. These energy resources include nuclear power (Chapter 27), hydraulic fracturing for shale gas extraction (Chapter 28), large-scale hydropower (Chapter 29), and wind and solar power (Chapter 30). This mix of nuclear, fossil, and renewable energy sources and their corresponding technologies not only does a reasonable job of capturing the energy landscape but also allows for the deconstruction of harms and opportunities surrounding their use.

One of the main premises of energy democracy is decarbonization. Szulecki and Overland (2020) even labeled decarbonization as one of the "ideal-typical understandings of energy democracy" (p. 2). As such, energy democracy social movements and the responding literature often focus on renewable energies as the best way to address environmental and social issues such as climate change and energy justice. These movements advocate a transition away from fossil fuels and other energy sources that not only contribute to greenhouse gas emissions but also have a poor track record of engaging with relevant publics democratically, often leading to issues of inequality, disproportionate impacts, and marginalization (Cha, 2020; de Onís, 2018).

The other understandings of energy democracy mentioned by Szulecki and Overland (2020) are process and a normative goal. Though the normative goal can vary by location, technology, and social desires and needs, process generally means that decision making will be carried out democratically, with thoughts toward justice and equality. As the Climate Justice Alliance (n.d.) astutely points out, justice and equality in the transition to renewables is not a guarantee. It is, however, an opportunity to resist the status quo currently practiced by many energy industries by encouraging shifts in power to other relevant stakeholders such as communities, consumers, and members of the public (Sweeney, 2012). Chapters in this section address both the well-known conflicts with traditional energy sources such as fossil fuels and nuclear (Chapters 28 and 27 respectively), as well as debunking assumptions that renewable energy inherently carries the guarantee of fairness and equity (Chapter 29). This section, however, also shows attempts at change through challenging preexisting paradigms of power and decision making (Chapter 30), thus reinforcing hope that society can take heed of the need for energy system transition and improve both environmental and social conditions.

This part of the book starts with a wake-up call to the risks of energy production. In Chapter 27, Kasperski and Kuchinskaya provide a powerful example of a regional attempt to come

DOI: 10.4324/9780429402302-31

back from nuclear disaster. Chernobyl is a well-known case study about the risks of nuclear power. The 1986 disaster has served both as a cautionary tale as well as a testbed for recovery. This chapter addresses the latter, evaluating two international efforts, ETHOS (1996–2001) and CORE (2003–2008), meant to empower communities with the knowledge and tools to promote safety and continue life in a radiated environment. Instead, researchers found a shallow attempt at knowledge production that left communities with a painful environmental legacy wanting.

Feldpausch-Parker and Batill-Bigler use Chapter 28 to showcase news media representations of the growing discontent with the oil and gas industry as it has sought alternative means to extract previously unreachable deposits during a growing climate crisis. Their analysis is focused on a technology referred to as hydraulic fracturing, also called hydrofracking or fracking for short. Using the Socio-Political Evaluation of Energy Deployment (SPEED) framework, they explored public debates occurring in news media over this controversial technology in the US states of Texas, Pennsylvania, New York, and Michigan. They found a disproportionate amount of negative coverage in all states, including Texas where the technology was first implemented on a commercial scale. Most discussions about the technology centered on legal, environmental, technical, and political systems, which demonstrate some connections to early articulations of energy democracy movements.

In Chapter 29, Finley-Brook describes multiple case studies throughout Latin America that showcase instances of both fast (e.g., murder, repression, death threats) and slow violence (e.g., racism, poverty, debt) related to the financing and construction of large-scale hydroelectric projects. These case studies, including a historic example from Guatemala and more recent examples in Panama, Brazil, and Ecuador, epitomize issues of extreme injustice and marginalization. Finley-Brook takes on an activist scholar perspective in her analysis, laying out a solid case for critique of the notion that renewable energy sources inherently carry a promise of greater democratic practice and thought toward community well-being.

In the final chapter of the section (Chapter 30), Horton, Hernandez, and Peterson address the US state of Hawaii's attempt to transition away from fossil fuels toward energy production through wind and solar. They focus on a key piece of legislation passed in 2008, known as the Hawaii Clean Energy Initiative. The goal of this initiative is to power the state's electric grid and transportation system 100% from clean energy by 2045. The researchers collected data from the 2016 Hawaii Power Summit, where they captured professionals' discourse about creating a more democratic energy system. Though Horton, Hernandez, and Peterson found challenges to this effort, they also saw opportunities for empowering a greater number of stakeholders in energy system decision making.

References

Cha, J. M. (2020). A just transition for whom? Politics, contestation, and social identity in the disruption of coal in the Powder River Basin. *Energy Research & Social Science, 69*, 101657. https://doi.org/10.1016/j.erss.2020.101657

Climate Justice Alliance. (n.d.). *Energy democracy*. Retrieved December 11, 2020, from https://climatejusticealliance.org/workgroup/energy-democracy/

de Onís, C. M. (2018). Energy colonialism powers the ongoing unnatural disaster in Puerto Rico. *Frontiers in Communication, 3*. https://doi.org/10.3389/fcomm.2018.00002

Sweeney, S. (2012). *Resist, reclaim, restructure: Union and the struggle for energy democracy*. Rosa-Luxemburg-Stiftungo.

Szulecki, K., & Overland, I. (2020). Energy democracy as a process, an outcome and a goal: A conceptual review. *Energy Research & Social Science, 69*, 101768. https://doi.org/10.1016/j.erss.2020.101768

27

ENERGY DEMOCRACY, NUCLEAR POWER, AND PARTICIPATORY KNOWLEDGE PRODUCTION ABOUT RADIATION RISKS

Tatiana Kasperski and Olga Kuchinskaya

Introduction

What makes energy democracy democratic is arguably not just decentralized, more participatory decision making that involves multiple stakeholders but also more democratic organization of the production of relevant knowledge. This chapter considers challenges related to knowledge production about the risks of such large-scale energy projects as nuclear power. It analyses the example of two participatory international projects, ETHOS (1996–2001) and CORE (2003–2008), meant to address the needs of several communities in Belarus affected by the fallout from the 1986 Chernobyl nuclear accident. We use the example of these projects to suggest two potential requirements or preconditions for energy democracy in the case of nuclear power: tools and places to bring the problem of radiation health effects out of nonexistence (make radiation and its health effects publicly visible), and opportunities for challenging underlying theoretical approaches, which in the case of radiation protection are often explicitly meant to be authoritative, technocratic, and unreceptive to local or alternative knowledge. We argue that energy democracy in a fuller sense of the word would require public engagement with "deeper," fundamental issues of the production of knowledge about radiation health effects.

As demonstrated by the introduction to this volume, the issue of "participation" is central to the theoretical project and practices of energy democracy. Our discussion of "participation" focuses on the relevant knowledge production necessary for decision making (such as, in the case of nuclear power, knowledge about radiation risks). Callon et al. (2001/2009) note that delegative democracy is insufficient for technoscientific issues because of the persistent problem of unanticipated consequences and issues that have not been or could not be accounted for. From this perspective, what questions have to be considered, what data have to be collected, and what knowledge has to be produced, and how—the discussion of all these and similar questions has to include various publics consistently and from the start, underscoring the importance of participatory forums (see also Kinsella, 2004). In short, knowledge that reflects different perspectives and interests is essential for adequate decision making on complex issues of energy democracy, and the production of such knowledge has to be participatory.

DOI: 10.4324/9780429402302-32

The questions of participatory knowledge production have been studied under the rubric of "citizen science" (Irwin, 2015; Kimura & Kinchy, 2016; Lave, 2012). In their review of literature on citizen science, Kimura and Kinchy (2016) "take issue with studies of citizen science that rate particular projects on their degree of participation or other virtues that are presumed to be inherent to the design of the project itself" (p. 348). For these authors, the key is not whether public participation is sufficiently thorough but the broader social context. For example, in her study of post-Fukushima citizen radiation monitoring organizations (CRMOs), Kimura (2018) observed that these organizations often refrained from telling their clients whether the levels measured for their food were dangerous. Kimura emphasized the broader conditions that affected the work of CRMOs in that context, including neoliberalism (in which managing risks becomes the affected individuals' own responsibility), along with the culture of scientism and of policing concerns about radiation risks as irrational and antivictims. As Kimura and Kinchy (2016) note, highly engaged, bottom-up participation does not guarantee being able to "challenge powerful institutions (such as the nuclear industry) or achieve desired outcomes" (p. 332).

This chapter considers the relationship between specific forms of public participation and the production of knowledge about nuclear risks. The issues of nuclear risks are central to deployment of nuclear energy production; in the very least, the public has to understand and accept nuclear risks. In this chapter, we consider the relationship between forms of public participation and the production of knowledge about nuclear risks from the example of communities affected by the 1986 Chernobyl nuclear accident. It addresses the context of nuclear power, though nuclear power has already been challenged as incompatible with the ideals of energy democracy (Burke & Stephens, 2018; Kinsella, 2012; Kinsella et al., 2015; Winner, 1980). The chapter also addresses the context of postaccident contamination rather than energy transition. Finally, the focus is on post-Chernobyl communities in Belarus, a country that cannot be described as democratic. Yet the participatory model considered in this chapter has been heralded as transferable to other, including more democratic contexts and countries in the event they face similar contamination circumstances. Even more important, the emphasized value of this model has been its participatory approach. This chapter therefore analyzes this model, the value of this form of participation, and its limitations for practicing and understanding energy democracy.

We revisit the relationship between public participation, knowledge production, and community action on the example of two participatory international projects, ETHOS[1] (1996–2001) and CORE (Cooperation for Rehabilitation of the living conditions in the Chernobyl-affected areas in Belarus). The projects were meant to address the needs of communities in Belarus affected by the fallout from the 1986 Chernobyl nuclear accident. Both projects aimed to educate the affected publics, empower them, and improve their living conditions by fostering a "practical radiological culture" consisting of everyday practices and skills that help protect oneself from radiation exposure (Hériard-Dubreuil et al., 2000). As discussed next, these top-down projects initiated by international organizations have been criticized for promoting a particularly limited remembrance of the accident and for advancing neoliberal approaches that individualized responsibility for radiation protection (Dupuy, 2006; Kasperski, 2020; Topçu, 2013). The framing of these projects was heavily affected by international nuclear organizations and sanctioned by the Belarusian government. Yet despite their numerous limitations, the projects provided tools and spaces to bring the problem of radiation health effects out of nonexistence and to make them more publicly visible, even if temporarily and in a limited way. Given the limited scope of this visibility, we argue that the projects played an important role in the context where the issues of Chernobyl were being erased. Yet these projects should not be uncritically transferred to other contexts because of their argued "participatory" value. In the context where affected populations cannot challenge the underlying conceptual approaches, the value of "participation" is limited.

The following sections outline the history of ETHOS and CORE, describe their conceptions of "practical radiological culture" and "participation" (Hériard-Dubreuil et al., 2000), and then discuss the value and limitations of participation in these and similar cases. We conclude by suggesting two potential requirements for energy democracy that the nuclear industry does not currently fulfill: first, providing tools and spaces for the public articulation of radiation risks and possible health effects, and, second, offering opportunities for challenging underlying conceptual approaches.

Postaccident context and the turn to "normalization" policies

On April 26, 1986, one of four reactors at the Chernobyl nuclear power station exploded. The power plant was located in the northern part of what was then the Soviet Republic of Ukraine, close to the Belarusian border. Chernobyl remains the largest civil nuclear accident to date (Josephson, 1999; Medvedev, 1990; Schmid, 2015). The explosion and resulting fire released massive amounts of radioactive materials into the environment. Radioactive plumes were carried by the wind, eventually depositing particles on the ground and durably contaminating vast territories in Europe, most notably in Ukraine, Belarus, and Russia. Close to 25% of the Belarusian territory was affected (International Atomic Energy Agency, 2006).

In the years following the explosion, the Soviet authorities sought to minimize and conceal the scale of the disaster, limit evacuations, and restrict the acknowledged spatial or temporal scope of the disaster (what areas were affected and how long the consequences are going to last). The public visibility of Chernobyl erupted in late 1988, partially as a result of the work done by the oppositional forces in Belarus and Ukraine and local protests (Kuchinskaya, 2014; see also Kasperski, 2020; Petryna, 2002). This was the period of extensive media attention to Chernobyl, establishment of several research institutes devoted to studying the consequences of Chernobyl, adoption of new, more sensitive radiation protection standards and then two Chernobyl-related laws on relocation and social protection of the affected populations. After the collapse of the Soviet Union at the end of 1991, attention in the state media in Belarus shifted to the enormous costs of the Chernobyl Program. Chernobyl was gradually reframed as an economic disaster. This meant that Chernobyl problems were largely the result of responses to the disaster, especially massive spontaneous and organized relocations from the most affected areas. Attempts at redefining state approaches to radiation protection were made in mid-1990s, and by the late 1990s the state policies explicitly shifted to *rehabilitating* the affected areas and restoring "normal" conditions of life through a combination of radiation protection, agricultural, and socioeconomic measures. By the late 1990s, the official media argued, "life overcame Chernobyl" or "We will not give up our land!" (Kuchinskaya, 2014, pp. 82–83).

It was in that context of normalization and rehabilitation that a team of French experts, funded by the European Commission, developed the approach behind ETHOS (Lochard, 2007). The "rehabilitation" in this case meant actively adapting to the consequences of the radioactive fallout and not necessarily forgetting the impact of the disaster and simply returning to the previous way of life. Radioactive contamination was envisioned as a particular type of environment that requires a different type of behavior (Lochard & Prêtre, 1995).

ETHOS grew out of the earlier international efforts. In 1991, the European Commission launched a series of projects to study the consequences of Chernobyl. The projects were carried out in cooperation with the governments of the Russian Federation, Ukraine, and Belarus in 1991–1996. As part of one of the joint study projects, the independent French research and study group Mutadis—a private consulting company created by Gilles Hériard-Dubreuil and specializing in the governance of risky activities and the sustainable development—carried out a study among the Belarusian and Ukrainian communities affected by the disaster. The first stage of that research meant to assess the social and psychological consequences of Chernobyl (Ukraine, 1992–1993); the second stage focused on living conditions in the contaminated territories and management of radiation risks (Belarus, Chechersk district, 1994).

During that investigation, the French experts noted a "generalized climate of anxiety" about the effects of the accident and the lasting contamination of the environment, the stress generated by the feelings of helplessness and loss of control in the face of unknown and invisible dangers, as well as mistrust toward scientific and medical experts and the authorities (Hériard-Dubreuil, 1994; Hériard-Dubreuil & Girard, 1997). Researchers sought to provide a nuanced analysis of postaccident stress. They criticized approaches characterized by "overpsychologization" of the postaccident situation or insisting on the "radiophobia" of the affected populations, the approach they attributed to the Soviet experts.[2] Mutadis researchers also criticized the psychological assistance measures for the local residents as implemented by UNESCO centers, which, in their opinion, could not lead to a lasting improvement in the psychological state of the local population. For Mutadis experts, such measures did not provide practical solutions to protect people against daily radiation risks and thus do something about the cause of their stress. The Belarusian state's information policies, implemented in a highly centralized manner, also had little effect, as the population had little confidence in the words of government experts (Girard & Hériard-Dubreuil, 1996).

Based on conclusions from their research, the French experts proposed a new "return to normality" approach (Lochard & Prêtre, 1995). They pointed out that "return to normality" could not mean a return to the conditions before the accident. They suggested not simply returning the affected territories back to normal functioning (as appeared to be the goal for the Belarusian government) but the rehabilitation of the "living conditions" on those lands that would remain radioactively contaminated for a long time. The new approach of "returning to normality" and "rehabilitation of the living conditions" was first implemented within the framework of the ETHOS project, which started in 1996. The nuances of interpreting what "rehabilitation" would mean did not matter as much for the Belarusian authorities; for them, ETHOS was contributing to the official strategy of rehabilitation or even "renaissance" of Chernobyl-affected lands (Kasperski, 2020).

"Rehabilitation of living conditions" and "practical radiological culture" in the ETHOS and CORE projects

ETHOS was coordinated by Mutadis and led by a multidisciplinary team of experts in radiation protection, risk management, and sustainable development representing several French organizations and scientific institutions. Most prominently, the team included experts from the Nuclear Protection Evaluation Centre (CEPN), a nonprofit organization created in 1979 that currently has three members, all of them major French nuclear organizations.[3] CEPN's chair, Jacques Lochard, later played a crucial role in dissemination of the ETHOS approach in Western Europe and in the post-Fukushima context. The first phase of the project took place between 1996 and 1998 in the village of Olmany (Stolin district), located in the southern area of Belarus. Six groups of local residents were created. With the support of the ETHOS team and the local authorities, the groups worked on projects to improve radiation protection: "the management of the radioactive safety of children by mothers," "the management of the radiological quality and marketing of meat," "the production of a video film by young people," "the development of a practical pedagogy for life in contaminated areas," "waste management and the preservation of the surrounding environment" (Rolevitch et al., 1999).

Several Belarusian scientific institutes were involved in the implementation of the second phase of the project, which took place between 1999 and 2001 in five localities in the Stolin district, including Olmany.[4] The objective of the second phase of the project was to create networks of local professionals—nurses and doctors, teachers, *kolkhoz* executives, and radiometrists—to support the dissemination of the approach developed under ETHOS-1. Those networks brought together nearly 80 local professionals (ETHOS, 2001, pp. 16–17).

In line with the earlier work by Mutadis, ETHOS experts insisted that the new approach had to take into account the impossibility of moving people from the contaminated territories. They considered it necessary to define an "acceptable level of protection" that would be sustainable in the long term, improve people's living conditions, and enable individuals to cope with radioactive risks without having to rely constantly on scientific and political authorities (Hériard-Dubreuil et al., 2000; Rigby, 2003).

Thus French and local experts worked closely with local populations and officials to develop and disseminate the "practical radiological culture" that has become the main pillar of ETHOS and subsequently of the CORE program. "Practical radiological culture" referred to a series of practices and rules for reducing one's radiation exposure through modifying family food consumption, farming (maintaining gardening plots and livestock), housekeeping, and recreation. Local residents, for example, had to learn to measure and interpret ambient levels of radioactivity in their homes, gardens, and in different parts of their village. They could also test various foodstuffs, especially privately grown. That testing, along with some basic knowledge of the properties of radioactive elements, would help them make their own decisions about their daily diet, places for recreation, and their farming. They also learned to manage highly radioactive ashes from woodstoves and fireplaces. The main purpose was to give local residents simple tools to protect themselves and reduce their radiation exposure.

Mothers and schoolchildren were among the most involved in the project. For example, mothers could learn how to manage an internal "contamination budget" over a year. The idea was not to prohibit consumption of contaminated products (such as wild mushrooms, the gathering of which is practically a national sport and which can have especially high levels of radioactivity). Rather, local residents—many of whom relied on their plots to support their families—would learn to combine more contaminated food products with relatively "clean" food in such a way as not to "exceed" the contamination budget, established at levels of 20,000 becquerels per year (or 50 becquerels per day) (Hériard-Dubreuil et al., 2000, p. 26–27). Contamination maps of homes, schools, or gardens were another example of a practical tool taught to mothers but also and especially to schoolchildren. The students learned the basics of mapping and measuring ambient radioactivity. They were then asked to create contamination maps of their nearby environments, identifying areas at risk, thus hopefully building a hold on the invisible danger (Pena-Vega, 2004).

The results of the ETHOS project formed the basis of the CORE program, launched in 2003 after a preparation phase, at the initiative of the Belarusian State Committee on Chernobyl. Similar to ETHOS, the CORE program sought to "return the sense of control to local people" and to "improve the living conditions of the inhabitants of selected districts by reaching out to the people themselves, helping them to contribute to formulating specific and common project proposals" (CORE, 2003, p. 2). Coordinators of the CORE program invited residents from four local districts, all with level of contamination at 15–40 curies per square kilometer, to submit project proposals on issues of health, economic development, radiological quality, and the memory of the Chernobyl disaster.

International partners of the CORE program included the United Nations agencies (UNDP, UN OCHA, UNICEF, UNESCO, FAO), OSCE, the World Bank, the European Commission, the Swiss Agency for Cooperation and Development, the European Union and its member countries through their diplomatic representations in Belarus, as well as such European NGOs as the French Association for Control of Radioactivity in the West (ACRO). Eclectic in terms of its participants and projects, the program aimed to provide methodological coordination. CORE's budget was limited to its coordination structures; partners responsible for each solicited project had to find their own funding. There were 146 projects registered under the CORE program with a total budget of €9 million (CORE, 2009, p. 12). The CORE program was completed in 2008.

The European Commission cofinanced a significant number of projects within the program, with the funds mainly committed through the Technical Assistance to the Commonwealth of Independent States (TACIS) program. In 2006, the Commission considered that its cooperation with the people and authorities of Belarus in the field of aid to the regions affected by the Chernobyl disaster could be considered successful (European Commission, n.d.). However, a large portion of the projects financed by the European Union, particularly those integrated into the CORE program, experienced serious difficulties during their implementation, which prompted some collaborating partners to abandon their efforts and return the subsidies received by the Commission. This was the case of the project Health Quality: Contribution to Reduction of Risks, in which Kasperski participated as a local coordinator between April 2005 and October 2006 (Kasperski, 2012).[5] The project was to be implemented as a project of "technical assistance," which would give it local tax exemptions for hiring local personnel or equipment purchases or donations (United Nations Development Project, 2004). Yet the framework of "technical assistance" was hard to enact in practice, and even routine tasks, such as receiving a shipment of donated computers, proved nearly impossible to accomplish in the context of the nontransparent bureaucracy of the Belarusian state.

Assessment of the projects

When international experts involved in ETHOS and CORE defended their "practical radiological culture" approach, they argued that the projects were helping people who had already chosen to live in radiologically contaminated territories. As Jacques Lochard, the head of the CEPN, put it in 2005, "What should we do? . . . If they [local residents] want to live there, it would be irresponsible not to give them the means to improve their living conditions" (as cited in Kempf, 2005). People living in the affected territories accumulated their doses by, for example, consuming wild mushrooms or produce from private gardening plots. ETHOS administrators argued then that, "Since he or she resides in a contaminated territory, it is the result of his or her actions, work as a farmer, behavior, life choices, that make him or her more or less affected" (ETHOS, 2001, p. 14). Educating individuals would enable them to choose practices that lessen radiation exposure. Another noted value was the shift from "a centralized approach to a participatory approach," which involved members of the "civil society" (Nuclear Energy Agency (NEA) 2006, p. 10–11). The projects presented an alternative to the official recommendations of the Belarusian state and its experts, often detached from the daily needs of the local residents.

"Participation" and "choice" were thus defined in contrast to being passive in the face of both the Chernobyl disaster *and* the state. The "practical radiological culture" meant to transform individuals from passive victims of the disaster and passive recipients of information and assistance from the state into informed, independent agents capable of coping with the challenges of living in the affected territories.

For the critics, the creation of the "practical radiological culture" implied that people would have to get over the "inhumanity" of having to adapt to living in the contaminated environment (Dupuy, 2006, p. 17). Furthermore, "participation," "empowerment," and the "choices" that could be made by local residents essentially corresponded to the introduction of neoliberal governance of radiation protection (Topçu, 2013). From this perspective, ETHOS and CORE served to transfer the work of managing chronic contamination to the affected individuals. What looked like assistance was a way to liberate the state from any disaster-related responsibilities. The outcomes would not be good: "[T]he promotion of empowerment in the context of radioactive contamination may imply a new form of abandonment of individuals by the state and the international bodies" (Topçu, 2013, p. 21). One might add that the emphasis on individual choice also sabotages possibilities for communal, democratic deliberation since there is no meaningful demos.

We agree with this critique of the projects. Indeed, "participation" and "empowerment" are suspicious terms in the context of an authoritarian regime (Kasperski, 2020, p. 276–277). Yet simply dismissing the projects as a neoliberal hoax (in the service of the regime) does not do justice to what the projects meant in their particular historical and cultural context in Belarus. By the end of the 1990s, the official media in Belarus barely mentioned Chernobyl; the government had long stopped considering Chernobyl as a primarily radiological problem (Kuchinskaya, 2014). In that context, ETHOS and especially the later and more expansive CORE program made Chernobyl more visible through their sheer operation as Chernobyl-related projects and their own publicity work, such as TV appearances of the head of CORE. Members of CORE, somewhat sheltered by its status as an international project, were more approachable and willing to answer questions than, for example, government officials. Even representatives of the Ministry of Health or the State Committee on Chernobyl recommended one of us to contact CORE to get more information or were more willing to talk about CORE than their own work (Kuchinskaya, 2014, p. 131).

CORE's relative structural openness mattered. Members of CORE were not only willing to describe their work, but they invited Kuchinskaya on their trip to the rural communities in southern Belarus. In personal conversations during that trip, some members of the team described their work as intermediaries between local residents and the international and state bodies. They perceived themselves accountable in front of the local communities: It was a question of looking them in the eye the next time they traveled to that area. And locally CORE provided at least some, though limited tools and communal spaces for articulation of possible Chernobyl-related problems and community needs.

That was not necessarily the goal of either of the projects. Yet local opportunities and effects can emerge even when they are not part of the project objectives; engagement creates opportunities for the kinds of engagement that go beyond the project framework envisioned by the project managers, though such excess of meaning is often sanitized and managed later (Michael, 2012). The local effects of ETHOS and CORE could and, we argue, did in some ways transcend the built-in limitations of the projects.

At the same time, the built-in limitations of the projects were significant, and they mattered in the long term. They also matter for our assessment of the transferability of these participation-based, "practical radiological culture" initiatives to other radiation-affected communities and contexts. The projects emphasized economic needs of the local communities and included limited discussion of radiation risks and no discussion of possible radiation health effects. During the meetings, Kuchinskaya observed on her trip with CORE, radiation levels in produce were mentioned rarely and mainly in the context of growing produce for sale rather than own consumption. Some local members of the CORE team commented to Kuchinskaya that the focus on economic aspects meant that community needs in Chernobyl-affected areas were not much different from needs in rural areas elsewhere.

The projects' emphasis on economic needs and downplaying of radiation protection echoed the position of the Belarusian government—and the international bodies. Established in 2003, the CORE program came in the wake of the United Nations report that framed Chernobyl as primarily a socioeconomic problem and argued that the government could not afford its current policies, including benefits and compensations to the affected populations, "health recuperation" programs, and radioprotective measures (United Nations Development Project & United Nations Children's Fund, 2002). That report followed a series of earlier reports and studies by the UN agencies—the International Atomic Energy Agency (IAEA), the World Health Organization (WHO), and the United Nations Scientific Committee on the Effects of Atomic Radiation (UNSCEAR)—that, to quote historian Kate Brown), "extended a great deal of effort not to know about the effects of the Chernobyl accident, to limit research and to contain judgement" (Brown, 2017, p. 2). The "practical radiological culture" was a Band-Aid kind of approach because the agencies recognized no radiation health effects in the general population (with the exception of the rise of thyroid cancer in children), attributing health problems among the affected populations to psychological or economic causes, and therefore saw no need for more systematic and expansive efforts at radiation protection.

In addition to inheriting a view of post-Chernobyl radiation risks as insignificant, the projects also inherited a narrow view of participation. "Participation" as promoted by the projects was limited, localized, and not meant to be built upon. Residents of select affected communities could propose small actionable projects that would have to fit within the overall scope of CORE, but they could not challenge or affect the framing of the CORE program itself. The projects did not presuppose any sustained knowledge production or collaboration between experts and nonexperts beyond top-down education of the affected populations. There was no feedback loop or mechanism for reflecting on the effectiveness of the projects. Any perspectives that emerged in local meetings were not meant to be documented. They were not meant to affect local, national, or international decision making. In short, the projects made no attempt at engaging the local populations for broader purposes; they appeared to see no point in doing that. In matters of nuclear power, regulatory agencies often see it as imperative that they provide the public with "authoritative and definitive" knowledge (Kuchinskaya, 2014, p. 126; see also, Boudia, 2007; Brown, 2017, 2019; Petryna, 2002; Stephens, 2002). This does not leave the public any other role besides having to be educated.

The projects were also limited in what areas they covered, lasted for only a few years each, and did not attempt any systematic, longer-term solutions. Zoya Trofimchik, the head of CORE, compared establishing "practical radiological culture" to learning to brush one's teeth, something that one learns to do habitually. Yet, as previously mentioned, mitigating one's exposure is more complex than that. Residents of the local rural communities were expected to learn to get the produce tested, follow particular agricultural techniques to reduce accumulation of radionuclides in produce, learn food preparation techniques such as boiling meat in salty water or processing milk with a skimming centrifuge. At the same time, in the words of one local resident, "Farmers never have enough time" (Kuchinskaya, 2014, p. 51). Mitigating radiation exposure is never ending work in a context in which local residents already lack resources. What the projects did not do is facilitate long-term solutions, such as infrastructural support for radiation protection activities or even integrated, sustained education on topics of radiation protection (that, as one local resident pointed out, should really be part of the local school curriculum) (Kuchinskaya, 2014, p. 53).

Finally, the projects also did not account for the broader political and socioeconomic factors that affected people's individual "choices." Two or more decades after Chernobyl, people contributed to their doses primarily by consuming contaminated food from private garden plots or wild mushrooms or game. Yet while people made their own doses, they did so not in circumstances of their own choosing. As Kuchinskaya (2014) argued elsewhere, what appears to be a problem of individual choice is a systematic, structural problem that results in greater radiation exposure for the most economically disadvantaged groups and that requires infrastructural solutions.

In short, the temporary nature of the projects and lack of integrated, infrastructural solutions meant the effects of the projects were limited. Though local independent radiation protection experts—experts from the Independent Institute of Radiation Safety (Belrad)—were optimistic about the CORE program when it started, they noted little improvement when testing local residents with a whole-body counter after the CORE program was completed (Kuchinskaya, 2014). The projects, meant to be short-term, fizzled away. They gave some visibility to the problems of the Chernobyl-affected areas, provided some tools and articulation spaces to select communities, but were also limited by their underlying dismissive conception of radiation health effects and their conception of participation, which did not include mechanisms for either producing communal, infrastructural solutions or challenging the underlying conceptions.

Conclusion

The participatory, "practical radiological culture" approach discussed here had an impact at the international level (Topçu, 2013). Its proponents successfully argued that, for example, the ETHOS project could serve as a foundation for developing strategies for disaster preparedness and postaccident management in other countries, including all European countries. Adapting that model was the aim of the SAGE project, funded by the European Commission in 1998–2002. Five scientific institutes from the United Kingdom, France, Germany, and Belarus cooperated to define "strategies and guidelines for the establishment of a practical radiological culture in the event of long-term radioactive contamination after a nuclear accident" (Lochard et al., 2005, p. 1).

One can find more examples of how it is implemented outside of Belarus and its original context. The SAGE final report noted that the existing emergency infrastructure in European countries focused only on the short term (Lochard et al., 2005, p. 17). In June 2005, to correct that lack of long-term vision, among other things, the French nuclear safety authority (ASN) set up the Steering Committee for the Management of the Post-accident Phase of a Nuclear Accident or Radiological Emergency (CODIRPA) precisely to develop a national doctrine for that case. CODIRPA had a planning group, later a commission, dedicated to the "long-term phase" following an accident. Chaired by Jacques Lochard from 2009 to 2012, that commission focused on "practical radiological culture" with the aim to identify the conditions necessary for the development of such culture among the French population, as well as education and health professionals. CODIRPA experts argued that, "to be effective from the emergency phase, the development of a practical radiation protection culture must be considered now," that is, before a potential accident, in order to allow people to "appropriate" and better understand this complex concept and its principles (CODIRPA, 2011, p. 5).

Training in "practical radiological culture" was also featured as part of PREPARE, an emergency preparedness project involving Mutadis and CEPN experts. The project was motivated by the Fukushima disaster, and it ran in Europe in 2013–2016 with the aim of "closing gaps identified in nuclear and radiological preparedness in Europe" (Schneider et al., 2017, p. 151).

PREPARE included a course, organized in collaboration with the Institute of Radiology in Belarus and "offering the unique opportunity to visit the Chernobyl contaminated areas and the local centers for developing a practical radiological protection culture" and to learn "from the inhabitants and local people engaged in the rehabilitation of living conditions" (Schneider et al., 2017, p. 154–155).

In 2013 Jacques Lochard—an economist specializing in the optimization of radiation protection, the head of CEPN from 1989 to 2016, and one of the promoters of ETHOS and CORE—also became vice chair of the International Commission on Radiological Protection, a powerful expert organization that makes recommendations about radiation protection. In 2011–2015, International Commission on Radiological Protection (ICPR) and the nongovernmental organization, Radiation Safety Forum Japan, initiated a series of 12 "dialogue meetings" with the local stakeholders in Fukushima Prefecture to identify the challenges related to the "rehabilitation of living conditions" in the affected territories (Tanigawa et al., 2016, p. 93S). The experts draw explicit connections to the ETHOS and CORE projects in Belarus and pointed out the importance of participation: "Involvement and empowerment of the affected populations is [sic] a way to provide them with the necessary elements to make informed decisions" (Gariel et al., 2018, p. 254).

One of the participants in the ICRP dialogue meetings was Ryoko Ando. Ando and a small study group of citizens from Iwaki and Suetsugi in the southern Fukushima found references to the ETHOS project in Belarus when browsing ICPR publications online in late 2011. They initiated contact with Jacques Lochard and ICPR, seeking similar assistance (Ando, 2016). They also created the website *Ethos in Fukushima* (ethos-fukushima.blogspot.com), which included Japanese translations of documents explaining the ETHOS approach and then video recordings of the ICRP dialogue meetings (Ando, 2016). While ICPR did not provide equipment for the Suetsugi citizens, their association with ETHOS "gave them something that a poor and depopulated rural region did not have: visibility" (Polleri, 2019, p. 220), and that visibility helped secure further funding and resources. Maxime Polleri's (2019) analysis suggests that, in the end, the collaboration reinforced the focus on individual responsibility and fell in line with the official discourse of normalizing life in the affected areas (see also Kimura, 2018).

The analysis of what the ETHOS and CORE projects meant in Belarus and what this approach could mean in other contexts highlights the following irony. On the one hand, it is important to recognize the value of the projects locally. The projects, especially CORE, sustained attention on the issues of Chernobyl in Belarus, even if in a limited way and only for a limited time. They also provided (limited) tools and communal spaces for articulating Chernobyl-related issues, including radiation protection issues on the level of local communities. Though the model of participation employed by the projects was very limited, they still created spaces that were relatively structurally open and that provided some opportunities for the articulation of issues and production of relevant knowledge. On the other hand, the projects were framed narrowly, and they did not allow for the development of infrastructural solutions, which, we argued, were needed. The emphasis on the individual choice at the expense of communal infrastructural solutions is problematic, especially given the multiple attempts to promote and expand this approach to other contexts and countries.

Furthermore, the projects lasted only several years each, and they focused on a very limited number of local communities. It is a drop in the bucket if one considers all similarly affected communities in Russia, Ukraine, and Belarus. Yet the positive results of the projects are presented in their final reports as if speaking for all the populations living in the contaminated territories. Estimations of the costs of making these projects sustainable are nowhere to be found. Instead, the limited local experience with establishing "practical radiological culture" is heralded as proof that it is possible and even acceptable, from the point of view of public health, for people to live in the radiologically contaminated environment.

Finally, we return to the question of "participation" and participatory forums as means for knowledge production in the context of energy democracy. While, as Kimura and Kinchy (2016) reminded us, more participation in itself does not guarantee that the communities will be able to challenge authorities or drive policy change, the actual procedures of participation—that is, the design of the project, including the timing and intensity of collaboration between experts and various publics—also matter (Callon et al., 2001/2009). In this case, participation appears to be limited by design. It is exactly the form of participation criticized by scholars as unlikely to benefit the publics: self-help, localized projects that solicit perspectives of local publics but do not allow for any impact on decision making and effectively leave local communities with their problems (Callon et al., 2001/2009, p. 188; Kimura, 2016; Kimura, 2018; Kimura & Kinchy, 2016, p. 348; Topçu, 2013). This fits a broader pattern: Nuclear agencies often see public perception of nuclear issues, including issues of radiation protection, as a matter of literacy, whereby to be educated is to agree with the regulatory agencies and their experts. Given that underlying assumption, including the public earlier and in more integrated ways appears to be unnecessary, even makes no sense, and might create political threats to the legitimacy of the nuclear industry. The underlying model of expertise and knowledge production puts nuclear energy at odds with the ideas of energy democracy (Feldpausch-Parker et al., 2019).

On the broader conceptual level, we argue that the discussion of energy democracy should pay attention to more democratic models of knowledge production, not just more democratic energy-related decision making. Participatory knowledge production is crucial for empowering local communities and ensuring the production of knowledge that is reflective of the needs of these communities. These questions of power and justice are central to the framework of energy democracy and the democratization of energy transition (Burke & Stephens, 2017; Feldpausch-Parker et al., 2019; Sovacool & Dworkin, 2014, 2015). Our view of participation is aligned with other approaches to participation and energy democracy that are informed by science and technology studies and that emphasize participation as an emergent, context-dependent phenomenon (e.g., Chilvers & Longhurst, 2016).

More specifically, we argue that more democratic forms of knowledge production require, first, tools and communal spaces for articulating key issues and concerns and, second, participation formats and broader social conditions that allow for engaging with underlying assumptions and conceptual approaches. The approach of participatory, "practical radiological culture" designed by international nuclear organizations for postaccident management limits the first and precludes the second. The key issue appears to be not participation but underlying assumptions among experts and policy makers about whether the public can actually contribute to knowledge production for any energy technology system, including nuclear systems.

Notes

1 The name of this program appears in project documentation in all capital letters; it is not an acronym.
2 On the history of international nuclear experts' studies of psychological effects of radiation exposure, see (Kuchinskaya, 2014, pp. 122–126).
3 CEPN's current three members are the French public electricity generating utility that operates French civil nuclear reactors (EDF), the Institute of Radiological Protection and Nuclear Safety (IRSN), and the French Alternatives Energies and Atomic Energy Commission (CEA). Previously it also included the French nuclear technology company AREVA.
4 Those included the Pinsk Regional Centre of the Radiological Research Institute, the Belarusian Research Institute for Soil Science and Agrochemistry, and the Brest State University.

5 The project was to be implemented in the Chechersk district with the goal of improving health of children in the district (as measured by routine checkups) and the local communities' access to information about radiation protection and to the system of monitoring radioactivity in food and the environment. The project was coordinated by the Institute for Radiological Protection and Nuclear Safety (IRSN), a French public body, with participation of French NGOs Médecins du Monde (Doctors of the World) and ACRO, the Belarusian NGO Institute for Radiation Safety "Belrad," and several Belarusian public health and radiological safety institutions.

References

Ando, R. (2016). Ethos in Fukushima and the ICRP dialogue seminars. *Annals of the ICPR*, *45*(2), 135–140. https://doi.org/10.1177/0146645316666759

Boudia, S. (2007). Global regulation: Controlling and accepting radioactivity risks. *History and Technology: An International Journal*, *23*(4), 389–406. https://doi.org/10.1080/07341510701527443

Brown, K. (2017). Blinkered science: Why we know so little about Chernobyl's health effects. *Culture, Theory and Critique*, *4*(58), 413–434. https://doi.org/10.1080/14735784.2017.1358099

Brown, K. (2019). *Manual for survival: A Chernobyl guide to the future*. W. W. Norton & Company.

Burke, M. J., & Stephens, J. C. (2017). Energy democracy: Goals and policy instruments for sociotechnical transitions. *Energy Research & Social Science*, *33*, 35–48. https://doi.org/10.1016/j.erss.2017.09.024

Burke, M. J., & Stephens, J. C. (2018). Political power and renewable energy futures: A critical review. *Energy Research & Social Science*, *35*, 78–93. http://doi.org/10.1016/j.erss.2017.10.018

Callon, M., Lascoumes, P., & Barthe, Y. (2009/2001). *Acting in an uncertain world: An essay on technological democracy* (G. Burchell, Trans.). MIT Press.

Chilvers, J., & Longhurst, N. (2016). Participation in transition(s): Reconceiving public engagements in energy transitions as co-produced, emergent and diverse. *Journal of Environmental Policy & Planning*, *18*(5), 585–607. https://doi.org/10.1080/1523908X.2015.1110483

Comité Directeur pour la gestion de la phase post-accidentelle d'un accident nucléaire ou d'une situation radiologique (CODIRPA). (2011, December). *Rapport du groupe de travail "Culture pratique de radioprotection en situation post-accidentelle"*. https://inis.iaea.org/collection/NCLCollectionStore/_Public/49/053/49053298.pdf?r=1&r=1

Cooperation for Rehabilitation (CORE). (2003). *Declaration of principles of the CORE programme "co-operation for re-habilitation of living conditions in Chernobyl affected areas in Belarus"*. CORE.

Cooperation for Rehabilitation (CORE). (2009). *Sotrudnichestvo dlia reabilitatsii uslovii zhizni v postradavshikh ot Chernobyl'skoi katastrofy raionakh Belarusi: 2004–2008*. Institut Radiologii.

Dupuy, J. P. (2006). *Retour de Tchernobyl: Journal d'un homme en colère*. Seuil.

ETHOS. (2001, Novembre 15–16). *La rehabilitation des conditions de vie dans les territoires contaminés par l'accident de Tchernobyl: La contribution de l'approche ETHOS. Actes du Séminaire international, Stolyn*. http://ethos.cepn.asso.fr/data/documents/stolyn/Actes_Stolyn_Verbatim.pdf

European Commission. (n.d.). *European neighbourhood and partnership instrument*. Country Strategy Paper 2007–2013 and National Indicative Programme 2007–2011. http://eeas.europa.eu/archives/docs/enp/pdf/pdf/country/enpi_csp_nip_belarus_en.pdf

Feldpausch-Parker, A. M., Endres, D., & Peterson, T. R. (2019). Editorial: A research agenda for energy democracy. *Frontiers in Communication*, *4*(53). https://doi.org/10.3389/fcomm.2019.00053

Gariel, J. C., Rollinger, F., & Schneider, T. (2018). The role of experts in postaccident recovery: Lessons learnt from Chernobyl and Fukushima. *Annals of the ICPR*, *47*(3–4), 254–259. https://doi.org/10.1177/0146645318756821

Girard, P., & Hériard-Dubreuil, G. (1996). Stress in accident and post-accident management at Chernobyl. *Journal of Radiological Protection*, *16*(3), 167–180. https://doi.org/10.1088/0952-4746/16/3/003

Hériard-Dubreuil, G. (1994). Un premier bilan des effets psychiques et sociaux de l'accident de Tchernobyl. *Radioprotection*, *29*(3), 363–376. https://doi.org/10.1051/radiopro/1994008

Hériard-Dubreuil, G., & Girard, P. (1997). Conditions de vie dans les territoires contaminés en Biélorussie 8 ans après l'accident de Tchernobyl. *Radioprotection*, *32*(2), 209–228. https://doi.org/10.1051/radiopro:1997107

Hériard-Dubreuil, G., Lochard, J., Girard, P., Guyonnet, J. F., Le Cardinal, G., Lepicard, S., Ollagnon, H., Pupin, V., Rigby, J., & Schneider, T. (2000). *The ETHOS project in Belarus 1996–1998: Synthesis of the*

major outcomes of the ETHOS research project on the rehabilitation of living conditions in contaminated territories affected by the Chernobyl accident. Mutadis.
International Atomic Energy Agency. (2006). *Environmental consequences of the Chernobyl accident and their remediation: Twenty years of experience*. International Atomic Energy Agency.
Irwin, A. (2015). Citizen science and scientific citizenship: Same words, different meanings? In B. Schiele, J. L. Marec, & P. Baranger (Eds.), *Science communication today—2015* (pp. 29–38). Presses Universitaires de Nancy.
Josephson, P. R. (1999). *Red atom: Russia's nuclear power from Stalin to today*. W.H. Freeman.
Kasperski, T. (2012). *La politique de la mémoire d'une catastrophe nucléaire: Les usages de l'accident de Tchernobyl en Biélorussie (1986–2008)* [Unpublished doctoral dissertation, Sciences Po].
Kasperski, T. (2020). *Les politiques de la radioactivité: Tchernobyl et la mémoire nationale en Biélorussie contemporaine*. Pétra.
Kempf, H. (2005, March 16). Leçons de survie à la radioactivité en Biélorussie. *Le Monde*. www.lemonde.fr/planete/article/2005/03/16/lecons-de-survie-a-la-radioactivite-en-bielorussie_401816_3244.html
Kimura, A. H. (2016). *Radiation brain moms and citizen scientists: The gender politics of food contamination after Fukushima*. Duke University Press.
Kimura, A. H. (2018). Fukushima ETHOS: Post-disaster risk communication, affect, and shifting risks. *Science as Culture*, *27*(1), 98–117. https://doi.org/10.1080/09505431.2017.1325458
Kimura, A. H., & Kinchy, A. (2016). Citizen science: Probing the virtues and contexts of participatory research. *Engaging Science, Technology, and Society*, *2*, 331–361. https://doi.org/10.17351/ests2016.99
Kinsella, W. J. (2004). Public expertise: A foundation for citizen participation in energy and environmental decisions. In S. P. Depoe, J. W. Delicath, & M. F. A. Elsenbeer (Eds.), *Communication and public participation in environmental decision making* (pp. 83–98). SUNY Press.
Kinsella, W. J. (2012). Environments, risks, and the limits of representation: Examples from nuclear energy and some implications of Fukushima. *Environmental Communication: A Journal of Nature and Culture*, *6*(2), 251–259. https://doi.org/10.1080/17524032.2012.672928
Kinsella, W. J., Andreas, D. C., & Endres, D. (2015). Communicating nuclear power: A programmatic review. *Annals of the International Communication Association*, *39*(1), 277–309. https://doi.org/10.1080/23808985.2015.11679178
Kuchinskaya, O. (2014). *The politics of invisibility: Public knowledge about radiation health effects after Chernobyl*. MIT Press.
Lave, R. (2012). Neoliberalism and the production of environmental knowledge. *Environment and Society: Advances in Research*, *3*(1), 19–38. https://doi.org/10.3167/ares.2012.030103
Lochard, J. (2007). Rehabilitation of living conditions in territories contaminated by the Chernobyl accident: The ETHOS project. *Health Physics*, *93*(5), 522–526. https://doi.org/10.1097/00004032-199501000-00002
Lochard, J., Croüail, P., Bataille, C., Fiedler, I., Voigt, G., Mercer, J., Nisbet, A., Sudas, A., Zaitzev, A., Zhukovskaya, L., Nesterenko, V. B., & Nesterenko, A. V. (2005). *Strategies and guidance for establishing a practical radiation protection culture in Europe in case of long-term radioactive contamination after a nuclear accident: Final report*. CEPN. www.iaea.org/inis/collection/NCLCollectionStore/_Public/43/009/43009221.pdf
Lochard, J., & Prêtre, S. (1995). Return to normality after a radiological emergency. *Health Physics*, *68*(1), 21–26. https://doi.org/10.1097/00004032-199501000-00002
Medvedev, Z. (1990). *The legacy of Chernobyl*. Blackwell.
Michael, M. (2012). "What are we busy doing?": Engaging the idiot. *Science Technology and Human Values*, *37*(5), 528–554. https://doi.org/10.1177/0162243911428624
Nuclear Energy Agency of the Organization for Economic Cooperation and Development. (2006). *Société civile et radioprotection: Les enseignements de Tchernobyl, 20 ans après. Un rapport du Comité de la radioprotection et de la santé publique (CRPPH)*. OECD. www.oecd-nea.org/rp/reports/2006/nea6171-tchernobyl.pdf
Pena-Vega, A. (2004). Le silence des cartes. Regards ethno-cartographiques sur l'accident de Tchernobyl. *Ethnologie Française*, *xxxiv*(4), 617–626. https://doi.org/10.3917/ethn.044.0617
Petryna, A. (2002). *Life exposed: Biological citizens after Chernobyl*. Princeton University Press.
Polleri, M. (2019). Conflictual collaboration: Citizen science and the governance of radioactive contamination after the Fukushima nuclear disaster. *American Ethnologist*, *46*(2), 214–226. https://doi.org/10.1111/amet.12763

Rigby, J. (2003). *Principes et processus à l'œuvre dans un projet d'amélioration de conditions de vie dans les territoires contaminées par la catastrophe de Tchernobyl-ETHOS I (1996–1998)* [Unpublished doctoral dissertation, University of Technology of Compiègne].

Rolevitch, I., Pachkievitch, V., Petroviet, V., Lepicard, S., Livolsi, P., Lochard, J., Schneider, T., Ollagnon, H., Pupin, V., Hériard-Dubreuil, G., Girard, G., Guyonnet, J. F., Le Cardinal, G., Monroy, M., Pena-Vega, A., & Rigby, J. (1999, April). Une démarche de réhabilitation des conditions de vie dans les territoires contaminés après Tchernobyl: Le projet ETHOS en Belarus. *Annales des Mines*, 5–11.

Schmid, S. D. (2015). *Producing power: The pre-Chernobyl history of the soviet nuclear industry*. MIT Press.

Schneider, T., Gering, F., Charron, S., Zhelezniak, M., Andronopoulos, S., Heriard-Dubreuil, G., Camps, J., & Raskob, W. (2017). Nuclear and radiological preparedness: The achievements of the European research project PREPARE . *Radiation Protection Dosimetry*, *173*(1–3), 151–156. https://doi.org/10.1093/rpd/ncw318

Sovacool, B. K., & Dworkin, M. H. (2014). *Global energy justice. Problems, principles, and practices*. Cambridge University Press.

Sovacool, B. K., & Dworkin, M. H. (2015). Energy justice: Conceptual insights and practical applications. *Applied Energy*, *142*, 435–444. https://doi.org/10.1016/j.apenergy.2015.01.002

Stephens, S. (2002). Bounding uncertainty: The post-Chernobyl culture of radiological protection experts. In S. M. Hoffman & A. Oliver-Smith (Eds.), *Catastrophe & culture: The anthropology of disaster* (pp. 91–112). School of American Research Press.

Tanigawa, K., Lochard, J., Abdel-Wahab, M., & Crick, M. J. (2016). Roles and activities of international organizations after the Fukushima accident. *Asia Pacific Journal of Public Health*, *29*(S2), 90S–98S. https://doi.org/10.1177/1010539516675699

Topçu, S. (2013). Chernobyl empowerment? Exporting 'participatory governance' to contaminated territories. In S. Boudia & N. Jas (Eds.), *Toxic world: Toxicants, health and regulation since 1945* (pp. 135–158). Pickering and Chatto.

United Nations Development Project. (2004). *International technical assistance to the republic of Belarus: Questions and answers*. Iunipak.

United Nations Development Project, & United Nations Children's Fund. (2002). *The human consequences of the Chernobyl nuclear accident: A strategy for recovery*. A Report Commissioned by UNDP and UNICEF with the support of UN-OCHA and WHO. www.by.undp.org/content/dam/belarus/docs/report-2002.pdf

Winner, L. (1980). Do artifacts have politics? *Daedalus*, *109*(1), 121–136.

28

A FRACKED SOCIETY

Multistate media analysis of hydraulic fracturing in the United States

Andrea M. Feldpausch-Parker and Katelind Batill-Bigler

Introduction

The political and production-based power held by the fossil fuel industry in the global energy sector, to many, is old news. What is relatively new, is publics' willingness to accept an old energy source being captured with new technologies in the face of a climate and social crisis where green energy sources are the new black. These international debates are occurring during a time of increased demand for public engagement in energy infrastructure decision making, sentiments commonly linked to the energy democracy movement. This movement, according to Burke and Stephens (2017), is a stakeholder call to action and a justification for "integrations of policies linking social justice and economic equity with renewable energy transitions" (p. 35). Though earlier scholars (see Giancatarino, 2012) defined energy democracy as a means to tackle climate and energy issues within a local grassroots context, energy democracy can be achieved at multiple scales from local to global, as noted in Part I of this book. Regardless, definitions across the literature point to a need to democratize energy systems at multiple scales and push toward just transitions to renewable energy sources as a means to address climate change and social equitability of energy production and consumption (Feldpausch-Parker et al., 2019).

In regard to global production and consumption, energy availability and geopolitical concerns over energy security are growing, and the development and deployment of emerging energy technologies have taken center stage. A modern extraction technology commonly known as hydraulic fracturing (HF), also referred to as hydrofracking or fracking, has allowed natural gas and oil production to become a potential solution for meeting increasing energy demands in the United States and Canada in North America as well as a handful of countries across Asia, Europe, and South America. In the United States, this technology has been linked extensively to energy independence from foreign producers and most recently to political rhetoric of Energy Dominance (Schneider & Peeples, 2018), with the United States having the second largest quantity of recoverable shale oil resources (behind Russia) and fourth largest quantity of shale gas resources globally (behind China, Argentina, and Algeria) (US EIA, 2014). Though HF is not a new technology—it has been widely used by the energy industry for more than 60 years to extract natural gas and oil from shale rock (Willie, 2011)—the use of high-volume HF coupled with horizontal drilling is new. During this process, vast amounts of water (from which the term "high-volume" is derived), combined with sand and chemicals,

DOI: 10.4324/9780429402302-33

are injected thousands of feet into the ground at high pressures to fracture rock in the oil- and gas-producing formations. Sand, or another "proppant," keeps the fractures open and allows the oil or gas to rise to the wellhead (US EPA, 2012). This technology has provided access to deep, abundant natural gas deposits in formations like the Barnett Shale play in northern Texas and the Marcellus Shale in the Northeastern United States. In this study we focus on high-volume HF solely in relation to natural gas extraction in the United States.

Controversy surrounding this technology has erupted around HF across the world. This is a technology, as Morrone et al. (2015) state, that "has the potential to fracture more than just shale; it may fracture communities as well" (p. 207). Even United Nations development experts have partially come out against the technology saying that fracking for natural gas specifically "produces cleaner energy than oil and coal, but it is not necessarily in the best interests of the world's poorest countries" (UN News, 2018). In the United States, deployment of the technology has offered economic stimulus for struggling rural areas in states such as Pennsylvania and has brought heavy industrial activity to urban areas in Texas. These conflicting impacts have captured the attention of researchers interested in public perceptions of HF (see Brasier et al., 2011; Boudet et al., 2014; Evensen et al., 2013; Theodori, 2009), finding mixed reactions among the public relative to location, degree of HF development, and a region's past experience with extraction technologies. Hydraulic fracturing has also piqued the interest of news media outlets that have the ability to further impact public perceptions of the extraction strategy. In this chapter, we investigate how regional newspapers portray HF at the state level as an indicator of public and elite discourse about the technology and its impact on communities where it is in practice. We focus on individual states, in part because the energy technology has historically been regulated at the state level. The ongoing controversy around HF has fueled a surge of media coverage that influences both emerging and continuing conversations surrounding not only the debate about fracking but also efforts to move away from fossil fuels overall.

In many regions, environmental hazards related to HF have sparked public concern and, in some cases, vehement opposition to HF and the natural gas industry. As Allen et al. (2019) note, "[R]esisting the power of large multinational fossil fuel energy companies that exacerbate inequities and disparities in energy" as well as "reclaiming the energy sector with more community and public control to redistribute benefits and risks" are two of the three reasons they identify as driving energy democracy and energy justice movements (p. 1). Claims of drinking water contamination, increased road traffic, and even seismic activity have been linked to HF in several US states (Anderson & Theodori, 2009; Christopherson & Rightor, 2011; Evensen et al., 2013; Wiseman, 2009). In response, calls for strict regulations have arisen across the country. Presently, shale gas accounts for approximately 23% of natural gas production in the United States and is estimated to grow to 49% by 2035 (US EIA, 2012). Given the projected rise in relevance to the country's energy portfolio, it is critical to identify and understand public attitudes; as in the case of other emerging technologies, public attitudes can affect related policy and regulation, as well as the viability of the technology itself (Boudet et al., 2014).

Numerous studies have been conducted under the premise that the news media create a public sphere for information transmission and debate and therefore serve as an adequate conveyance to capture and analyze public discourse surrounding an issue (Habermas, 1989; Hoffman & Slater, 2007). Carvalho (2007) also asserts that as "a forum for the discourses of others and a speaker in their own right, the media have a key part in the production and transformation of meaning" (p. 224). We characterize the public discourse about this unconventional drilling technology by analyzing HF coverage in newspapers from four states (Texas, Pennsylvania, New York, and Michigan) where HF technology is present at varying levels of deployment. The analysis identifies the frequency and article type of HF coverage and differences between newspapers within and between states; as well as the presence of social system–level dialogue (Stephens et al., 2009) and their perceived risks and benefits related to HF deployment. We begin by briefly reviewing the literature on media analysis studies with a particular focus on risk perceptions and energy democracy.

Public perceptions of fossil fuel production

While HF is considered an unconventional energy technology, its application is directly linked to the extraction and production of oil and natural gas—both of which are traditional fossil fuels. Researchers have long investigated public attitudes about the use of fossil fuels and, more recently, the impact of fossil fuels on the climate. Oil and natural gas extraction consistently demonstrate a unique pattern of expansion and contraction in relation to supply and demand described as boom-bust cycles (Forsyth et al., 2007), with a mixed bag of economic and social impacts (Jacquet, 2009). As it relates to natural gas drilling, boom-bust discussion has expanded in recent years as HF activity across various states has dramatically increased (Brasier et al., 2011). As Gopalakrishnan and Klaiber (2014) summarized, "Shale exploration is likely to produce both winners and losers in a community" (p. 44). The so-called winners can include individuals who own subsurface mineral rights, as well as the people and municipalities who can benefit from increased tax revenue during drilling activity. The losers generally include homeowners who can experience increased road traffic, water contamination, and lowered property values.

As more states allow the extraction of natural gas using HF, the quantity of research focusing on public perceptions of the technology has also increased. According to a Pew Research Center for the People and the Press (2012) poll, only 26% of Americans polled had heard much about the HF issue, while 37% had heard a little, and another 37% said they had heard nothing at all. Of those who were aware of the technology, 52% supported its implementation, whereas 35% opposed it. Surveys (Boudet et al., 2014; Davis & Fisk, 2014; Jacquet, 2012; Rabe & Borick, 2011; Theodori, 2009; Wynveen, 2011) and interviews (Brasier et al., 2011) have been used to gauge public attitudes, levels of support and/or opposition, as well as perceptions of specific impacts of HF activity and opinions of various policy and regulation options.

The majority of HF perception research has focused on particular regions close to HF activity. The findings, however, are mixed. For example, Rabe and Borick (2011) found through telephone surveys that Pennsylvania residents believe that HF brings more benefits than problems to the state. Yet when Brasier et al. (2011) evaluated public perceptions of HF in the Marcellus Shale through interviews with key informants in four counties (three in Pennsylvania and one in New York State), interviewees reported both positive and negative impacts from HF operations.

Impacts of news media on public perceptions of energy and energy democracy

While few public opinion surveys on environmental issues have incorporated the exact influence of media, three main findings that relate to the issue have been consistent: (1) the public is highly interested in knowing about developing environment/science/technology issues; (2) mass media serve as a primary source of information about such issues; (3) other factors—firsthand experience, local knowledge, and access to direct sources of information—influence just how that information is processed and perceived (Hansen, 2010). Risk communication research suggests media coverage can affect public risk perception and acceptance of emerging technologies through agenda setting (Boudet et al., 2014). Because of the major implications of energy security and availability, related issues tend to be quite visible in daily media, and the archiving nature of newspapers makes them a notable tool serving as a record of ongoing discourse. News media thus serves as a source for researchers investigating public sentiment and discourse about said technologies.

With the goal of understanding how climate change considerations are inserted into conversations around carbon capture and storage technology, Feldpausch-Parker et al. (2013) studied how newspapers in four states captured CCS discourse and which *social function systems* (see Stephens et al., 2009) were emphasized. The analysis allowed for understanding of the evolution and fluctuation of discussion topics over time, related issues of greatest salience, as well as which risks and benefits of CSS were discussed most often in the public sphere. Boyd and Paveglio (2014) also examined CCS technology, finding that media tended to influence public opinions on issues relating to science, policy, and risk. Media analysis of energy technologies has also been used in conjunction with other methods to assess regional differences in wind energy deployment. In an effort to capture the sociopolitical context that affects these variations, Fischlein et al. (2010) conducted interviews with key energy policy stakeholders and also studied related newspaper coverage. Identifying risks and benefits, the authors found significant differences in how wind energy was framed in newspapers; however, among policy stakeholders, the majority generally touted wind as a positive alternative energy option. Stephens et al. (2009) examined wind technology through media analysis in three states, including Texas, finding a correlation between frequency of newspaper articles focused on wind energy in a state and the degree of controversy among state residents around the technology. This observation is supported by Mazur's (1981) argument that the amount of media coverage surrounding a specific scientific technology is related to the level of public opposition to that technology.

To date, analysis of media coverage of HF has been mainly negative, focusing on environmental risks including water contamination (Boudet et al., 2014). In a broader analysis of HF regulation and policy, Davis and Hoffer (2012) found in a content analysis of newspapers in three states with commercial HF operations a general rise in HF media coverage from 2009 to 2011. While over half of the stories published included some reference to the possible risks of HF, the authors also noted the frequent mention of specific local impacts brought on by the drilling technology. Regulatory issues were also handled differently by the newspapers. Evensen et al. (2013) conducted a similar content analysis of newspaper articles published in New York and Pennsylvania in areas close to HF activity. They found that the most frequently mentioned environmental impact of HF was on water quality. Jobs and land leasing revenue were common economic impacts; social impacts—such as noise levels or crime rates—were rarely covered. Environmental and social impacts were framed negatively, while the majority of economic impacts received a positive presentation. The authors also conducted interviews with the journalist who wrote the most articles about HF-related topics. The journalist described coverage as generally reactive to local events such as town hall meetings and indicated that reports were largely a reflection of the public discourse.

In addition to exploring perceptions of energy sources and technologies, media also play a role in the discourses of energy democracy, or, as Chilvers and Pallett (2018) argue, energy democracies, considering that energy systems are dependent upon sociopolitical, technical, and geographic contexts. Not only do articles showcase relevant stakeholders (Augoustinos et al., 2009; Walker et al., 2019), researchers are also able to discover gaps in public and elite discourses, as Antal and Karhunmaa (2018) noted in media coverage of Germany's energy transition efforts. Their study points out a heavy focus in coverage on energy supply and the technoeconomic while giving little attention to energy demand and issues related to democratic practice. This reflects the observations of energy democracy scholars that the technological aspects of energy transitions receive more attention, without giving proper weight to social and cultural considerations that may pose significant barriers to renewable energy adoption (Chilvers & Pallett, 2018; Miller et al., 2013). In this chapter, we demonstrate how energy discourses reflected in news media about HF in fact do mirror discourses present in energy democracy movements, including but not limited to addressing issues of justice and inequalities present in fossil fuel energy extraction processes as well as showcasing the leverage of power by industry as well as other stakeholders seeking to influence decision-making processes. These concepts of justice, power, and participation are reflected in Feldpausch-Parker et al.'s (2019) framework for energy democracy research.

Theoretical framework

Stephens et al. (2008) developed a comprehensive framework for the sociopolitical evaluation of energy deployment (SPEED) systems. It was developed with the awareness that there is an increasing need for a "transition in society's energy infrastructure" (p. 169) and that, despite a growing urgency, the implementation of new energy technologies has been slow. SPEED acknowledges that existing energy systems and emerging technologies are linked and that public perceptions influence deployment decisions.

These concepts resonate strongly with Luhmann's (1989) theory of social systems, which Stephens et al. (2008) determined provided a useful structure for analyzing and categorizing the risks and benefits present in public discourse in relation to energy technology deployment and application (i.e., Stephens et al., 2009). SPEED was adapted from Luhmann's (1989) subsystems for empirical research in order to examine the social-economic factors involved with energy-related decision making. Luhmann (1989) argues that modern industrialized society is constructed of function subsystems that suggest potential societal responses to the existing environment. Subsystems include but are not limited to: economy, science, law, politics, religion, and education. Luhmann asserts that society communicates through these subsystems in response to environmental stresses or conflicts such as anthropogenic climate change. The revised system categories include technical, economic, environmental, health/safety, political, legal, and aesthetic. Within each of those system categories are factors pointing to the risks or benefits related to the deployment of an energy technology, such as HF.

The SPEED framework also integrates the value of state-level analysis as it relates to energy technology development. Stephens et al. (2008) recognize that state-level sociopolitical variation can play a crucial role in determining the success of an emerging technology. Given the geographic and demographic diversity throughout the United States, and the fact that most often state-level governments will be the first to evaluate, permit, and regulate new technologies, understanding state-level differences is worthy of continued attention. By using the SPEED framework to analyze newspaper articles as a proxy for public and elite discourse, we can develop a better understanding of the driving forces behind technology deployment trends and a possible push away from fossil fuels as a dominant source of energy.

Methods

We examined the discourse surrounding HF by analyzing three newspapers published within each of four states: New York, Pennsylvania, Texas, and Michigan. Each state was selected based on its location, level of HF activity, and history of energy extraction technology. Texas has served as the center of the country's oil and gas business and is the state in which high-volume HF was first implemented. In Pennsylvania, HF has been in commercial practice since 2005

(Harper, 2008). The state offers a premier example of both the economic stimulus that is associated with high-volume HF activity, as well as environmental damage. Oil and gas companies have been using traditional, vertical HF in Michigan for the last 60 years (Michigan Department of Environmental Quality, n.d.), and most wells drilled in the recent past incorporate horizontal drilling technology as well. Of the four states, New York is unique, in that it eventually banned high-volume HF after instituting a moratorium on the technology in 2008. In 2011, the New York Department of Conservation (DEC) received approximately 80,000 comments on the issue, an indication of the high level of public involvement during the period of study (DEC, 2015). In December 2014, the state's acting Health Commissioner made public his intention to support a ban on HF in New York, based on the results of a years-long health impact survey of the technology. A statewide ban soon followed. In 2020, the New York governor announced legislation to make this ban permanent.

Articles were selected from the (1) largest regional outlet based on circulation; (2) state capital's newspaper; and (3) the newspaper in the region most affected by HF (Feldpausch-Parker et al., 2013; Table 28.1).

Three academic databases were used to collect newspaper articles: ProQuest, LexisNexis, and Access News Bank. Coverage was analyzed during a five-year period from January 1, 2008, through December 31, 2012. The start date corresponds with the emergence of the HF discussion in New York State. Articles were retrieved based on searches using a combination of the following keywords: "hydraulic fracturing," "hydrofracking," "fracking," and "shale drilling." In order to qualify for inclusion, one or more of these keywords had to appear in either the headline or first 200 words of an article. We implemented this strategy in an attempt to ensure that all articles were directly focused on HF technology or a closely related issue (i.e., disposal of HF wastewater or chemical disclosure of HF fluids). All main news articles, features, editorials, op-eds, and letters to the editor were included to capture all content included in the media discourse.

We conducted a content analysis to identify the presence or absence of perceived indicators of risk and benefit associated with HF. The core of the coding included the codification of the social systems described by the SPEED framework. We developed a codebook in order to ensure accurate and consistent coding throughout the project; the codebook outlines the seven subsystems defined in the SPEED framework and sets out clear guidelines establishing appropriate indicators within each category.

Results

Results are organized into five subsections: (1) frequency, (2/3) presence/absence of categories, and (4/5) presence/absence of risks and benefits.

Table 28.1 Newspapers analyzed by state

Newspaper type	*New York*	*Pennsylvania*	*Texas*	*Michigan*
Highest circulation	*New York Times*	*Philadelphia Inquirer*	*Houston Chronicle*	*Detroit Free Press*
State capital	*Albany Times Union*	*Harrisburg Patriot News*	*Austin American-Statesman*	*Lansing State Journal*
Closest to HF activity	*Syracuse Post-Standard*	*Scranton Times-Tribune*	*Dallas Morning News*	*Hillsdale Daily News*

Publication frequency

While publication frequency varied greatly, all 12 newspapers reported on HF during the five-year period from 2008 to 2013, resulting in a total of 1,815 articles. Newspapers in New York (774) had the highest frequency, followed by Pennsylvania (590), Texas (430), and Michigan (21; Figure 28.1). Differences in frequency also existed within states, where regional newspapers close to HF activity in New York (*The Post-Standard*) and Pennsylvania (*The Times-Tribune*) published the most HF articles. In Texas and Michigan, the highest circulating newspapers (*Houston Chronicle* and *Detroit Free Press*, respectively) reported most on HF in the state.

As Figure 28.1 illustrates, coverage was low and inconsistent across all states during 2008 and 2009. New York continued to show the highest rates of reporting throughout the entire five-year period. This is largely attributed to the number of letters to the editor published in the state (see Figure 28.2), especially in the regional and capital newspapers.

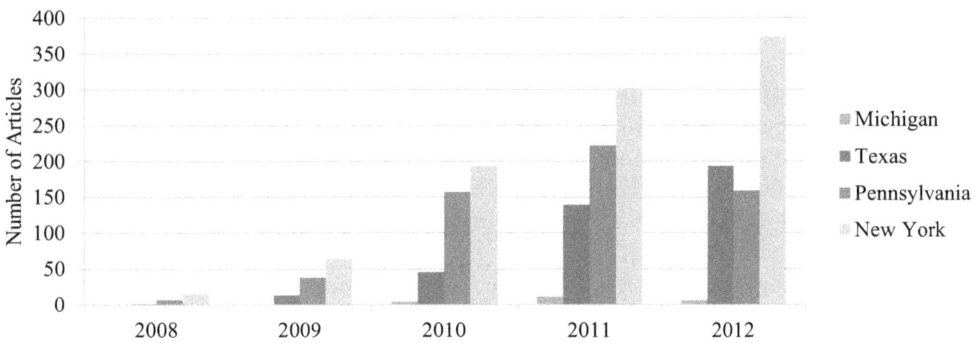

Figure 28.1 Number of articles reporting on HF from January 1, 2008, to December 31, 2012, by state

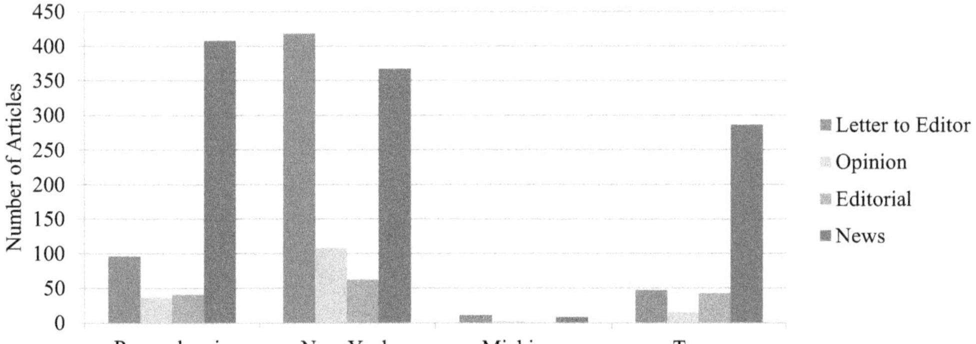

Figure 28.2 Published HF articles by state, reported as number of articles of each article type during five-year period

These letters to the editor offer unique representation of the level of public engagement in certain states.

Presence/absence of categories

Legal discourse appeared most frequently in publications from Pennsylvania and Michigan (24% and 21%, respectively; Figure 28.3). Topics in this category included calls for stricter regulation of HF or the drilling industry (including proposed or passed legislation) and also any incidents of drillers being fined for regulatory violations. This became increasingly common discourse in Pennsylvania and an especially frequent topic in editorials where the editorial staff presented a skeptical opinion of HF.

Environment was the second most frequently coded category in Pennsylvania and Michigan. References to a threat (or a specific case) of water contamination appeared in the majority of these articles. HF wastewater disposal was also an especially common topic in Pennsylvania, where the state's geologic profile prevented construction of injection wells.

In New York and Texas, environmental and legal discussions appeared equally, accounting for approximately 20% of coded occurrences. In New York there was significant coverage of HF moratoriums and bans—and especially whether bans instituted by towns could supersede a state ruling on allowing high-volume HF to move forward in the state. Texas newspapers focused (negatively) on potential regulatory oversight from the Environmental Protection Agency (EPA). In New York, a noteworthy amount of the environmental discourse involved calls for further scientific inquiry into HF and its potential effects. In Texas, coverage focused on events in the Northeast with frequent references to public water concerns. Air pollution was also a major environmental issue in Texas especially in areas of urban drilling.

Political reporting—focused on political support for or opposition to HF activity—was the third most reported category in New York and Pennsylvania. Political activism was also a frequently covered topic and included in this category. In both states, HF supporters and opponents were active participants in demonstrations regarding proposed HF policies.

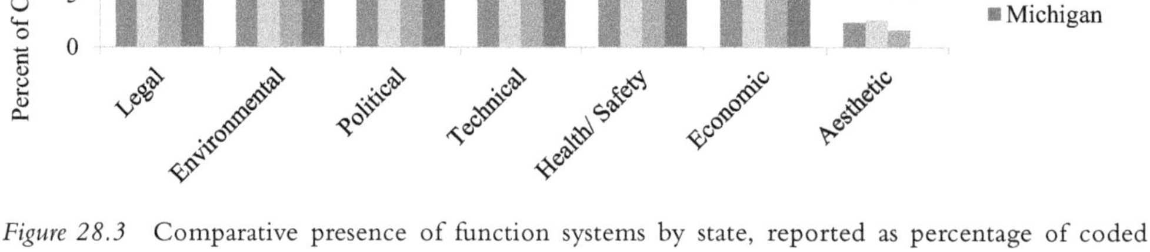

Figure 28.3 Comparative presence of function systems by state, reported as percentage of coded occurrences

In New York, a substantial amount of coverage was focused in later years on the lack of progress by state leaders to decide whether to allow HF within the state. Letters often indicated New Yorkers' skepticism that Gov. Andrew Cuomo would actually rely solely on scientific data to make a decision, as he repeatedly vowed to do in interviews with journalists. Two governors held office in Pennsylvania during the time period studied. While Gov. Ed Rendell was often touted as a proponent of HF and the natural gas industry, he was not subjected to the same level of disapproval as his successor, Gov. Tom Corbett, who was often labeled as a "friend" of the natural gas industry. In Texas, coverage was always clear about the position of the state's administration with respect to HF and energy production. Gov. Rick Perry was often quoted declaring support for the unconventional drilling technique and as having confidence in its safe operation.

In Texas and Michigan, however, coverage of the technical aspects of HF was the third most frequently reported social system. Often, Texas news reports were published about HF technology being "safe" but also emphasized that poor well construction was more likely to lead to water contamination than HF activity itself (both occurrences were coded as technical discussions).

Health/safety issues followed in frequency of coverage for all states. This category was covered most frequently in Pennsylvania (14%) and least frequently in Michigan (7%). By far the most common health/safety-related issue reported on was contamination of drinking water.

Aesthetic discourse was consistently the least represented category. Most of the aesthetic conversation involved concerns about HF noise and fears that rural towns would be turned into "industrial zones" by HF activity.

Presence/absence of risks and benefits

Analysis of the risks and benefits presented in HF discourse in each state reveal mixed sentiment over HF technology but also consistent trends. There are substantially more instances of coded risks than benefits in the legal, political, environmental, health/safety, and aesthetic categories (Figure 28.4).

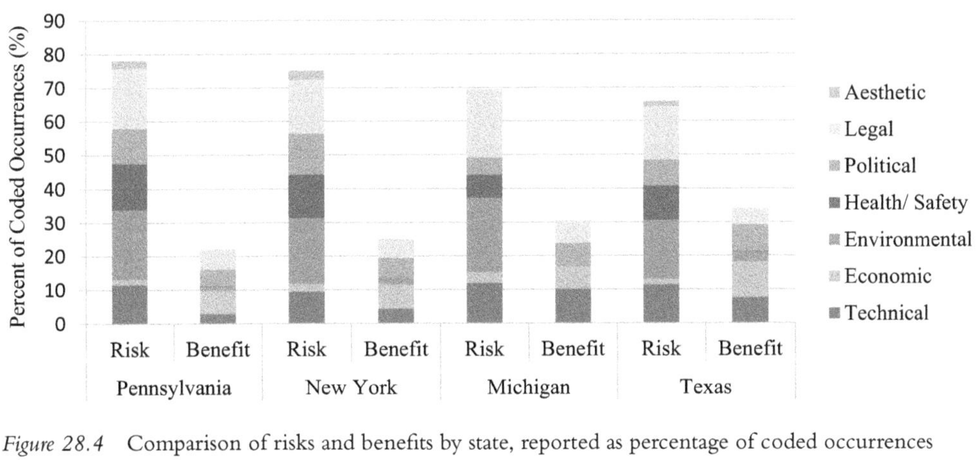

Figure 28.4 Comparison of risks and benefits by state, reported as percentage of coded occurrences

While coded legal and political risks far outnumbered the coded benefits, these two categories saw the most discussion overall. In all four states, the number of articles incorporating legal risks outweighed articles incorporating legal benefits three to one. Again, these risks were predominately focused on HF regulations or oversight. The political category includes a wider range of topics. New York and Pennsylvania consistently reported more risks (68% and 67%, respectively) such as backlash against lawmakers supporting HF activity and reporting of anti-fracking protests. Keep in mind, risk refers to any factor that might undermine the eventual or continued deployment of HF. Political coverage was evenly divided in Texas. In Michigan, however, political benefits actually outweighed risks. These findings were the result of coded benefits about drillers' lobbying efforts and connections made in coverage between HF and achieving energy independence in the United States.

The technical discourse in Texas was also closely divided between coverage of risks and benefits. Discourse included frequent assurances about HF technology being reliable and "safe." HF was also associated with providing a "bridge fuel" (i.e., natural gas) as development continued for future renewable energies. Alternatively, poor well construction and the volume of water required in HF were regularly identified as technological risks.

Economics was the only function system for which newspapers consistently reported more on HF benefits than on risks (Figure 28.4). Texas had the strongest coverage of economic benefits from HF of any of the four states. Related coverage included job creation and especially reinvigoration of supporting and surrounding businesses. Job creation was by far the most often reported economic benefit in New York, in which the majority of any future HF drilling would take place in economically depressed counties along the Pennsylvania border. In Pennsylvania, in addition to jobs, improved business opportunities and increased revenue accompanying Marcellus Shale development were the focus of significant coverage.

Economic risks, however, occurred more frequently in Michigan and New York, where concerns about diminished property values and repair costs for roads worn down by industry trucks were common fodder for discussion. Again, articles were extremely limited in Michigan, and only two articles presented economic risks.

Discussion

This content analysis enabled clarification of many complex and conflicting discussions that have dominated the HF debate in a handful of US states, thus highlighting the perceived benefits and drawbacks of implementing a new technology for fossil fuel energy exploration. Through media coverage, these arguments were placed into the greater public discourse to be considered by various publics (i.e., accepted or dismissed) and used to further conversations of whether to accept and support the technology or protest its implementation through public participation processes and/or informal rallies. Through published letters to the editor and as interviews in news articles, energy stakeholders were also able to engage in a public forum to voice opinions, furthering discussions of the technology. Though how an issue is covered by media does not equate to a general public acceptance or rejection of HF or stand in place of a formal public participation process, it does serve as representations of public and elite sentiments on an issue that can inform future stakeholder participation in decision making. News media serve an important role in showcasing discourses linked to the future of energy infrastructure generally, if not efforts in the practice of energy democracy (e.g., highlighting instances of push-back against powerful interests in attempts to change the trajectory of energy production away from fossil fuels). These news stories and reports thus legitimize energy democracy narratives, though not in name, within the eyes of a more generalized public, further bringing debates surrounding HF into the public sphere for deliberation. In this section, we therefore link our findings to media and energy literatures and determine how it plays into the energy democracy movement and related scholarship.

Publication frequency

Three out of the four states examined in this study presented generous amounts of HF media coverage. Applying Mazur's (1990) quantity of coverage theory (QCT), which argues that the amount of media coverage surrounding a specific technology is directly related to the level of public opposition against that technology, we would be justified in claiming that there is a substantial amount of opposition to HF in New York, Pennsylvania, and even Texas. Such claims are supported by our findings, which show that in all four states and in every category (with the exception of economic), reporting focused much more heavily on the perceived risks of HF than on perceived benefits. Additionally, New York's relatively extreme number of letters to the editor (accounting for 54% of the state's total articles) seem to support survey data collected by Stedman et al. (2012) suggesting that New Yorkers were more likely to engage in public participatory activities related to HF than their Pennsylvania neighbors. This likelihood was later realized due to the record-making levels of public participation in public comment periods and the sheer number of rallies and counter rallies in New York before the decision to ban the technology was made. Though the energy democracy movement was only gaining traction when debates of HF were at their highest in these states, the movement's tenets of democratizing the energy sector through greater participation in energy decision making were already happening in New York.

Again, when applying QCT, Michigan's lack of HF coverage would suggest that there is extremely limited controversy in the state. HF in Michigan is regarded as a well established technology, and high-volume HF has not been distinguished from the conventional, vertical technique by the state's Department of Environmental Quality (DEQ) (Michigan Department of Environmental Quality, n.d.). This may account for the high occurrence of coded technical benefits within the state, which would include claims that the technology is safe, reliable, and sufficiently regulated.

Texas has been developing the Barnett Shale play for natural gas since 2001 (TCEQ, 2012), and yet only one article was published on HF in 2008 and 13 the following year. In fact, the degree of coverage in Texas—especially in the *Houston Chronicle*—seemed to be closely linked to publication frequency in New York and Pennsylvania. This could suggest that the public was not aware of the expanded HF activity in Texas prior to events in the Northeast (i.e., large-scale fracking opposition efforts) or had not associated the negative health effects with HF technology until linkages were made in the Northeast and were heavily circulated throughout news media.

With regard to states' past energy production, it is worth noting that Pennsylvania also has an extensive history. The state has been a longtime producer of coal, a fossil fuel that comes with its own set of risks. Occasionally, opinion articles and letters to the editor made reference to the environmental damage caused by coal mining, and authors demonstrated a desire to avoid a similar fate from HF drilling. This finding resonates with Brasier et al.'s (2011) study of risk perceptions of HF in the Marcellus Shale, which demonstrated a strong tendency for interview subjects to draw parallels between the two extraction technologies. New York has a fairly limited energy production history, though natural gas production is not new to the state.

Presence/absence of social systems

Discussion of legal matters accounted for more news coverage than any other category across all states. This was not unexpected, especially in the Northeast where high-volume HF was considered new and uncertain, and articles supported the idea that establishing sufficient regulations for HF was the states' highest priority when considering the technology. HF was the subject of intense controversy in Pennsylvania and even more so in New York. The level of coverage attributed to this topic area, we believe, is accurately representative of the ongoing concerns about what degree of regulatory control is necessary to ensure safety. The topic of HF regulations has already captured the attention of academics. In his growing work on HF processes in the Midwest, Theodori (2012) assessed the HF regulatory differences between Colorado and Texas. He found that Colorado's regulatory approach was aimed at stronger environmental protection initiatives than in Texas. That analysis mirrors our findings with respect to Texas, where, while concerns were raised within (especially urban communities) about the sufficiency of current safety parameters, most regulatory concerns about environmental quality speculated whether state control of HF would be overtaken by federal regulation, though this never came to fruition.

Environmental indicators were the second most commonly focused area of coverage for Pennsylvania and Michigan and appeared just as frequently as legal topics in New York and Texas. When Stephens et al. (2009) examined wind technology deployment in Texas, they found a lack of emphasis on environmental issues, interpreting that trend at least partly due to the state's long history of energy production as a prominent economic driver. The same tendency was not observed in the present media analysis. In fact, Texas newspapers reported significantly on environmental issues related to HF. In the results, a substantial amount of environmental discourse in the Marcellus Shale region (New York and Pennsylvania) involved concerns about water contamination and references to specific cases of well contamination. The most prominent of such cases occurred in Dimock, Pennsylvania, where a group of residents claimed that rampant HF activity led to their wells being contaminated (McKay et al., 2011). These findings most likely reflect a growing desire by Americans to move away from fossil fuels, in part because of climate change but also because of issues of social justice (Burke & Stephens, 2017). They also align with inequalities associated with issues of energy justice, an aspect of the energy democracy movement which addresses "disparities among socio-economic groups in terms of impacts from energy infrastructure, and access to energy and energy decision-making power" (Allen et al., 2019, p. 2). These inequalities are most evident through the presentation by media outlets of communities suffering disproportionately from the extraction process.

Presence/absence of risks and benefits

The potential risks of HF activity are more apparent in media coverage than the benefits. This is true in all four states. There are some intriguing contradictions when comparing risks and benefits within certain categories, most notably the environmental and economic function systems. Reports of environmental risk were prolific in articles, as well as in letters and opinion articles by HF opponents. As described in earlier sections, environmental risks included reference to water/air pollution, wastewater disposal, earthquakes, and impacts to wildlife. The prominence of public concern regarding water contamination is also present in another media analysis conducted by Evensen et al. (2013). Additionally, articles often cited findings by Howarth and Santoro (2011), which argued that HF contributes to natural gas having a larger carbon footprint than coal. The study's findings were reported in 8 of the 12 newspapers analyzed. Before and after this report was released, however, many HF supporters argued that natural gas is a much cleaner fuel and its expanded use could help the country lower its greenhouse gases levels. In this unique case, HF is a fossil fuel–driven technology marketed as simultaneously environmentally harmful and helpful.

The economic category also presented an interesting juxtaposition. While most often this coverage focused on perceived benefits, it also included economic drawbacks, especially in the Northeast. In New York, many articles cited the extensive job opportunities that the industry has made available in other states. In contrast, dissenting articles suggested these were exaggerated claims and that most jobs would be filled by out-of-state workers. Reporting in New York and Pennsylvania also included the frequent mention of indirect costs to communities located near HF activity. Those costs include road repairs, increased housing rates, and decreased property values. Many opinion articles and letters in New York also presented concerns that the "boom" phase of natural gas production would quickly shift to a "bust," leaving once quaint, rural towns gutted industrial zones, devoid of the people or industry necessary to support the new infrastructure. Texas newspapers were much less likely to emphasize this "bust" phase or other economic risks, with the exception of road repair costs. With the exception of our Texas findings, these patterns seem to reiterate the findings of multiple studies focused on the boom-bust cycles of oil and natural gas extraction and of HF in particular (Gopalakrishnan & Klaiber, 2014; Jacquet, 2009). By this measure, the findings suggest that the news media can offer a true and accurate representation of the documented risks of boom-bust cycles in various regions of the United States due to HF.

As in Fischlein et al.'s (2010) wind power study, we found that Texas discourse included more of both positive and negative political perspectives than other states. More specifically, we found an equal division between the occurrence of political risks and benefits in Texas news coverage. While political coverage in Michigan—another state where HF has been seemingly less controversial—is also balanced, coverage in the Marcellus Shale states profoundly emphasized political *risks*. The majority of these coded occurrences involved dissatisfaction with pro-HF policy makers and also frustration or distrust with the legislative process or the ability of state agencies to sufficiently regulate the drilling industry. Many of these sentiments thus serving as the precursor if not driver for the energy democracy movement.

Conclusion

Media analysis of HF coverage exposed prominent conversations surrounding this controversial energy technology, which has been framed as both a fossil fuel technology and a bridging technology between fossil fuels and renewables. Even so, media coverage revealed inconsistencies that have potentially fueled pubic wariness of the technology, which, in some parts of the United States, eventually led to community/regional bans of HF as well as statewide bans in Vermont (2012), New York (2014), Maryland (2017), and Washington State (2019). This study presents a unique opportunity to examine media portrayal of a technology and how these findings ultimately reflected both later public participation and decision outcomes in the states studied.

From our findings we saw increasing coverage as well as opposition to HF natural gas extraction, with states impacting the dialogue of other states (i.e., Pennsylvania and New York media coverage impacting that of Texas). There were also a high number of letters to the editor represented in our analysis, demonstrating the use of news media as a platform for discourse in the public sphere. Though the energy democracy movement was just ramping up when we had stopped our data collection for this study, it is unsurprising to find similarities between the antifracking movement and that of the larger and more all-encompassing energy democracy movement, most notably fears of environmental and social impacts related to fossil fuel extraction. Other media analyses of renewable energy coverage have shown more favorable portrayals of those technologies (see Feldpausch-Parker et al., 2015; Fischlein et al., 2014).

This analysis offers a useful foothold for further research about media, public perceptions of energy systems, and the energy democracy movement. There are still numerous questions to be answered including determining media's direct influence on public engagement in decision-making processes as well as further investigation into the relationship between states' energy extraction histories (potentially playing a prominent role in Texas and Pennsylvania's acceptance of HF) and receptivity to future energy technology deployment strategies. Additionally, the role of social media is also important in understanding demographic differences in where people seek out information about science and technology as well as how they use such platforms for engaging in the public sphere surrounding energy systems ranging from extraction to consumption. Our findings and the additional questions they pose, however, demonstrate that media have an important role to play in energy democracy.

References

Anderson, B., & Theodori, G. (2009). Local leader's perceptions on energy development in the Barnett shale. *Southern Rural Sociology*, *24*(1), 115–129.

Antal, M., & Karhunmaa, K. (2018). The German energy transition in the British, Finnish and Hungarian news media. *Nature Energy*. https://doi.org/10.1038/s41560-018-0248-3

Allen, E., Lyons, H., & Stephens, J. C. (2019). Women's leadership in renewable transformation, energy justice and democracy: Redistributing power. *Energy Research & Social Science*, *57*, 101233.

Augoustinos, M., Crabb, S., & Shepherd, R. (2009). Genetically modified food in the news: Media representations of GM debate in the UK. *Public Understanding of Science*. https://doi.org/10.1177/0963662508088669

Boudet, H., Clarke, C., Bugden, D., Maibach, E., Roser-Renouf, C., & Leiserowitz, A. (2014). "Fracking" controversy and communication: Using national survey data to understand public perceptions of hydraulic fracturing. *Energy Policy*, *65*, 57–67.

Boyd, A. D., & Paveglio, T. B. (2014). Front page or "buried" beneath the fold? Media coverage of carbon capture and storage. *Public Understanding of Science*, *23*(4), 411–427.

Brasier, K., Filteau, M., McLaughlin, D., Jacquet, J., Stedman, R., & Kelsey, T. (2011). Residents' perceptions of community and environmental impacts from development of natural gas in the Marcellus Shale: A comparison of Pennsylvania and New York cases. *Journal of Rural Sciences*, *26*(1), 32–61.

Burke, M. J., & Stephens, J. C. (2017). Energy democracy: Goals and policy instruments for sociotechnical transitions. *Energy Research & Social Science*, *33*, 35–48.

Carvalho, A. (2007). Ideological cultures and media discourses on scientific knowledge: Re-reading news on climate change. *Public Understanding of Science*, *16*(2), 223–243.

Chilvers, J., & Pallett, H. (2018). Energy democracies and publics in the making: A relational agenda for research and practice. *Frontiers in Communication*. https://doi.org/10.3389/fcomm.2018.00014

Christopherson, S., & Rightor, N. (2011). How should we think about the economic consequences of shale gas drilling? Working paper series: A comprehensive economic impact analysis of natural gas extraction in the Marcellus shale. *Journal or Town & City Management*, *2*(4), 1–20. www.greenchoices.cornell.edu/downloads/development/marcellus/ Marcellus_SC_NR.pdf

Davis, C., & Fisk, J. (2014). Energy abundance or environmental worries? Analyzing public support for fracking in the US. *Review of Policy Research*, *31*(1), 1–16.

Davis, C., & Hoffer, K. (2012). Federalizing energy? Agenda change and the politics of fracking. *Policy Science*, *45*(3), 221–241.

Evensen, D. T., Clarke, C. E., & Stedman, R. C. (2013). A New York or Pennsylvania state of mind: Social representations in newspaper coverage of gas development in the Marcellus Shale. *Journal of Environmental Studies and Sciences*, *4*, 65–77.

Feldpausch-Parker, A. M., Burnham, M., Melnik, M., Callaghan, M. L., & Selfa, T. (2015). News media analysis of carbon capture and storage and biomass: Perceptions and possibilities. *Energies*, *8*, 3058–3074. https://doi.org/10.3390/en8043058

Feldpausch-Parker, A. M., Endres, D., & Peterson, T. R. (2019). Editorial: A research agenda for energy democracy. *Frontiers in Communication*. https://doi.org/10.3389/fcomm.2019.00053

Feldpausch-Parker, A. M., Ragland, C. J., Melnick, L. L., Chaudhry, R., Hall, D. M., Peterson, T. R., Stephens, J. C., & Wilson, E. J. (2013). Spreading the news on carbon capture and storage: A state-level comparison of US media. *Environmental Communication: A Journal of Nature and Culture*, 7(3), 336–354.

Fischlein, M., Feldpausch-Parker, A. M., Peterson, T. R., Stephens, J. C., & Wilson, E. J. (2014). Which way does the wind blow? Analyzing the state context for renewable energy deployment in the United States. *Environmental Policy and Governance*, *24*(3), 169–187.

Fischlein, M., Larson, J., Hall, D., Chaudhry, R., Peterson, T., Stephens, J. C., & Wilson, E. J. (2010). Policy stakeholders and deployment of wind power in the sub-national context: A comparison of four US states. *Energy Policy*, *38*(8), 4429–4439.

Forsyth, C., Luthra, A., & Bankston, W. (2007). Framing perceptions of oil development and social disruption. *The Social Science Journal*, *44*(2), 287–299.

Giancatarino, A. (2012). *Energy democracy: Supporting community innovation*. The Center for Social Inclusion. www.centerforsocialinclusion.org/new-report-energy-democracy-supporting-community-innovation/

Gopalakrishnan, S., & Klaiber, H. A. (2014). Is the shale energy boom a bust for nearby residents? Evidence from housing values in Pennsylvania. *American Journal of Agricultural Economics*, *96*(1), 43–66.

Habermas, J. (1989). *The structural transformation of the public sphere: An inquiry into a category of Bourgeois society* (T. Burger, Trans.). MIT.

Hansen, A. (2010). *Environment, media, and communication*. Routledge.

Harper, J. A. (2008). The Marcellus Shale: An old "new" gas reservoir in Pennsylvania. *Pennsylvania Geology*, *38*(1), 2–15.

Hoffman, L., & Slater, M. (2007). Evaluating public discourse in newspaper opinion articles: Values-framing and integrative complexity in substance and health policy issues. *Journalism and Mass Communication Quarterly*, *84*(1), 58–74.

Howarth, R., & Santoro, R. (2011). Methane and the greenhouse-gas footprint of natural gas from shale formations. *Climatic Change*, *106*, 679–690.

Jacquet, J. (2009). *Energy boomtowns & natural gas: Implications for Marcellus Shale local governments & rural communities* (NERCD Rural Development Paper No. 43). Northeast Regional Center for Rural Development, Pennsylvania State University.

Jacquet, J. (2012). Landowner attitudes toward natural gas and wind farm development in northern Pennsylvania. *Energy Policy*, *50*, 677–688.

Luhmann, N. (1989). *Ecological communication* (J. Bednarz, Jr., Trans). The University of Chicago Press.

Mazur, A. (1981). Media coverage and public opinion on scientific controversies. *Journal of Communications*, *31*(2), 106–115.

Mazur, A. (1990). Nuclear power, chemical hazards, and the quantity of reporting. *Minerva*, *28*, 294–323.

McKay, L. K., Johnson, R. H., & Salita, L. A. (2011). Science and the reasonable development of Marcellus Shale natural gas resources in Pennsylvania and New York. *Energy Law Journal*, *32*(1), 125–143.

Michigan Department of Environmental Quality (DEQ). (n.d.). Office of oil, gas, and minerals. *Five Facts about Hydraulic Fracturing*. www.michigan.gov/documents/deq/Five_facts_about_HF_1_page_455379_7.pdf

Miller, C. A., Iles, A., & Jones, C. F. (2013). The social dimensions of energy transitions. *Science as Culture*, *22*, 135–148. https://doi.org/10.1080/09505431.2013.786989

Morrone, M., Chadwick, A. E., & Kruse, N. (2015). A community divided: Hydraulic fracturing in rural Appalachia. *Journal of Appalachian Studies*, *21*(2), 207–228.

New York Department of Environmental Conservation (DEC). (2015). *Final supplemental generic environmental impact statement on the oil, gas and solution mining regulatory program*. www.dec.ny.gov/docs/materials_minerals_pdf/findingstatehvhf62015.pdf

Pew Research Center for the People and the Press. (2012). *As gas prices pinch, support for oil and gas production grows*. www.people-press.org/2012/03/19/as-gas-prices-pinch-support-for-oil-and-gas-production-grows

Rabe, B. G., & Borick, C. (2011). *Fracking for natural gas: Public opinion on state policy options*. The Center for Local, State and Urban Policy (CLOSUP), Gerald R. Ford School of Public Policy. University of Michigan.

Schneider, J., & Peeples, J. (2018). The energy covenant: Energy dominance and the rhetoric of the aggrieved. *Frontiers in Communication*. https://doi.org/10.3389/fcomm.2018.00005

Stedman, R., Jacquet, J., Filteau, M., Willits, F., Brasier, K., & McLaughlin, D. (2012). Environmental reviews and case studies: Marcellus shale gas development and new boomtown research: Views of New York and Pennsylvania residents. *Environmental Practice*, *14*(4), 382–393.

Stephens, J. C., Rand, G. M., & Melnick, L. L. (2009). Wind energy in US media: A comparative state-level analysis of a critical climate change mitigation technology. *Environmental Communication*, *3*(2), 168–190.

Stephens, J. C., Wilson, E. J., & Peterson, T. R. (2008). Socio-political evaluation of energy deployment (SPEED): An integrated research framework analyzing energy technology deployment. *Technological Forecasting and Social Change*, *75*(8), 1224–1246.

Texas Commission on Environmental Quality (TCEQ). (2012). *Texas in Vanguard of nationwide oil and gas energy boom*. Natural Outlook.

Theodori, G. (2009). Paradoxical perceptions of problems associated with unconventional natural gas development. *Southern Rural Sociology*, *24*(3), 97–117.

Theodori, G. (2012). Public perceptions of the natural gas industry: Data from the Barnett Shale. *Energy Sources, Part B: Economics, Planning, and Policy*, 7(3), 275–281.

UN News. (2018). *Not so fast on fracking, UN agency tells developing countries*. https://news.un.org/en/story/2018/05/1010571

US Energy Information Administration (EIA). (2012). *Annual energy outlook 2012 early release overview*. www.eia.gov/forecasts/aeo/er/pdf/

US Energy Information Administration (EIA). (2014). *Shale oil and shale gas resources are globally abundant*. www.eia.gov/todayinenergy/detail.php?id=14431

US Environmental Protection Agency (US EPA). (2012). *Study of the potential impacts of hydraulic fracturing on drinking water resources: Progress report*. Retrieved February 2, 2013.

Walker, C., Alexander, A., Doucette, M. B., Lewis, D., Neufeld, H. T., Martin, D., Masuda, J., Stefanelli, R., & Castleden, H. (2019). Are the pens working for justice? News media coverage of renewable energy involving indigenous peoples in Canada. *Energy Research & Social Science*. https://doi.org/10.1016/j.erss.2019.101230

Willie, M. (2011). Hydraulic fracturing and "spotty" regulation: Why the federal government should let states control unconventional onshore drilling. *Brigham Young University Law Review*, *5*, 1743–1781.

Wiseman, H. (2009). Untested waters: The rise of hydraulic fracturing in oil and gas production and the need to revisit regulation. *Fordham Environmental Law Review*, *20*, 115.

Wynveen, B. (2011). A thematic analysis of local respondent's perceptions of Barnett shale energy development. *Journal of Rural Sociology*, *26*(1), 8–31.

29
LATIN AMERICAN HYDROPOWER SACRIFICE ZONES

Mary Finley-Brook

Introduction

Across the globe scholars have identified instances where unequal power creates unjust distribution of harm in both renewable and fossil fuel energy (Finley-Brook & Holloman, 2016; Finley-Brook & Jordan, 2019). Large hydroelectric projects in particular spur land dispossession and disproportionately target Indigenous Peoples with criminalization of opposition and repression (Del Bene et al., 2018). When violence occurs, impunity follows, a pattern documented since the infamous Guatemalan Chixoy dam massacres and scorched Earth resettlements in the 1980s (Global Witness, 2019; International Rivers, 2016; Lynch, 2019). Such violence related to hydro conflict can occur in different forms, including structural (i.e., colonialism, internal colonization), territorial (i.e., landgrabbing, privatization of customary territories), and climate (i.e., methane leaks, disinformation) violence. "Slow" violence (*sensu* Nixon, 2010) tied to structural inequality, undergirds "fast" violence in rapid dramatic events, such as protester assault or dam failure. This chapter focuses on the slow violence of structural racism, poverty, and debt that facilitates human rights violations that occur during permitting and the resulting sacrifice zones formed during impoundment and flooding.

Recognizing prior limitations in cross-case analysis for the study of slow violence, this work envisions a comprehensive framework to categorize patterns. While cautioning against generalizations, important trends deserve attention. First, illustrative case studies demonstrate that, without impacted residents risking their lives to call out predatory actors (Lakhani, 2020), there is no documentation of the violence to marginalized populations. Second, since large dams often have unwavering support from top politicians, extreme power differentials result in grave harm for local residents, particularly marginalized groups.

Utility infrastructure planning and development require scrutiny from government oversight agencies, watchdog groups, and stakeholders, especially rate payers (Finley-Brook & Thomas, 2011; Millikan, 2014). An assumption of benevolence on the part of utilities providing services can be common; yet their return on investment (ROI) is often predatory. Limited oversight and closed-door interactions have been known to facilitate corruption. Legislators often provide advantages to utilities; for example, government officials may grant monopoly status or allow quasi-monopolies with little competition among a few large suppliers. The profitability of new construction may encourage unnecessary infrastructure buildout, with rate payers assuming

DOI: 10.4324/9780429402302-34

the costs. Assumptions that energy transmission is critical to the economy and that all projects therefore support essential "public" interests can thus encourage repression of opponents. Self-interested politicians echo and reinforce industry positions, potentially undermining public support for anti-dam efforts (Fearnside, 2017b; Vieira, 2020). Motivated by returns on investment from new construction or by earnings from loan repayment, foreign developers often design excessively large dams that run at partial capacity (Lozano, 2020), and well financed public relations campaigns support industry agendas to construct new dams, even when there are less expensive or greener options.

One reason hydro conflict must be part of the conversation around energy democracy is that Latin America remains unprepared for "just" transition to clean energy, a process requiring independent verification to eliminate coercion and reduce pollutants. *Energy justice*, like broader environmental justice, necessitates attention to *distributive* justice (i.e., fair allocation of costs and benefits), *procedural* justice (i.e., institutionalizing accessible and participatory decision making from planning through implementation, management, waste disposal and decommissioning), and *recognition* justice (i.e., respect for and inclusion of diverse stakeholders, including historically marginalized populations) (Finley-Brook & Holloman, 2016). It also requires the elimination of energy poverty and the disproportional energy burden, where the poor experience the greatest harm.

Included in the justice umbrella of concerns related to energy systems, energy colonization is a spatial process creating high costs in the project area while exporting benefits. Energy megaprojects often cause disproportionate harm to people of color and low-income communities (Del Bene et al., 2018), and pro-dam carbon mitigation schemes often receive foreign investment from areas where energy emissions remain high, instigating criticisms of carbon colonialism. For example, colonialism imposed by big dams causes ecological degradation and resettlement in impounded areas while exporting electricity and profit (Finley-Brook & Thomas, 2011). Some climate change mitigation advocates argue that mega dams are preferable to fossil fuels because they have lower carbon emissions (Erlewein, 2014). This, however, ignores the fact that dam building can release methane, a potent greenhouse gas.

Many Latin American countries face obstacles to energy transition because of inadequate safeguards before construction (Garzón, 2018). For example, unplanned water releases from large dams expose equity gaps in emergency notification and evacuation protocols with the most vulnerable populations not receiving notice or assistance to move from harm's way. Poor households also face intense hardship after damage to crops or homes. As dams burst, or as water is released to prevent dam failure, the dangerous buildup of mercury and heavy metals can poison downstream communities reliant on fishing and agriculture (Pestana et al., 2019).

In addition to community safety and justice concerns, dams are also susceptible to a changing climate. Large dams are vulnerable to climate disruption in various ways: (1) melting alpine glaciers create excess seasonal flows; (2) extreme weather like hurricanes cause temporary flooding; and (3) higher ambient temperatures translate into lower fluvial levels, which means less water flows over dams. Hydroelectricity is insecure for months of the dry season each year in many Latin American locations, and water levels will be substandard for increasingly long periods (Carvajal et al., 2017). Recent dam planning has often relied on out-of-date projections without consideration of climate impacts. Climate disruption also feeds into Latin America's debt treadmill. Before countries even pay off construction loans, hydroelectric facilities become vulnerable to increased uncertainty regarding year-round water levels, leading to recommendations to expand fossil fuel backups, justifying additional loans (Higgins, 2020).

Contributions to research on energy democracies

By corralling water flow and creating industrial landscapes, large hydro projects appeal to technocratic and ethnocentric worldviews (Atkins, 2017, 2019; Teräväinen, 2019; Warner et al., 2017) that venerate human control of nature (Boelens et al., 2019; Nixon, 2010; Sneddon, 2015). Energy democracy research connects theory and praxis to move beyond oppressive and destructive relationships with Earth and other humans (Feldpausch-Parker et al., 2019). Sustainable energy movements reject patterns of "accumulation by dispossession" (*sensu* Harvey, 2005) promulgated by powerful, moneyed interests.

Sustainable energy transition requires transparency and accountability. This can be facilitated by cross-case analysis of hydropower conflicts by (1) exposing persistent cracks in institutional architecture that encourage corruption and impunity and (2) highlighting the role of multiscale civil society (nongovernmental organizations [NGOs], labor unions, faith-based organizations, charitable foundations, and more) to name and shame for change. Diverse anti-dam coalitions often enable groups living at sites of energy consumption to connect with populations residing in sites of energy production (Fearnside, 2017b; Vieira, 2020).

Despite increased awareness, construction of exploitative, repressive, and even dangerous dams (e.g., likely to collapse, overflow, or fall short of necessary output) continues, with developers now claiming to embrace environmental and social responsibility (International Rivers, 2019). Thus big dams highlight the limitations of market-oriented sustainability commitments, such as certified emission reductions (CERs) in the Clean Development Mechanism (CDM), with which industrialized nations continue to pollute after purchasing pollution offsets tied to dam building in less developed areas (Finley-Brook & Thomas, 2011). Market-based reforms like the Equator Principles, formed by financial institutions who voluntarily committed to social responsibility, extend investor willingness to take harmful risks (Eshet, 2017; Testa et al., 2018). To move beyond vacuous "virtual signaling" and greenwashing, energy research benefits from critical and unconventional investigative methods. These include: (1) sustained effort to hear from people who are being silenced or erased; (2) unflinching commitment to go beyond media or public relations veneer; and (3) willingness to follow the money trail to identify and call out conflicts of interest. The existing concentrations of power and resources in the energy sector suggest a need for cooperative scholarly and grassroots activism to enable equitable transition to renewable energy that is good for people and the planet.

Methods

While small-scale run-of-the-river hydro plants may cause low or mitigatable harm, multidam cascades and large dams with impoundment often create significant negative costs (Athayde et al., 2019; Fearnside, 2009). In the 1990s, my electrical grid's tie to Native struggles seeded awareness of dam displacement: Hydro Quebec's James Bay dams supplied renewable electricity to my Vermont home, yet flooding elevated methyl mercury levels in the system that forced Indigenous Cree villages to stop their consumption of fish and other staple wild foods (Gagnon & Desbiens, 2018). I joined electricity consumers and Cree activists to deter Hydro Quebec's Great Whale expansion in 1994.

My work builds on a tradition of activist scholarship in geography, anthropology, political ecology, environmental justice, and other fields (see Hale, 2001, 2006; Pain, 2003; Tickell, 1995). Although incorporating top-down policy review and bottom-up activism can expose deep contradictions, activist-researchers have the potential to improve both policy and praxis

(Chatterton, 2008; Hale, 2006). Rigorous, transparent, and participatory methodologies pair productively with equity-driven grassroots and NGOs' movements promoting epistemic justice (i.e., fair access to knowledge). Accessible, accurate information is a foundation of energy democracies at all governance scales from local to global.

Energy justice research builds from knowledge coproduction incorporating perspectives from impacted communities in addition to findings from scholarly researchers, state agencies, funders, the private sector, and watchdog organizations. Other methodological considerations include (1) the need for a long view spanning decades, as events from one point in time may not adequately represent broader processes, and (2) identifying important intersections between ecological and social factors. Narrowly assessing rivers in terms of energy maximization abstracts electricity from other uses (Gerlak et al., 2020). For example, dams may energize industry, so local populations experience cumulative impacts, whether from water use or pollution (Table 29.1).

Because place-based understanding is imperative, investigation of energy violence has frequently relied on "deep dives" into particular cases (e.g., Johnston, 2005, 2010; Lakhani, 2020). I utilize four illustrative cases from Brazil, Guatemala, Honduras, and Panama for this analysis. These case studies exhibit (1) reliable peer-reviewed research as well as legal and technical records, which is often missing from rural conflicts, and (2) years of attempts by multiscale (local, national, international) organizations to protect human rights and promote sustainability. It is my hope that prominent examples of hydro failure will attract global attention and spur reform. The next subsections provide relevant concepts and background information important to the case studies that follow later in this chapter.

Collateral damage and green grabbing

Large dams are monuments of inequality even in Western democracies (Sneddon, 2015). Intensification of insecurity following dam construction is common for vulnerable households

Table 29.1 Sample of hydroelectricity's socioecological intersections

Trigger	*Impact*
Reservoirs	• Community displacement • Haven for invasive plant species • Harmful algal blooms
Impoundments	• Declines in migratory fish • Increased methane from decomposing vegetation • Increased mineral concentrations in reservoir (i.e., buildup of methyl mercury) impacting downstream communities after water release
Dewatering	• Fish kills that amplify food insecurity and biodiversity loss
Extreme weather	• Dam bursts or emergency water discharges causing fatalities or property damage
Associated infrastructure (i.e., transmission lines, substations, roads, housing for foreign workers)	• Deforestation or coastal disruption • Marginalized groups lose resource access
Associated expansion of ports or industry, such as mines	• Additional community displacement • Resource and water use competing with local needs • Cumulative emissions and pollution

(Richter et al., 2010; WCD, 2000). Project developers create sacrifice zones (i.e., areas that are nonrecoverable due to irreparable harm), treating host communities as collateral damage. A stated goal is to displace the fewest number per unit of electricity (Ledec & Quintero, 2003), but reservoir development often requires relocation. The line between voluntary and involuntary resettlement blurs when impacted families perceive they have no choice (Jordan, 2008).

In an attempt to obscure their actions, media offices of hydro companies tend to spin tales of stewardship based on indicators such as waste recycling or novel turbine technology. Notably, public relations budgets often surpass the amounts allocated for economic assistance to impacted communities. Human rights violations are justified with development frames premised upon racism and ethnocentrism. For example, in Panama, staff created narratives in which the relocation of Indigenous Peoples became an opportunity to improve housing conditions with modern "sanitary" cement structures for nuclear families (Finley-Brook & Thomas, 2010, 2011). And restrictions on customary subsistence farming were transformed into watershed conservation. Meanwhile, police and military forces imposed spatial and social control to silence Indigenous dissent.

Dam conflict often extends from internal colonization where central governments subordinate ethnic regions through structural violence. In these cases, dam developers grab and privatize customary common property, and regulatory systems fail to account for harm to local subsistence economies. For example, traditional practices like agriculture or building homes and canoes from wood become illicit to protect forests and water resources for hydroelectricity (Finley-Brook & Thomas, 2010, 2011).

Although evidence of widespread harm in the World Commission on Dams (WCD) (2000) report contributed to a shift away from support for large dams (Gerlak et al., 2020), they were repopularized a decade later with promotion of hydroelectricity as a carbon mitigation strategy (Finley-Brook & Thomas, 2010, 2011; Millikan, 2014). The largest funder of large dams, the World Bank, embraced hydro as climate friendly, only later reversing this stance (Gerlak et al., 2020; Siciliano et al., 2019). Dam building can be categorized as green grabbing, or claiming resources through the effective manipulation of ineffective environmental policy when dams are justified to reduce dependency on fossil fuels (Fearnside, 2016, 2017a, 2017b; Finley-Brook & Thomas, 2011; Millikan, 2014). However, dams contribute to climate change as vegetative decomposition following impoundment produces methane, with emissions particularly high in tropical ecosystems (Fearnside, 2009, 2016). The International Finance Corporation (IFC, 2015)) recommends harvesting timber before flooding, yet firms infrequently extract logs due to either cost or regulatory or legal constraints on felling timber.

Accountability in the hydro sector

Established in 1998, the WCD was a trisectoral network that included members of governments, civil society, and business. Their two-year process established principles and guidelines for future dam building. Vast harm documented in testimonies from impacted communities worldwide provided convincing proof of extensive oppression throughout the hydro sector (WCD, 2000). The commission estimated large dams displaced between 40 and 80 million people and concluded that the majority of large dams failed to fulfill expectations (Dingwerth, 2005). Findings elevated concerns that social and ecological consequences received only marginal consideration during planning. In the international spotlight and faced with the weight of evidence, industry promised to change.

In 2006, the International Hydropower Association (IHA) released sustainability protocols, a series of best practices. These developed into a voluntary Hydropower Sustainability Assessment

Protocol (Hydropower Sustainability Governance Committee, n.d.). Nevertheless, negative impacts continued. With support for large dams slipping among donors, in 2017 IHA promised further improvement (Del Bene et al., 2018). Nonetheless, detailed assessment of project implementation suggests that transparency continues to be selective, and accountability for damage is poor (International Rivers, 2019).

Although the World Bank pulled back its support for large hydro, dam building continues with private and bilateral funds (Siciliano et al., 2019). Since 2015, a majority of international big dam construction occurred with Chinese sponsorship (Gerlak et al., 2020; International Rivers, 2019). An illustrative case discussed later shows how Chinese financing spurred economic leakage from Ecuador with subcontracts for Chinese contractors, supplies, and workforce (Ellis, 2018; Vallejo et al., 2018).

Latin America, like parts of Asia, relies heavily on big hydro, even as climate change makes tropical dams less reliable (Gerlak et al., 2020). Over the past decade, the Brazilian government has been a major supporter of Amazonian hydro development. Recent projects like Belo Monte and Coca Codo Sinclair (CCS), covered later in the chapter, used outdated stream flow data during design, putting the structural integrity of projects at risk (Garzón & Castro, 2018). Deforestation from Amazonian wildfires, now surging with climate change, also threatens water availability. Sustainable hydroelectricity requires recognition of multiple water uses and a new climate normal in watershed management (Almeida et al., 2019); unfortunately, politics and greed often corrupt planning decisions. Although new Latin American dam construction has slowed in recent years, hundreds of hydro projects remain under review in Argentina, Brazil, Ecuador, Peru, and Uruguay (Gerlak et al., 2020). While support for new large dams is less certain, stalled or failed hydro projects have a tendency to resurface (Atkins, 2019; Finley-Brook & Jordan, 2019).

Case study applications

Historic analysis for this chapter starts in Guatemala, where the Chixoy dam demonstrates decades of injustice. This case shows how development agencies are resistant to reform where even minor reparations took decades (Munzer, 2019). Following the historic Chixoy case, three contemporary examples demonstrate ongoing conflict. The most violent dams involve assault, murder, torture, or rape (WCD, 2000; Finley-Brook & Thomas, 2010, 2011; Lakhani, 2020). Hydroelectric energy has a low mortality rate per kilowatt-hour compared to fossil fuels (Conca, 2012); however, nonlethal conflicts create economic hardship and stress. A multidimensional screen (Table 29.2) addresses six areas of violence: physical-emotional, ecological, structural, economic, territorial, and political.

Dam Massacres in Guatemala

Initiated in 1975 during a civil war, Chixoy dam, now called Pueblo Viejo Hydroelectric, remains a major power source for Guatemala. Feasibility studies stated the valley was uninhabited and there was no consultation with the 1,500 Indigenous Maya Achi living in the impact zone (Munzer, 2019). Table 29.3 demonstrates the extent of the subsequent violence. Excluding massacres, similar negative repercussions happened with each example presented in this chapter.

Over the course of this project, Chixoy's original loans of $365 million ballooned to $825 million. The World Bank's evaluation of dam resettlement recorded failure to create social safeguards and gross violations in contractual obligations, yet the bank did not correct this

Table 29.2 Multidimensional violence screen

Type of violence	*Sample indicators*
Physical-emotional	• Murder • Repression with use of force • Torture • Trauma and stress • Death threats
Ecological	• Water scarcity or pollution • Technological hubris foments environmental disasters • Toxic- or high-risk hotspots
Structural	• Racism and ethnic violence • Sexism and gender violence • Rural–urban inequality • Global North–Global South inequality • Internal colonization
Territorial	• Loss of property rights or dispossession • Land maldistribution • Privatization or misappropriation of commons
Economic	• Cost/benefit maldistribution • Livelihood threats • Energy poverty regardless of residential proximity to infrastructure
Political	• Constraints to participation • Misuse of power to advance misinformation • Lack of transparency and accountability

Table 29.3 Chixoy dam violence

Type of Violence	*Example*
Physical-emotional	• Homicide of 445 children, women, and men in five massacres • Intimidation, threats, fear, stress, and trauma • Villages destroyed by displacement
Ecological	• Methane from decomposition of flooded vegetation • Impoundment-altered water cycle • Fish and biodiversity loss
Structural	• Racism • Low-income villages targeted • Debtor nation with high foreign debt
Economic	• Disrupted local livelihoods • Ballooned constructions costs • Energy poverty without access to electricity
Political	• Power and resources not shared between urban and rural areas • Impunity • Slow, partial indemnification of damages

Sources: Colajacomo & Chen (1999); Johnston (2005); Goodland (2010); Lynch (2019); Munzer (2019)

situation when issuing a second loan in 1984. Although there were never World Bank reparations for damages, in 1996 Bank officials announced it had fulfilled its obligations. Rather than take responsibility for Chixoy harm, the Guatemalan government promoted privatization of the facility.

Impacted populations organized with NGOs to demand a response from the state (Aguirre, 2014; Johnston, 2004, 2005, 2010). In 2005, Maya communities petitioned the Inter-American Human Rights Commission (IACHR), but it took until 2012 for the Inter-American Court on Human Rights (IACtHR) to pass down a decision requiring compensation and restoration. Finally, in 2014, following international pressure, the Guatemalan president announced $154.5 million in reparations to 33 dam-impacted communities (Aguirre, 2014). When the state breached this agreement, the World Bank and Inter-American Development Bank made reparations conditional for loans in 2015. Response from the state and from donors continued to be weak and inadequate. In 2019, Maya protested again. In 2020, Chixoy-impacted villages experienced insecurity and neglect during flooding events in the midst of the COVID-19 pandemic.

Contemporary illustrative cases

Analysis of high conflict dams in Panama, Brazil, and Ecuador demonstrates key similarities to the preceding case (Table 29.4). Violence and displacement contributed to controversy, delays, and lawsuits in each case. Developers and international financiers contend that they support ethical operations, yet each project violates human rights.

Green grabbing and Panamanian climate violence

Panama underwent a dam building spree following the onset of the Clean Development Mechanism (CDM) of the United Nations Framework Convention on Climate Change (UNFCCC) (Finley-Brook & Thomas, 2010; Jordan, 2008). Barro Blanco received CDM verification in 2011, even though the consultation process contradicted stakeholder rules (Finley-Brook & Jordan, 2019; Hofbauer, 2017). Construction processes violated international human rights norms as well as Panama's *comarca* law granting self-government rights (Finley-Brook & Jordan, 2019; Pérez et al., 2016).

Table 29.4 Contemporary case studies

Country (cost)	*Dam (build dates)*	*Developer*	*Key donors and investors*
Panama (*US$78.3 million*)	Barro Blanco (*2014–2017*)	Generadora del Istmo, S.A. (GENISA)	Dutch Development Finance Company, Central American Economic Integration Bank, German Investment Corporation
Brazil (*US$7.8 billion*)	Belo Monte (*2011–2018*)	Norte Energia	Brazilian Development Bank, among many others
Ecuador (*US$2.25 billion*)	Coca Coda Sinclair (*2010–2016*)	Sinohydro	China's Export–Import Bank

Police responded to dam opponents blocking the Pan-American Highway in 2012 with tear gas and artillery, injuring protestors (Jordan, 2018). Three protestors later died, including a 16-year-old boy shot at close range by a police officer (Finley-Brook & Jordan, 2019). Police jailed 171 people, who reported physical, verbal, and sexual abuse during custody.

Pushback from civil society and Indigenous leadership in Panama attracted international support, and petitions gathered thousands of signatures. Nevertheless, donors and developers refused to change course, dispersing project financing even after the UN Special Rapporteur for Indigenous Peoples visited the affected area in 2013 and documented that the Indigenous Peoples directly affected by the project had not given free, prior, and informed consent (Jordan, 2018).

Impacted populations filed a complaint with Dutch donors asking the agency to investigate and withdraw funds (SOMO/Both Ends, 2014). They were unsuccessful even though the Dutch development agency FMO should have demonstrated social responsibility as a signatory to the Equator Principles (EPs). The EPs encapsulate a voluntary risk management strategy for the finance agencies that back the majority of the world's large infrastructure projects, like dams. Although their purpose is to promote environmental, social, and governance investing that seeks social and ecological benefits along with positive returns, evidence proliferates that EP market mechanisms do not effectively protect human rights (e.g., BankWatch, 2017, 2018; Eshet, 2017; Hofbauer, 2017).

The Panamanian state failed to adequately address human rights violations tied to Barro Blanco (Jordan, 2018; Pérez et al., 2016). In 2015, national courts temporarily halted construction due to noncompliance with the environmental impact assessment and inadequate agreements with affected communities. Developers paid fines of $775,000 and dam construction resumed, with military and police suppressing opposition. In 2016, police officers shot tear gas and rubber bullets at Indigenous Ngäbe protesting involuntary relocation. The filling of the dam reservoir and forced resettlement occurred soon after (Jordan, 2018).

After years of discussion, the CDM cancelled verification and the right to sell greenhouse gas emissions offsets in 2016, with the dam 95% complete (Pérez et al., 2016). The UN never advocated for reparations for the Ngäbe (Hofbauer, 2017), preferring to focus on improving future operating procedures (personal communication, Panama City, 2014). Grievance mechanisms with EP lenders were also absent. Since social responsibility commitments from funders are voluntary, negatively impacted communities are generally abandoned without clear mechanisms to address damage. BankTrack (2019) lists Barro Blanco as one of several disastrous EP projects, alongside Belo Monte dam in Brazil, the next case study.

Battle for Belo Monte

Developers broke ground in 2011 on Belo Monte, one of the world's largest hydro projects. During early planning, the World Bank encouraged Brazil to fast-track Amazonian dams (Banco Mundial, 2008), a position the Bank would later backtrack. A lack of necessary safeguards became clear during construction. Dozens of national and international campaigns produced petitions criticizing human rights violations and demanding reforms. Belo Monte displaced 20,000 Indigenous Peoples according to the state; Amazonian organizations reported the number of displaced to be much higher at 50,000 (Fearnside, 2017b).

When the Belo Monte reservoir filled in 2015, the dam was a monument to corruption. The Brazilian Development Bank (BNDES) created a loophole to increase its budget to be able fund Belo Monte construction (Latrubesse et al., 2017): the state assumed 77.5% of costs under a consortium called *Norte Energia* (BankTrack, 2017) and lost billions of taxpayers' funds on the megadam (Higgins, 2020).

Belo Monte opponents faced political persecution or retaliation (Fearnside, 2017a, 2017b; Millikan, 2014). The pro-dam lobby functioned as a "hydro-mafia" (Atkins, 2019). Brazil's so-called Car Wash corruption scandal involved many ministries backing Belo Monte. Misuse of BNDES funds and other illegal strategies extended to development projects in more than a dozen countries as far away as Angola; government officials involved in the crime ring subsequently served jail time in multiple countries.

In spite of severe consequences for speaking out and even several suspicious deaths, protests citing negative impacts erupted among local river communities, Indigenous groups, and environmental organizations. Yet tightly controlled pro-dam messages in Brazilian media misinformed the public; official rhetoric asserted that the dam was necessary for national development, that it would lower energy prices and reduce blackouts, and that repercussions for Amazonian tribes and ecosystems were minimal (Fearnside, 2017a, 2017b).

More than 300 lawsuits emerged, several citing failure to provide promised reparations. Brazil's Public Federal Ministry and National Indian Foundation sued dam developers for ethnocide. Nonetheless, the Brazilian state justified the dam as too essential to face constraints from the courts (Millikan, 2014). Officials disrupted advancement of lawsuits in the courts various times with a "security suspension," using legal code held over from the military dictatorship.

While national officials failed to block the controversial project, protests drew international attention (Atkins, 2017, 2019; Fearnside, 2017a, 2017b). Although protests initially garnered support via global events and millions of petition signatures, state actors skillfully crafted counternarratives criticizing international meddling and suggesting outsiders desired Brazilian development to fail as a means to undermine competition (Atkins, 2017).

As with Panama's Barro Blanco dam, voluntary investor commitments to social responsibility proved insufficient to deter finance. In addition to BNDES, three Brazilian banks financed the Belo Monte despite previous commitment to the EPs (BankTrack, 2019). Another EP signatory, the US-based bank JPChase Morgan, supported Belo Monte alongside other international insurers, investment funds and pension funds (BankTrack, 2018). Two agencies of the World Bank Group financed additional loans to BNDES during Belo Monte construction lending additional support to the corrupt agency (Fearnside, 2017a). Controversy over finance to Barro Blanco exposed limitations to EP transparency and risk management as well as the continued lack of effective grievance mechanisms, as occurred earlier in Panama.

In 2018, after the Belo Monte dam was completed, the Brazilian government briefly proclaimed an intention to stop building large dams in the Amazon, stating costs were too high (Branford, 2018). Leaders later reversed this position, and President Jair Bolsonaro (2019–present) ran on a platform of Amazonian development, including large dams. Yet threats to Belo Monte's integrity and functioning are extreme. Uncontrolled fires and rampant deforestation of up to 80% in the Xingu basin from 2018 to 2019 reduced water flow to the point that in 2019 Norte Energia declared a water emergency (Higgins, 2020). Belo Monte's earthen structure was at risk because of low water levels. Meanwhile, the dam failed to provide the electrical supply promised by dam owners.

Dam construction has exacerbated water shortages, putting local inhabitants' subsistence needs and regional agribusinesses in conflict with new industry. For example, one of the largest electricity purchasers is a Canadian gold mine established after the Belo Monte dam dewatered a section of river, thus creating an opportunity to exploit gold deposits previously considered unextractable (Bratman & Dias, 2018). Further, as water scarcity became dire with deforestation and uses for mining, to the point it even threatened the function of the dam, prompting Chinese investors, who built transmission lines associated with Belo Monte, to propose a new fossil fuel plant alongside the hydro facility (Higgins, 2020). Failure to deliver on climate change mitigation promises also plagues the final case, the Coca Coda Sinclair dam in Ecuador.

Coca Coda = dam disappointment

The intent of the Coca Coda Sinclair (CCS) dam, the largest energy project in Ecuador, was to reduce dependence on unsustainable oil development. The state even relocated a pipeline to make way for hydro infrastructure. President Rafael Correa's stated objective was to move away from historic dependence on imperial powers while producing surplus renewable energy for export to neighboring countries (Aidoo et al., 2017; Ellis, 2018; Garzón & Castro, 2018); however, CCS created a debt trap with unfavorable terms (Vallejo et al., 2018). The Ecuadorian government's regulatory stipulation requiring self-financing, which was put in place in order to increase stringency on project finance, backfired as it eliminated all contenders except the Chinese. Without competitors bidding for CCS contracts, Chinese developers put forth a number of requirements privileging their partners in supply, construction, and operation. Ecuador had to renounce standard operating procedures, reinforcing Chinese advantage (Siciliano et al., 2019).

China promotes their involvement in Latin America dams as South–South aid and appeals to nationalist desires in countries seeking neoliberal alternatives (Ellis, 2018; Freeman, 2017; Warner et al., 2017). Ironically, terms of trade with China are often exceedingly exploitative (Ellis, 2018; Purcell & Martinez, 2018). The line between private sector and state is blurry in Chinese projects, and they often foment foreign control with allied Chinese actors churning out a range of profit generation mechanisms (Gerlak et al., 2020). With CCS, a Chinese consortium held 92% ownership and later obtained even more when South American companies that initially held a small portion divested.

CCS benefits leaked from Ecuador (Ellis, 2018), as approximately 75% of the construction materials originated in China (Garzón & Castro, 2018). CCS loans from the Export–Import Bank of China (CHEXIM) were $1.7 billion with an interest rate of 7% over 15 years. Ecuador paid China with oil (Aidoo et al., 2017), which meant China got 80% of Ecuador's oil production in 2018. To achieve that production, Ecuador initiated oil extraction from the previously protected Yasuni National Park (Garzón, 2018; Garzón & Castro, 2018), which had been the focus of an international campaign to pay Ecuador to leave oil in the ground (Finer et al., 2010). Although many believed this unique proposal held promise for mitigating climate change, that promise evaporated when Yasuni was opened for oil extraction to finance CCS construction (Kingsbury et al., 2019).

Chinese developers working abroad tend to use nonbinding requirements for reporting and defer to in-country systems for ex ante and ex post environmental and social impact assessments, which usually have weaker safeguards than protocols used by multilateral lenders (Freeman, 2017; Gallagher & Yuan, 2017). These weak standards can contribute to delays and cost overruns, as occurred in CCS (Garzón & Castro, 2018). After prolonged conflicts during contract negotiations, there were six orders of suspension and a warning of contract termination during construction (Garzón & Castro, 2018).

CCS construction demonstrated unfulfilled promises (Lozano, 2020). Chinese developers promised technology and knowledge transfer to Ecuadorians but did not deliver the necessary software or engineering details (Garzón & Castro, 2018; Vallejo et al., 2018). Poor communication from Chinese project managers riddled regional development. Local inhabitants acquired debt to open hotels and restaurants when they believed workers would need housing and food; subsequently, Chinese developers built camps where most lived and ate. Local inhabitants sought temporary jobs in dam construction to pay debt, only to have to take new loans when they returned home to restock livestock and prepare lands. Poor treatment of workers created unrest during construction, similar to other Chinese projects in Ecuador (Vallejo et al., 2018). There were 92 labor claims filed, as well as civil lawsuits and cases for damages (Garzón & Castro, 2018).

CCS, which moved forward despite an active volcano and previous earthquakes deterring earlier plans to build a dam in this location (Casey & Krauss, 2018), exhibits a critical level of insecurity (Garzón & Castro, 2018; Lozano, 2020). Developers relied on poor environmental impact assessments and out-of-date water data. During and after construction, criticisms surfaced about the use of substandard steel and inadequate welding (Casey & Krauss, 2018; International Rivers, 2019). Thousands of large and small cracks emerging in the wall of the dam after completion created apprehension about the quality of the workmanship and the overall longevity of the project. Design flaws allow an excess of sand and silt to enter machinery that may harm vital equipment over time.

CCS runs at less than half its capacity after causing blackouts when tested at higher production levels (Casey & Krauss, 2018). Annual output from CCS is also constrained by a lack of water. The dam may have zero output during the dry season (Carvajal et al., 2017), requiring seasonal alignment with other electricity sources. This signifies a return to dependence on fossil fuels or additional investments in other forms of renewable energy, even after spending billions of dollars on CCS.

Discussion

The construction of large hydro projects often exemplifies slow violence punctuated with fast violence. In stark contrast to core objectives of energy democratization, these cases demonstrate structural inequality. While dams do not necessarily promote violence, safeguards are insufficient to prevent cases such as those presented here. Problems surface from the earliest phases of planning through implementation, and patterns suggest impediments for responsible future decommissioning, a long-term challenge beyond the scope of this research.

In these cases, developers obscured the shortcomings of hydro, including climate vulnerability. They justified dams as alternatives to fossil energy and alleged support for clean energy and climate mitigation, yet scientists argue that after flooding low-elevation tropical forests, dams like Belo Monte would need to operate for decades at optimal conditions to reduce emissions compared to fossil fuels (Fearnside, 2009). The cases exemplified both climate violence and green grabbing with slow violence. This was epitomized in Ecuador where the CCS dam was justified as a means to end a dependency on oil and in Panama where the Barro Blanco dam was justified through the CDM. Reservoirs also have displaced agroforestry lands and local subsistence uses, such as in Panama. Impoundments create sacrifice zones where total future restoration is unlikely, as in Brazil where villages were flooded and the government was charged with ethnocide. After Belo Monte construction, placement of a gold mine in the course of the Xingu River means restoring water flow would be even more challenging.

Patterns of fast violence extend beyond assassination to occupational violence, as seen in Guatemala with homicides, disappearances, and torture, in Panama with homicides and police brutality, and in Ecuador that saw 16 worker deaths, 13 of which occurred in a tunnel with known structural instability. Chinese development in Ecuador created unsafe, exploitative labor conditions and contributed to discord not only with CCS but also at other Chinese project sites. The Ecuadorian state then used political persecution of labor unions and workers to quell social unrest (Vallejo et al., 2018). Thus, the core practice of energy democracy, where impacted populations contribute to planning and decision making, seldom occurs with large dams.

Findings and recommendations

Three main conclusions from this cross-case analysis suggest recommendations consistent with democratic principles. First, since marginalized ethnic groups continue to experience grave risks, dam developers must be trained in and adhere to protocols for Free, Prior, and Informed Consent (FPIC) found in the United Nations Declaration on the Rights of Indigenous Peoples (UNDRIP) (UN, 2007). UNDRIP was ratified by the countries under assessment and nearly all national governments more than a decade ago. Energy democracy advocates trained in environmental justice, including recognition justice, would be more adept at centering the rights of marginalized groups and amplifying antiracism work. The Jemez Principles for Democratic Organizing notes affected people should speak for themselves (SNEEJ, 1996), and allies can demand spaces for this to occur and support access to participatory decision-making processes.

A second important finding, after following the money trail, is the importance of reforming banks and creating sanctions for investments tied to human rights violations. Even after Chixoy, the World Bank showed interest in supporting Belo Monte in Indigenous territories without proper consultation. Later, when the Bank put in place additional safeguards and announced it would no longer finance large dams, this encouraged bifurcation of funding. Countries like Brazil who sought to continue without oversight looked for sources with fewer regulatory demands (Wörsdörfer, 2015). This same split exists throughout the energy sector. Repressive energy projects will continue until finance institutions reform. Bilateral and multilateral agencies supported the conflictive hydro case studies covered in this research and other cases not discussed here, even after evidence of human rights violations emerged. At least 65 institutions from 12 countries took part in Belo Monte construction (Finley-Brook & Jordan, 2019). Broad consortiums disperse risk in capital-intensive infrastructure projects, but they also obfuscate assigning responsibility for harm. Transparency and accountability measures must improve.

The third and related finding is that environmental, social, and governance commitments must have teeth. Without oversight and accountability mechanisms, voluntary schemes like EPs allow funders to exaggerate commitments and facilitate companies claiming undeserved wins using misleading public relations campaigns. Exploitation will continue without laws and clear oversight—including consequences for harm. Approaches without third-party verification and enforcement are nothing short of abusive, given the continuation of mistakes across decades. For example, EP institutions are supposed to require stakeholder engagement, grievance mechanisms, monitoring, and reporting, yet EP investors failed to adhere to due diligence obligations in Barro Blanco and Belo Monte (BankWatch, 2017, 2018; Hofbauer, 2017). In fact, a majority of EP projects that BankTrack assessed in 2020 lacked stakeholder engagement processes or grievance mechanisms (BankTrack, 2020). In fact, research suggests institutions may back even riskier projects after institutionalizing EP (Eshet, 2017).

When energy project damage occurs in host communities, developers have sought to avoid reputational damage by deflecting responsibility. Even after recording grievances in proper channels, EP violations go unpunished due to the ability of violators to obscure damages. For example, a common technique is selective social responsibility campaigns where developers distract from harm in impacted communities with media campaigns highlighting symbolic achievements, like charitable giving of scholarships to disadvantaged youth. Without systemic review, insidious deception continues, such as firms highlighting their economic support for building a new school, without first mentioning that the involuntary relocation for dam construction made use of the old school no longer viable. Implying strong practices without sanction for violations simply encourages such deception and disinformation (Testa et al., 2018).

Conclusion

This cross-case analysis included four powerful examples of how deception and disinformation riddle hydropower and argues that tokenism and superficial symbolic gestures lacking substance and oversight threaten, rather than serve energy democracy. A lack of participatory procedures continues in dam building with few improvements across the two decades since the WCD report. Absence of independent verification persists, even in locations with widespread knowledge of governmental corruption and conflicts of interest.

Although dangerous projects face pressure from impacted communities, who often receive support from national and international civil society, power inequities obstruct just resolution of concerns. Illustrative cases demonstrate a need for simultaneous top-down reforms involving shifts across the energy sector and bottom-up movements to demand accountability. Otherwise, people who speak out may be assassinated, as occurred with Berta Caceres in Honduras and with other lesser known Latin American dam opponents (International Rivers, 2016).

It is important to acknowledge that renewable energy projects can fall into the same traps as fossil fuels. An international movement for energy democracy demanding distributive, procedural, and recognition justices must create the groundwork to move beyond false solutions, such as justifying large tropical dams for climate change mitigation. Abuse of power by industry-captured or corrupt government officials, or by investors seeking interest on loans, impedes the transition to clean, participatory, and equitable energy. Many sustainable energy advocates working to end fossil fuels are unaware of the problems with large dams. Whether across locations or types of energy, solidarity requires education to expose how developers with profit motives often play energy justice movements against one another. Lessons from these and other cases show that energy democracy requires not only epistemic justice, with access to accurate and honest social and ecological information, but also the power to punish actors who promote disinformation to shift attention away from human rights violations. In sum, democracy necessitates both education and accountability.

References

Aguirre, M. (2014, November 17). Healing begins for the Maya Achi people of Guatemala. *International Rivers*. www.internationalrivers.org/blogs/233-0

Aidoo, R., Martin, P., Ye, M., & Quiroga, D. (2017). Footprints of the dragon: China's oil diplomacy and its impacts on sustainable development policy in Ecuador and Ghana. *International Development Policy/ Revue Internationale de Politique de Développement*, *8*(1).

Almeida, R. M., Shi, Q., Gomes-Selman, J. M., Wu, X., Xue, Y., Angarita, H., Barros, N., Forsberg, B. R., García-Villacorta, R., Hamilton, S. K., & Melack, J. M. (2019). Reducing greenhouse gas emissions of Amazon hydropower with strategic dam planning. *Nature Communications*, *10*(1), 1–9.

Athayde, S., Duarte, C. G., Gallardo, A. L. C. F., Moretto, E. M., Sangoi, L. A., Dibo, A. P. A., Siqueira-Gay, J., & Sánchez, L. E. (2019). Improving policies and instruments to address cumulative impacts of small hydropower in the Amazon. *Energy Policy*, *132*, 265–271.

Atkins, E. (2017). Dammed and diversionary: The multi-dimensional framing of Brazil's Belo Monte dam. *Singapore Journal of Tropical Geography*, *38*(3), 276–292.

Atkins, E. (2019). Disputing the 'national interest': The depoliticization and repoliticization of the Belo Monte Dam, Brazil. *Water*, *11*(1), 103.

Banco Mundial. (2008). *Licenciamento Ambiental de Empreendimentos Hidrelétricos no Brasil*. http://siteresources.worldbank.org/INTLACBRAZILINPOR/Resources/Brazil_licenciamento_SintesePortugueseMarch2008.pdf

BankTrack. (2017, October 3). *Belo Monte*. www.banktrack.org/project/belo_monte_dam#about
BankTrack. (2018, November 19). *Barro Blanco*. www.banktrack.org/project/barro_blanco_dam_project
BankTrack. (2019). *Equator principles disaster projects*. www.equatorbanksact.org/disaster_projects
BankTrack. (2020). *Trust us, we're equator banks*. www.banktrack.org/download/trust_us_were_equator_banks_briefing_paper/200810_trust_us_were_equator_banks_briefing_paper.pdf
BankWatch. (2017, June). *2017 report on European code of conduct on partnership implementation in CEE countries*. https://bankwatch.org/sites/default/files/briefing-ECCP-29Aug2017.pdf
BankWatch. (2018, October 15). *EBRD renewable investments finally matched its fossil fuel investments in 2017—so why is the bank's draft Energy Strategy still fixated with gas?* https://bankwatch.org/publication/ebrd-renewable-investments-finally-matched-its-fossil-fuel-investments-in-2017-so-why-is-the-bank-s-draft-energy-strategy-still-fixated-with-gas
Bnamericas. (n.d.). *Company profile: Norte energia*. www.bnamericas.com/en/company-profile/norte-energia-sa-norte-energia
Boelens, R., Shah, E., & Bruins, B. (2019). Contested knowledges: Large dams and mega-hydraulic development. *Water, 11*(3), 416.
Branford, S. (2018, January 3). Brazil announces end to Amazon mega-dam building policy. *MongaBay*. https://news.mongabay.com/2018/01/brazil-announces-end-to-amazon-mega-dam-building-policy/
Bratman, E., & Dias, C. B. (2018). Development blind spots and environmental impact assessment: Tensions between policy, law and practice in Brazil's Xingu river basin. *Environmental Impact Assessment Review, 70*, 1–10.
Carvajal, P. E., Anandarajah, G., Mulugetta, Y., & Dessens, O. (2017). Assessing uncertainty of climate change impacts on long-term hydropower generation using the CMIP5 ensemble—the case of Ecuador. *Climatic Change, 144*(4), 611–624.
Casey, N., & Krauss, C. (2018, December 24). It does not matter if Ecuador can afford this dam. China still gets paid anyway. *The New York Times*. www.nytimes.com/2018/12/24/world/americas/ecuador-china-dam.html
Chatterton, P. (2008). Demand the possible: Journeys in changing our world as a public activist-scholar. In K. Mitchell (Ed.), *Practising public scholarship: Experiences and possibilities beyond the academy* (pp. 73–79). Wiley-Blackwell.
Colajacomo, J., & Chen, C. (1999). The Chixoy dam: The Maya Achi' genocide: The story of forced resettlement. *Indigenous Affairs*, 3–4.
Conca, J. (2012, June 10). How deadly is your kilowatt? We rank the killer energy sources. *Forbes*. www.forbes.com/sites/jamesconca/2012/06/10/energys-deathprint-a-price-always-paid/#3608503e709b
Del Bene, D., Scheidel, A., & Temper, L. (2018). More dams, more violence? A global analysis on resistances and repression around conflictive dams through co-produced knowledge. *Sustainability Science, 13*(3), 617–633.
Dingwerth, K. (2005). The democratic legitimacy of public-private rule making: What can we learn from the world commission on dams? *Global Governance: A Review of Multilateralism and International Organizations, 11*(1), 65–83.
Ellis, R. E. (2018). Ecuador's leveraging of China to pursue an alternative political and development path. *Journal of Indo-Pacific Affairs, 1*(1), 79–104.
Erlewein, A. (2014). The promotion of dams through the clean development mechanism: Between sustainable climate protection and carbon colonialism. In M. Nüsser (Ed.), *Large dams in Asia: Contested environments between technological hydroscapes and social resistance* (pp. 149–168). Springer.
Eshet, A. (2017). Sustainable finance? The environmental impact of the 'equator principles' and the credit industry. *International Journal of Innovation and Sustainable Development, 11*(2–3), 106–129.
Fearnside, P. (2009). Las hidrolétricas de Belo Monte e Altamira (Babaquara) como fontes de gases de efeito estufa. *Novos Cadernos NAEA, 12*(2), 5–56.
Fearnside, P. M. (2016). Planned disinformation: The example of the Belo Monte Dam as a source of greenhouse gases. In L. R. Issberner & P. Léna (Eds.), *Brazil in the Anthropocene* (pp. 141–158). Routledge.
Fearnside, P. M. (2017a). Belo Monte: Actors and arguments in the struggle over Brazil's most controversial Amazonian dam. *DIE ERDE—Journal of the Geographical Society of Berlin, 148*(1), 14–26.
Fearnside, P. M. (2017b). Brazil's Belo Monte Dam: Lessons of an Amazonian resource struggle. *DIE ERDE—Journal of the Geographical Society of Berlin, 148*(2–3), 167–184.
Feldpausch-Parker, A., Sprain, L., Endres, D., & Peterson, T. R. (Eds.). (2019). Editorial: A research agenda for energy democracy. *Frontiers in Communication, 4*, 1–8. https://doi.org/10.3389/fcomm.2019.00053

Finer, M., Moncel, R., & Jenkins, C. N. (2010). Leaving the oil under the Amazon: Ecuador's Yasuní-ITT Initiative. *Biotropica, 42*(1), 63–66.
Finley-Brook, M. (2017). Hydropower's fluid geographies. In B. Soloman & K. Calvert (Eds.), *Handbook on the geographies of energy* (pp. 119–133). Edward Elgar Publishing.
Finley-Brook, M. (2019). Extreme energy injustice and the expansion of capital. In D. Paley & S. Granovsky-Larsen (Eds.), *Organized violence and the expansion of capital* (pp. 23–47). University of Regina Press.
Finley-Brook, M., & Holloman, E. L. (2016). Empowering energy justice. *International Journal of Environmental Research and Public Health, 13*(9), 926.
Finley-Brook, M., & Jordan, O. (2019). Energy transition and uneven development. In J. Cupples, M. Palomino-Schalsha, & M. Prieto (Eds.), *Routledge handbook of Latin American development* (pp. 446–457). Routledge Press.
Finley-Brook, M., & Thomas, C. (2010). Treatment of displaced indigenous populations in two large hydro projects in Panama. *Water Alternatives, 3*(2), 269.
Finley-Brook, M., & Thomas, C. (2011). Renewable energy and human rights violations: Illustrative cases from indigenous territories in Panama. *Annals of the Association of American Geographers, 101*(4), 863–872.
Freeman, C. P. (2017). Dam diplomacy? China's new neighbourhood policy and Chinese dam-building companies. *Water International, 42*(2), 187–206.
Gagnon, J., & Desbiens, C. (2018). Mapping memories in a flooded landscape: A place reenactment project in Pessamit (Quebec). *Emotion, Space and Society, 27*, 39–51.
Gallagher, K. P., Kamal, R., Jin, J., Chen, Y., & Ma, X. (2018). Energizing development finance? The benefits and risks of China's development finance in the global energy sector. *Energy Policy, 122*, 313–321.
Gallagher, K. P., & Yuan, F. (2017). Standardizing sustainable development: A comparison of development banks in the Americas. *The Journal of Environment & Development, 26*(3), 243–271.
Garzón, P. (2018). Implicaciones de la relación entre China y América Latina. Una mirada al caso ecuatoriano. *Ecología Política*, (56), 80–88.
Garzón, P., & Castro, D. (2018). China-Ecuador relations and the development of the hydro sector. *Building, 24*, 37–57.
Gerlak, A. K., Saguier, M., Mills-Novoa, M., Fearnside, P. M., & Albrecht, T. R. (2020). Dams, Chinese investments, and EIAs: A race to the bottom in South America? *Ambio, 49*(1), 156–164.
Global Witness. (2019). *Enemies of the state?* www.globalwitness.org/fr/campaigns/environmental-activists/enemies-state/
Goodland, R. (2010). The World Bank versus the World Commission on Dams. *Water Alternatives, 3*(2), 384–398.
Hale, C. R. (2001). What is activist research? *Social Science Research Council, 2*(1–2), 13–15.
Hale, C. R. (2006). Activist research v. cultural critique: Indigenous land rights and the contradictions of politically engaged anthropology. *Cultural Anthropology, 21*(1), 96–120.
Hall, A., & Branford, S. (2012). Development, dams and Dilma: The saga of Belo Monte. *Critical Sociology, 38*(6), 851–862.
Harvey, D. (2005). *A brief history of neoliberalism*. Oxford University Press.
Higgins, T. (2020, January 17). Belo Monte boondoggle: Brazil's biggest, costliest dam may be unviable. *MongaBay*. https://news.mongabay.com/2020/01/belo-monte-boondoggle-brazils-biggest-costliest-dam-may-be-unviable/
Hofbauer, J. A. (2017). Operationalizing extraterritorial obligations in the context of climate project finance: The Barro Blanco case. *Journal of Human Rights and the Environment, 8*(1), 98–118.
Hydropower Sustainability Governance Committee. (n.d.). www.hydrosustainability.org/assessment-protocol
IFC (International Finance Corporation). (2015). *Hydropower: A guide for developers and investors*. www.ifc.org/wps/wcm/connect/906fa13c-2f47-4476-9476-75320e08e5f3/Hydropower_Report.pdf?MOD=AJPERES&CVID=kJQl35z
IHA (International Hydropower Association). (n.d.). *Our story*. www.hydropower.org/our-story
International Rivers. (2011). *Documenting Belo Monte's Dam risks: Buyers beware*. www.internationalrivers.org/resources/documenting-belo-monte-dam-s-risks-buyer-beware-3428
International Rivers. (2016, June 20). *Murdered for their rivers: A roster of fallen dam fighters*. www.internationalrivers.org/resources/murdered-for-their-rivers-a-roster-of-fallen-dam-fighters-11499

International Rivers. (2019). *Watered down: How do big hydropower companies adhere to social and environmental policies and best practices?* www.internationalrivers.org/sites/default/files/attached-files/watered_down_eng_full_report_web_0.pdf

Johnston, B. R. (2004). Reparations for dam-displaced communities? Report from the Chixoy Dam legacy issues meeting, July 25, 2003, Pacux, Guatemala. *Capitalism Nature Socialism, 15*(1), 113–119.

Johnston, B. R. (2005). *Volume two: Chixoy Dam legacy issues study*. Center for Political Ecology. www.internationalrivers.org/wp-content/uploads/sites/86/2020/06/Chixoy-Dam-Legal-Issues-Study-vol-2.pdf

Johnston, B. R. (2010). Chixoy Dam legacies: The struggle to secure reparation and the right to remedy in Guatemala. *Water Alternatives, 3*(2).

Jordan, O. (2008). I entered during the day, and came out during the night: Power, environment, and indigenous peoples in a globalizing Panama. *Tenn. Journal of Law, Economics & Policy, 4*, 467.

Jordan, O. (2018). The privatization of environmental discourse: Clean development and indigenous territoriality in western Panama. *ABYA-YALA: Revista sobre acesso á justiça e direitos nas Américas, 2*(1), 140–168.

Kingsbury, D. V., Kramarz, T., & Jacques, K. (2019). Populism or petrostate? The afterlives of Ecuador's Yasuní-ITT initiative. *Society & Natural Resources, 32*(5), 530–547.

Lakhani, N. (2020). *Who killed Berta Caceres? Dams, death squads, and an Indigenous defender's battle for the planet*. Verso.

Latrubesse, E. M., Arima, E. Y., Dunne, T., Park, E., Baker, V. R., d'Horta, F. M., Wight, C., Wittmann, F., Zuanon, J., Baker, P. A., & Ribas, C. C. (2017). Damming the rivers of the Amazon basin. *Nature, 546*(7658), 363.

Ledec, G., & Quintero, J. D. (2003). *Good dams and bad dams: Environmental criteria for site selection of hydroelectric projects*. The World Bank.

Lozano, G. (2020, February). Can Latin America's first AIIB member learn from past mistakes? *China Dialogue*. https://chinadialogue.net/en/business/11903-can-latin-america-s-first-aiib-member-learn-from-past-mistakes/

Lynch, B. D. (2019). What Hirschman's hiding hand hid in San Lorenzo and Chixoy. *Water, 11*(3), 415.

Millikan, B. (2014). The Amazon: Dirty dams, dirty politics and the myth of clean energy. *Tipití: Journal of the Society for the Anthropology of Lowland South America, 12*(2), 134–138.

Munzer, S. R. (2019). Dam(n) displacement: Compensation, resettlement, and indigeneity. *Cornell International Law Journal, 51*, 823.

Nixon, R. (2010). Unimagined communities: Developmental refugees, megadams and monumental modernity. *New Formations, 69*(69), 62–80.

Pain, R. (2003). Social geography: On action orientated research. *Progress in Human Geography, 27*(5), 649–657.

Pérez, B. F., Hofbauer, J. A., Mayrhofer, M., & Calzadilla, P. V. (2016). Rethinking the role of development banks in climate finance: Panama's Barro Blanco CDM project and human rights. *Law, Environment and Development Journal, 12*, i.

Pestana, I. A., Azevedo, L. S., Bastos, W. R., & de Souza, C. M. M. (2019). The impact of hydroelectric dams on mercury dynamics in South America: A review. *Chemosphere, 219*, 546–556.

Purcell, T. F., & Martinez, E. (2018). Post-neoliberal energy modernity and the political economy of the landlord state in Ecuador. *Energy Research & Social Science, 41*, 12–21.

Richter, B. D., Postel, S., Revenga, C., Scudder, T., Lehner, B., Churchill, A., & Chow, M. (2010). Lost in development's shadow: The downstream human consequences of dams. *Water Alternatives, 3*(2).

Siciliano, G., Del Bene, D., Scheidel, A., Liu, J., & Urban, F. (2019). Environmental justice and Chinese dam-building in the global South. *Current Opinion in Environmental Sustainability, 37*, 20–27.

Sneddon, C. (2015). *Concrete revolution: Large dams, cold war geopolitics, and the US bureau of reclamation*. University of Chicago Press.

SNEEJ (Southwest Network for Environmental and Economic Justice). (1996). *Jemez principles for democratic organizing*. www.ejnet.org/ej/jemez.pdf

SOMO/Both Ends. (2014). *Complaint filed against Dutch development bank FMO for funding the Barro Blanco dam in Panama*. www.banktrack.org/show/news/complaint_filed_against_dutch_development_bank_fmo_for_funding_the_barro_blanco_dam_in_panama_1

Teräväinen, T. (2019). Negotiating water and technology—competing expectations and confronting knowledges in the case of the Coca Codo Sinclair in Ecuador. *Water, 11*(3), 411.

Testa, F., Boiral, O., & Iraldo, F. (2018). Internalization of environmental practices and institutional complexity: Can stakeholders pressures encourage greenwashing? *Journal of Business Ethics*, *147*(2), 287–307.

Tickell, A. (1995). Reflections on 'activism and the academy'. *Environment and Planning D: Society and Space*, *13*, 235–237.

UN (United Nations). (2007). *United Nations Declaration on the Rights of Indigenous Peoples*. https://undocs.org/A/RES/61/295

UNDP (United Nations Development Program). (2015). *Social and environmental policies and procedures*. www.undp.org/content/dam/undp/library/corporate/Social-and-Environmental-Policies-and-Procedures/UNDP-Social-Environmental-Screening-Procedure-1January2015.pdf

Vallejo, M. C., Espinosa, B., Venes, V., Lopez, V., & Anda, S. (2018). *Esquivando estándares de desarrollo sustentable: Estudios de casos en proyectos hidroeléctricos del Ecuador*. www.bu.edu/gdp/files/2018/10/GEGI-GDP-Backgrounder-Ecuador.pdf

Vieira, D. M. (2020). The discourse and coordination among advocacy coalitions: The case of Belo Monte. *RAUSP Management Journal*, *55*(1), 86–99.

Warner, J. F., van Dijk, J. H., & Hidalgo, J. P. (2017). Old wine in new bottles: The adaptive capacity of the hydraulic mission in Ecuador. *Water Alternatives*, *10*(2), 332–340.

WCD (World Commission on Dams). (2000). *Dams and development: A new framework for decision-making. The report of the world commission on dams*. Earthscan.

Wörsdörfer, M. (2015). Equator principles—now and then. In M. Schwartz, D. Comer, & H. Harris (Eds.), *The ethical contribution of organizations to society* (pp. 121–160). Emerald Group Publishing Limited.

30

POSTCARDS FROM THE FUTURE

A case study in Hawaii's transition to wind and solar energy

Cristi Choat Horton, Nicolas Hernandez and Tarla Rai Peterson

Introduction

In 2018, approximately 9 million people travelled to the Aloha State to walk some of the largest white sand beaches in the United States ("Hawai'i Visitor Statistics," 2019), admire one of Hawaii's longest waterfalls ("50 Fun Facts," 2015), and hike through tropical rain forests ("The Hawaiian Islands," 2019). Those experiences, captured by postcards, were sent home to friends and family. However, tourists are not the only ones sending postcards. For a little over a decade, national and international attention has focused on Hawaii's transition from a predominantly fossil fuel–based energy system to a 100% renewable energy system. Both observers and active participants describe Hawaii as a "postcard from the future" (Grimley & Farrell, 2015) that sheds light on both the promises and the challenges of transitioning to a renewable energy future.

Surrounded by the Pacific Ocean, Hawaii's eight islands are 2,390 miles (3,846 km) from California and 3,850 miles (6,196 km) from Japan, making the State the most isolated population center on Earth and accessible only by air or water ("50 Fun Facts," 2015). Due to its location, Hawaii experiences its fair share of imports, with crude oil consistently ranked at the top of Hawaii's imports ("State Imports for Hawaii," 2019). The 2008 energy crisis brought into sharp focus Hawaii's susceptibility to fluctuating oil prices. As oil prices hit $150 per barrel, Hawaiians were paying $4 per gallon at the fuel pumps ("Celebrating 10 Years of Success," 2018). In 2015, oil and coal accounted for 67.3% and 15.1%, respectively, of Hawaii's energy use ("Hawaii Energy Facts & Figures," 2018). In 2016, the State's transportation sector (air, ground, marine) accounted for 63% of Hawaii's petroleum use, and electric power accounted for 25%, while industrial, commercial, and residential made up the rest ("US Energy Information Administration," 2019). Since a large percentage of Hawaii's electricity comes from imported fuel oil, the state's electricity costs fluctuate with the often volatile crude oil prices, at times causing electricity costs to be more than double the mainland's electricity costs (Grimley & Farrell, 2015).

Concerns about this volatility led to an effort to shift Hawaii to a 100% renewable energy system. This chapter examines how energy professionals tasked with developing the new system discursively construct energy democracy. First, we review the energy democracy movement's call for a more democratic energy system. We then offer the 2016 Hawaii Power Summit as a case for examining Hawaii's energy transition, focusing on energy professionals' discourse about

DOI: 10.4324/9780429402302-35

the transition. Finally, we discuss how our findings provide insight into potential challenges and opportunities for empowering a broader suite of participants to become active contributors in decisions about energy systems.

A call for a democratically designed renewable energy future

Hawaii's situation supports Sweeney's (2012) argument that energy democracy is, in part, a response to an energy emergency: overdependence on fossil fuels that not only threatens the climate but also both economic and energy security. While energy democracy remains a loosely defined concept (Burke & Stephens, 2018), a nascent definition comes from the Lausitz Climate Camp's proposal that:

> everybody is ensured access to sufficient energy. Energy production must thereby neither pollute the environment nor harm people. More concretely, this means that fossil fuel resources must be left in the ground, the means of production need to be socialised and democratised, and that we must rethink our overall attitude towards energy consumption.
>
> *("Camp Booklet," 2012)*

This definition emphasizes local control and prudent consumption of accessible, sustainable clean energy. Szulecki (2018) describes energy democracy as an "ideal political goal," wherein "citizens are the recipients, stakeholders (as consumers/producers [prosumers]) and accountholders of the entire energy sector policy" (p. 35). Sweeney (2012) contends that energy democracy "is a decisive shift in power towards workers, communities and the public," where fewer decisions regarding energy production and consumption are left to elite groups (p. ii).

Although these definitions vary, they all indicate that the movement challenges expectations that energy governance should be limited to elites. Building from a loss of confidence in existing energy governance systems and spurred by desires to transition to local or regional-based energy systems and decentralized technologies (Burke & Stephens, 2017), the energy democracy movement advocates for increased citizen control over energy decisions. Citizen control potentiates a sustainable energy future that embraces energy efficiency and renewable energy (Weinrub & Giancatarino, 2015). Sweeney (2012) suggests that development of sustainable energy democracy requires (1) resistance to fossil fuel corporations' agenda that promotes risky extraction methods and opposes climate change mitigation efforts; (2) reclamation of the energy economy from market mechanisms that favor corporate interests to enable small companies or local communities to lower energy costs and improve energy services; and (3) restructuring of global energy supply chains to allow more localities to benefit from the upgrade of renewables in the form of job creation and other economic benefits.

Although some critique energy democracy as a quasi-utopian ideal that is impossible to operationalize, others suggest criteria and indicators to guide its development (Burke & Stephens, 2017; Szulecki, 2018; Welton, 2018). Szulecki (2018) suggests that criteria for energy democracy should be citizen centered and include popular sovereignty, participatory governance, and civic ownership. Welton (2018) offers similar criteria including increased participatory decision making; consumer choice in energy purchasing decisions and energy generation, storing, and selling; and more local community control over energy resources. Burke and Stephens (2017) integrate a discussion of US policy instruments for advancing renewable energy transitions based on Sweeney's (2012) resist-reclaim-restructure goals. They articulate various intended outcomes resulting from policy implementation and argue that "collectively these policies offer the possibility to provide synergies and enhance effectiveness for achieving integrated energy democracy goals and outcomes" (p. 38). The Hawaii Clean Energy Initiative (HCEI) provides an ongoing example of how such policies may simultaneously potentiate energy democracy and climate change mitigation.

Hawaii's call for a renewable energy future

Spurred by an overdependence on fossil fuels, the State of Hawaii partnered with the US Department of Energy (DOE) to launch the HCEI. The initiative set ambitious goals for Hawaii to minimize dependence on fossil fuels for electricity and transportation. Driven by an overall goal of 100% clean energy by 2045, HCEI set out five objectives: (1) identifying new infrastructure needed for a clean energy economy; (2) promoting innovation in clean energy technologies, financing, and policy to aid the State's transition to clean energy; (3) developing and diversifying Hawaii's economy to benefit from sustainable energy policy; (4) establishing an open-source learning model to support other island communities with similar clean energy goals; and finally (5) creating a workforce with "green" skills to transition Hawaii into an energy-independent future ("Hawaii Clean Energy Initiative," 2019). By meeting these objectives, Hawaii would become the first state in the United States to generate all of its electricity from indigenous renewable resources ("Celebrating 10 Years of Success," 2018).

Praised by Hawaiian US Sen. Brian Schatz as an initiative that spurred innovation, reduced emissions, and cut energy costs, the HCEI is a result of a collaborative effort between the State of Hawaii and the DOE to "create a holistic, locally driven strategy to advance Hawaii's enterprising clean energy agenda by pooling the ideas, resources, and capabilities of a diverse team of stakeholders" ("Celebrating 10 Years of Success," 2018, p. 2). Hawaii's clean energy policy agenda has created regulatory reform and influenced tax policy and clean energy financing. As Hawaii moves toward its 2045 goal, the Hawaii Public Utility Commission (PUC), Hawaiian Electric Companies (HECO), Kauai Island Utility Cooperative (KIUC), third-party service providers, and other stakeholders strive to meet HCEI objectives. By 2018, 27% of the state's electricity sales were generated from solar and wind. Distributed solar photovoltaic (PV) was the number one renewable energy source and 60 utility-scale renewable energy projects had been established ("Celebrating 10 Years of Success," 2018).

In 2016, the Hawaii Power Summit (Summit) convened in Honolulu, Hawaii, to address the State's progress toward the 2045 goal. Key energy professionals discussed capital requirements, technology innovation, and policy challenges for attaining the 100% renewable mandate. Hawaii's governance of energy resources is unique in the need to decide whether a uniform policy is needed for all islands or whether individual islands should design separate policies to optimize their own resources. Finally, energy professionals explored novel possibilities for consumers in a distributed energy system ("Hawaii Power Summit 2016," 2018). The three-day event included a preconference workshop that examined system issues connected with high penetration of solar and wind and a two-day conference that addressed opportunities and challenges related to Hawaii's progress of achieving renewable portfolio standards (RPS) mandates and reducing consumer's electricity costs.

The 2016 Summit provided an opportunity to examine energy professionals' discourse about renewable energy as a means of transitioning from a centralized fossil fuel–based energy system to a decentralized system that could enable more democratic practices. Our primary purpose in analyzing this discourse was to discover what opportunities and challenges these energy professionals understand as most central to the success of their program and how that understanding contributes to and limits the possibilities they envision for energy democracy. Because

Sweeney's (2012) vision of energy democracy and the goals of the HCEI appeared so complementary, we used his tripartite path toward energy democracy via resistance, reclamation, and restructuring as our analytical guide (Sweeney, 2012).

Data collection and processing

We collected data[1] from the preconference workshop and 16 conference sessions for a total of 17 sessions and 39 presenters. We grouped presenters into three categories: utility, policy, and technical. Responsibilities of the utility professionals center on electricity generation and distribution for sale, usually in a regulated market. Responsibilities of the policy professionals centered on developing and implementing policy related to renewable energy. Technical professionals were responsible for providing expertise regarding the installation and operation of electrical and energy storage and transmission systems and to devise ways to reduce energy consumption.

Panel discussions and individual talks were digitally recorded to capture the in situ discourse that is rarely found in formalized texts such as government documents and professional publications (Middleton et al., 2015). This enabled us to include jokes, stories, and other informal comments about transitioning to a renewable energy system (Horton et al., 2016; Middleton et al., 2015). The recorded presentations were transcribed to produce verbatim texts, which the research team reviewed for accuracy. Text was separated into individual sentences, assigned a unique label, reviewed for labeling accuracy, and then entered into NVivo 10 qualitative software (QSR International, Doncaster, Victoria, Australia).

To capture utterances related to resistance, reclamation, and restructuring, which Sweeney (2012) identified as central to energy democracy, we conducted a word search of all inputted sentences. We supplemented this search with terms gleaned from Burke and Stephens (2017), Szulecki (2018), and Hawaii's clean energy initiative literature ("Hawaii Clean Energy Initiative," 2019; "Hawaii Energy Facts & Figures," 2018) (see Table 30.1). Our search produced 1,975 sentences.

The energy democracy movement advocates for change in an energy sector that is not only comprised of complex generation, transmission, and distribution technologies but is an intricate web of cultural, economic, and political practices that impact the environment. We used the SPEED framework (Stephens et al., 2008, 2013) to examine the intricacies of energy democracy objectives expressed in each sentence. We chose this framework because its theoretical derivation from Luhmann's (1989) theory of social systems highlights the simultaneous independence and intertwined nature of cultural, economic, legal, technical, and other functions operating within modern societies. As such, it provided a coding protocol that guided us to an expansive vision of our informants' ways of giving meaning to Hawaii's energy system transition. Further, it discouraged us from limiting our analysis to one or two of the most commonly recognized social functions of such transitions, such as economics and technology.

We coded for culture, economics, environmental science, policy, and technology. The same sentence was coded in multiple categories if it fit more than one. Each was further coded with a positive (opportunities) or negative (challenges) tone. For example, "mandating the 100% RPS benefits Hawaii by promoting sustainable economic development" was coded under policy and economic categories with a positive tone. To reduce preconceived bias, each code and tone was considered independently. Preliminary analysis conducted during the coding process highlighted certain opportunities and costs related to transitioning to a renewable energy future. We then used an inductive approach guided by grounded theory (Strauss & Corbin, 1994) to further examine our texts. Finally, we explored how categories related to each group of professionals (utility, policy, and technical).

Table 30.1 Sweeney's energy democracy objectives and key terms used for coding

Resist	*Reclaim*	*Restructure*
Cap-and-dividend	Cap-and-dividend	Cap-and-dividend
Carbon tax	Carbon tax	Cooperatives
Clean Energy Initiatives (CEI)	Cooperatives	Community-based renewable energy
Hawaii Clean Energy Initiative	Community-based renewable energy	Customer grid supply (CGS)
Decarbonization	Customer grid supply (CGS)	Customer self-supply (CSS)
Low carbon energy	Customer self-supply (CSS)	Distributed energy
Distributed energy	Feed-in tariffs	Distributed generation
Distributed generation	Green subsidies	Microgrids
Renewable energy standards	Microgrids	Prosumer
Renewable portfolio standards (RPS)	Net metering	Public bonds
	On-bill financing	Solar energy
	Prosumer	Solar photovoltaics
	Public bonds	PV
	Solar energy	Sustainable energy utilities
	Solar photovoltaics	Wind energy
	PV	Wind turbine
	Sustainable energy utilities	
	Wind energy	
	Wind turbine	

Note: This table lists key terms used for the word search. The table headings list Sweeney's (2012) energy democracy objectives. The terms used to operationalize these objectives were drawn from a review of Burke and Stephens (2017), Szulecki (2018), "Hawaii Clean Energy Initiative: Goals and objectives," 2019, and "Hawaii Energy Facts & Figures," 2018. Because these terms may describe operationalization of more than one of the objectives, they may be repeated in more than one column.

Transitioning to a renewable energy future: opportunities and challenges

Our informants shared the presumption that successful transition to a renewable energy future centered on economics, policy, and technology to navigate the uncharted course to 100% renewable energy by 2045. Although only the policy-oriented professionals demonstrated strong self-identification as energy democracy advocates, all groups implicitly connected concepts central to energy democracy such as localized control, participatory governance, and self-determination, with implementation of the HEIC. Discourse about how the HEIC and energy democracy might operate in a mutually supportive fashion, including how energy professionals imagined themselves interacting with other participants, differed widely. Across all of their differences, however, these energy professionals introduced and elaborated on resistance to centralized authority, reclamation of a more direct relationship between energy users and the system, and restructuring to compose a new energy system (Sweeney, 2012).

Technical opportunities

Participants indicated that innovative technology was a primary means to move the electric system and consumers closer to Hawaii's renewable energy goals. One policy-oriented energy professional commented: "We need to keep the doors open for the new technologies. . . . The fact that we now have better options, . . . the generation that we decide to move forward with today, is largely going to determine the distribution systems of the future" (WP235). Another policy professional specified, "We're trying to implement real-time distributed energy resources to try to use the tools of both central plants, utility type of scale, as well as distributed energy resources and generation tools in different innovative ways" (WP228).

Energy storage technology was a common topic. As a utility-oriented participant explained, storage technology is needed "to take care of excess energy during the day" (WP232). Storage technology is an asset for a consumer-sided system that "is designed for customers with rooftop PV systems and energy storage" (WP228). Technically oriented participants stressed the importance of storage technologies to compensate for renewable energy's variability that impacts grid stability, explaining that battery storage can solve frequency support issues and noting that options such as "pumped hydro, compressed air, energy flywheels" (WP225) already exist.

Technology discussions often explored available options for integration of solar technology into the existing grid. Participants advocated that "there could be a feature where you do have microgrid developments" (WP263) and added that an intelligent grid was needed to manage or optimize "all the resources" (WP251). Both technical and policy professionals echoed the sentiment that an "integrated grid. . . [is] really important for us to achieve 100% renewable" (WP 228).

Participants prioritized technology to ensure reliability and resilience of the energy system. Utility professionals stressed the importance of having "devices, resources, storage [for] . . . the right amount of generation, load, storage, grid support services" (WP263) to assure system reliability. Participants also acknowledged that new infrastructure for wind and solar must be built and "the standby fleet, or . . . the generation fleet" must be modernized (WP260). Policy makers added that maintenance of current technology could optimize its utilization and maximize its length of service time. One encouraged the utilities to "get going on modernizing the generation system on each island" (WP226). Reliability and resilience were repeatedly emphasized by utility professionals as they talked about the technological components of a renewable energy system. One participant summarized the importance of system resiliency: "Nobody will care that we are at 100% renewables if we are in Day 21 of an island-wide power outage. . . . Is this energy system that we are planning today, going to be resilient enough that it can withstand damage?" (WP263). The expectation was that weather events and other unpredictable damage will occur and that utilities are required to recover quickly.

Policy opportunities

Although our participants saw technology as an important component to meeting the 100% RPS in 2045, they acknowledged that policy played a significant role in establishing common objectives for Hawaii's renewable energy transition. One utility professional commented:

> When we signed the 2008 Clean Energy Agreement, it was major step forward. I think that all parties, all policymakers, utility folks would agree that that was probably one of the most significant steps in this evolution of Hawaii's move towards clean energy. Because for the first time, you'd had at a very high policy level, everybody saying, this is what Hawaii needs.
>
> *(WP263)*

Our energy professionals indicated that governing bodies should continue to play key roles in guiding Hawaii to 100% renewable energy by 2045. One explained that Hawaii's PUC has advanced the transition to renewable energy sources by issuing "several orders and major dockets in 2014 and 2015 to outline the strategic roadmap for Hawaii's utilities in the future" (WP228). Another participant expanded on the PUC's role, explaining that HECO companies create power supply improvement plans (PSIPs) that describe actions they will take to achieve the 100% RPS by 2045. Our technical professionals referenced governing entities such as Federal Energy Regulatory Commission (FERC) and North American Electric Reliability Corporation (NERC), which develop and require the maintenance of safe, reliable, economically efficient, secure energy and which establish reliability standards for bulk power.

Policy was described as providing guidelines for necessary changes to conventional practices. Utility professionals, noted for example, that policy changes caused them to "rethink" current policies established prior to the RPS. One participant explained, "DR [demand response] policies came up before the 100 percent RPS," which requires utilities to approach "customer sided, behind the meter assets" differently (WP248). He added that "the commission's guidance, [provides] a full spectrum of ancillary services" (WP248). Other utility professionals discussed changing practices of power purchase agreements (PPAs) dealing with curtailment issues. In the past, utilities typically dealt with an oversupply of solar or wind by curtailing their electricity output. However, as one participant explained, utilities have taken:

> a huge step in agreeing to [pay the] . . . risk-adjusted price of PPAs [and there's compensation] for excess energy curtailment. It's still a must-take PPA. But what our vision is for the future is what we're calling a dispatchable PPA. And that's what we need, we think, to get to 100%.
>
> *(WP232)*

This participant was indicating that favorable policies can encourage utilities to collaborate with both consumers and regulators to develop creative options for both storage and use of electricity in ways that mitigate some of the waste. The agreement to pay a "risk-adjusted price" exemplifies taking a step that directly resists standards and norms developed within the centralized fossil fuel–dependent system. Favorable policies can ensure that participating utilities do not inadvertently incur free-rider costs when other utilities elect to maintain established practices. Finally, utility-oriented energy professionals commented that small changes, such as billing procedures, reflected steps taken to accommodate the 100% RPS. One participant explained how their utility engaged with customers regarding a time-of-use (TOU) rate plan. "The Commission directed us to provide a shadow bill to our customers who signed up for this program for the first six months" (WP251). Following the PUC's recommendation, the utility provided a bill where consumers could see what they would have paid under the old rate schedule in contrast to the rate schedule of an interim time-of-use program. This gave consumers access to more information and functioned as an informal invitation to become more involved in decisions about energy.

Technology-oriented energy professionals remarked that policy was crucial for coordination among utilities and third-party administrators of energy conservation and efficiency programs. "Energy efficiency programs are overseen by a third-party administrator," explained one participant, who also noted that, "given the increasing convergence between demand response and targeted energy efficiency, the Commission has observed the need for the companies to coordinate closely" (WP247). Coordinated efforts assist and prepare customers to participate in demand response programs "[which enable reaching the] goal of a reduction of 4,300 gigawatt hours by 2030" (WP249).

Economic opportunities

Participants acknowledged the need for economic benefits of transitioning to renewable energy to balance out the costs, emphasizing that the "price tag of getting to 100% renewables [should be] as low as possible" (WP258). They justified this claim with their assumption that economics constrained technology adoption. As one argued, "economics will dictate when microgrids become viable; it will be just a sheer financial decision" (WP231). Another stressed the importance of helping consumers understand the value behind choosing a technology, explaining, "[Y]ou have to be able to explain the value proposition. And the value proposition that I put in front of these end-use sites are cost, cost stability, and energy supply security. . . . I can help a customer know exactly what his future is going to look like" (WP245).

Both utility- and technology-oriented energy professionals were optimistic that technology such as PV panels and geothermal storage was becoming more affordable. Although they foresaw continued need to mitigate new technology costs, they framed mitigation as an economic opportunity. One anticipated that the capital would "come from primarily three areas: The utilities as one, independent power producers and VER [Verified Emission Reduction] aggregators, [and] the last is customers through investing and distributing energy resources" (WP258). Another suggested, "[I]ncentives such as an interim time-of-use program" (WP251) that encourages customers to shift their electricity usage from an expensive high-peak period to a less expensive midday period. They also discussed tax credits that would incentivize consumers to install renewable energy technologies and to decrease their income tax through solar investments.

Participants saw further opportunity in direct financial aid, such as the Green Energy Money Saver (GEMS) loan program, which finances solar PV and other clean energy. Its objective is to make energy improvements more accessible to multifamily rental projects, nonprofits, single-family residential renters and homeowners (low to moderate income), and small businesses. One participant expressed cautious optimism: "Right now there's about $144 million in a fund that we need to spend on clean energy here in Hawaii. . . . I have my fingers crossed that the PV-plus storage program notification for GEMS, mean[s] that you could use GEMS in order to get these installations in place" (WP243).

Participants acknowledged that the cost of the grid, which is influenced by energy pricing, was an important consideration when transitioning to renewable energy sources. One noted that "there's an opportunity to have one of the lowest-cost grids in the world" (WP250). Another participant added that Hawaii needs "to explore and pursue cost-effective opportunities to increase the operational flexibility of our grid" (WP247). A policy-oriented participant suggested that establishing a value for services such as energy storage and smart inverters could decrease costs, and recommended "look[ing] through production cost modelling [to determine] how they can lower system cost" (WP 226).

One participant observed that utilities' business strategies were changing "to align the business model with customer expectations" (WP226). Another elaborated on this idea with a decoupling example. The practice of decoupling designates the revenue necessary to cover known costs and allows electricity rates to change as consumption changes to meet designated revenue goals. As he explained, decoupling has:

> two main components, . . . the revenue balancing account. . . . The other component . . . is called the revenue adjusted mechanism. . . . Using the decoupling

> mechanism . . . ensured that the benefits of decoupling were helping to achieve the desired outcome [to meet customers' needs] . . . while a public utility is required to have a reasonable opportunity to earn a fair financial return.
>
> *(WP255)*

Policy-oriented energy professionals summed up economic opportunities by discussing Hawaii's promising economic and energy future, wherein "renewables will be required in a manner that's beneficial to Hawaii's economy" (WP264). One explained, "by mandating 100% RPS by 2045, Hawaii not only reduces the cost of electricity but creates new clean energy jobs and promotes sustainable economic development" (WP228), while another rejoiced that "[we] empower our island's families to make smart energy choices" (WP 252).

Environmental opportunities

A small number explicitly noted links between renewable energy and environmental quality. One observed that "the electric utility slice of our overall greenhouse gas emissions and oil consumption is shrinking, [and Hawaiians will reap] the environmental preservation benefits of 100% renewable energy" (WP262). Others mentioned that "one of the reasons why we like things like wind and solar is because there is an emissions benefit" (WP227).

Technical challenges

Although our participants were enthusiastic about technology, they also acknowledged technical challenges. As one acknowledged, the hardest challenge "may not be 100% *renewable* but 100% *variable* resources. . . . As we start moving toward this more distributed, more variable resource power system, there's a ton of upgrades that need to be made to the distribution network, to transmission" (WP227). The need to develop technologies that respond productively to the variability of renewable resources was a generalized concern.

They described grid stability and storage as two related challenges, focusing on grid disruptions due to solar PV generation that involve curtailment when "the system is unstable, or there's a risk to the grid" (WP232). Some participants expressed anxiety about off-grid microgrids operated by nonexperts; "when you do have off-grid microgrids, they do not have the ability anymore to contribute to the bigger grid in terms of keeping things in balance" (WP263). Participants also discussed generation and its impacts on existing grids. As one explained, "[O]nce you start getting circuit after circuit, having surplus power like this, there reaches a point where you have too much power for the particular grid to handle during peak solar generation hours" (WP238). Others observed that "distributed generation can cause voltage fluctuations that some loads really can't tolerate" (WP227) and that "in a low-flexibility grid, meaning not a lot of demand response, maybe the generation isn't very flexible. Transmission capacity is stressing the system" (WP243).

Participants pointed out one technical challenge they viewed as unique to Hawaii: islanded grids. One explained that since "each island is an islanded grid . . . we can't export excess generation if we need to. And, our frequency deviations mean a whole lot more to us here on our smaller islanded grid" (WP251). Another noted that Hawaiians, living in the world of distributed energy, find "it is harder than ever to plan for transmission" (WP235). Currently, the islands' transmission grids are not interconnected, so each island must modernize its own grids for possible integration of consumers' microgrids. One participant offered the University of Hawaii's research facility as an illustration of the wider problem:

> It has a transmission cable from the main island, but it's getting deteriorated. They're looking at rebuilding it . . . and putting in renewables and storage . . . to make it independent. It's a 5-megawatt grid with a lot of PV on it. And it's had a lot of issues with its system stability.
>
> *(WP226)*

In this context, energy storage is especially problematic. As one participant asked, "[W]hat do you do with that excess energy? You got to put it in storage" (WP 260). Another added, "The wind variability is such that it's hard to build a battery system big enough to capture it" (WP226).

Policy challenges

Utility-oriented participants discussed challenges associated with planning and executing rapid policy changes to meet new requirements. They expressed frustration with policy that determined how quickly to phase out fossil fuels. As one explained:

> As far as we're concerned, 100% in 2045 means we don't have fossil fuel anymore in our generation group. . . . Prior to having 100% in 2045, we were focused on what's it takes to get to 40% by 2030. . . . What we're trying to do here right now is come up with as reasonable cases, scenarios, of generation and energy resource mixes, which would get you to that 100% in 2045. [We want] to make sure there is adequate time on the technology side, the system side, the timing side, and of course, the project implementation to have a fair shot of succeeding.
>
> *(WP263)*

Although this example indicates aggravation with policy that moves too rapidly, our technology-oriented participants found outdated policy that makes it difficult to push technological innovations forward especially frustrating. One participant explained, "[W]e've adopted a 2008 national electric code in Hawaii. So, from a design standpoint, it's hard to design storage on some of these new, improved, and technological improvements, based on a 2008 code" (WP230). From this person's perspective, the greatest policy challenge came from lagging behind the technologically possible.

If there was any agreement regarding policy challenges, it centered around the difficulty of navigating cumbersome bureaucracy. They argued incentives were subject to bureaucratic decisions that irrationally privilege one type of tax credit over another. Speaking about the Investment Tax Credit, one protested that, "Speaker Ryan wants to reduce the corporate tax rate to 15 percent. The only way he can do that is to get rid of virtually all the tax incentives out there" (WP 241). Our technology experts observed that for Hawaii to continue forward movement toward 100% RPS by 2045, complementary policy for sectors such as transportation is needed, but "it's not as clean as RPS, where the utilities got on board. . . . You know, the topic of 100% renewable energy and transportation is complicated by the fact that you've got large groups of decision-makers" (WP265). And utility professionals argued for "a real common-sense definition for renewables, things that can be renewed [such as] solar, wind, wave, [but also consider] scenarios where fossils could, through the life cycle, be considered renewable down the line" (WP265). Beyond the obvious fears associated with any change, statements such as these illustrate how important it is for stakeholders to recognize themselves as contributing partners in determining the rules of the game.

Economic challenges

Participants pointed out numerous economic challenges. Some relate to funding for new projects. One participant said, "[W]hether it's rooftop solar or solar farm or a wind farm, they all require capital" (WP258). Another noted that "energy storage [is] another 1 to 2 billion dollars" (WP260). One utility professional added, "[T]here's a lot more expense that [microgrid operators] have to put into designing their system, whether it's the control system or the back-up storage" (WP263). They acknowledged it is "hard to see that far ahead to 2045 to know the capital required to get to 100% renewable energy because there will be changes in technologies, costs, customer behavior" (WP258).

Policy-oriented energy professionals described challenges in terms of cost-effectiveness of renewables or use-of-rate plans. One participant cautioned that "what might be cost-effective early on as you're moving through your higher and higher penetrations of renewables might not be as cost-effective or the most optimal thing to do as you get higher in penetrations" (WP 226). In other words, a rate plan that is economically viable when renewable energy is providing less than 20% of the electricity for a given locale may not be economically viable when renewable energy is providing more than 50% of the electricity for that locale. Another suggested that "the impact of TOU [time of use] on low-income [consumers] is something to be careful of, to anticipate, and monitor" (WP253). The assumption was that, since these consumers rarely have upgraded, energy-efficient appliances, this customer type could not benefit from TOU rates.

Technology-oriented energy professionals explained that failure to make thoughtful assessments about technology could be disastrous, claiming that "if Hawaii makes poor buying decisions about energy storage resources, you'll end up with one of the most expensive grids on the planet" (WP 250). Others focused on technical complexity of valuing electricity in the proposed system: "[Y]ou have a new number that shows up sometimes every five minutes. A real-time market will change the price of energy every five minutes. It's a little bit harder to narrow down what the value is" (WP225). They also noted the importance of interconnections among policy, technology, and economic functions in statements such as, "I don't think you're going to bring in the capital that's required to get to 100% renewables unless HECO is being allowed to put assets in place . . . and recover through the rate-based" (WP225), and "[W]e're very used to high reliability, high availability grid. If you want that same level of availability from a PV plant, then the relative cost is going to be rather high" (WP227). Statements such as these suggest that availability and reliability must be intricately linked to an affordable energy system before stakeholders will buy into renewable energy.

Environmental challenges

When these energy professionals raised environmental concerns, they focused on interactions between humans and their environment. One participant reminded everyone that "we [can] have natural disasters . . . that disrupt an island . . . or a grid on an island" (WP 228); while another noted that grid resilience is "especially [important] with respect to natural . . .

catastrophes" (WP259). They also claimed that, although the electricity side of the renewable transition was reducing emissions, other sectors lagged. As one participant observed:

> [I]t's critical to capture the transportation. And set some sort of goal. We have 1.1 million vehicles in Hawaii. The good news is we just passed the threshold of 5,000 electric vehicles. But we have a long way to go to make our ground transportation and train transportation renewable.
>
> *(WP 265)*

Mentions of environmental challenges tended to result more from inadequate responses to the clean energy initiative rather than from the initiative itself.

Discussion

The discourse at the Summit centered around the goal of restructuring the State's energy system to replace conventional means of generation, distribution, and transmission with a renewable energy system that promotes sustainable development and "empower[s] our island's families to make smart energy choices" (WP 252). The notions of empowering people "to make smart energy choices" and contributing to sustainability resonates with concepts seen throughout the energy democracy literature (Burke & Stephens, 2017; Cantarero, 2020; Feldpausch-Parker et al., 2019). Although our respondents explicitly emphasized economic, policy, and technology-oriented dimensions of sociotechnical change, our decision to code for the full suite of sociotechnical dimensions identified in the SPEED framework enabled us to recognize that they implicitly connected economics, policy, and technology to others. For example, they attributed many of the economic, policy, and technical challenges to cultural divisions between energy insiders and outsiders; and although not emphasized, they loosely connected proposed changes to environmental benefits. Because utility companies provide infrastructure and other services at the operational scale, their transformation was recognized as immediately critical, and interconnected with all other transition components. Economic discourse emphasized the importance of ensuring financial profit to encourage participation from electricity and technology providers, as well as incentivizing consumer choice. These energy professionals framed Hawaii's postcard from the future as a tropical paradise where independent sociotechnical social functions simultaneously contributed to and constrained efforts to resist, reclaim, and restructure a democratically governed and sustainable energy system.

Policy-oriented energy professionals

These self-described "leaders in the energy space" (WP252) were the most explicitly supportive of Sweeney's (2012) suggested path to energy democracy. For them, the HCEI's resistance to the fossil fuel–based energy system is the initial step toward economic prosperity, and policy is key to ensuring that prosperity is shared in a democratic manner that promotes the well-being of all Hawaiians. They are committed to replacing imported oil as the state's primary fuel for generating electricity and agree that technology to capture and store renewable energy is key to transitioning away from fossil fuel dependency. They express confidence that with advanced customer-side technology and modernization of the utility side, less need for fossil fuels will exist. They see policy as basic to reclamation of the energy system by people focused on stabilizing energy costs and improving energy services. For example, they note that the HCEI was the springboard for implementing new programs to give all low-income, single-family, and multifamily customers access to renewable energy, programs that assure equal access to the benefits of renewables and opportunities for participation in community-based renewable facilities. As these professionals reflect upon their progress toward restructuring, however, they admit, "[T]here is still a long way to go" (WP241) to achieve accessible, affordable energy for all Hawaiians.

These professionals acknowledge that complex and sometimes unlikely coalitions will be crucial to enable restructuring Hawaii's energy future. They praise Hawaiians for their resistance to fossil fuels and seeking 100% renewable energy, energy cost savings, and security benefits. They offer increased use of diverse renewables (wind, solar, geothermal, and hydropower) and some utility companies' efforts to increase renewables on the system as indicators that reclamation of Hawaii's energy economy is occurring. Policy energy professionals are encouraged that agencies, such as the National Renewable Energy Laboratory, offer resources to estimate impacts from installing low-carbon energy technologies. However, they pointed out that others impede reclaiming and restructuring of Hawaii's energy system. Some utility companies, for example, have opposed distributed generation and net metering. They also noted that, although the diversity of energy consumers means increased opportunities for consumer involvement in decisions, this is not a panacea for restructuring. Even a term so apparently straightforward as "stability" may be interpreted differently by each person. For some energy consumers, stability means that they can continue behaving as they have behaved for the past decade and that even suggesting they should change the time of day they do laundry is an affront. For others, stability means that their electricity cost and respiratory health (due to lower emissions/Kw) will remain the same, so long as they are willing to change the time of day when they wash that laundry. Policy-oriented professionals are understandably wary of alienating either group.

Utility-oriented energy professionals

Utility-oriented participants interpreted Sweeney's (2012) pathway to energy democracy somewhat differently than policy-focused professionals. Although they generally supported the idea of resisting the political power of the fossil fuel industry, they expressed fear of the uncertainty associated with change in general and, more immediately, with the intermittency of renewable energy sources. They worried that the speed of transition would lead to costly mistakes when it came to reclaiming and especially to restructuring the energy system. They emphasized the importance of diverse energy sources in support of resisting, reclaiming, and restructuring the energy system. When utility professionals talked about 100% in 2045, they expressed fear that completely removing fossil fuels from the mix will cause negative repercussions to daily operations. Not surprisingly, their focus on daily operability leads them to prefer an incremental resistance plan that initially includes fossil fuels in the energy mix, while transitioning toward renewable energy. Affordability is the criterion for reclamation that concerns them the most. They accept responsibility for minimizing the immediate costs of shifting from fossil fuels to renewable sources, linking it directly to reliability. They emphasize the challenges of resisting and reclaiming rather than restructuring Hawaii's energy system, perhaps because they believe a "best fit" for the future energy system is yet to be discovered (WP263).

Utility professionals acknowledge that Hawaii's transition to a renewable energy future depends on a vast array of diverse entities. Residential and commercial consumers are transitioning away from fossil fuels and are electing solar installations. Policy makers' efforts to push through the 2008 Clean Energy Agreement provided a significant step in Hawaii's transition to a renewable-based system. However, these professionals are concerned that policy makers may not fully comprehend the complexities involved in reclaiming Hawaii's energy economy. They express concern that technological innovation may not keep pace with policy mandates, which will negatively impact reliability. For example, in stating that the initial mandate is 30% by 2020, then shifting to 40% by 2030, utility operators worried that the 10% increase was driven more by wishful thinking than by technological advances. Additionally, consumers building off-grid microgrids concerned utility energy professionals because it complicates operations, which means the system incurs more expense.

Technology-oriented energy professionals

This high energy group expressed confidence that technological innovations for harnessing renewable energy are already driving the resist, reclaim, restructure objectives of energy democracy. Technology-oriented professionals were enthusiastic about innovations that enable Hawaii to resist the fossil fuel–based energy system. However, they also expressed caution about the complete elimination of fossil fuels. Some argued for managing the risk of overreliance on renewables by creating an adaptable system that can use many kinds of fuels, including biofuel, diesel, and liquid natural gas (LNG). From their perspective, the energy system cannot be reclaimed without maintaining (and potentially increasing) reliability and affordability, and they joined utility professionals in identifying operational flexibility as essential to success. In fact, many of their efforts focus on maintaining operational flexibility. They use computer modeling to plan and diagnosis events, such as extreme weather, that would disrupt power generation and transmission. They strive to develop or improve energy storage technologies that not only will be cost-effective and equitable for all customers but will increase penetration of renewable energy into the transportation, as well as the electricity sector. They emphasized the need for measurable milestones along the path toward 100% renewables and were concerned that policy makers might forget about the need to adequately test new technologies before counting on them to reliably shift distributed generation to transmission.

Hawaii's transition to a renewable energy future was seen as a cooperative effort among numerous interested parties, such as collaboration on grid design and public outreach across policy, utilities, technology experts, and consumers. Scholars from a variety of backgrounds have noted the strong links between renewable energy and the decentralized organizational configuration needed to compose a more democratic energy sector (Cantarero, 2020; Feldpausch-Parker et al., 2019). Energy professionals involved in HCEI demonstrated this connection as they spoke of how resistance to the old, centralized model begins with renewable energy financial incentives for consumers and energy storage companies. They attribute past progress toward reclaiming Hawaii's energy economy to policies and programs that evaluate the effectiveness of new technologies and that reward both developers and users of those that score well. They also note the benefits of Hawaii's PUC encouraging coordination across diverse energy organizations. Technically oriented participants were unique in their insistence that policy and regulatory entities must ensure that a restructured energy system is comprised of affordable and reliable technologies installed by licensed entities to protect consumers' investment and the renewables industry's integrity.

Across all three groups, energy professionals' discourse focused on solar and, to a lesser extent, wind energy opportunities and challenges, both of which can be configured to support a relatively decentralized energy system. The energy democracy movement challenges energy professionals to join in questioning generally accepted notions about both the appropriate mix of fuels and organizational structures for future energy systems (Burke & Stephens, 2017; Feldpausch-Parker et al., 2019; Sweeney, 2012; Szulecki, 2018). These professionals focused on their belief that, to be sustainable, democratic energy systems need to provide affordable and reliable energy, while also enhancing opportunities for individuals and communities to transcend the traditional gulf between consumer and producer, taking on the role of prosumer. At the same time, the system should continue to serve individuals who have no desire to take on such an active role.

Hawaii Power Summit participants were almost ten years into a 37-year goal of 100% renewables by 2045. They demonstrated cognizance of, if not comfort with the synergistic impacts of decisions regarding Hawaii's move toward renewable energy and expressed awareness that their transition requires resisting the current fossil fuel–based energy system as a prelude to reclaiming and then restructuring an energy system that will be sustainable and secure.

Conclusion

This case provides insight into the challenges of developing and practicing more democratic energy systems: In this "postcard from the future," policy drives the shift to renewable energy, but there is concern that policy makers may not have ensured that technological innovation will keep pace with policy demands. Individuals who focused on utility operations and on technology were wary about grid instability that may be triggered without maintaining a mix of diverse energy sources, including fossil fuels. Although policy makers expressed a less targeted concern with maintaining fossil fuels as part of the energy mix, they shared a desire to maintain stability.

Participants generally considered economics to be the most powerful sociotechnical function system and vacillated between how much they believed policy pushes the transition versus how much technology innovation pulls it forward. Coding with SPEED helped us avoid ignoring sociotechnical functions that were not stressed, such as culture. While not emphasized by energy professionals, culture was the bond that sutured economics, policy, and technology together in particular ways. These three sociotechnical functions were emphasized as the most central to Hawaii's energy transition. And, coding for culture encouraged us to explore how energy professionals differentiate what *they* think they should be doing from what *others* expect them to do, which helped us understand how they imagined energy democracy could (and could not) be practiced.

If energy democracy relies on resistance, reclamation, and restructuring, the most contentious issue seems to be that some energy professionals feel they are being rushed into decisions that require more careful thinking. They generally seemed comfortable with the idea of resisting the old, centralized approach to energy but expressed discomfort with determining what should be reclaimed and by whom. Finally, they feel pressure to restructure before they have had sufficient time to reclaim the human and natural resources needed for restructuring. This case suggests that energy democracy does not necessarily grow out of initiatives for clean energy systems, which is why rubrics such as Sweeney's (2012) resist, reclaim, restructure approach offer valuable guidance for future energy initiatives that seek to enable more democratic energy futures.

Note

1 The Institutional Review Board at the University of Texas at El Paso (771749–1) approved data collection.

References

50 Fun Facts About the 50th State (2015). https://governor.hawaii.gov/wp-content/uploads/2015/01/50-Fun-Facts-about-the-50th-State.pdf

Burke, M. J., & Stephens, J. C. (2017). Energy democracy: Goals and policy instruments for sociotechnical transitions. *Energy Research & Social Science*, *33*, 35–48.

Burke, M. J., & Stephens, J. C. (2018). Political power and renewable energy future: A critical review. *Energy Research & Social Science, 35*, 78–93.
Camp Booklet, Lausitzer Climate and Energy Camp 2012: Structure, Workshops, and Useful Information. (2012). Retrieved August 30, 2019, from https://archiv.lausitzcamp.de/wp-content/uploads/Lausitzcamp-2012_Infobroschere-EN-2012-08-06-Web.pdf
Cantarero, M. M. V. (2020). Of renewable energy, energy democracy, and sustainable development: A roadmap to accelerate the energy transition in developing countries. *Energy Research & Social Science, 70.*
Celebrating 10 Years of Success: Hawaii Clean Energy Initiative 2008–2018. (2018). Retrieved August 24, 2019, from www.hawaiicleanenergyinitiative.org/wp-content/uploads/2018/01/HCEI-Brochure_FINAL_Jan2018.pdf
Feldpausch-Parker, A. M., Endres, D., & Peterson, T. R. (2019). Editorial: A research agenda for energy democracy. *Frontiers in Communication, 4.*
Grimley, M., & Farrell, J. (2015, October 6). *Hawai'i at the energy crossroads.* Institute for Local Self-Reliance. https://ilsr.org/report-renewable-hawaii/
Hawaii Clean Energy Initiative: Goals and Objectives. (2019). Retrieved June 22, 2019, from www.hawaiicleanenergyinitiative.org/about-the-hawaii-clean-energy-initiative/goals-and-objectives/
Hawaii Energy Facts & Figures. (2018). https://energy.hawaii.gov/wp-content/uploads/2018/06/HSEO_2018_EnergyFactsFigures.pdf
Hawaii Power Summit 2016: Prioritizing the Challenges Ahead. (2018). Retrieved June 23, 2019, from https://www.euci.com/event_post/1216-hawaii-power-summit/
Hawai'i Visitor Statistics Released for 2018. (2019). Retrieved August 22, 2019, from https://www.hawaitourismauthority.org/media/2972/december-2018-visitor-statistics-press-release-final.pdf
The Hawaiian Islands. (2019). Retrieved August 22, 2019, from https://www.gohawaii.com/
Horton, C. C., Peterson, T. R., Banerjee, P., & Peterson, M. J. (2016). Credibility and advocacy in conservation science. *Conservation Biology, 30*(1), 23–32.
Luhmann, N. (1989). *Ecological communication.* University of Chicago Press.
Middleton, M., Hess, A., Endres, D., & Senda-Cook, S. (2015). *Participatory critical rhetoric: Theoretical and methodological foundations for studying rhetoric in situ.* Lexington Books.
State imports for Hawaii. (2019). Retrieved August 15, 2019, from https://www.census.gov/foreign-trade/statistics/state/data/imports/hi.html
Stephens, J. C., Peterson, T. R., & Wilson, E. J. (2013). Socio-political evaluation of energy deployment (SPEED): A framework applied to smart grid. *UCLA Law Review, 61*, 1930–1961.
Stephens, J. C., Wilson, E. J., & Peterson, T. R. (2008). Socio-political evaluation of energy Deployment (SPEED): An integrated research framework analyzing energy technology deployment. *Technological Forecasting and Social Change, 75*(8), 1224–1246.
Strauss, A., & Corbin, J. M. (1994). *Basics of qualitative research: Grounded theory procedures and techniques.* Sage Publications, Inc.
Sweeney, S. (2012). *Resist, reclaim, restructure: Union and the struggle for energy democracy.* Rosa-Luxemburg-Stiftungo.
Szulecki, K. (2018). Conceptualizing energy democracy. *Environmental Politics, 27*, 21–41.
US Energy Information Administration. (2019). Retrieved August 24, 2019, from https://www.eia.gov/
Weinrub, A., & Giancatarino, A. (2015, June). Toward a climate justice energy platform: Democratizing our energy future. *Local Clean Air Energy Alliance & Center for Social Inclusion*, 1–22. https://community-wealth.org/content/toward-climate-justice-energy-platform-democratizing-our-energy-future#:~:text=Toward%20a%20Climate%20Justice%20Energy%20Platform%3A%20Democratizing%20Our%20Energy%20Future,-Al%20Weinrub%20and&text=This%20new%20report%20from%20the,systems%20from%20corporations%20to%20communities
Welton, S. (2018). Grasping for energy democracy. *Michigan Law Review, 116*, 581–644.

31
PART V RESPONSE

Andrea M. Feldpausch-Parker

Introduction

Social science and humanities scholars have been researching energy systems for a long time, focusing on elements such as the boom and bust cycles of energy resource extraction (e.g., Lehman & Kinchy, 2021; Measham et al., 2019), justice (e.g., Endres, 2009; Klassen & Feldpausch-Parker, 2011; Sovacool et al., 2017), energy policy contexts (e.g., Fischlein et al., 2014), impacts from energy colonialism on marginalized communities (e.g., de Onís, 2018; Endres, 2012), public participation (e.g., Kinsella, 2004, 2020), public perceptions of energy technologies (e.g., Feldpausch-Parker et al., 2015; Peterson et al., 2015), and much more. The research featured in this part of the handbook examines and extends the trajectory of social science and humanities research on social and cultural tensions within energy democracy movements, especially as they relate to transitioning away from fossil fuels to more sustainable energy sources.

Themes

As noted in the Part V introduction (Chapter 26), the chapters in this section address energy sources and technologies from across the energy and geographic landscape. Not surprisingly, the presence of stakeholder tensions was a universal factor across resource/technology types. These tensions, however, related to a willingness (or not) to resist traditional paradigms of technocracy, demonstrations of power, low levels of public participation, and marginalization and other issues of justice. For instance, Kasperski and Kuchinskaya's chapter (Chapter 27) described a default to a technocratic approach of top-down information and technology dissemination to address radiation health effects from two international programs seeking to empower impacted communities from the Chernobyl disaster. Instead of empowering these communities, the programs minimally engaged them in reclaiming their land by providing them with tools to manage their own contamination levels (i.e., technical assistance). Kasperski and Kuchinskaya argue that this was at best a short-term solution that also left victims having to manage their own health following an energy catastrophe.

In Chapter 28, Feldpausch-Parker and Batill-Bigler demonstrated a growing dissatisfaction in the United States with the oil and gas industry by examining media coverage of hydraulic fracturing across different states. Their study demonstrated not only a growing concern with

DOI: 10.4324/9780429402302-36

environmental issues such as groundwater and surface water contamination as well as climate change but also a growing willingness by members of the public to engage in energy systems. Case in point: New York media coverage was full of anti- and profracking rallies, record-setting public comment periods, and political maneuvering. After a few-years-long fight, fracking was eventually banned in the state. They also found that events in New York influenced discourses in other states, demonstrating a traction in the fight against fossil fuel extraction but also a call to engage in civil action.

Finley-Brook's Chapter 29 addressed the truly horrific side of energy systems. While a plethora of studies talk about the dark side to fossil fuels and nuclear (including studies mentioned at the beginning of this chapter), renewables can and, in some cases, do follow similarly destructive paradigms. Large-scale dam projects across the globe have been linked to violations of human rights. Finley-Brook's chapter provides stunning examples of horribly skewed power dynamics, greed, and injustice. Her examples from Guatemala, Panama, Brazil, and Ecuador show a complete disregard for public involvement in decision making and institutionalization of fast and slow violence in the name of sustainable energy production.

In the last chapter of the section (Chapter 30), Horton, Hernandez, and Peterson's study of energy transitions in the US state of Hawaii serves as an example of struggles but also of hope. They found a continued heavy focus among energy professionals on economics, policy, and technology, though implicit connections were made to other factors critical to just and equitable energy transitions. This finding aligns with many energy democracy scholars' and practitioners' critiques of energy system transitions in that social and cultural context is often ignored. Even with this glaring issue, the Hawaii case study shows promise in that energy professionals were seeking ways to resist the fossil fuel–dominated, centralized model of energy systems and to replace it with a more decentralized model that relies on locally secured renewable energy resources.

Future directions

Tensions are an inevitability in any decision-making space. Conflict scholars would even argue that conflict can be a productive component of those processes (Feldpausch-Parker & Peterson, 2020). As Mouffe (2000) states: "What is specific and valuable about modern liberal democracy is that, when properly understood, it creates a space in which this confrontation is kept open, power relations are always being put into question and no victory can be final" (p. 15). With this in mind, it would be valuable to both energy democracy scholarship and practice to showcase increasingly diverse examples of energy decision making that allow for a plurality of voices and for the presence of dissent and deliberation. Essentially, more examples of energy democracy in practice, such as those in Part 6 of this book, and at various scales of governance, such as in Part I, are needed to guide navigation of these tensions productively and in ways that benefit both people and Earth.

References

de Onís, C. M. (2018). Energy colonialism powers the ongoing unnatural disaster in Puerto Rico. *Frontiers in Communication, 3.* https://doi.org/10.3389/fcomm.2018.00002

Endres, D. (2009). From wasteland to waste site: The role of discourse in nuclear power's environmental injustices. *Local Environment, 14*(10), 917–937.

Endres, D. (2012). Sacred land or national sacrifice zone: Competing values in the Yucca Mountain controversy. *Environmental Communication, 6*(3), 328–345.

Feldpausch-Parker, A. M., Burnham, M., Melnik, M., Callaghan, M. L., & Selfa, T. (2015). News media analysis of carbon capture and storage and biomass: Perceptions and possibilities. *Energies, 8*, 3058–3074.
Feldpausch-Parker, A. M., & Peterson, T. R. (2020). Conflict in wildlife science and conservation. In N. J. Silvy (Ed.), *Techniques for wildlife investigations and management* (8th ed., Vol. 2, pp. 82–92). Johns Hopkins University Press.
Fischlein, M., Feldpausch-Parker, A. M., Peterson, T. R., Stephens, J. C., & Wilson, E. J. (2014). Which way does the wind blow? Analyzing the state context for renewable energy deployment in the United States. *Environmental Policy and Governance, 24*(3), 169–187.
Kinsella, W. J. (2004). Public expertise: A foundation for citizen participation in energy and environmental decisions. In S. P. Depoe, J. W. Delicath, & M. F. A. Elsenbeer (Eds.), *Communication and public participation in environmental decision making* (pp. 83–98). SUNY Press.
Kinsella, W. J. (2020). Extracting uranium's futures: Nuclear wastes, toxic temporalities, and uncertain decisions. *The Extractive Industries and Society*, 7(2), 524–534.
Klassen, J. A., & Feldpausch-Parker, A. M. (2011). Oiling the gears of public participation: The value of organisations in establishing Trinity of Voice for communities impacted by the oil and gas industry. *Local Environment, 16*(9), 903–915.
Lehman, J., & Kinchy, A. (2021). Bringing climate politics home: Lived experiences of flooding and housing insecurity in a natural gas boomtown. *Geoforum, 121*, 152–161.
Measham, T. G., Walton, A., Graham, P., & Fleming-Munoz, D. A. (2019). Living with resource booms and busts: Employment scenarios and resilience to unconventional gas cyclical effects in Australia. *Energy Research & Social Science, 56*. https://doi.org/10.1016/j.erss.2019.101221
Mouffe, C. (2000). *The democratic paradox*. Verso.
Peterson, T. R., Stephens, J. C., & Wilson, E. J. (2015). Public perception of and engagement with emerging low-carbon energy technologies: A literature review. *Energy & Sustainability, 2*. https://doi.org/10.1557/mre.2015.12
Sovacool, B. K., Burke, M., Baker, L., Kotikalapudi, C. K., & Wlokas, H. (2017). New frontiers and conceptual frameworks for energy justice. *Energy Policy, 105*, 677–691.

PART VI

Energy democracies in practice

32

ENERGY DEMOCRACIES IN PRACTICE

Introduction

Stephanie L. Gomez and Danielle Endres

Part VI of the handbook introduces a series of case studies that highlight energy democracies in practice. The chapters in this section demonstrate how energy democracies can act like a verb, implying that energy democracy is an ongoing active process that is enacted through on-the-ground struggles in myriad contexts. Each chapter in this section analyzes and describes current practices of energy democracy as well as imagines possible next steps to actively transform energy systems in line with principles of democracy. While a variety of examples are included in these chapters—from specific campaigns, such as carbon neutrality pledges (Chapter 33), to university partnerships (Chapter 34), to the relationships between the Global North and Global South (Chapter 35), to Indigenous Nations (Chapter 36)—there are many more possibilities for energy democracies in action.

This section's focus is valuable because it grounds conceptualizations and theorizations of energy democracy in particular places, moments, contexts, and situations. As Chilvers and Pallet (2018) note, energy democracy does not look the same in every situation. They resist approaches to energy democracy that rely on fixed and preformed meanings of democracy and participation, instead advocating for analysis of how "different forms of energy democracy and their publics are *made*, constructed, and co-produced" (p. 3). Hence, a focus on energy democracies recognizes the plurality of ways of engaging publics, democratic principles, energy systems, and modes of participation. Specific enactments of energy democracy have different audiences, situations, contexts, political structures, assumptions about democracy, models of participation, and more. The chapters in this part of the handbook lay out four different ways of enacting energy democracy that are linked to the places and people involved.

This part of the handbook advances an understanding of energy democracies as embodied and emplaced. Embodiment recognizes that energy democracy does not happen without people taking actions to participate in energy decisions that affect them. Energy democracies are fundamentally lived through the power of people who are putting their bodies on the line toward struggling for more just, equitable, and democratic systems of energy transition. As embodied phenomena, a focus on energy democracies allows for analysis of the power dynamics, systems of oppression, and policies that treat people inequitably in relation to structures of racism, sexism, classism, ableism, and more.

Emplacement couples with embodiment to recognize that people are always situated in place. Place is a semibounded location with material and symbolic elements that allow it to be

DOI: 10.4324/9780429402302-38

recognized as distinct from other locations or broader notions of empty space (Endres & Senda-Cook, 2011). Like embodiment, places are diverse and inextricably linked to power dynamics, systems of oppression, and normative expectations. Places cannot be unconnected from people who make, contest, and remake meaning in those places. The Dakota Access Pipeline (DAPL) water protectors exemplify the notion of a multiplicity of embodied and emplaced struggles for energy democracy. The contours of the DAPL controversy reflect the people (Dakota Indigenous people) and place (Dakota sacred land) in which the struggle is taking place and cannot be theorized without attention to the specificities of people and place.

An energy democracy research agenda needs to be responsive to the wide variety of local instantiations of embodied and emplaced energy democracies. Indeed, the variance in energy democracies in practice makes the creation of a unified framework or theory of energy democracy difficult (see Szulecki & Overland, 2020). While the framework presented in the introduction of this handbook does seek to situate ED at the intersection of justice, participation, power, and technology, it does so with a deliberate recognition that the mix of these elements will be different for each embodied and emplaced enactment of energy democracy (see Chapter 1).

In Chapter 33, McKasy and Yeo focus on coalition-building in the state of Utah through community support for carbon-neutral pledges. They assess residential stakeholders' attitudes toward government carbon-neutral pledges in an effort to understand perceived risk and support of such pledges. In doing so, they analyze the role that coalition-building plays in energy democracy and in creating progressive climate change initiatives. Ultimately, they argue that carbon-neutral pledges exhibit key hallmarks of energy democracy and should be further employed in efforts to fight climate change.

Moving to a broader scale, in Chapter 34, Pérez-Lugo, Ortiz García, and Orama Exclusa focus on the role that universities can and should play in on-the-ground energy democracy work. Locating their study in Puerto Rico, the authors explore the role that the National Institute for Energy and Island Sustainability (INESI) at the University of Puerto Rico played before, during, and after the September 2017 hurricanes. They push back against the hands-off "ivory tower" position that universities often take toward community-based action. Instead, they argue that universities should be responsible for driving collaborations with the community that center energy democracy and justice.

In Chapter 35, Campbell, Cloke, and Brown interrogate Euro-centric assumptions about the Global South, particularly in regard to low-carbon energy transitions. They study the historical development and contemporary role that the UK Low Carbon Energy for Development Network (LCEDN) plays in energy research, as well as their partnership with other communities of practice, including NGOs, businesses, and policy communities that are all dedicated to energy justice and low-carbon transitions. They argue that the LCEDN demonstrates that true energy justice is only possible with decentralized, localized systems. Ultimately, Campbell et al. conclude that energy democracy should strive for distributed systems of governance in order to achieve greater equity and push back against the disempowerment inherent in centralized structures of energy governance.

In Chapter 36, Bessette et al. refer to energy democracy as a social movement, aimed toward creating energy policy that is just, equitable, and inclusive. Key to this movement is energy sovereignty, which refers to individuals and communities having the right to make decisions about their energy policy; unfortunately, US energy policy actively works against energy sovereignty by centralizing control over energy production and consumption. The authors focus their analysis on the Michigan Community Anishinaabe and Rural Energy Sovereignty project (MICARES), which centers on the Anishinaabe teachings of the medicine wheel and aims to facilitate energy sovereignty for Indigenous Nations and rural communities. They use four case studies from two locations in Michigan and two Indigenous communities to highlight the strategies used by MICARES in pushing for greater energy sovereignty and identify the capabilities and constraints of each.

References

Chilvers, J., & Pallett, H. (2018). Energy democracies and publics in the making: A relational agenda for research and practice. *Frontiers in Communication*, *3*, 1–16. https://doi.org/10.3389/fcomm.2018.00014

Endres, D., & Senda-Cook, S. (2011). Location matters: The rhetoric of place in protest. *Quarterly Journal of Speech*, *97*(3), 257–282. https://doi.org/10.1080/00335630.2011.585167

Szulecki, K., & Overland, I. (2020). Energy democracy as a process, an outcome and a goal: A conceptual review. *Energy Research & Social Science*, *69*(101768), 1–14. https://doi.org/10.1016/j.erss.2020.101768

33

CARBON-NEUTRAL PLEDGES

Public opinions, opportunities, and challenges for energy democracy

Meaghan McKasy and Sara K. Yeo

Introduction

In June 2017, President Donald Trump announced his intent to leave the landmark Paris Agreement, an international accord pledging to reduce greenhouse gas emissions and address mitigation, adaptation, and financial issues related to climate change (Shear, 2017). While individuals and businesses expressed discord with this action through civic climate marches (Fandos, 2017) and public advertisements (Winston, 2017), perhaps some of the most interesting reactions occurred in government. American politicians at various levels of government, particularly cities, were bolstered by citizen feedback and pledged to still adopt the agreement (Mindock, 2017). While cities only occupy 2% of the surface area on the planet, they contribute to two-thirds of global energy use and release over 70% of carbon emissions (Intergovernmental Panel on Climate Change, 2018). As such, the response of cities to climate change is both symbolically critical and substantively influential for addressing global carbon emissions.

Municipal governments across the United States have intensified their efforts to reduce and offset their emissions through many approaches. One such example is a carbon-neutral pledge. According to the United Nations Framework Convention on Climate Change (UNFCCC), carbon-neutral pledges are actions that organizations, businesses, and governments take to measure and report greenhouse gas emissions, reduce their overall carbon footprint, and offset the remaining emissions (UNFCCC, n.d.a). In the United States, over 90 cities have adopted clean energy or carbon-neutral goals (Sierra Club, n.d.). While this seems minor compared to the 187 countries that have ratified the Paris Agreement to date (UNFCCC, n.d.b), without national leadership on climate change in the United States, cities and citizens play a crucial role.

Since the United States indicated its intention to withdraw from the Paris Agreement, citizen pressure on local governments exemplifies the initiative for improved opportunities for public participation both in political arenas and in energy creation and distribution. Increasingly, citizens strive to be involved in energy-related decision making as more than just consumers; instead, they are *prosumers* who have an active role in energy-related decisions (Cozen et al., 2018). While this involvement is transpiring globally, the democracies of the Northern hemisphere offer clear examples of participatory energy efforts (Stephens et al., 2015). This phenomenon of energy democracy deprivatizes the energy economy, engages citizen

 DOI: 10.4324/9780429402302-39

stakeholders, and dramatically increases renewable energy production to reduce carbon emissions (Sweeney, 2014).

There are many instances in which the energy democracy movement can be observed, such as electric utility reform efforts or government regulation of energy production (Kunze & Becker, 2012). Municipal government carbon-neutral pledges are another example of energy democracy; community residents unite through deliberation and cumulative voting, demonstrating inclusivity in the energy decision-making process through equitable distribution of control. In this way, tenets central to energy democracy (justice, power, and participation) are also foundational to carbon-neutral pledges (Feldpausch-Parker et al., 2019). Further, carbon-neutral pledges address what many climate change mitigation strategies omit: the social and cultural change necessary for effective policy change (Ehrhardt-Martinez et al., 2015). Too often, organizations, governments, and reports identify climate change adaptation and mitigation strategies as technological impediments, ignoring the social aspects necessary for successful transition (e.g., equitable representation of stakeholders from diverse backgrounds [e.g., gender, ethnicity]). An inclusive approach is distinctly contrary to the energy realm that is traditionally male dominated and excludes women and minorities (Stephens, 2019). Energy democracy elevates inclusiveness by unifying diverse stakeholders through activism and involvement in community-oriented energy governance (Fairchild, 2017). Accessibility and addressing the concerns of communities are identified prerequisites for any successful energy policy in a democracy (Burke, 2018; Chilvers & Pallett, 2018; Teron & Ekoh, 2018).

While a solitary carbon-neutral pledge may appear to have minimal effects, the cumulative influence of a multiscale approach (individual, organizational, and governmental pledges) could have a global impact. Each pledge is tasked with addressing climate change and limiting global temperature rise. Individual pledges allow people the opportunity to track their emissions, understand where their emissions come from, and improve their personal actions. For instance, the UNFCCC has a citizen climate pledge where anyone can hold themselves accountable for reducing their carbon footprint (UNFCCC, n.d.a). Business pledges may proactively address issues related to extreme weather events, create innovative jobs, and take actions that make economic sense. In 2019, Amazon made headlines by pledging to be carbon-neutral by 2040, partially in response to corporate employees' criticism that the company lacked climate leadership (Selyukh, 2019). Finally, while municipal and national government pledges not only symbolically show decisive bold steps, they also create the legal standards to instigate concrete change. As of 2019, New York, New Mexico, Washington, and the District of Columbia required carbon-free energy by the turn of the century, while Maine and Nevada pledged similar voluntary goals (Swartz et al., 2020). These actions influence all individuals and organizations within their territories.

While previous research on the publics' role in energy democracy has examined municipal outreach efforts in areas such as Washington, D.C. (Teron & Ekoh, 2018), research has yet to explore citizen attitudes toward and involvement in discourses for combined action such as government carbon-neutral pledges. An assessment of individual attitudes about carbon-neutral pledges is important for several reasons. First, the number of municipal governments participating in carbon-neutral pledges is growing, often due to citizen mobilization efforts, normalizing the action among governments and putting pressure on corporations within their district. Second, carbon-neutral pledges allow publics to engage in and take ownership of the energy decision-making process through public comment periods and voting. Finally, carbon-neutral pledges account for the unique social, political, and economic complexities faced by substate governments by taking a more concentrated approach with city and county administrations.

This exemplifies a key component of energy democracy: integrating social change with a transition to renewable energy (Burke & Stephens, 2017).

Carbon-neutral pledges take citizens' wants, needs, and lifestyles into account at a microlevel, and therefore their attitudes play a pivotal role in reducing emissions through successful climate mitigation and adaptation policies. In this chapter, we analyze public attitudes about carbon-neutral pledges in the state of Utah as these attitudes provide insight toward pledges as a strategy of energy democracy. We first provide an overview of the current energy landscape and changes that have allowed for the development of mitigation strategies such as carbon-neutral pledges. We then provide a close examination of the components of energy democracy, focusing on how carbon-neutral pledges embody its principles. Then, using a case study, we examine residential stakeholders' perceptions of carbon-neutral pledges for municipal governments in Utah, questioning how these actions may inspire future efforts to address climate change. While carbon-neutral pledges in Utah represent energy democracy in action, we also offer commentary on the potential limitations of this strategy. This chapter discusses the implications of these pledges, including equitable citizen involvement in energy production and dissemination and strategies for other governments and organizations aiming to enact carbon-neutral pledges. We argue that carbon-neutral pledges represent the key tenets of energy democracy: participatory action, social justice, and a clean energy future. Further, this research explores the dynamic social, political, and economic discourses that led to carbon-neutral pledges in Utah and outlines some of the roles that cities and citizens play in countering climate change.

Background

The current energy and political landscapes

Responding to climate change is one of the most demanding sociopolitical problems of our time (Giddens, 2009). The International Renewable Energy Agency (IRENA) estimates that warming above 2°C will lead to climate damage costs of US$1.5–3.3 trillion a year in 2050 (IRENA, 2017). Organizations such as the National Academies of Sciences, Engineering, and Medicine (NASEM) in the United States and Intergovernmental Panel on Climate Change (IPCC) promote reducing emissions with resilient energy technologies and advocate that an appropriate resolution requires collective action through adaptation and mitigation efforts. While adaptation refers to changes in natural or human systems to withstand the existing and forthcoming effects of climate change, mitigation efforts focus on reducing emissions or improving the sinks of greenhouse gases, i.e., carbon sequestration methods (Klein et al., 2005). Corporations and governments across the globe must commit to decarbonization strategies in hopes of limiting warming to the 1.5°C goal set in 2016 by the UNFCCC in the Paris Agreement (Rogelj et al., 2016).

After the turn of the century, there was an increased interest in renewable energy projects. In the United States, following the 2008 election, President Obama set lofty goals and heavily subsidized renewable energy production, bringing many new projects onto the power grid (Wesoff, 2011), the interconnected network that electric utility companies use to balance the supply and demand of energy through powerlines, switches, and transformers (Fang et al., 2012). The Obama administration also implemented the Clean Power Plan, which required individual states to meet specific standards with respect to the reduction of carbon dioxide emissions. Similarly, while renewable portfolio standards exist in several countries such as the United Kingdom, Poland, Chile, Sweden, and Italy, they often differ in the aggressiveness of targets and strategies by area (European Parliament and Council of 11, 2018).

A substantial decrease in the cost of solar power and growing consumer interest in green energy stimulated a dramatic rise in solar adoption at the residential, community, and utility scales during the last decade in the United States (Muro & Saha, 2016). Energy utility companies developed networks of two-way communications to coordinate their customers' energy use and production, while simultaneously making the power grid more efficient, secure, and reliable. This interactivity in energy creation and dissemination allowed for greater customer engagement, leading to more informed customers and increased energy efficiency (Langheim et al., 2014). In recent years, solar was second only to natural gas for newly constructed or permitted energy generating capacity in the United States (Fox-Penner, 2014). In 2018, the total capacity for solar was estimated to be about 64.2 GW, or enough to power 12.3 million homes (Wood Mackenzie Limited, 2019). These rapid changes to the energy mix presented challenges with distribution but also displayed the national commitment to a decarbonized energy infrastructure.

However, following the 2016 presidential election, climate and energy discussions in the United States were transformed. When President Donald Trump pledged to leave the Paris Agreement, national climate action in the United States stalled or even regressed (Jotzo et al., 2018). The lack of leadership from the world's largest emitter of greenhouse gas emissions raised concerns internationally about reaching global emissions targets and additional countries rescinding on their own pledges (Jotzo et al., 2018). Following President Trump's announcement, former President Obama stressed that mayors and local governments had to be the leaders of future American climate action (Madhani, 2017; Smith, 2017). In June 2017, almost 400 US mayors, representing over 68 million Americans, committed to uphold the climate change goals set in the Paris Agreement (Madhani, 2017), emphasizing the power of local governments to effect change. Cities around the country pledged to uphold the nonbinding agreement by intensifying their efforts to decrease emissions and create a clean energy economy.

Energy democracy

Energy democracy is a holistic concept that incorporates the tenets of democratic public participation with a transition to a distributed, clean energy economy. Previous research has described energy democracy as a decarbonized energy transformation with grassroots civic action, meaning it involves persons in energy decision making while emphasizing the importance of reducing fossil fuel use and increasing distributed renewable energy systems (Fairchild, 2017). Advocates of energy democracy envision an energy future in which citizens are active participants in publicly owned utility companies and renewable energy cooperatives. Within energy democracy, consumers move from a passive role to an active role in a decentralized energy environment. These *prosumers* participate in energy-related decisions, from production to consumption (Giancatarino, 2012). As such, energy democracy emphasizes participation, mutual control, collective benefits, and equitable decision making, which results in less energy consumption and the promotion of renewable energy (Kunze & Becker, 2012).

Since energy democracy calls for participatory communication and broader civic engagement in energy systems change (Stephens, 2017), the notion of democracy is key. Democratic engagement is the foundation of modern liberal societies around the world and can be seen in the cooperation necessary for resolving environmental conflicts (Stephens et al., 2015). Within a democracy, citizens exercise their right to vote by electing representatives with the goal of giving power to the majority, while still protecting the rights of the minority. Energy democracy embodies this by giving citizens a voice and the ability to evaluate the legitimacy of energy decisions made by leaders (Clarke, 2017).

This thriving civic role in the energy landscape aligns with a broader effort to address human-caused threats to ecological systems and the failure of social institutions to confront these threats (Cox, 2007). Though energy democracy has been compared to other terms such as "energy justice" and "environmental democracy," it is its own unique movement. Recent attempts to merge energy matters, public engagement, and social and political landscapes have resulted in a burgeoning energy democracy research agenda (Endres et al., 2016; Feldpausch-Parker et al., 2019). Within this agenda, energy democracy is conceptualized as geographically distributed, publicly accessible, environmentally conscious, and equitable (Becker & Naumann, 2017). Scholarship has examined energy democracy in municipal sustainability planning (Teron & Ekoh, 2018), in electric utility company communication efforts (McKasy & Yeo, 2018), and in community energy projects (van Veelen, 2018). In the political arena, Szulecki (2018) identifies that "[g]overnance in energy democracy should be characterized by wide participation of informed, aware, and responsible political subjects, in an inclusive and transparent decision-making process relating to energy choices, with the public good as its goal" (p. 35). Szulecki further emphasizes that public, private, and collective means are all necessary for civic empowerment and autonomy. This cooperative effort of governance may be observed in small municipalities, such as cities.

The role of cities

Despite only occupying a tiny fraction of the planet's surface area, cities use more than two-thirds of global energy and create the most carbon emissions (Day et al., 2018). It is critical to understand cities' carbon emissions, since despite their large impact—or perhaps because of it—cities have the potential to be much greener (Mitchell et al., 2018). For example, New York City claims to be North America's greenest city since its residents have a much smaller carbon footprint than other American cities and rural areas due to the availability of public transit, lower energy use, and proximity of necessities (Owen, 2009). As urban areas continue to grow exponentially, they are both a problem for and a solution to climate change. In fact, research indicates that city action alone could help the United States reach 40% of their Paris Agreement goal (C40 Cities, 2016). Together, cities create economies of scale with purchasing power to build more sustainable and resilient communities.

The decline of national climate action in the United States stimulated several city-centered organizations and environmental networks both in the United States and internationally, such as C40 Cities (www.c40.org), We Are Still In (www.wearestillin.com), and the Global Covenant of Mayors for Climate & Energy (www.globalcovenantofmayors.org). Many of these organizations have participation standards, such as committing to a specific quantified target, developing a reduction plan, and reporting progress toward their goals (Day et al., 2018). Currently, only 45% of US cities with populations of 75,000 or higher participate in a climate change organization (Lusk, 2018). But the lack of national leadership in the United States and vocal citizens are creating pressure for mayors and cities to lead climate action and decrease global carbon emissions. This pressure exists globally as well. Edmonton, Alberta, once promoted as the oil capital of Canada, recently hosted the IPCC's 2018 Cities & Climate Change Conference, highlighting its evolution as a city leading efforts to address climate change. It is notable that most examples exist in Europe and North America since few cities in the Global South have committed to reducing emissions (Bansard et al., 2017).

This pressure for cities to act is critical since addressing climate change remains a politicized issue (Ballew et al., 2019). Research indicates that 84% of US mayors believe in anthropogenic climate change, however, that number is divided by partisan ideologies, with 95% of

Democrats and 50% of Republicans believing that climate change is due to human activities (Einstein et al., 2019). Yet there is hope for bipartisan climate change mitigation efforts. For example, former New York City Mayor and Republican Michael Bloomberg[1] and Democratic California Governor Jerry Brown collaborated to found America's Pledge, an organization that unites public and private sectors with the goal of reducing emissions and acting on the goals of the Paris Agreement in order to keep the United States in a leadership position for addressing climate change. America's Pledge works with and quantifies the actions of cities, states, businesses, and other stakeholders committed to reducing greenhouse gas emissions. Most of the world's population lives in urban areas, and so the choices made by cities like New York can impact the entire world (Bloomberg & Pope, 2017). Therefore, understanding how much, where, and why cities emit carbon remain key science research goals (Mitchell et al., 2018).

A carbon-neutral pledge is a common action for a city responding to climate change. As previously stated, when organizations, businesses, or governments make a carbon-neutral pledge, they commit to measuring and reporting greenhouse gas emissions, making efforts to reduce their emissions and to offset any remaining emissions (Flagg, 2015). In addition to the emissions reductions, these pledges act as public signals of mayors' environmental commitments, which may fuel the ambition of others, create peer pressure, and provide an opportunity for non-national stakeholders to share best practices. Carbon-neutral pledges are a unique approach to reducing greenhouse gases. For a pledge to be successful, it concurrently needs both bottom-up citizen mobility for the effort and top-down political leadership. The citizens may be the impetus that puts pressure on the government to support a pledge, but the political leader must propose the act for a popular vote.

Carbon-neutral pledges are more abstract than actions like carbon taxes that impose a fee on carbon-based fuels, forcing individuals to pay the *true* price for the full consequences of emissions (Metcalf & Weisbach, 2009). A tax puts strong pressure on the operating costs of energy development, and this monetary disincentive could stimulate a clean energy transition. Meanwhile, a carbon-neutral pledge acts as a reward system that would be weighted toward capital costs, such as the purchase of an electric vehicle. As a carbon-neutral pledge is a vast goal, it may involve multiple clean energy actions such as increased energy efficiency or community solar to help reach the overall target. One necessary component for any successful policy is citizen support, which is critical not only to enact a policy but also to secure citizen commitment for proper implementation with necessary individual and societal changes. As Szulecki (2018) noted, governance in energy democracy requires individuals to be accountable and knowledgeable political subjects to make informed energy decisions. Therefore, it is important to examine citizens' understanding of carbon-neutral pledges and attitudes toward public participation in the energy arena.

Carbon-neutral pledges in Utah: a case study

In the United States, 81% of citizens live in urban areas (US Census Bureau, 2010). In Utah, three-quarters of the state's population lives along the Wasatch Front, an urban region stretching along the Wasatch Mountain Range in the north-central part of the state, with the capital Salt Lake City at the central point. The population of Utah is projected to more than triple by 2060 (from 2 million to almost 7 million), with most of that growth occurring along the Wasatch metropolitan corridor (World Population Review, 2019). With rapid urban population growth occurring in a concentrated area of a predominantly rural state, cities and counties experiencing the most growth will face increased political pressure to address concerns such as housing, transportation, and environmental issues like water supply, air quality, and climate change (Gochnour, 2018).

On August 30, 2018, 16 mayors from cities across Utah wrote an opinion editorial in a local newspaper, *Deseret News*, calling for bipartisan actions to address climate change (Utah Cities and Mayors, 2018). At the time of this study in 2018, four communities in Utah, each with unique advantages and challenges, pledged to be carbon-neutral by 2032: Salt Lake City, Park City, Moab, and Summit County.[2] Though these communities are considered liberal hubs in a generally conservative state, climate change threatens their populations in ways that could more broadly harm the state's economy. Park City (the largest city in Summit County) and Moab are predominantly tourist towns; they face the unique challenge of seasonal and large population changes with the influx of tourists. For example, though Park City has 8,500 year-round residents, it becomes a metropolitan city for two weeks every winter during the Sundance Film Festival. Almost 125,000 people attended the 2018 festival (Hattem, 2018). Meanwhile, Moab is the gateway to two of Utah's five national parks, which are large contributors to the state's economy (Leaver, 2016) and susceptible to changes in the climate. Additionally, the commercial ski industry is a major draw for the state; the capitol, Salt Lake City, was selected as the US nominee for the 2030 Winter Olympic Games (Winslow, 2018). Yet climate change is projected to negatively impact the ability of certain locations, including Salt Lake City and Park City, to host the Winter Olympics due to warming temperatures and rising snow elevations (Scott et al., 2015). With the susceptibility of these vital economic industries, Utah may seem like an obvious place for carbon-neutral pledges to emerge. However, the politically conservative nature of the State's population makes the unique dynamics of these energy discourses thought provoking.

This tumultuous time in climate change policy could result in a widening gap in public opinion and attitudes about environmental policies. Recent research indicates that the majority of Americans believe in climate change; however, there is less agreement about cause and solutions (Leiserowitz et al., 2018). Due to this belief consensus, research must focus on adaptation and mitigation solutions (as suggested by Pan et al., 2019), such as carbon-neutral pledges. Currently, there is little understanding of public attitudes toward carbon neutrality and how public opinion may inform proposed policies to reach these goals. Municipal representatives in Utah repeatedly expressed that, despite making these pledges, governments had little knowledge of citizen understanding of carbon-neutral pledges. Salt Lake County's Sustainability Manager stated, "We take these steps and make these sacrifices, but honestly right now we don't really know if our constituents support our actions or see the benefits" (A. Yoder, personal communication, April 9, 2018). On this path toward carbon neutrality, it is important to gain an understanding of what factors may influence individuals' beliefs, attitudes, and subsequent formation of public opinion about climate change adaptation and mitigation efforts.

Therefore, an online survey was conducted in the fall of 2018 to examine these questions and decipher public opinion of carbon-neutral pledges in the state of Utah. The survey of 942 individuals was collected to represent the demographics of Utah. For the quota sample to align with the gender, ethnicity, and geographic distribution of the state, the median age was approximately 50 years, which was older than the average age of 30 years (World Population Review, 2019). Further aligning with demographics of the state, most respondents self-identified as white, Republican, and holding a college degree. Relative to topics such as climate change or drought, participants reported not feeling very well-informed about carbon-neutral pledges. Regardless, after reading a definition, respondents reported high levels of support for carbon-neutral pledges made in local government and elsewhere.

The predominantly optimistic response regarding carbon-neutral pledges in Utah presents a useful analysis for governments undertaking or considering pledges, particularly in politically conservative areas. According to the data, respondents perceived many benefits to carbon-neutral pledges, including decreased human impacts on the environment and reliance on fossil fuels and making the planet safer for future generations. In accordance, participants perceived relatively few risks, generally disagreeing with the idea that carbon-neutral pledges may hurt the economy or place unnecessary pressure on local communities. Most Utahns ranked carbon-neutral pledges occurring in their local area as more important than those happening in other places, indicating that proximity was a key factor for positive opinions. When asked questions pertaining to their own self-efficacy or government effectiveness for carbon-neutral pledges, results indicated that while individuals were likely to make personal changes, they also believed that governments play integral roles. Interestingly, individuals speculated that local governments were more likely than the federal government to take action against climate change (e.g., making a carbon-neutral pledge).

These findings indicate that carbon-neutral pledges are considered a generally positive action in Utah due to the high levels of support and perceived benefits. In Utah, over half of the state population identifies with the predominant faith group, the Church of Jesus Christ of Latter-Day Saints (LDS). In general, Utah's population identifies as Republican (54%). However, within the LDS faith in Utah, that number rises to 76% identifying as Republican (Canham, 2018; "Party Affiliation," 2014). Typically, individuals who identify as Republican hold more politically conservative views. Within the survey, identifying as politically conservative was negatively related to support, benefits, and importance of carbon-neutral pledges, along with efficacy measures, while positively associated with perceived risks of carbon-neutral pledges. While political ideology and religiosity are often closely related, interestingly, identifying as strongly religious was only significantly related to increased perceptions of risk—religiosity did not correlate with perceptions of benefits or support for carbon-neutral pledges.

These responses are provocative when viewed through the lens of energy democracy: While individuals recognize the importance of their personal actions, they also exert pressure on local governments for climate-related action. Much research has taken the term "prosumer" to mean those who are active in their personal electric energy generation. Yet, the foundational understanding of energy democracy is about participatory communication and broader civic engagement in energy systems change (Stephens, 2017). Therefore, in recent years scholars have worked to deepen the understanding of prosumer to include individual interest and involvement in other forms. For instance, a prosumer is not necessarily a person who produces energy in an attempt to separate from the power grid but rather a person who is an active participant in the power grid, e.g., through home solar ownership or monitoring with smart meters (Stephens et al., 2015). Carbon-neutral pledges offer the opportunity for greater energy involvement and inclusivity for interested prosumers. Not everyone can put solar on their homes or pay to offset their carbon footprint, but they can vote to ensure the production of clean energy rather than fossil fuels, thereby participating in decisions about energy production and dissemination. Individuals exercise their rights through the democratic system: specifically, by electing representatives who act equitably for a clean energy future by listening to the requests of their constituents and voting for actions such as carbon-neutral pledges.

This democratic involvement in the energy system was visible during the 2019 Utah State Legislative Session when the legislature passed House Bill 411 (H.B. 411), a bill that enacted the Community Renewable Energy Act in the Public Utilities Code, which required the Utah Public Service Commission to define rules, rates, and expectations in order to aid communities setting goals toward a net-100% renewable electricity portfolio (Community Renewable Energy Act, 2019). This landmark legislation was heralded as a bipartisan breakthrough for Utah's clean energy goals and first-of-its-kind piece of legislation due to the collaboration between communities and investor-owned utilities (Vasic, 2019). Within this act, municipalities had until the end of 2019 to make a renewable energy pledge and qualify for the benefits offered by H.B. 411 (Community Renewable Energy Act, 2019). So by the close of 2019, less than one year after the study described in this chapter took place, 24 cities and counties in the state of Utah had carbon-neutral pledges (Stevens, 2020). All entities involved in the legislation highlighted the importance of city, state, and community individuals working together toward a common goal. While Park City, Salt Lake City, Moab, and Summit County cleared the path in 2017, it took years of community negotiations with the local utility company, Rocky Mountain Power, to negotiate the Community Renewable Energy Act (Beebe, 2019). Once passed, this legislation opened the flood gates for other communities to follow suit, resulting in more than a quarter of the state's population potentially receiving net-100% renewable energy by 2030 (Stevens, 2020).

The clear shortcoming to this proactive clean energy advancement is that of the 24 communities with carbon-neutral pledges in Utah, 21 exist in urban areas. While those communities who pledged as a result of the Community Renewable Energy Act will transition to 100% renewable energy, those who did not will be left to pay for Rocky Mountain Power's costly coal-based infrastructure (Beebe, 2019). These coal-fired power plants are anticipated to be in working condition until past 2040. Yet legislation such as H.B. 411 will potentially make Rocky Mountain Power shorten the timelines for power plant retirement (Vasic, 2019), meaning the economic burden of these aging coal-fired power plants will be predominantly on the shoulders of rural Utahns.

The stark energy future facing rural Utahns mirrors their own understanding of carbon-neutral pledges. When questioned about carbon-neutral pledges, individuals who self-identified as living in rural areas reported significantly lower levels of support, perceived benefits, and importance of carbon-neutral pledges. Additionally, they reported lower levels of self-efficacy or the belief that personal actions would influence other individuals or government action toward addressing climate change through carbon-neutral pledges. They also reported lower levels of understanding of carbon-neutral pledges, presumably since discourses related to these decisions often take place in urban areas where local government is centralized. Yet not all constituents live in an urban hub; they do not have the same opportunities for engagement. An equitable energy system is one that represents citizens *and* fairly disseminates both the costs and benefits of energy services (Sovacool & Blyth, 2015). So, while carbon-neutral pledges involve citizens in localized democratic energy engagement, the repercussions of these actions may lead to greater energy injustice by differentially affecting individuals in urban areas relative to those in rural areas. This is an important consideration for other cities and counties around the country that made or are considering making carbon-neutral pledges.

Discussion and future research

As shown by the dramatic rise in carbon-neutral pledges in the state of Utah, communities want local action to address climate change. This case study of carbon-neutral pledges in Utah indicates that these actions are taking place at locations beyond liberal cities. Despite the presence of conservative religious and political identities and values, carbon-neutral pledges were viewed favorably as an approach to mitigate the effects of climate change. Yet while the findings of this study reinforced the notion that individuals are actively engaged in energy issues and see real benefits of local government action, they also indicated that there is a stark difference geographically based on urban versus rural perceptions of carbon-neutral pledges. Our survey shows that urban Utahns were more likely to support carbon-neutral pledges. This finding was accentuated by the subsequent pledges that occurred in urban Utah communities in 2019.[3] The divergent nature of citizens' comprehension of climate action may aid in the replication or advancement of pledges in other areas with similar geographic layouts: an urban hub and vast stretches of rural space.

The finding that geographic location was key to perceptions of carbon-neutral pledges echoes the aftermath of the 2016 US presidential election. America's urban and rural areas have a long history of a spatial divide in political attitudes and voting patterns. Liberals lean toward living in dense urban areas, while conservatives tend to live farther apart in rural parts of the country (Dimock et al., 2014). Despite the numerous polls that predicted Hillary Clinton would be victorious in the 2016 election, and although she won the popular vote, Donald Trump won the presidency. Trump's victory was even more decisive in rural areas, where he received 62.3% of the vote (Monnat & Brown, 2017). Studies repeatedly show that even when controlling for other demographics such as race, sex, and political ideology, physical geography itself is a strong predictor of political choice. Even though rural voices only account for approximately 15% of the US population (Monnat & Brown, 2017), they are important participants for future elections and energy decisions.

A common comment from rural voters during the 2016 election was that they felt as though their communities and economic hardship were being ignored by the rest of America (Monnat & Brown, 2017). Similarly, surveyed rural Utahns in this study reported lower levels of belief in government effectiveness. Rural communities are often overlooked in campaigns since they are not densely populated, and therefore outreach requires much more determined effort. Strategists ascertain that individuals in rural areas often pay less attention to specific policies and are more aware of culture, references, and tone (Evich, 2016). Accordingly, one of Trump's strategies for the 2016 election was to appeal to voters culturally in rural areas. This approach offers a lesson for advocates of energy democracy: Find ways to connect with rural citizens for greater energy inclusivity. Some of the main criticisms of carbon-neutral pledges are that they may hurt the economy and job creation while transferring benefits abroad to other countries (Sampathkumar, 2017). Therefore, these are key points to address in rural areas. For instance, studies show that perceptions of climate change among rural farmers vary considerably; however, farmers who identified the effects of climate change on the agricultural economy were more likely to support government mitigation actions (Arbuckle et al., 2013). Consequently, advocates for carbon-neutral pledges may find conversations centered on economic benefits to be more successful with rural citizens than those targeting environmental benefits.

Despite the limitation of an urban–rural divide in the state of Utah, this case study highlights the advantages of carbon-neutral pledges as an energy democracy strategy and informs the way that future research into carbon mitigation strategies may branch out beyond heavily populated areas. It is important to consider rural voices and perspectives, not just in the United States but internationally as well. The idea should not be to *win* over individuals who may disagree with more liberal-minded actions but rather to have an open dialogue and participation. By advocating for rural engagement, polices ensure greater understanding, if not support. It may be that future research should work to overcome the urban-centric focus of carbon-neutral pledges and seek rural momentum to place pressure on their local communities to take similar actions. Or perhaps a carbon-neutral pledge is a strategy uniquely situated to assist in urban cities, and rural areas should seek alternative solutions. The results of this study are instructive for those seeking election or lobbying for carbon-neutral pledges, for those composing energy and environmental narratives to reach beyond individuals in urban centers and seek out rural voices that may offer a differing idea or opinion, and for individual citizens to know that through their communal actions they can foster climate change policy. The strategic use of carbon-neutral pledges in

Utah and elsewhere contributes to the ongoing conceptualization of energy democracy as a unique term with a burgeoning research agenda and encourages future scholarship to elucidate the role of cities and citizens in a participatory energy future both in the United States and abroad.

Notes

1 Michael Bloomberg switched his political party registration from Republican to Democrat for his 2020 presidential campaign (Stewart, 2019).
2 This number rose to 24 communities by early 2020.
3 The 2019 pledges that took place in Utah were Ogden, Emigration Township, Millcreek, West Valley City, Kearns, Holladay, West Jordan, Bluffdale, Cottonwood Heights, Salt Lake County, Alta, Coalville, Oakley, Kamas, Francis, Castle Valley, Ivins, and Springdale.

References

Arbuckle, J. G., Prokopy, L. S., Haigh, T., Hobbs, J., Knoot, T., Knutson, C., Loy, A., Mase, A. S., McGuire, J., Morton, L. W., Tyndall, J., & Widhalm, M. (2013). Climate change beliefs, concerns, and attitudes toward adaptation and mitigation among farmers in the Midwestern United States. *Climatic Change*, *117*(4), 943–950. https://doi.org/10.1007/s10584-013-0707-6

Ballew, M. T., Leiserowitz, A., Roser-Renouf, C., Rosenthal, S. A., Kotcher, J. E., Marlon, J. R., Lyon, E., Goldberg, M. H., & Maibach, E. W. (2019). Climate change in the American mind: Data, tools, and trends. *Environment: Science and Policy for Sustainable Development*, *61*(3), 4–18.

Bansard, J. S., Pattberg, P. H., & Widerberg, O. (2017). Cities to the rescue? Assessing the performance of transnational municipal networks in global climate governance. *International Environmental Agreements: Politics, Law and Economics*, *17*(2), 229–246. https://doi.org/10.1007/s10784-016-9318-9

Becker, S., & Naumann, M. (2017). Energy democracy: Mapping the debate on energy alternatives. *Geography Compass*, *11*(8), e12321.

Beebe, L. (2019, December 18). In Utah, the path to 100% clean energy is clear. *Sierra Club*. www.sierraclub.org/articles/2019/12/utah-path-100-clean-energy-clear?utm_source=instagram&utm_medium=social&utm_campaign=readyfor100&utm_content=201912_Utah100Path

Bloomberg, M., & Pope, C. (2017). *Climate of hope: How cities, businesses, and citizens can save the planet*. St. Martin's Press.

Burke, M. J. (2018). Shared yet contested: Energy democracy counter-narratives. *Frontiers in Communication*, *2*, 22.

Burke, M. J., & Stephens, J. C. (2017). Energy democracy: Goals and policy instruments for sociotechnical transitions. *Energy Research & Social Science*, *33*, 35–48. https://doi.org/10.1016/j.erss.2017.09.024

C40 Cities. (2016). *Deadline 2020: How cities will get the job done*. ARUP. www.arup.com/perspectives/publications/research/section/deadline-2020-how-cities-will-get-the-job-done

Canham, M. (2018, December 9). Salt Lake County is now minority Mormon, and the impacts are far reaching. *The Salt Lake Tribune*. www.sltrib.com/religion/2018/12/09/salt-lake-county-is-now/

Chilvers, J., & Pallett, H. (2018). Energy democracies and publics in the making: A relational agenda for research and practice. *Frontiers in Communication*, *3*, 14.

Clarke, C. E. (2017, July 13). *Incorporating citizen engagement in energy development decision*. Energy Democracy Symposium.

Community Renewable Energy Act, HB 411, State of Utah, 2019 General Session. (2019).

Cox, R. (2007). Nature's "crisis disciplines": Does environmental communication have an ethical duty? *Environmental Communication*, *1*(1), 5–20.

Cozen, B., Endres, D., Peterson, T. R., Horton, C., & Barnett, J. T. (2018). Energy communication: Theory and praxis towards a sustainable energy future. *Environmental Communication*, *12*(3), 289–294.

Day, T., Gonzales-Zuñiga, S., Nascimento, L., Höhne, N., Fekete, H., Sterl, S., Hans, F., Warembourg, A., Anica, A., & van Breevoort, P. (2018). *Climate opportunity: More jobs; better health; liveable cities*. C40 Cities. https://c40-production-images.s3.amazonaws.com/researches/images/71_ClimateOpportunity_Full_online_%283%29.original.pdf?1536323452

Dimock, M., Doherty, C., Kiley, J., & Oates, R. (2014). *Political polarization in the American public: How increasing ideological uniformity and partisan antipathy affect politics, compromise and everyday life*. Pew Research Center 12.

Ehrhardt-Martinez, K., Rudel, T. K., Norgaard, K. M., & Broadbent, J. (2015). Mitigating climate change. *Climate Change and Society: Sociological Perspectives*, 199–234.

Einstein, K. L., Glick, D. M., Palmer, M., & Fox, S. (2019). *Menino survey of mayors*. Boston University Initiative on Cities. www.voanews.com/a/survey-says-mayors-view-climate-change-as-pressing-urban-issue/4220963.html

Endres, D., Cozen, B., Barnett, J., O'Byrne, M., & Peterson, T. (2016). Communicating energy in a climate (of) crisis. *Annals of the International Communication Association*, *40*(1), 419–447.

European Parliament and Council of 11. (2018). *Directive on the promotion of the use of energy from renewable sources*. Official Journal of the European Union.

Evich, H. B. (2016, November 13). Revenge of the rural voter. *Politico*. https://politi.co/2BAKpR6

Fairchild, D. (2017). *Energy democracy: Advancing equity in clean energy solutions*. Island Press.

Fandos, N. (2017, April 29). Climate march draws thousands of protesters alarmed by Trump's environmental agenda. *The New York Times*. www.nytimes.com/2017/04/29/us/politics/peoples-climate-march-trump.html

Fang, X., Misra, S., Xue, G., & Yang, D. (2012). Smart grid—the new and improved power grid: A survey. *IEEE Communications Surveys Tutorials*, *14*(4), 944–980. https://doi.org/10.1109/SURV.2011.101911.00087

Feldpausch-Parker, A. M., Endres, D. E., & Peterson, T. (2019). A research agenda for energy democracy. *Frontiers in Communication*, *4*, 53.

Flagg, J. A. (2015). Aiming for zero: What makes nations adopt carbon neutral pledges? *Environmental Sociology*, *1*(3), 202–212. https://doi.org/10.1080/23251042.2015.1041213

Fox-Penner, P. (2014). *Smart power anniversary edition: Climate change, the smart grid, and the future of electric utilities*. Island Press. https://books.google.com/books?hl=en&lr=&id=phQ-BAAAQBAJ&oi=fnd&pg=PR6&dq=utilities+net+metering+change&ots=goXHxSj3oM&sig=VyJ3ZzrszDa7ovxGwAs0ZNKWB14

Giancatarino, A. (2012). *Energy democracy: Supporting community innovation*. The Center for Social Inclusion. www.centerforsocialinclusion.org/wp-content/uploads/2012/02/Energy-Democracy-Supporting-Community-Innovation.pdf

Giddens, A. (2009). *The politics of climate change*. Polity Press.

Gochnour, N. (2018, December 21). *Utah's population is booming*. Utah Business. www.utahbusiness.com/utahs-population-is-booming/

Hattem, J. (2018, July 17). Study finds dramatic increase in 2018 Sundance attendance. *AP News*. https://apnews.com/cd4a1c90441949b18d34a16e3a3a8ed7

Intergovernmental Panel on Climate Change. (2018). *Global warming of 1.5°C*. www.ipcc.ch/report/sr15/

IRENA. (2017). *Stranded assets and renewables: How the energy transition affects the value of energy reserves, buildings and capital stock*. International Renewable Energy Agency (IRENA). www.irena.org/menu/index.aspx?mnu=Subcat&PriMenuID=36&CatID=141&SubcatID=3870

Jotzo, F., Depledge, J., & Winkler, H. (2018). US and international climate policy under President Trump. *Climate Policy*, *18*(7), 813–817. https://doi.org/10.1080/14693062.2018.1490051

Klein, R. J. T., Schipper, E. L. F., & Dessai, S. (2005). Integrating mitigation and adaptation into climate and development policy: Three research questions. *Environmental Science & Policy*, *8*(6), 579–588. https://doi.org/10.1016/j.envsci.2005.06.010

Kunze, C., & Becker, S. (2012). *Energy democracy, practice and visions*. Büro für eine demokratische Energiewende.

Langheim, R., Skubel, M., Chen, X., Maxwell, W., Peterson, T. R., Wilson, E., & Stephens, J. C. (2014). Smart grid coverage in US newspapers: Characterizing public conversations. *The Electricity Journal*, *27*(5), 77–87.

Leaver, J. (2016). *The state of Utah's travel and tourism industry* (p. 18). Kem C. Gardner Policy Institute, The University of Utah.

Leiserowitz, A., Maibach, E., Roser-Renouf, C., Rosenthal, S., Cutler, M., & Kotcher, J. (2018). *Climate change in the American mind: March, 2018*. Yale Program on Climate Change Communication.

Lusk, K. (2018, October 24). *Mayors: Taking action on climate*. CO2-USA Workshop. http://sites.bu.edu/co2usa/files/2018/11/02_Lusk_Mayors-on-Climate_KLusk_10-24-18.pdf

Madhani, A. (2017, December 5). Obama praises mayors as "new face" of leadership on climate change in Trump era. *USA Today*. www.usatoday.com/story/news/2017/12/05/obama-trump-paris-agreement-chicago-climate-summit/923865001/

McKasy, M., & Yeo, S. K. (2018). A comparative case study of electric utility companies' use of energy democracy in strategic communication. *Frontiers in Communication*, *3*, 10.

Metcalf, G. E., & Weisbach, D. (2009). The design of a carbon tax. *Harvard Environmental Law Review*, *33*, 499.

Mindock, C. (2017, June 17). Paris agreement: US states and mayors fight climate change after Donald Trump pulls US out of deal. *The Independent*. www.independent.co.uk/news/world/americas/us-politics/paris-climate-change-agreement-mayors-states-american-hawaii-salt-lake-donald-trump-pulls-us-out-a7795426.html

Mitchell, L. E., Lin, J. C., Bowling, D. R., Pataki, D. E., Strong, C., Schauer, A. J., Bares, R., Bush, S. E., Stephens, B. B., & Mendoza, D. (2018). Long-term urban carbon dioxide observations reveal spatial and temporal dynamics related to urban characteristics and growth. *Proceedings of the National Academy of Sciences*, *115*(12), 2912–2917.

Monnat, S. M., & Brown, D. L. (2017). More than a rural revolt: Landscapes of despair and the 2016 presidential election. *Journal of Rural Studies*, *55*, 227–236.

Muro, M., & Saha, D. (2016). *Rooftop solar: Net metering is a net benefit*. Series Paper 91. www.brookings.edu/research/rooftop-solar-net-metering-is-a-net-benefit/

Owen, D. (2009, October 26). The greenest place in the US may not be where you think. *Yale E360*. https://e360.yale.edu/features/greenest_place_in_the_us_its_not_where_you_think

Pan, Y., Opgenhaffen, M., & Van Gorp, B. (2019). Negotiating climate change: A frame analysis of COP21 in British, American, and Chinese news media. *Public Understanding of Science*, *28*(5), 519–533. https://doi.org/10.1177/0963662518823969

Party Affiliation Among Adults in Utah—Religion in America: US Religious Data, Demographics and Statistics. (2014). *Pew Research Center*. www.pewforum.org/religious-landscape-study/

Rogelj, J., den Elzen, M., Höhne, N., Fransen, T., Fekete, H., Winkler, H., Schaeffer, R., Sha, F., Riahi, K., & Meinshausen, M. (2016). Paris agreement climate proposals need a boost to keep warming well below 2 °C. *Nature*, *534*(7609), 631–639. https://doi.org/10.1038/nature18307

Sampathkumar, M. (2017, June 2). Mike Pence says climate change is just an issue for the left. *The Independent*. www.independent.co.uk/news/world/americas/us-politics/mike-pence-climate-change-us-vice-president-paris-agreement-issue-left-for-some-reason-donald-trump-a7769081.html

Scott, D., Steiger, R., Rutty, M., & Johnson, P. (2015). The future of the Olympic Winter games in an era of climate change. *Current Issues in Tourism*, *18*(10), 913–930. https://doi.org/10.1080/13683500.2014.887664

Selyukh, A. (2019, September 19). Amazon makes "climate pledge" as workers plan walkout. *NPR*. www.npr.org/2019/09/19/762336929/amazon-makes-climate-pledge-as-workers-plan-walkout

Shear, M. (2017, June 1). Trump will withdraw US from Paris climate agreement. *The New York Times*. www.nytimes.com/2017/06/01/climate/trump-paris-climate-agreement.html

Sierra Club. (n.d.). *100% commitments in cities, counties, & states*. www.sierraclub.org/ready-for-100/commitments

Smith, M. (2017, December 5). 4 takeaways from a gathering of mayors on climate change. *The New York Times*. www.nytimes.com/2017/12/05/us/climate-change-mayors-chicago.html

Sovacool, B. K., & Blyth, P. L. (2015). Energy and environmental attitudes in the green state of Denmark: Implications for energy democracy, low carbon transitions, and energy literacy. *Environmental Science & Policy*, *54*(Supplement C), 304–315. https://doi.org/10.1016/j.envsci.2015.07.011

Stephens, J. C. (2017, July 12). *Strengthening resilience by redistributing power: Energy democracy, gender, diversity, and inclusion in the renewable energy transition*. Energy Democracy Symposium.

Stephens, J. C. (2019). Energy democracy: Redistributing power to the people through renewable transformation. *Environment: Science and Policy for Sustainable Development*, *61*(2), 4–13. https://doi.org/10.1080/00139157.2019.1564212

Stephens, J. C., Wilson, E. J., & Peterson, T. R. (2015). *Smart grid (r)evolution*. Cambridge University Press. http://books.google.com/books?hl=en&lr=&id=LAREBgAAQBAJ&oi=fnd&pg=PR11&dq=info:ts4t20qMVWwJ:scholar.google.com&ots=bpLw7AkfxN&sig=pwXHJ3rw_rtxVRAdJNgI5jC6sds

Stevens, T. (2020, January 13). At least 24 Utah cities, counties pledge to use renewable energy by 2030. *The Salt Lake Tribune*. www.sltrib.com/news/politics/2020/01/13/least-utah-cities/

Stewart, E. (2019, November 8). Michael Bloomberg's 2020 presidential campaign and policy positions, explained. *Vox*. www.vox.com/policy-and-politics/2019/11/8/18221878/michael-bloomberg-2020-president-campaign-democrat-policies

Swartz, K. E., Kuckro, R., Klump, E., Tomich, J., & Anchondo, C. (2020, January 3). 5 energy fights to watch in 2020. *E&E News*. www.eenews.net/stories/1061978867

Sweeney, S. (2014). Working toward energy democracy. In *State of the world 2014* (pp. 215–227). Island Press.

Szulecki, K. (2018). Conceptualizing energy democracy. *Environmental Politics*, *27*(1), 21–41. https://doi.org/10.1080/09644016.2017.1387294

Teron, L., & Ekoh, S. S. (2018). Energy democracy and the city: Evaluating the practice and potential of municipal sustainability planning. *Frontiers in Communication*, *3*. https://doi.org/10.3389/fcomm.2018.00008

UNFCCC. (n.d.a). *Climate neutral now pledge*. United Nations Climate Change. https://unfccc.int/climate-action/climate-neutral-now/i-am-a-company/organization/climate-neutral-now-pledge

UNFCCC. (n.d.b). *Paris agreement—Status of ratification*. https://unfccc.int/process/the-paris-agreement/status-of-ratification

US Census Bureau. (2010). *Urban and rural classification and urban area criteria*. www.census.gov/geo/reference/ua/urban-rural-2010.html

Utah Cities and Mayors. (2018, August 30). When it comes to climate solutions, there's nothing like what Utah can offer. *Deseret News*. www.deseretnews.com/article/900029921/guest-opinion-when-it-comes-to-climate-solutions-theres-nothing-like-what-utah-can-offer.html

van Veelen, B. (2018). Negotiating energy democracy in practice: Governance processes in community energy projects. *Environmental Politics*, *27*(4), 644–665.

Vasic, H. (2019, April 17). Rocky mountain power moves forward with 2030 renewable energy goal as feasibility questions linger. *KUER*. www.kuer.org/post/rocky-mountain-power-moves-forward-2030-renewable-energy-goal-feasibility-questions-linger#stream/0

Wesoff, E. (2011, March 30). *Obama speaks on energy in the US*. www.greentechmedia.com/articles/read/obama-on-energy-in-the-u.s

Winslow, B. (2018, December 14). Salt Lake City chosen as US nominee for 2030 (or beyond) Winter Olympics. *Fox News*. https://fox13now.com/2018/12/14/salt-lake-city-chosen-as-u-s-nominee-for-2030-or-beyond-winter-olympics/

Winston, A. (2017, May 31). US business leaders want to stay in the Paris climate accord. *Harvard Business Review*. https://hbr.org/2017/05/u-s-business-leaders-want-to-stay-in-the-paris-climate-accord

Wood Mackenzie Limited. (2019). *US solar market insight*. Solar Energy Industries Association. / us-solar-market-insight

World Population Review. (2019, June 5). *Utah population 2020*. http://worldpopulationreview.com/states/utah-population/

Yoder, A. (2018, April 9). *Interview with Salt Lake County sustainability manager* [Personal communication].

34

BEYOND THE IVORY TOWER

Exploring the role of universities toward sustainable energy transitions in postdisaster environments

Marla Pérez-Lugo, Cecilio Ortiz García, and Lionel Orama Exclusa

Introduction

According to emergent literature in in the field, energy democracy is both an academic concept as well as a grassroots movement (Burke & Stephens, 2017). As an academic concept, it examines and describes the ways in which power relations influence the distribution of energy services, with the political economy of energy decision-making processes currently centered on fossil fuels. As a grassroots movement, it focuses on demanding the necessary changes in policies. However, it also demands changes in the politics of energy to promote transitions to renewable energy sources including considerations of justice for workers, communities, and other vulnerable sectors of civil society (Burke & Stephens, 2017).

In theory, universities are repositories and generators of local knowledge. Therefore, they should be in a privileged position to bridge the gap between the conceptual discussion around energy democracy and its practical and policy applications. On one hand, they provide the state-of-the-art education and research for technological innovation. According to the NCES (National Center for Education Statistics, 2019), in the United States there are fewer than 4,600 accredited universities and colleges. In total, they impact more than 20 million students and 4 million faculty and nonfaculty employees. These universities are key actors in the US research and development landscape. To this point, research conducted by the European Union about renewable energy research in the United States suggests that states that spearhead research and development (such as California, Colorado, Illinois, and Texas) are characterized by a high degree of spatial concentration of university-based research groups, centers, and institutes (ENRICH in the USA, 2018). Another nine states, according to this analysis, stand out due to the presence of leading university research groups, research centers, and industry clusters: Idaho, Kentucky, Maine, Massachusetts, Michigan, New Mexico, New York, North Carolina, and Tennessee (ENRICH in the USA, 2018).

On the other hand, universities are considered important economic and political actors at the local and the regional levels. In fact, with the advent of the "global university," their impact goes well beyond national boundaries (Erkkilä, 2014). Research conducted in Europe and Africa suggests that universities provide the space for the collaborative multisectorial capacity building necessary for effective governance and sustainable technological implementation

DOI: 10.4324/9780429402302-40

(Brennan et al., 2004). In the United States, many university presidents are aware of the influence that their institutions have in their learning landscapes. They utilize the term "anchor institutions" to describe the relationship between their academic organizations and their social, cultural, and physical ecosystems. For example, Dr. Drew Gilpin Faust and John L. Hennessy, presidents of Harvard and Stanford respectively, say:

> Universities uniquely bring together a wealth of intellectual resources across fields, an abundance of creativity and collaborative energy across generations, an opportunity to convene key actors on neutral ground, a commitment to serving society in ways that privilege objective evidence and rigorous analysis and the dedication to pursuing powerful long-term solutions without becoming subservient to near-term economic interests or partisan political concerns.
>
> *(Harvard University Center for the Environment, 2014)*

These statements point to universities as having the potential to promote and facilitate sustainable, just, and democratic energy transitions by training highly skilled labor, advancing R&D, influencing local and regional policy, and promoting institutional change. However, universities face substantial barriers of cultural, social, economic, and political nature that can hinder their capacity to promote just, democratic, and sustainable futures. For example, budgetary cuts at the state and local levels can inhibit the university's ability to innovate in its curriculum, reorganize necessary resources in ways that reflect the complexities of energy issues, and ultimately shift their attention toward fiscal survival instead of effectively tackling what are perceived as politically hot issues (Barr & Turner, 2013). Also, the disciplinary fragmentation of academia hinders the institutional capacity to incubate inter- and transdisciplinary approaches by imposing a hierarchy of power relations between disciplines, diminishing collaboration and convergence of science, knowledge, and technology (Darbellay, 2015).

Lastly, political regimes are capable of shifting universities' institutional goals from generating knowledge and inserting it into decision-making processes to maintain the political regime in place by capturing their administrative apparatus (Chege, 2009). For example, in universities that are funded by the state, board members are usually appointed by the governor, or at least they have a direct impact on the selection of its members (Woodhouse, 2015). According to Chege (2009), state-funded universities are particularly vulnerable to the influence of ruling political parties when changes in the leadership are tied to political cycles. This hinders institutional memory, stable succession mechanisms, and the university's capacity to effectively serve as community anchor (Chege, 2009).

While not an exhaustive list, these structural conditions significantly lower the capacity of universities to be agents of change toward energy sustainability, justice, and democracy. The purpose of this chapter is to propose that if universities desire to fulfill their mission as agents of change, they need to use their resources innovatively, collaboratively, and in transdisciplinary convergence. In essence, they must go beyond the so-called ivory tower model. This is particularly important when dealing with wicked problems such as energy transitions, disasters, community vulnerability, and resilience (Brown et al., 2010).

The ivory tower, or the traditional academic approach, assumes that the purpose of universities is to create new knowledge by observing reality in an unbiased, detached, and neutral way (Moore, 2004). This expression originally comes from the Bible and means purity. However, in recent times the expression is more related to unworldly seclusion, an individual development that happened in isolation apart from the rest of society. In the university environment, it translates to the highly stratified and segmented social structure that universities have developed since medieval times and their tendency to divorce the creation of knowledge from day-to-day experiences and needs.

However, as many theorists of modernity and postmodernity have argued, the problems of contemporary societies are not "tamed" but "wicked." "Tamed" problems, although complicated, enjoy a common societal understanding and can be fixed by one discipline, either in the laboratory setting or in consultancy (Rittel & Webber, 1973). An example of a "tamed" problem could be the electrification of a rural region or national territory. If all of the main stakeholders agree that it is good and necessary, engineering, as a distinct academic discipline and a practice, can provide the tools and expertise to get the job done.

Rittel and Weber (1973) argue that contemporary societal problems, such as sustainability, resilience, or energy justice and democracy, are "inherently different from the problems that scientists and perhaps some classes of engineers dealt with in the past" (p. 160). These problems are inherently "wicked." They are value laden, multiperspectival, socially constructed, and not prone to a single solution. Wicked problems can be distinguished from problems in the natural sciences, which are definable and separable and may have solutions that are findable; the problems of governmental planning—and especially those of social or policy planning—are ill defined, and they rely upon elusive political judgment for resolution. Therefore, issues like sustainability, resilience, just transitions, and energy democracy do not lend themselves to a laboratory setting, where variables can be controlled and one disciplinary expertise can design an optimal solution to the problem. "At best, they are only re-solved—over and over again" (Rittel & Webber, 1973, p. 160) by bringing all stakeholders to the table through the development of collaborative networks or extended peer communities (Funtowicz & Ravetz, 1994).

We describe, as a case study, the institutional origins of the University of Puerto Rico's National Institute for Energy and Island Sustainability (INESI) and its role in promoting an energy transition toward a more sustainable, just, and democratic electric system for Puerto Rico. We also analyze its role in bringing the basic principles of energy democracy to the public discussion during the response, recovery, and reconstruction of Puerto Rico's electrical system after the catastrophic impact of Hurricanes Irma and Maria. These events triggered the collapse of Puerto Rico's electrical system and the deaths of more than 2,975 US citizens (Santos-Burgoa et al., 2018). We focus on INESI's attempts to insert local knowledge in the response, recovery, and reconstruction processes as well as the creation of an interuniversity collaborative network of mainland universities that traveled to Puerto Rico in the aftermath of Maria to rethink the ways in which universities intervene in communities during disasters. The chapter ends with a reflection on the lessons learned about university–community interactions in postimpact environments and the need for developing a new ethic to guide them.

The UPR's National Institute for Energy and Island Sustainability

The National Institute for Energy and Island Sustainability (Instituto Nacional de Energía y Sostenibilidad Isleña, INESI) was born in 2015 as the only interdisciplinary and intercampus entity of the University of Puerto Rico (UPR) (MPRV, 2015). It was conceived as a collaborative platform that identifies and interconnects faculty, research efforts, education, and service initiatives across the 11 campuses of the UPR system that could contribute to an integrated and applied discussion of energy issues in Puerto Rico (Garcia et al., 2014). INESI's main goals also included reconceptualization of the UPR's organizational structure to address not only the conceptual, structural, and political barriers to interdisciplinary and intercampus collaboration but also the insertion of UPR human resources in energy policy processes. INESI's third goal was to promote multisectoral capacity building of all energy stakeholders to increase their participation in the public energy policy and sustainability processes of Puerto Rico (Instituto Nacional de Energía y Sostenibilidad Isleña, 2019).

Until then, only a limited number of local stakeholders were considered as having a "stake" in Puerto Rico's energy policy decisions (Instituto Nacional de Energia y Sostenibilidad Isleña, 2015c). Those stakeholders were mainly the Puerto Rico Electric Power Authority (PREPA), government actors, and representatives of powerful economic interests such as the Puerto Rico Manufacturers Association (O'Neill-Carrillo & Rivera-Quinones, 2018). At the same time, energy issues were considered as being in the purview of the engineering academics at the UPR (O'Neill-Carrillo et al., 2008a, 2008b). University of Puerto Rico, mostly in its Mayaguez Campus (UPRM), was already recognized as a powerhouse for the formation and training of Hispanic engineers. In 2013, the American Society for Engineering Education (ASEE) ranked UPRM among the first 50 universities at the national level in relevant categories such as the number of Hispanic Tenured/Tenure Track Faculty in Engineering and the number of Engineering Bachelor's Degrees Awarded to Hispanics (Yoder, 2012). At the local level, UPRM was and still is the main engineering school in Puerto Rico, and it has trained most of the personnel in charge of the Puerto Rico Electric Power Authority (PREPA) since the 1950s.

In fact, the relationship between UPRM and PREPA was multilevel. Organizationally, an assembly line–like structure developed between the UPRM's School of Engineering and PREPA. UPRM provided the valuable human resources and knowledge capital for the public electricity monopoly. At the policy level, for decades, an iron triangle of interests was formed around energy issues in Puerto Rico. This impenetrable coalition was able to keep energy discourse within a technoeconomic realm, thus limiting other sociopolitical aspects. The Professional Association of Engineers, the PR Manufacturers Association, prominent island economists, and some of the most prestigious law offices in the island kept an airtight lock on every discussion regarding a possible transformation of the current system.

PREPA had a prominent role in keeping this iron triangle together, dominating the island legislature and keeping Puerto Ricans outside of all decision-making processes, as mere consumers or clients (Instituto Nacional de Energia y Sostenibilidad Isleña, 2015c). For example, it was not until 2012 that consumers could elect representatives to serve in PREPA's board of directors (El Nuevo Dia, 2012). It even invoked national security reasons behind highly secretive energy decision-making processes. This symbiotic relationship, among other factors, placed the UPR, particularly the Engineering Departments, at the center of the archipelago's energy history and of the territory's industrial and economic development.

Pivotal events in 2014 provided powerful blows to PREPA's monopoly on PR's energy policy arena. First, PREPA entered an agreement with its bondholders to restructure the public corporation's debt (Caribbean Business, 2015). This involved hiring a restructuration officer to spearhead that process (Merced & Corkery, 2014). In that sense, Lisa Donahue and Alix Partners (the restructuration officer hired by PREPA's board and her company) became the symbol of a debilitated PREPA, having to share the strong hold it had over PR's energy policy processes with other actors. On the coattails of Lisa Donahue's introduction, various sectors of the Puerto Rican society started to demand a new energy regulatory framework, based on transparency, a counterbalance for the powers of the monopoly utility and citizen participation in energy policy (Instituto Nacional de Energia y Sostenibilidad Isleña, 2015c). That opened the door for a more inclusive and interdisciplinary approach to energy issues within the university. INESI was proposed as an intercampus collaborative platform, building upon smaller preexisting initiatives for interdisciplinary work on energy such as the UPRM's Tropical Institute for Energy, Society and the Environment (ITEAS) (Instituto Tropical de Energía, Ambiente y Sociedad, 2009).

In 2014, in response to the legislative initiative to dramatically change the energy policy landscape in Puerto Rico, the UPR administration commissioned a faculty committee to compile a catalog of its human resources and initiatives related to energy throughout the university system. A steering committee for what eventually became INESI performed in-depth interviews and focus groups with known researchers, web page reviews, transactions and proceedings reviews, and online surveys across the 11 campuses (García et al., 2014). They collected information from faculty including descriptions of their activities, initiatives, and projects relevant to the energy issue, a list of collaborators inside and outside their campus, their perception of the barriers imposed by the UPR system in their individual efforts and interdisciplinary and other collaborations, and their perception of the connection between their activities and the public discussion of the energy issues of Puerto Rico. The results, as later described in the proposal to create the INESI, suggested that more than 166 faculty in the UPR system were working in areas relevant to energy issues. The academic areas to which these faculty belong are extremely varied but include natural and social sciences, humanities, and engineering. The study also identified a total of 51 entities within the system, including laboratories, centers, and institutes, many of which were at the forefront of energy research and sustainability at the international level.

The researchers also identified the cultural and structural barriers to interdisciplinary and intercampus collaborations within the UPR system. Among the most important are the following: (1) the lack of understanding of the interdisciplinarity of wicked problems; (2) collaboration among disciplines and among campuses was not encouraged; (3) a generalized perception that so-called soft scientists (social sciences and humanities) received less administrative support when compared to their colleagues in the hard sciences (natural sciences and engineering); (4) lack of communication and coordination between initiatives; (5) competition among the different campuses and departments as if they were different institutions; (6) lack of research space and reliable infrastructure; and (7) highly bureaucratic and obsolete administrative processes. These barriers are very consistent with the literature cited at the beginning of this chapter with regard to challenges to interdisciplinary work in the academy. Also, several agreements were identified among the participant faculty members: (1) that the UPR had to make energy a top priority, opening spaces for informed public discussion; (2) the need to coordinate efforts and resources with government agencies, municipalities, and the private sector; and (3) the urgency to recognize and counter the political capture of the university administrative structure and processes by the two main political parties. INESI was designed to overcome these barriers and to bring the much needed change within the UPR culture and structure.

INESI was based on three main conceptual pillars: a sociotechnical approach to energy systems, the wickedness of energy transitions, and the need for convergence when organizing around energy issues (Garcia et al., 2014). The first conceptual pillar refers to the notion that energy systems are sociotechnical due to their multiple layers that include both physical infrastructures and social components such as values, beliefs, practices, and power dynamics (Verbong & Geels, 2007). The second addresses the multiplicity of values, visions, and interests, as well as the high levels of uncertainty associated with the transformations of electric systems, that make energy transitions wicked problems (Valkenburg & Cotella, 2016). Several authors have indicated that innovative ways of organization are needed to effectively tackle these types of problems. More heterarchical, collaborative, polycentric structures with an emphasis on knowledge sharing and adaptive learning are some of the features mentioned in the literature (Cummings et al., 2013). The third conceptual pillar, the need for convergence, is what the National Science Foundation (NSF) (n.d.) has called one of the ten big ideas for this century.

According to NSF, convergence calls for interdisciplinary approaches to issue-based initiatives (National Science Foundation, n.d.). INESI tried to operationalize convergence by redirecting the UPR's fragmented disciplinary structure to one based on collaboratively addressing some of the most pressing issues ailing the Puerto Rican society today: its energy and sustainability issues (Garcia et al., 2014). In essence, INESI recognized that not one single discipline has the wherewithal to solve the island's energy issues. It also recognized that a hierarchical organizational structure was not conducive to collaboration and the creation of much needed extended peer communities.

It was precisely its foundation on the concept of convergence, but in an indirect way, that made INESI a tool for achieving energy democracy for Puerto Rico. The INESI founders quickly discovered that, while much has been written about the concept of convergence in areas such as biotechnology and the allied health sciences (Sharp et al., 2016), no blueprint really existed for the operationalization of convergence in energy and sustainability areas. Out of necessity, INESI started with the creation of a catalog of resources in energy and sustainability within the UPR system. After contacting all the human resources identified in the foundational research, INESI published a catalog with the profiles of all of those who wanted to participate in this initiative. The catalog included 90 resources, representing 23 academic disciplines such as public administration, law, political science, social work, sociology, psychology, biology, climatology, and oceanography, among others (Instituto Nacional de Energia y Sostenibilidad Isleña, 2015d).

That meant that for the first time in Puerto Rico, INESI was including issues and basic concepts of energy democracy that were seldom associated with what was perceived as a purely technical issue in the public discussion through the specific inclusion of academics in other areas apart from engineering. Through the introduction of INESI's catalog of resources, a severe blow was dealt to the monopoly of power that natural sciences and engineering disciplines have consistently demonstrated when dealing with socioecological or sociotechnical systems. By including, for example, community psychologists, bioethicists, as well as humanists and artists, a more heterarchical and democratic organization of energy knowledge was achieved. The inclusion of social sciences and humanities in INESI's catalog of resources in energy and sustainability served as a guarantee that the best scientific, academic, and human talent of UPR was going to support, in a transdisciplinary way and through research, training and servicing the energy transformation efforts.

Later, INESI was able to mobilize those resources for the creation of working groups around the development of research reports and policy tools and participation in multisectoral discussions. INESI's platform was designed to achieve multilevel interconnectivity. At the local and community levels, the institute developed projects around issues regarding the food/energy/water nexus, energy insecurity, and sustainable management of common property resources. At the institutional and governance levels, INESI created Puerto Rico's Energy Stakeholder Forum (PRESF) (Instituto Nacional de Energía y Sostenibilidad Isleña, 2019), as a space for multisectoral collaborative deliberation and capacity building.

One of PRESF's main successes was the development of a vision for Puerto Rico's electrical system based on the conceptual premises of energy justice and democracy (Instituto Nacional de Energía y Sostenibilidad Isleña, 2015a). Representatives of about ten grassroots and community-based organizations participated in a two-month process to develop a vision based on the following fundamental values: democracy, justice, solidarity, responsibility, happiness, and sustainability (Instituto Nacional de Energía y Sostenibilidad Isleña, 2015b). The result of this exercise was the following vision: Our electrical system is both social and physical, sustainable and innovative and is able to anticipate and assimilate social, climatic, and market changes, as well as consumption patterns. It maximizes the use of local and renewable energy sources to achieve a Puerto Rico that is resilient, prosperous, just, democratic, sustainable, and happy. This vision, continues the document, "is based on a public participation scheme that is transparent, inclusive, integrative, ample and effective, with respect towards all sectors of the Puerto Rican society." With this vision statement, the PRESF recognized that electrical systems do not have intrinsic value. Instead, the PRESF stated that their value lay in the services that they provide and their effectiveness in helping a society to reach a long-term vision of a more just and democratic society.

INESI's heterarchical structure clashed with the highly hierarchical and compartmentalized essence of the University of Puerto Rico. Its foundational values of justice and democracy through active multisectoral participation also clashed with the academic model of the ivory tower, prevalent in the UPR system. However, those precise characteristics allowed INESI to insert UPR faculty into policy processes in highly political and contested policy processes, such as the conceptualization of Puerto Rico's Energy Commission and the Puerto Rico Electric Power Authority's Integrated Resources Plan (Proceso evaluacion Plan Integrado de Recursos AEE, 2015). It also allowed the institute to respond immediately after the impact of Hurricanes Irma and Maria in September of 2017, after the total collapse of Puerto Rico's electrical system, when INESI served as a tool for energy justice and democracy by fact-checking the government's narrative and by bringing solar technology to neglected communities.

INESI after Irma and María

Hurricanes Irma and Maria impacted Puerto Rico on September 6 and 20 of 2017, respectively (Holpuch, 2017; Johnson et al., 2017). The impact of those two hurricanes on Puerto Rico's electrical system was catastrophic, triggering the collapse not only of the electric transmission and distribution systems but also all other interdependent systems such as the communications network and water distribution and treatment. According to Johnson et al. (2017), Hurricane Irma passed north of Puerto Rico as a Category 5 storm "killing at least 12 people, leaving thousands of others homeless and plunging more than 1 million residents of Puerto Rico into darkness" (para. 1). More than 56,000 people were left without potable water. Hurricane Maria, according to Kwasinski et al. (2019) triggered the collapse of the electrical system, destroying close to 2,416 miles of transmission lines and leaving Puerto Rico's population without power for almost a year. See Figure 34.1 for the trajectory of Hurricane Maria through the main island of Puerto Rico and its impact in the electrical system. Almost immediately, authorities announced that more than 200,000 communities were not going to be reconnected to the electrical grid during the reconstruction process due to cost/benefit considerations (Instituto Nacional de Energia y Sostenibilidad Islena, 2017). This has brought PR the dubious honor of having suffered the longest blackout in US history and the second longest ever recorded worldwide (Castro-Sitiriche et al., 2018).

Researchers from George Washington University and the Harvard School of Public Health have calculated the death toll associated with the impact of these powerful storms between 2,975 and 4,645. Research by Santos-Burgoa et al. (2018) and Kishore et al. (2018) suggests that there were striking differences in the socioeconomic conditions of those who died and those who did not, with the poor, the elderly, and those living in remote rural areas being the most vulnerable. The Center for Investigative Journalism has suggested that most of these deaths were related to the lack of electrical service (Centro de Periodismo Investigativo, 2018). At the same time, research conducted by INESI affiliates has demonstrated a relationship between deaths and energy insecurity. According to Castro-Sitiriche et al. (2018), when measured in consumer

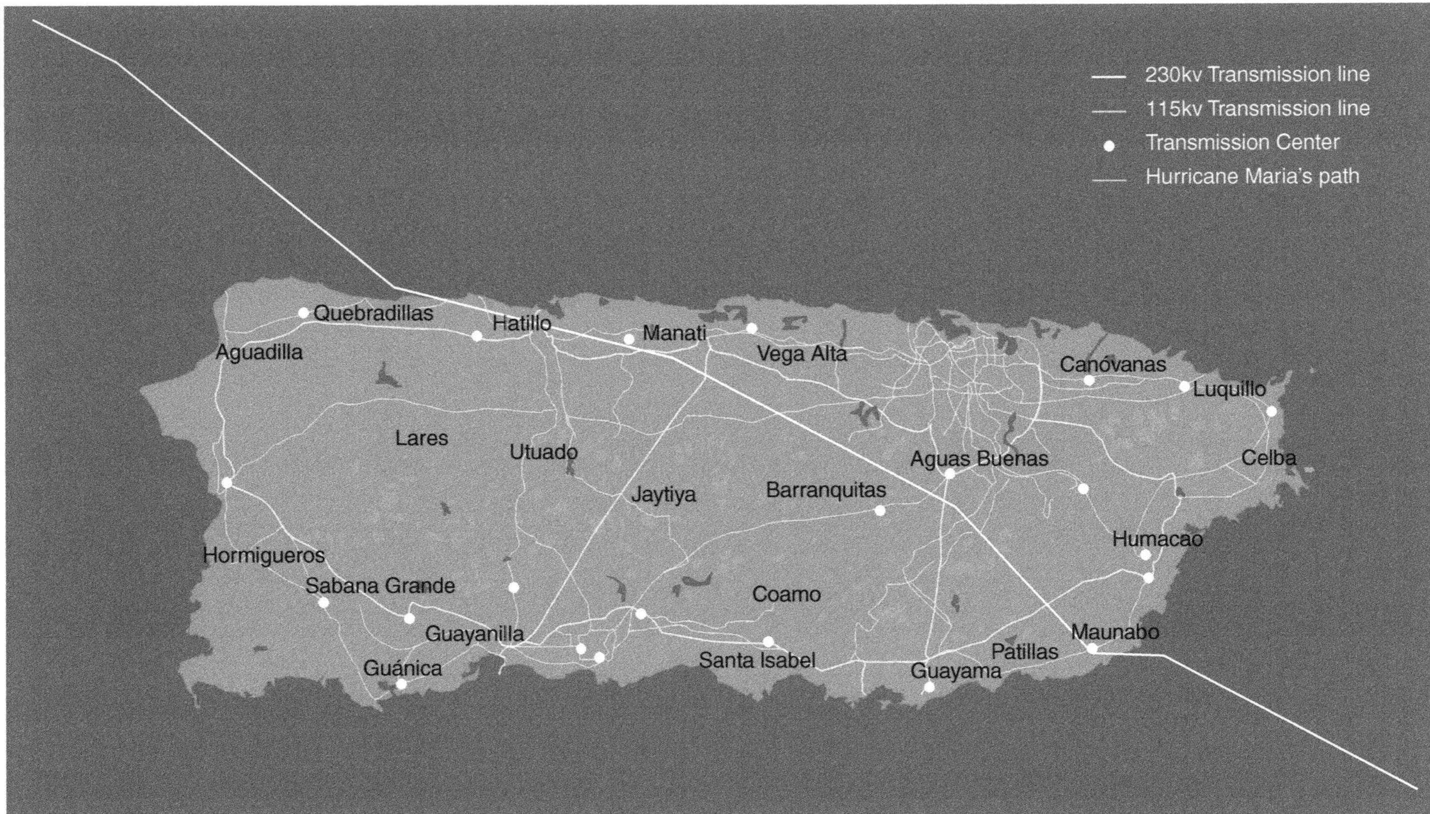

Figure 34.1 Hurricane Maria's trajectory through the main island of Puerto Rico and the damage to the electrical system.

Source: Figure by Jan Cordero for INESI

hours of lost electric service (CHoLES), almost a third of the total CHoLES (900 million) was due to the last 200,000 customers that were reconnected to the grid from day 156 to day 329 after Hurricane Maria. The estimates also indicate that slightly more than a third of the total CHoLES (1,041 million) were due to the 300,000 customers that were reconnected to the grid from day 71 to day 156 after the hurricane. Those customers that were reconnected last were located mainly in remote and economically depressed areas of the mountain region.

Admittedly, Irma and Maria also had devastating impacts on the UPR, and while not all 11 campuses were impacted equally, important physical infrastructures such as libraries, student housing, research labs, and access roads were severely affected (López-Alicea, 2017). To this day, the university, as well as Puerto Rico, are still in reconstruction mode, with the magnifying effect of the dramatic budget cuts brought by the Fiscal Oversight and Management Board (FOMB) created by Congress's Puerto Rico Oversight, Management and Economic Stability Act (PROMESA) (Rosario, 2019).

However, it was the UPR's centralized and hierarchical administrative structure that was actually limiting its ability to effectively integrate its resources into the response and recovery processes. INESI's human resources were ready to engage just one week after Hurricane Maria. A group of INESI faculty developed the blueprint for the reconstruction of the collapsed electrical system toward energy justice, resilience, and sustainability. Yet several attempts to approach the island's central government and the office of the UPR's president were met with deaf ears. Additionally, there was no receptiveness among important players in the response and recovery effort such as PREPAs, the US Corps of Engineers, and FEMA, at a time where INESI's wide spectrum of resources was desperately needed. At the same time, substantial contracts were being granted to unexperienced foreign companies for the system's reconstruction. Only months after our efforts, major scandals with regard to such contracts (Whitefish, Cobra, etc.) surfaced through the media (Newkirk, 2017; The Washington Post, 2017).

Nevertheless, INESI did find a strong receptiveness at the municipal level, through the networks of students and faculty. INESI's expertise flowed toward rural mountainside municipalities who had to deal with high levels of vulnerability in their communities and a total disregard by the central government. For example, because of the lack of communications system, mayors from the mountain region had to travel daily for hours through blocked or destroyed roads to the capital city of San Juan to meet with central government officials to personally deliver news about their towns' conditions. Many times, after making those trips, according to several mayors who were collaborating with INESI faculty, they encountered a disorganized emergency management apparatus that was many times unable to even listen to their reports. Members of INESI's steering committee themselves made that trip to the Center of Emergency Operations at the Convention Center Pedro Rosselló in the capital city of San Juan, only to find a resort-like atmosphere (in a luxurious hotel with working air conditioning and mechanical stairs while the rest of the population lacked electric service and running water) and a sense of lack of urgency among state and federal officials.

At the request of several of those mayors from municipalities in the mountains, INESI resources focused on providing immediate relief by creating modular solar-powered community resilience hubs called "oasis of light" that included charging stations for small electronics, a small refrigerator to store temperature-sensitive medication, and much needed illumination for common areas (Noticiasprtv, 2017). Soon the example of Jayuya (as seen in Figure 34.2) was replicated in the towns of Caguas, Aibonito, Utuado, and Villalba (Instituto Nacional de Energia y Sostenibilidad Isleña, 2018) with similar results.

Figure 34.2 Nocturnal image of the Oasis of Light in Jayuya

Source: Captured by Jose Miranda

INESI's role in addressing immediate community needs was even recognized first by the US Department of Energy (DOE) and then by the Puerto Rico's Legislature. The DOE, in their 2018 report titled "Energy Resilience Solutions for the Puerto Rico Grid," stated that:

> in conjunction with federal recovery and mitigation efforts, INESI could provide the foundation on which to firmly establish Puerto Rico as a Center of Excellence on distributed grid operations, and could provide both the supply of interdisciplinary engineers and policymakers Puerto Rico will need and the expertise other island and remote grid systems will need in their transition to a distributed, resilient electricity sector.
>
> *(US Department of Energy, 2018)*

On February 20, 2019, the PR House of Representatives also recognized INESI for its important role working for the communities' well-being and sustainable energy future as part of their celebrations for the day of social justice (Universidad de Puerto Rico, 2019).

At the same time, INESI compiled and disseminated much needed information about the post-Maria situation through webinars, presentations, meetings, briefings, and testimonies in front of mainland actors (such as the US Congress) with the purpose of combating knowledge gaps (or "manufactured ignorance") generated by government communications. "Manufactured ignorance," also called "negative knowledge," refers to the knowledge gaps created purposively for strategic reasons (Rappert, 2012). Two crude examples of manufactured ignorance in Hurricane Maria's aftermath are the official accounting of deaths and the sudden changes in the way in which PREPA reported progress during the reconstruction process. The Government of Puerto Rico kept the official number of deaths related to Hurricane Maria at 64 for a year, even in the face of media reports and academic research suggesting a much higher number (Barclay et al., 2018). It also switched from reporting in press conferences the "percentage of clients reconnected" to reporting on the "percentage of generation recovered." In fact, even the DOE questioned at some point the veracity of the electricity data provided by PREPA and stopped reporting progress in its electronic page (Ferris, 2018).

Perhaps one of the longest lasting impacts of INESI's work in the post-Maria period was its role as an interconnector with US mainland universities that were doing research, humanitarian aid, or service/learning trips in Puerto Rico. More than 30 universities, according to the accounts of INESI's steering committee members, landed on the main island of Puerto Rico in the months after the events. While these efforts were commendable and necessary, a few organizational deficiencies began to emerge almost immediately. For example, some highly charismatic communities received multiple visits from many universities, while other localities equally in need saw none. Furthermore, sometimes members of the same universities did not know that other members of the same institution were undergoing interventions of their own just miles away from each other. In a more operational sense, due to the rush of getting to Puerto Rico as quickly as possible, some of these teams did not emphasize training the volunteers to ethically intervene in communities that were dramatically impacted by a disaster and that were culturally and socioeconomically different from them. The result was a cultural shock that sometimes hindered (and negatively affected) the very purpose of their trip.

Similar issues also emerged even outside the disaster areas. At least five universities in the mainland generously offered Puerto Rican students free or discounted tuition so that they could continue their studies when the conditions of the UPR's physical and social infrastructure were not conducive (Pérez, 2017). In at least two of them, according to participant student's accounts, conflict arose when the host universities were requesting paperwork such as official transcripts in a business-as-usual matter. They were also charging for housing costs, fees, medical insurance, or meal plans to students who lost their homes and other possessions and were fleeing a disaster-ravaged island without electricity, running water, communications, or even working traffic lights.

Due to those organizational deficiencies identified by INESI members, the institute's steering committee, in collaboration with various colleagues from the School for the Future of Innovation in Society (SFIS) at Arizona State University, proposed the formation of an interuniversity convergence collaborative platform. That platform was named Resilience Through Innovation in Sustainable Energy for Puerto Rico (or the RISE-PR Network). The purpose of RISE-PR was to interconnect and coordinate resources, knowledge, and practice in a collaborative and interactive process of capacity building to address, through culturally appropriate research, collaborations, and interventions, the collapse of Puerto Rico's critical electrical infrastructure. RISE-PR was also intended to serve as a pathway to the integration of communities and other local stakeholders in these decisions, helping Puerto Ricans develop (not to impose or dictate through decontextualized technology transfer) a better understanding of what a sustainable and resilient society would look like.

RISE-PR organized two virtual workshops addressing these points. In these workshops, local experts shared their knowledge about the characteristics of Puerto Rico's electrical system and the impacts of Hurricanes Irma and Maria on that physical infrastructure. The workshops also explored convergences and divergences among participants through brief presentations about their experiences in the Puerto Rican archipelago after the Hurricanes. More than 50 researchers and students from prestigious institutions, such as State University of New York, University of Texas at El Paso, Northeastern University, Northwestern University, Massachusetts Institute of Technology, Tufts University, SUNY-ESF, University of Minnesota, and University of Puerto Rico, participated in these encounters, as along with representatives from local and US-based nongovernmental organizations, the PR Department of Puerto

Rico's Economic Development Department, and the USDA–Forest Service. These workshops served as a spark for new groups coming to Puerto Rico and more interaction and collaboration between US-based and local institutions. Moreover, several invitations were made for INESI to participate in conferences, workshops, and dialogues in universities across the mainland. Even in an environment of severe budget cuts brought about PROMESA's fiscal control board on the UPR, INESI became a prophet of innovation toward the redesign of university–community interventions from self-reliance due to institutional neglect to empowerment based on collaborative partnerships.

A third workshop was held in-person at the end of June 2018 in Mayaguez, Puerto Rico, for three days (Dirks et al., 2018). Although it was hosted by University of Puerto Rico, the workshop was designed and funded by University of Minnesota's School of Architecture. The Fletcher School of Diplomacy at Tufts University also provided the assistance of a summer intern to help with the logistics. With the active participation of 90 representatives from 26 universities, from both Puerto Rico and the mainland, and 14 representatives from local and federal governments, as well as community-based organizations, the workshop developed roundtable discussions on issues such as the relationships between universities and communities; the macropolitical, -economic, and -cultural context of resilience; the relationship between resilience and the built environment; and students' experiences in disasters. The workshop also included presentations by communities in Puerto Rico who narrated their firsthand experiences after the hurricanes. All these roundtable discussions fed into a unified discussion on what were the pressing issues that universities as a sector must address in the climate change era and how to organize an interuniversity collaborative platform to addresses these issues.

The results of this event can be summarized in three main agreements (Dirks et al., 2018). The first one is that research with human subjects in extreme operating environments is qualitatively different from similar research in nondisaster circumstances. In disasters, communities experience collective trauma, and by definition, they suffer from the discontinuation of their daily activities and lack of the institutional support they need for fulfilling their most basic needs (Britton, 1986). In fact, some workshop participants proposed that "informed consent," which is the most basic premise of the ethics of research with human subjects, is impossible under disaster conditions. The second agreement was that communities were not empty recipients of help. Communities have history, norms, social structure, and their vulnerability to the impact of natural hazards has been politically, economically, and socially constructed. The third agreement was that universities must be cognizant about the impact of their interventions, which can go well beyond the disaster geographical area. The example that was brought up in the discussion was the displacement of students and the lack of resources that universities put behind their well-intentioned initiatives to host students that have been displaced by disaster.

In summary, participants of the Puerto Rico RISE(ing) workshop came to the realization that in the era of climate change, universities could not continue doing business as usual. Based on that understanding, RISE-PR would be a polycentric, multisectoral, and multidimensional network of universities, government units, private and nonprofit sectors, and communities brought together by the need to develop the necessary relationships for shared adaptive learning, knowledge transfer, and partnership building toward community resilience. The RISE Network would use but at the same time transcend the case of Puerto Rico to develop a new ethic for universities' process in disasters and other extreme operating environments in four main areas: student mobility, peer-to-peer relations, university–community interactions, and the use of academic/scientific knowledge to decrease and mitigate vulnerabilities.

Lessons learned: universities as agents of change for energy democracy

To summarize, we can extract several lessons from INESI's experience. The first is that a new type of university that is transdisciplinary, issue based, and open to collaborative capacity building is desperately needed to address wicked problems such as energy democracy and justice. Incentivizing STEM is not enough. Incorporating the social sciences and the humanities in energy policy discussions opens the door for expanding not only the public conversations on energy issues but also the type of stakeholders that are welcome at the proverbial table where energy policy decisions are made. In the end, energy transitions toward sustainability, resilience, democracy, and justice are not about technological innovation but a change in the way in which the power to decide energy futures is distributed across the population.

Another lesson learned is that relationship building is essential to innovation processes. At the same time, innovation processes need innovation. Spaces like INESI's Energy Stakeholder Forum are essential to social change. To do that, the experience of INESI suggests that a clear internal policy shift is required inside universities themselves. Universities need to change the way in which they relate to other regional stakeholders, including government, industry, and community leadership. However, universities often do not feel comfortable dealing with "wicked problems" because the relationships they are accustomed to establishing with other stakeholders are usually framed as politically adverse, hierarchically determined, and unidisciplinary in nature. In fact, there is a growing literature on academic and scientific entrepreneurship that documents the tight relationships that have developed between universities and industry in the last few decades and the movement of academia from generating new knowledge for the sake of knowledge itself to the commercialization of knowledge and technology (Thursby & Thursby, 2002).

This new type of relationship will be construed sometimes as political and biased, but universities need to get into a realm that goes beyond the controllable environments of the classroom, the laboratory, or consultancy work. To address wicked problems, universities need to provide the space for multisectoral collaborative capacity building. This means that universities need to grapple with the unbalanced power relations that keep energy injustice and the lack of energy democracy a norm. This requires the support of the university leadership, by providing legitimacy and recognition to those efforts, even when academics and scientists confront the government's narrative by fact-checking and pointing out "manufactured ignorances."

The fight for energy democracy requires that universities engage in deep transformations and to develop pathways that enable such a transition. INESI became an instrument for change and democratic capacity development for energy decision making in two main fronts. It impacted intrauniversity relations with the development of the resource catalog forcing the institution to officially recognize the interdisciplinary nature of its expertise in energy and sustainability issues. It also influenced extrauniversity relations by creating the stakeholder forum, which forced the university to embrace its role as neutral convener to attract the multiple sectors of Puerto Rican society to a discussion on energy issues in a more collaborative setting.

The disaster in the aftermath of Hurricanes Irma and Maria and the struggle of INESI's scientists and academics show that local universities should be considered critical infrastructure. They can provide essential local knowledge not only about the technical conditions but also the contextual social, political, and cultural factors that impact them. Disasters, although usually framed in terms of their negative impacts, are also windows of opportunity that universities can seize not only to decrease existing vulnerabilities in their surrounding communities but also to promote transitions to more sustainable futures. Universities share a very powerful incentive to think about the common good: They, as well as their surrounding communities, will have to live with the consequences of the decisions made through the reconstruction process. That means that they should be included in not only disaster response, relief, and recovery but also in the routine policy processes that create disaster vulnerabilities. In fact, a growing body of literature suggests that excluding local knowledge makes the reconstruction process after a disaster more painful and longer lasting (Nelson et al., 2007).

Universities in Puerto Rico and the mainland still face the enormous challenge to coproduce knowledge in more distributionally and procedurally just and democratic ways. More importantly, a new ethic for disaster research and other interventions in the context of socioecological and sociotechnical systems is desperately needed.

Conclusion

If you are looking for a silver bullet or a quick fix in the path toward energy democracy, the case of INESI is not it. The development of the National Institute for Energy and the Environment of the University of Puerto Rico took over a decade of scientific deliberation, interdisciplinary confrontations, and bold and innovative experimentation to achieve what could be considered mildly successful outcomes. Electric power is more than just electrons flowing through wires. It is political power. Electricity is also a matter of life or death, even more so in remote or isolated areas, both geographically and socially. The mutually agreed on recognition of INESI researchers and academics of this principle made the effort worthwhile.

When Hurricanes Irma and Maria triggered the collapse of PR's electrical system, this cascaded into the collapse of other systems that provided basic and essential services to the population such as water, communications, food, health, education, public safety, and transportation. As a consequence, almost 3,000 people died, and thousands were left in the dark for a year. In the context of Puerto Rico, decentralized electricity generation, bringing it closer to the consumption point and avoiding extensive and complex transmission and distribution lines, would save people's lives. However, that is not a technological innovation. It is a governance innovation. It is a consequence of efforts toward energy democracy. If universities want to fulfill their mission as agents of change toward a better society, they need to rethink their ivory tower model and develop a more engaged, transdisciplinary, and convergent model for the pursuit of justice and democracy in decision-making processes.

Energy democracy, as a growing social and academic movement, has the potential for redistributing power to the people (Stephens, 2019), and universities need to become more relevant and entrepreneurial through actively participating and enabling wide public participation in energy policy processes. The integration of energy justice and equity issues into curricula, research, and service endeavors is crucial for universities to play the role of agents of change. To concentrate on the one-dimensional technoeconomic paradigm will continue to serve as a barrier for the integration of important social, cultural, and political dimensions of energy transitions. Emphasis needs to be placed on the development of different types of relationships between universities and communities locally, regionally, and globally. In fact, a new ethic of university–community relations should emerge, creating a more collaborative environment based on respect, trust, and collaborative capacity building.

References

Barclay, E., Fernández-Campbell, A., & Irfan, U. (2018, September 20). Hurricane Maria: 4 ways the storm changed Puerto Rico—and the rest of America: The worst disaster of the Trump administration (so far). *Vox*. www.vox.com/2018/9/20/17871330/hurricane-maria-puerto-rico-damage-death-toll-trump

Barr, A., & Turner, S. E. (2013). Expanding enrollments and contracting state budgets: The effect of the great recession on higher education. *The Annals of the American Academy of Political and Social Science, 650*(1), 168–193.

Brennan, J., King, R., & Lebeau, Y. (2004). *The role of universities in the transformation of societies: An international research project.* Centre for Higher Education, Research and Information.

Britton, N. R. (1986). Developing an understanding of disaster. *The Australian and New Zealand Journal of Sociology, 22*(2), 254–271.

Brown, V. A., Harris, J. A., & Russell, J. Y. (2010). *Tackling wicked problems through the transdisciplinary imagination.* Earthscan.

Burke, M. J., & Stephens, J. C. (2017). Energy democracy: Goals and policy instruments for sociotechnical transitions. *Energy Research & Social Science, 33*, 35–48.

Caribbean Business. (2015, December 24). *Prepa reaches restructuring deal with two insurers.* https://caribbeanbusiness.com/prepa-rsa-agreement-extended-until-dec-23-as-it-continues-to-seek-consensus-on-term-sheet/

Castro-Sitiriche, M., Gomez, J., & Cintrón, Y. (2018, November). *The longest power blackout in history and energy poverty.* International Conference on Appropriate Technology.

Centro de Periodismo Investigativo. (2018, September 14). *Los muertos de Maria.* http://periodismoinvestigativo.com/2018/09/los-muertos-de-maria/

Chege, M. (2009). The politics of education in Kenyan universities: A call for a paradigm shift. *African Studies Review, 52*(3), 55–71.

Cummings, S., Regeer, B. J., Ho, W. W., & Zweekhorst, M. B. (2013). Proposing a fifth generation of knowledge management for development: Investigating convergence between knowledge management for development and transdisciplinary research. *Knowledge Management for Development Journal, 9*(2), 10–36.

Darbellay, F. (2015). Rethinking inter-and transdisciplinarity: Undisciplined knowledge and the emergence of a new thought style. *Futures, 65*, 163–174.

Dirks, L., Hansen, L., Yace, J. A., Heid, A., Mans, J., Graffy, E., Garcia, C. O., & Lugo, M. P. (2018). *PUERTO RICO RISE(ing) workshop summary.* Instituto Nacional de Energía y Sostenibilidad Isleña. https://inesi.upr.edu/puerto-rico-riseing-workshop-summary/

El Nuevo Dia. (2012, May 18). A la Junta de la AEE Agustín Irizarry y Juan Rosario Serán los nuevos representantes ciudadanos. *El Nuevo Dia.*

ENRICH in the USA. (2018). *A study on us R&D related to energy: Renewable energies.* European Network of Research and Innovation Centres and Hubs—European Commission.

Erkkilä, T. (2014). Global university rankings, transnational policy discourse and higher education in Europe. *European Journal of Education, 49*(1), 91–101.

Ferris, D. (2018, February 20). Hurricane Maria Puerto Rico's grid recovery, by the numbers. *E&E News.* www.eenews.net/stories/1060074219

Funtowicz, S. O., & Ravetz, J. R. (1994). Uncertainty, complexity and post-normal science. *Environmental Toxicology and Chemistry: An International Journal, 13*(12), 1881–1885.

Garcia, C. O., Lugo, M. P., & Exclusa, L. O. (2014). *Propuesta para la creacion del Instituto Nacional de Energía y Sostenibilidad Isleña.* Propuesta, Universidad de Puerto Rico, Administración Central.

Harvard University Center for the Environment. (2014). *How universities can tackle climate change.* https://environment.harvard.edu/news/general/how-universities-can-tackle-climate-change

Holpuch, A. (2017, September 20). Hurricane Maria leaves Puerto Rico in total blackout as storm batters island. *The Guardian.* www.theguardian.com/world/2017/sep/20/life-threatening-winds-and-floods-hit-puerto-rico-as-maria-makes-landfall

Instituto Nacional de Energía y Sostenibilidad Isleña. (2015a). *Declaraciones de visión para Puerto Rico y su sistema eléctrico.* https://inesi.upr.edu/declaraciones-de-vision-para-puerto-rico-y-su-sistema-energetico/

Instituto Nacional de Energía y Sostenibilidad Isleña. (2015b). *Foro de constituyentes.* https://inesi.upr.edu/old/constituyentes.php

Instituto Nacional de Energía y Sostenibilidad Isleña. (2015c). *Informe de Estado de Situación Energética de Puerto Rico 2015.* Oficina Estatal de Politica Publica Energética de Puerto Rico.

Instituto Nacional de Energía y Sostenibilidad Isleña. (2015d). *Recursos.* https://inesi.upr.edu/old/resources.php

Instituto Nacional de Energía y Sostenibilidad Isleña. (2017). *Informe de Estado Energético para Puerto Rico 2017.* Instituto Nacional de Energía y Sostenibilidad Isleña de la Universidad de Puerto Rico.

Instituto Nacional de Energia y Sostenibilidad Isleña. (2018, April 11). *Oasis de Luz*. www.youtube.com/watch?v=QXHDTttJHT8
Instituto Nacional de Energía y Sostenibilidad Isleña. (2019). Universidad de Puerto Rico. https://inesi.upr.edu/foro-de-constituyentes/
Instituto Tropical de Energía, Ambiente y Sociedad. (2009). Universidad de Puerto Rico-Mayagüez. http://iteas.uprm.edu/
Johnson, A., Arkin, D., Cumming, J., & Karins, B. (2017, September 7). Hurricane Irma skirts Puerto Rico, leaves 1 million without power. *NBC News*. www.nbcnews.com/storyline/hurricane-irma/hurricane-irma-skirts-puerto-rico-lashing-it-powerful-winds-flooding-n799086
Kishore, N., Marqués, D., Mahmud, A., Kiang, M. V., Rodriguez, I., Fuller, A., & Maas, L. (2018). Mortality in Puerto Rico after Hurricane Maria. *New England Journal of Medicine*, *379*(2), 162–170.
Kwasinski, A., Andrade, F., Castro-Sitiriche, M., & O'Neill-Carrillo, E. (2019). Hurricane Maria effects on Puerto Rico electric power infrastructure. *IEEE Power and Energy Technology Systems Journal*, *6*(1), 85–94.
López-Alicea, K. (2017, September 24). Serios daños en el sistema de la Universidad de Puerto Rico. *El Nuevo Dia/Locales*. www.elnuevodia.com/noticias/locales/nota/seriosdanosenelsistemadelauniversidaddepuertorico-2360112/
Merced, M. J., & Corkery, M. (2014, September 4). Puerto Rico hires supervisor for utility restructuring. *New York Times*. https://dealbook.nytimes.com/2014/09/04/puerto-rico-said-to-hire-alixpartners-to-lead-restructuring-of-power-authority/
Moore, J. (2004). Living in the basement of the ivory tower: A graduate student's perspective of participatory action research within academic institutions. *Educational Action Research*, *12*(1), 145–162.
MPRV. (2015). *Nace el Instituto Nacional de Energía y Sostenibilidad Isleña*. www.miprv.com/nace-en-la-upr-el-instituto-nacional-de-energia-y-sostenibilidad-islena/
National Center for Education Statistics. (2019). https://nces.ed.gov/
National Science Foundation. (n.d.). *NSF's 10 big ideas*. www.nsf.gov/news/special_reports/big_ideas/
Nelson, M., Ehrenfeucht, R., & Laska, S. (2007). Planning, plans, and people: Professional expertise, local knowledge, and governmental action in post-hurricane Katrina New Orleans. *Cityscape*, 23–52.
Newkirk, V. R. (2017, November 3). The Puerto Rico power scandal expands. *The Atlantic*. www.theatlantic.com/politics/archive/2017/11/puerto-rico-whitefish-cobra-fema-contracts/544892/
Noticiasprtv. (2017, December 21). *Oasis de luz para los ciudadanos de Jayuya*. http://noticiasprtv.com/oasis-luz-los-ciudadanos-jayuya/
O'Neill-Carrillo, E., Frey, W., Ortiz-García, C., Irizarry-Rivera, A. A., Pérez-Lugo, M., & Colucci-Rios, J. A. (2008a). Advancing a sustainable energy ethics through stakeholder engagement. In *2008 IEEE energy 2030 conference* (pp. 1–6). IEEE.
O'Neill-Carrillo, E., Irizarry-Rivera, A., Colucci-Rios, J., Pérez-Lugo, M., & Ortiz-García, C. (2008b). Sustainable energy: Balancing the economic, environmental and social dimensions of energy. In *2008 IEEE energy 2030 conference* (pp. 1–7). IEEE.
O'Neill-Carrillo, E., & Rivera-Quinones, M. A. (2018). Energy policies in Puerto Rico and their impact on the likelihood of a resilient and sustainable electric power infrastructure. *Centro Journal*, *30*, 147–171.
Perez, J. J. (2017, October 2). 14 universidades de Estados Unidos recibirán estudiantes de Puerto Rico. *El Nuevo Dia Digital*. www.elnuevodia.com/noticias/locales/nota/14universidadesdeestadosunidosrecibiranestudiantesdepuertorico-2362458/
Proceso evaluación Plan Integrado de Recursos AEE, CEPR-AP-2015-0002. (2015). *Comisión de Energía de Puerto Rico*. AEE.
Rappert, B. (2012). States of ignorance: The unmaking and remaking of death tolls. *Economy and Society*, *41*(1), 42–63.
Rittel, H. W., & Webber, M. M. (1973). Dilemmas in a general theory of planning. *Policy Sciences*, *4*(2), 155–169.
Rosario, F. (2019, April 11). Siguen los aumentos y recortes en la UPR. *Periódico Primera Hora*. www.primerahora.com/noticias/gobierno-politica/nota/siguenlosaumentosyrecortesenlaupr-1336763/
Santos-Burgoa, C., Sandberg, J., Suárez, E., Goldman-Hawes, A., Zeger, S., Garcia-Meza, A., & Andrade, E. (2018). Differential and persistent risk of excess mortality from Hurricane Maria in Puerto Rico: A time-series analysis. *The Lancet Planetary Health*, *2*(11), e478-e488.
Sharp, P., Jacks, T., & Hockfield, S. (2016). Capitalizing on convergence for health care. *Science*, *352*(6293), 1522–1523.

Stephens, J. (2019). Energy democracy: Redistributing power to the people through renewable transformation. *Environment: Science and Policy for Sustainable Development*, *61*(2), 4–13.

Thursby, J. G., & Thursby, M. C. (2002). Who is selling the ivory tower? Sources of growth in university licensing. *Management science*, *48*(1), 90–104.

Universidad de Puerto Rico. (2019, February 20). *INESI recibe reconocimiento de la cámara de representantes* (A. C. Oficina de Prensa, Productor). www.upr.edu/inesi-recibe-reconocimiento-de-la-camara-de-representantes/

US Department of Energy. (2018, June 20). *Energy resilience solutions for the Puerto Rico grid*. www.energy.gov/sites/prod/files/2018/06/f53/DOE%20Report_Energy%20Resilience%20Solutions%20for%20the%20PR%20Grid%20Final%20June%202018.pdf

Valkenburg, G., & Cotella, G. (2016). Governance of energy transitions: About inclusion and closure in complex sociotechnical problems. *Energy, Sustainability and Society*, *6*(1).

Verbong, G., & Geels, F. (2007). The ongoing energy transition: Lessons from a socio-technical, multi-level analysis of the Dutch electricity system (1960–2004). *Energy Policy*, *35*(2), 1025–1037.

The Washington Post. (2017, October 30). *The Whitefish contract in Puerto Rico shows the real cost of bad government*. www.washingtonpost.com/news/posteverything/wp/2017/10/30/the-whitefish-contract-in-puerto-rico-shows-the-real-cost-of-bad-government/

Woodhouse, K. (2015, July 7). New push for trustee training. *Inside Higher Ed*. www.insidehighered.com/news/2015/07/07/states-explore-required-training-university-board-members

Yoder, B. L. (2012). *Engineering by the numbers*. American Society for Engineering Education.

35

LOW-CARBON ENERGY DEMOCRACY IN THE GLOBAL SOUTH?

Ben Campbell, Jon Cloke, and Ed Brown

Introduction

Academics and practitioners who advocate renewable energy solutions to energy poverty and lack of access in the Global South frequently meet the response that prioritizing low-carbon energy condemns the Global South to permanent energy inequality and lower quality of life. This chapter confronts glaring global inequalities in energy policy. The ideas of energy democracy as both a normative project of transition and as a field of empirical practice, face remarkably different dialogical engagements, challenges, and frictions in the context of work in the Global South than in conversations about relationships between energy and democracy in late industrial societies.

We agree with Stirling (2014) that versions of history conferring an "exceptionalist" role to energy in development are misplaced. Certainly, the forms of democracy that emerged in North Atlantic polities were strongly conditioned by collective interests configured through fossil fuel infrastructures (Mitchell, 2011). Following Moore (2016), it was the exploitation of "cheap nature" by higher-income countries that fueled the onset of the Anthropocene (or, as he prefers, Capitalocene). Now we face the need to address the unequal basis for climate vulnerabilities together with the effects of financial crises (2008 and Coronavirus), exacerbating global inequalities. This requires rethinking energy politics as social agency that creates division (between North and South as well as within societies) as much as it creates connection.

It is frequently the default assumption in technoeconomic policy that simply providing energy supply will lead to improved living conditions. We argue that excessively privileging the role of energy in Global South development oversimplifies the dynamic relationships among economic growth, climate change, energy transitions, and political entitlements to infrastructural services (Brown et al., 2018; Campbell et al., 2016; Castán Broto, 2019; Cloke et al., 2017; Ockwell et al., 2019). Recent analyses question the focus on centralized energy which continues to frame most Global South governments' approaches to energy policy (Ikejemba et al., 2017; Lilliestam & Hanger, 2016). This questioning is gradually bringing states, societies, communities, households, and persons to confront and reinvent themselves in economic conditions exacerbated by climate change and in the knowledge of the structural limits of national grids (Bakke, 2016).

DOI: 10.4324/9780429402302-41

In this general context of environmental and political-economic transformation, growing conflict, and international refugee movements, it is clear that sustainable energy agendas in the Global South will require thoroughly different sorts of thinking about democratic actors in the Anthropocene. Perspectives on energy that traditionally focused on physical provision of infrastructure need to consider how to approach the political arenas of utility consumers and prosumers, how to deepen participatory energy citizen connectivity to renewable energy infrastructures, and how people themselves are cultural agents of innovation in everyday practices of energy livelihoods.

Szulecki (2018) observes that democratic planning around energy has tended toward an expert enclave from which society and social science were excluded, yet much of the pressure for extending electrification came from grassroots decentralized innovation and local public pressure (Nye, 1990). Given a new disposition to view energy in a broader context of diverse sociotechnical actors and imaginaries, we consider the Euro-centric energy legacy to be a poisoned gift, in contrast to which other "worldings" of energy are possible and necessary (Bonneuil & Fressoz, 2017; McNeill & Engelke, 2014).

To expand current work on social energy systems, this chapter presents findings and analyses from seven years of work over a range of research programs, partnerships, and network-building emanating from the UK Low Carbon Energy for Development Network (LCEDN).[1] The LCEDN has sought to explore the complexity of low-carbon transitions in the Global South, promote approaches that critically extend skills and capabilities for practices of low-carbon democracy, and contribute to a more inclusive conception of what it means to be energy poor.

Following this brief introduction, the chapter introduces the LCEDN and the range of initiatives it has been involved in. The focus then moves to aspects of the LCEDN program that have reflected most closely on the challenges facing enhanced democratic participation in new kinds of social energy systems across the Global South. We subsequently review existing patterns of power within energy systems across the Global South and the identification of the energy disenfranchised and opportunities for their empowerment. We finish by exploring some ways in which LCEDN research has attempted to analyze existing power structures and democratize possible energy futures.

Building the LCEDN community

The LCEDN was founded in 2012 via UK government funding[2] to address what was seen as a lack of cohesion to research and innovation within the spheres of energy and international development, particularly regarding technocentrism dominating transdisciplinary forms of research practice. In response, there was a strong emphasis in LCEDN on community building to ensure sustainability in the types of transitions envisaged, as well as a focus on promoting greater integration of research into the wider community of stakeholders working on energy access issues.

The work of the LCEDN has primarily been about the 2.5 billion people in the Global South who depend significantly on biomass and for whom decentralized, community energy projects represent a substantial form of empowerment over their current situations (particularly in locations such as Sub-Saharan Africa). But empowerment does not arise through "mere" electrification; energy initiatives need to be situated in the lived realities of those communities and have potential for enhancing livelihoods.

The LCEDN has gathered expertise across disciplines involved in renewable energy research in the developing world, to address poverty and gender inequality issues in the Global South. The LCEDN has come to constitute an extended community of practice linked globally to other communities of practice formed by NGOs, businesses, and policy communities committed to low-carbon transitions and energy development justice. This work supports an agenda of acquiring evidence from case studies, using critical participatory approaches, and researching grassroots and municipal initiatives, thereby identifying weaknesses in the standard depoliticized discourse of energy-for-development.

The LCEDN's attention to energy decentralization thus far has revealed the critical importance of formal and informal institutions for effective and appropriate energy services, to counter the waste inherent in centralized systems and the political economy of incumbent energy supply and distribution mechanisms. Low-carbon energy transitions imply decentralized governance structures working with broader sociocultural change, particularly but not exclusively toward gender-equitable coping strategies, to reverse existing infrastructural priorities that exacerbate disempowerment. As one example, in the process of research into low-carbon possibilities in areas of off-grid and under-the-grid (i.e., underperforming and unreliable grid service) social realities, one of the LCEDN's central purposes has been explicitly connecting social relations with energy use, systems, and innovation and making these connections visible to the interdisciplinary community of energy and development scholars. This connection between social relations and energy often challenges entrenched cognitive formations of how to go about engineering solutions to perceived problems, and instead focuses on finding out how energy services feature in the priorities relevant to people's lived worlds where water, food, health, security, and disempowerment are frequently prominent concerns.

Transition in paradigms of everyday practice to new pathways of low-carbon energy systems in the Global South is not so much about offering greater technological choices but about transforming power relations and power norms. Low-carbon democracy involves participation by all members of society in finding ways that energy systems might improve livelihoods, enable sustainable local economies, and become part of communities' institutional self-governance. Renewable energy technologies (especially in minigrids) bring opportunities for decision making in ways that central grid dependency cannot, given that centralized grid dependency can even actively prevent decision making. The crucial change in power relations concerns inclusive participation in the areas of energy production, distribution, and exchange, as well as knowledge development in how to mobilize action for sociotechnical change (see Feldpausch-Parker et al., 2019).

One of the most important themes for the LCEDN community has been a focus on governance in relation to energy, which has opened up to social scrutiny all manner of technical and organizational alternatives to be shared in broader contexts of gender, class, and ethnic difference. This work has suggested that political assumptions and discretionary choices are hidden in supposedly technical solutions that are frequently advanced under umbrella assumptions about "efficiency" rather than in terms of either interrogating different understandings of efficiency or equitable access outcomes (see the following SONG project example). This is especially notable in examining off-grid interventions where the agenda for low-carbon transitions currently depends on grafting technocratic, Western-centric assumptions about what constitutes efficiency are imposed on the masses of the Global South poor, referred to as Bottom-of-the-Pyramid (BoP), which is setting them up to fail. BoP communities have their own (different and variegated) understandings of efficiency. Energy democracy, however, both can and should take many different forms, and the LCEDN has been able to participate in or support a variety of forms in the Global South:

1 *Understanding Sustainable Energy Solutions*: The LCEDN played an overall facilitating role in USES and an initiating role in two USES projects (READ and SONG); all of the

USES projects examined very different aspects of decentralization and democratization of energy innovation, ranging from using renewable energy to "support[ing] rural community business models for low and renewable energy input into optimized food processing" (RE4FOOD—Energy Efficient Rural Food Processing Utilising Renewable Energy to Improve Rural Livelihoods) through to capacity-building in local authorities (SAMSET—Supporting Sub-Saharan Africa's Municipalities with Sustainable Energy Transitions). USES was focused on five themes: (1) energy systems and decentralized use; (2) solar; (3) bioenergy; (4) urban contexts and transport; and (5) energy efficiency. Central to each project was engaging with sociocultural energy use by individuals, households, and communities and analyzing the premise that technologies should be shaped as much around the social as the technical.

2 *SONG*: The SONG (Solar NanoGrid) project undertook to analyze the "best practices involved in making solar/PV harmonize with the interests of low-income communities and households" ("Solar Nano-Grids," n.d.). These aims were determined from extensive community consultations, focus group discussions, semistructured interviewing, and interviewing key community actors; the actual technology was not installed until three phases of these consultations had been undertaken, complete with setting up democratically elected village energy committees (VECs). Community technicians were trained to maintain and repair the solar equipment, but chief among the achievements of the SONG project has been the continuity of discussion and decision-making processes by the VECs (one at least now dominated by women—see Figure 35.1), the local NGO partner Sustainable Community Development Services SCODE, and the UK SONG team through WhatsApp. All decision making, including the recent installment and setting up of an industrial-size egg

Figure 35.1 Village Energy Committee in Lemolo B community, Nakuru County, Kenya, with the solar hub in the background.

Source: Photo by Jon Leary

incubator as a community business in the solar hub, is derived from this ongoing, tripartite conversation.

3 *The MiniGrid Game*: The LCEDN first engaged with the Minigrid Game developed by Energy Action Partners[3] in Kathmandu in 2018.[4] The Minigrid Game revolves around a set of laptops and a wi-fi hotspot and was initially designed to demonstrate to community members the advantages and limitations of decentralized solar grids. As a learning tool, it has extraordinary potential for teaching households and communities not only about the technical issues relating to minigrids but also how to address these issues as a community of stakeholders. The LCEDN was fortunate in being able to link Energy Action Partners to colleagues from UK Smart Villages[5] for the most recent round of the UK government Energy Catalyst funding call, resulting in an award sufficient to support the next round of game development through deployment in communities in Somaliland.

These are concrete examples of attempts to practice transdisciplinary and participatory energy research for Global South realities, yet the idea of low-carbon energy transition and the apparatus of twentieth-century energy access thinking is open to more critical examination (Bonneuil & Fressoz, 2017), especially in Euro-centric top-down templates for development impact through technology transfer (Harriss-White, 2018). By contrast, versions of transition are under way in some grassroots communities, which open up new possibilities for emerging energy citizenships, innovating with new skilled practices alongside claims for deliberative spaces through low-carbon energy alliances; previous work by the authors has promoted discussion of these possibilities under the label "ethnoengineering."[6] These take energy out from under the control of a technocratic elite and a disembedded ethics, effectively resituating energy transitions as taking place alongside multiple other transitions: agrarian, urban, digital, conflict-and-climate related.

Locating lines of power

Before continuing with how sociotechnical innovation in renewables can play a role in democratizing energy futures (and the role of academic research), such as the LCEDN initiatives previously noted, comes consideration of the barriers to such a transformation. If one asks whether renewable energy technologies *intrinsically* have greater potential for democratic control, the answer has to be no, but the consensus is that off-grid renewables at least are recognizably more socially embedded (Ahlborg, 2017; Hughes, 2017). This means there can be greater sociotechnical potential for democratic control of renewable energy production/consumption, particularly in the Global South. The lines of energy power and sociopolitical power are more clearly visible to the unconnected and excluded.

There can be no automatic assumption regarding democratic potential of low-carbon energy technologies, which are as vulnerable to elite capture as any other form—for instance, when landowners install a microhydro system not benefitting the landless or when large dams displace people, inundating their land. Energy access initiatives of all types are embedded in social structures of inequality, revealing the questionable validity of the term "common" if small-scale renewables are assumed to be a common property resource management system. Another common example is where a commercial venture arrives in a community and does not engage with the community in the design of a renewable system and sets unrealistically high charges and/or unreasonable payment systems, leaving the community poorer than they were at the start of the project.

Communities living in precarious conditions are, of course, not only to be found in off-grid settings. In rapidly urbanizing contexts, particularly where precarity is stimulated by environmental or political crises, informal settlers are regularly present on waterway edges or on public lands. Their entitlement to live where they do is contested by municipal authorities, yet these people are meanwhile productive contributors to city economies. Though their rights in property-based, citizenship frameworks are not recognized, they have needs for basic services, including energy. Some purchase illicit connections at two or three times the normal rate for illicit metered supply, or others pay three times the unit rate of grid connection for a solar home system (Campbell & Adamu, 2019).

If a certain access struggle with global/local energy regimes is characteristic of grid-based and fossil fuel systems from the twentieth century's legacy, how differently can we conceive of the ways that the current energy poor in the Global South are perceiving energy in their lives? The domination of the centralized grid creates a deep sense of unequal citizenship among those denied connection but also creates a cultural space of unplugged powerlessness as compared to the possibilities that low-carbon agendas could offer for legitimacy, community welfare, and environmental and technology justice.

Who are the energy disenfranchised? Formal and informal political spaces

The relationships among citizens, state, and political norms and institutions have markedly different characteristics in the Global South, with consequences for how energy mediates differentiation in the public sphere and inhibits grassroots transformational capacity. In *The Politics of the Governed*, the Indian historian Partha Chatterjee (2004) explains the limited value of Western Enlightenment notions of democratic citizenship and rule of law for understanding the dynamics of collective action and claims-making in countries like India, where marginalized people manage survival only by circumventing the law and managing to garner support from the powerful when necessary through extralegal means of occupation, protest, and appeals to diffuse moral entitlements.[7]

Castán Broto et al. (2017) provide an excellent review of the kinds of questions about the energy needs of the urban and periurban poor, which complement Chatterjee's theorization and critique default responses to energy provision in urban environments.

> Problems in slums are "invisible" when government officials do not acknowledge their needs or even their existence. On the one hand, local governments may lack capacity to respond to the needs of informal settlements. On the other hand, urban development practices regularly ignore or misrepresent their existence. . . . Access to electricity in such conditions, for example through off-grid systems, may have transformative impacts both in terms of directly improving the lives of people at the household level, and enabling them to be recognized as urban citizens through the provision of services.
> *(p. 778)*

Understandings of the needs of poor urban women and men and of their priorities such as for clean cooking, and how both grid and off-grid electricity could service domestic, entrepreneurial, and community activities, have hardly encroached into dominant energy policy arenas, even as grid extensions and off-grid developments have accelerated. For example, the levels of electricity consumed by new consumers of grid extension programs are notoriously low, and many low-income customers cannot afford connection payments, even at highly subsidized rates, while attempts to encourage productive uses among off-grid energy consumers remain at best partially successful. This stems from government perspectives assuming that physical access to electricity bestows living standard improvement without the need for support activities.

In one of the authors' research projects, a neighborhood leader from an informal settlement in Kathmandu drew attention to challenges in registering eligibility for a line connection despite the appointment of a new pro-poor head for the Nepal Electricity Authority who pronounced that all Nepali citizens should have access to light.

> We were asked to bring the verification from the ward office regarding landlessness for connecting electricity to our households. The ward office which itself hasn't recognized us as landless, cannot in any case agree to provide the verification of landlessness.
>
> *(Lama, n.d.)*

Inadequate energy democracy for urban informal settlers reinforces their more general struggle to find accountability and just treatment in infrastructure provision. Their exclusion has been configured in the growth of "Third World" slums, as theorized as the "Myth of Marginality" in the 1970s (Perlman, 1975). Davis (2006) and McNeill and Engelke (2014) chart the history of informal urban expansion, making it clear that, rather than being an unfortunate condition of infrastructural falling-short, the shanty towns actually have been integral to the viability of city economies.

Countering marginalization and seeking equality of respect feeds informal settlers' collective struggle. Against the slur of being accused of stealing electricity, a community spokesperson quoted by Das and Walton (2015) says:

> Sir ji, how can you call us thieves? If you don't give us electricity on the grounds that we are not an authorized colony—and people naturally need electricity—a man wants to run a fan, his little children are burning in the heat—he will get electricity with whatever means—then why call him a thief?
>
> *(p. 48)*

By critical pressure on authorities and politicians to take heed of the plight of the urban poor in Delhi, Das and Walton (2015) argue that:

> democratic politics have been deepened by the participation of the poor. After all, it is because they have put political labor into going to courts, insisting that the law take into account what the constitutional provision of the right to life actually means, or their active participation in asking how city life is to be made viable that democracy has taken shape.
>
> *(p. 53)*

Attending to these articulations by informal settlers who are predicted to quadruple in numbers by 2050, they deconstruct the politics and ethics of energy. The slum dwellers call to account the abstract notion of energy as an assumed matrix of modern life that is disembedded from ethics, and point to the necessary democratic participation in deliberating public spaces for reinventing energy in the social distress of the Anthropocene.

Intensified research into democratic dimensions of energy programs has to address the invisibility of energy poverty among the urban poor, as well as the relationship of rural-to-urban contexts. It is often assumed that this division corresponds to presences and absences of markets, infrastructures, communications, and access to state institutions, and that the asymmetry of development facilities for empowering citizens can be overcome by providing greater energy access. The reality of existing energy provision is often well below the claims of energy access programs, however, when only minimal supplies of electrification, or liquid propane gas access, to restricted numbers of users has actually been achieved. Furthermore, a gendered invisibility of the energy poor, in both urban and rural situations, links to a lack of accounting for the biomass energy work of social reproduction by the unpaid labor of women and girls, which leads to male bias in setting development priorities (Nathan et al., 2018).

Nonetheless, there is an affinity of interests that create spaces of sociotechnical possibility and produce considerable allure for community-based solutions. We claim, moreover, that in a world beset by the urgencies of anthropogenic global warming working in contradiction to the need for universal energy access, community-based decentralized energy initiatives are the only means to bridge these contradictory processes and at the same time imbue energy access with elements of democracy and empowerment. Caution is needed about calls for democracy, empowerment, and new politics of energy knowledge and practice, if the capacity for grassroots innovation are overstated. Contesting inherency once again, decentralized community energy initiatives will not inspire a social revolution by themselves but do need to be part of emerging inclusive civic responses.

Notable for the attention that has been paid to inventiveness among urban poor in India is the phenomenon of *jugaad* or everyday bricolage. Kaur (2016) explains the affinity that the fetishization of self-reliance has with the interests of the neoliberal state, in maintaining that welfare and state intervention hinder India's intellectual capital of innovation. Elsewhere, South African urban development leads Redfield and Robins (2016) to argue that democratic sensibilities of its citizens have "become tightly tethered to popular demands for access to state services, technologies and infrastructure" (p. 145). Furthermore, the indignity of the conditions of living of the poor, with flimsy toilet facilities or inadequate energy provision, activates deep-seated historical resonances.

> To live adjacent to a grid, and yet not enjoy its benefits, vividly renders continuing racialized inequity in material terms. In this sense, for South Africans, the relative "modernity" of service delivery is less an abstract conceptual dispute than a continuing political issue.
>
> *(Redfield & Robins, 2016, p. 152)*

Harvey and Knox (2015) discuss "the promise of infrastructure," where energy access as a normative expectation of everyday life creates a modern infrastructural gaze that conversely sees other territories as places where "there is no infrastructure." This risks recreating neocolonial gazes of an infrastructural *terra nullius*, voided of existing livelihood "services" performed by diverse human and nonhuman networks of low-tech knowledge and practice. This renders invisible spaces of recognition for community actors to register precedence for livelihood practices and claims to territory-making (Latour, 2017). Distributed energy governance needs to build on those kinds of informal rights and the movements for alliance-making that enlivens "deep democracy" (Appadurai, 2002) in contested spaces. The worlding of low-carbon energy transitions in emergent locations of need with unequal but complementary capacities for innovation will not come from exogenous blueprints for sustainable energy but via intermediate collaborations within for socioecological capacities for action.

The LCEDN's gathering in of renewables research and case studies across the Global South shows the importance of amplifying how grassroots activities and networks are doing things on a local basis, beyond the radar of metropolitan research. Dipti Vaghela (2017) for instance, has documented the proliferation of small decentralized energy initiatives, especially microhydro and biomass gasification in Myanmar, working in a culture of research communication far from the internet frenzy associated with research circulating in the Global North. Smits and Bush's (2010) study of Lao PDR by way of contrast demonstrates the gulf of misunderstanding in policy circles over the extent and scale of horizontally expanding numbers of pico-hydro installations set against central policy obsession with foreign investment for large-scale dams at the national level and solar power as the RE-de-jour.

In the Anthropocene, visions of orderly sustainable energy transitions need to be seen realistically as purposive ideologies encountering conflict, in which "fallacies of control" (Stirling, 2014) have to be moderated. The "reckless Anthropocene" (de Waal, 2018) and resistance to disassembly from unsustainable energy system interests (Bridge, 2018) are the reality of doing of politics in sub- and supranational new territories of energy. People's relational capacity to act becomes the premise for energy politics in institutional networks and in relationships for distributed energy potential. This requires better knowledge of people's engagement with energy practice and discourse that includes the role of infrapolitics (Scott, 2012) with people's "mechanical intelligence" undergirding vernacular livelihoods, as much as it does citizens' capacity to act in public spheres of technological democracy.

In the following section, we explore some pathways of research that have attempted to reboot the intellectual and practical alignment of low-carbon energy transition work and to discuss a growing body of work attempting sociotechnical research collaborations. Within a space of intersectoral research dialogues between interdisciplinary academics from North and South, local community members, NGOs, "pro-poor" donor-supported research programs, some of the work that has been conducted since 2012 under the remit of the Low Carbon Energy for Development Network has shown glimpses of the way toward (and the challenges to) low-carbon democracy.

Extending low-carbon energy Development via a UK network

The space that opened up for interdisciplinary energy work in the UK was initiated by the then Department of Energy and Climate Change, which was followed up by the Engineering and Physical Sciences Research Council, and Department for International Development who funded and managed the program of 13 projects on Understanding Sustainable Energy Solutions (USES—see preceding discussion) from 2013 onward. This section looks to how those projects helped inform subsequent energy research policy pathways.

The USES program was designed in a climate change–oriented context of seeking "new ways of doing policy" (Brown et al., 2018, p. 112). Its overall aim was to achieve an:

> improved understanding of clean energy options and opportunities for developing countries; improved understanding of the social, market and political economy aspects of scaling sustainable energy access for poor people; strengthened developing country research capacity on clean energy and improved access to practical and policy relevant knowledge on the challenges and opportunities for sustainable energy solutions.
>
> *(Brown et al., 2018, p. 121)*

Working across the USES projects and auditing their impact for beneficiaries and for other researchers, important lessons were learned about communicative processes. A different order from the technology transfer model, the "dialogical production of impact" takes time, needs adequate funding, and requires broad-based stakeholder involvement:

> It takes time to build relationships and platforms, to share knowledge and develop impact, and it is these relational conditions of possibility for an impact that need much better recognition in participatory ownership of project objectives. Research does not automatically lead to impact, so considerable resources need to be devoted to creating opportunities, nurturing relationships and influencing policy strategically.
>
> *(Brown et al., 2018, p. 122)*

One of the projects was Renewable Energy and Decentralisation (READ), working with local government actors in Kenya and Rwanda to engage with processes of decentralization and bringing attention to the fact that there are "surprisingly few studies of the relationship between political decentralization and energy issues in the Global South" (Brown et al., 2015, p. 36). Even where formal decentralization policies exist, a UNDP and WHO (2009) study found only four cases that explicitly addressed energy decentralization. The common storyline is not local governments picking up "new roles and responsibilities in relation to energy but rather it is frequently of local governments struggling to even maintain inherited infrastructures rather than actively seek to expand their role" (Brown et al., 2015, p. 39). The report advised that "local people . . . play a more active role in articulating local solutions to the challenges which they face including those relating to energy" but added "this assumes that political decentralization is accompanied by the encouragement and facilitation of local consultation and priority setting which is, of course, not always the case" (Brown et al., 2015, p. 36). Examples where people at local levels had become notably engaged in planning and implementing energy projects have been especially recognized in Bangladesh, Nepal, and Mali (Brown et al., 2015, p. 37), while the general situation in the Global South is of regional and local authorities having no "direct legislated responsibility for meeting the energy needs of their citizens" (Brown et al., 2015, p. 38).

To address the lack of policy engagement, hybridizing the two concepts of energy literacy and energy efficiency, a new term, "energy proficiency," is proposed by the READ project members Batchelor and Smith (2014):

> energy proficiency is the degree to which local authority officials are fluent with the nature, role, and socio-political context of energy production systems in their nation and region, and can obtain, process, understand, evaluate and act on energy information to provide sustainable and efficient energy for their communities.
>
> *(p. 21)*

The project material effectively communicates how decentralization has to be seen as processual. The project's final report concludes more strongly that the process is political and will have defenders and challengers:

> The what and where of political/energy decentralization is only the starting point (the rhetoric)—questions of agency (who is involved, what factors are at play), process (how is it being pursued/achieved—i.e. through what mechanisms), and specific interests (why are they doing it) are ultimately the key to understanding what is possible/not possible.
>
> *(Brown et al., 2015, p. 45)*

Drawing from USES project examples, Cloke et al. (2017) turn the usual deficit model concerning energy literacy on its head. Instead of being about informing and educating unknowledgeable communities and publics about technology and maintenance, they suggest that a reverse information dialogue needs to take place educating "energy technology developers and project implementers about the project community's livelihood needs and aspirations in which energy plays a key enabling role" (p. 267).

Broader lessons about planning for renewable energy infrastructures according to the geographical distribution of people and landscapes suitable for renewables were drawn from a comparison of Kenya and Ghana in the Green Growth Diagnostics project of the USES program. Nonstandard solutions adapted to local circumstances of need and possibilities for development of energy services are part of a new way of energy planning (Pueyo, 2018, p. 96), counter to the agenda-driven, deterministic external influence of donors and corporations in countries in the Global South (Newell & Phillips, 2016).

Who should be involved in developing and participating in democratic planning processes has relevance to Silver and Marvin's (2017) attention to urban energy governance in the SAMSET (Supporting Sub-Saharan Africa's Municipalities with Sustainable Energy Transitions) project on municipal level energy policy making in South Africa, Ghana, and Uganda.

> Global North–anchored understandings of what constitutes an urban energy network face huge challenges concerning energy service operation where there are large numbers of energy users depending on biomass, and where the urban poor (particularly women and female-headed households) have problems regularizing property relations.
> *(Silver & Marvin, 2017, p. 854)*

The post-/colonial and capitalist legacies of social difference play out in urban spatial configurations and their energy regimes with "ongoing production (and circulation) of inequalities and injustice through and across energy systems together with sites and processes that produce divergent visions of future energy transition" (Silver & Marvin, 2017, p. 856). The SAMSET project aimed to "design, test, and evaluate a knowledge exchange framework to facilitate the implementation of an effective sustainable energy transition in Africa's Sub-Saharan urban areas" ("Supporting Sub-Saharan Africa's Municipalities," n.d.), and the challenges rather than solutions were most prominent in the findings.

The research networks facilitated by the USES program went on to find new spaces for skills development and partnership work in the Transforming Energy Access program, and latterly the Modern Energy Cooking Services initiative. This last program consists in a refusal to accept the marginalization of cooking from central energy sector attention in Africa especially, that marginalization being a prime example of antidemocratic energopatriarchy at work. The first improved cookstoves (*magan chulhas*) in India (dating from the 1930s) were promoted mainly on the basis of making the lives of women easier, and the issue of reducing the irritant effects of smoke was also a selling point. However, for decades following this innovation, women and their lived realities were excluded from design processes after clean cooking became a fixture in development practices.

This technocratic regime reduced women to victim status, allowing them little or no agency. Far from just being victims of more primitive stoves, however, women can be active agents in defense of the three-stone stoves passed down to them, which are sanctified by custom and belief. Women accrue power through cooking, fire, firewood, and the multiple meanings of cooked food in ways that have nothing to do with fuel efficiency. Kitchen location, control of food and purchases, and the centrality of the woman's role as nourisher of the family revolve around the cookstove, which is an important nexus of struggle in a woman's daily relationships.

Conclusion

Fossil fuel energy systems have been central to the expansion of extractive global economic relations over the last 200 years and to the forms of democratic representation struggled for by men and women in the capitalist Global North. That system is reorientating itself after peak oil austerity and the coronavirus, with the Anthropocene reaching public consciousness and giving rise to "green new deal" political pressure. With the increasing availability and decreasing cost of renewables, a novel territory for sociotechnical collaboration, for a gender-equitable and ethical economy, and for the relocalization of energy decision making around climate resilient systems governed by informed energy citizenships becomes more than a possibility.

The continuing hegemony of technocentric approaches to energy remains a severe impediment, as a depoliticized view of modern energy infrastructures configured as objective delivery systems to abstract users. This hegemonic discourse remains detached from conditions of reciprocally lived worlds of relation and locality and perpetuates development in terms of catch-up and leapfrog that present transitions as manageable substitutions of infrastructure, without appreciating the obvious normative revolution required in order to keep below a two-degree increase in global warming.

That normative revolution will need wholescale reevaluation of economic relations and living relationally on land, by water, and with energy. Renewable technologies have yet to be absorbed into and by actor networks without the unjustly asymmetrical competition of subsidized fossil fuel technomass. The neoliberal myth further suggests that economic and technological change can come about by individual choices in aggregate rather than (to date) by decisions in and between powerful institutions well practiced at democratic exclusion.

A low-carbon rather than high-carbon energy democracy will need to emerge from new forms of democratic energy literacy. Diverse kinds of knowledge and skill have to be developed out of local renewable energy landscapes. Community-led microhydro and community electrification examples already demonstrate hybrid innovations between local autonomous traditions of skill in technical practice and governance institutions with new contexts of utility stakeholding and IT-mediated service regulation. The democratization of innovation and learning in renewables are fundamental to a low-carbon democratic culture and to the unpicking of expert enclaves effected by energopatriarchy.

While community scales of appropriating low-carbon energy technology can only be part of a more complex approach to reducing inequality and advancing people's relational capacity to act, they are no panacea for the greater set of global challenges.

> The literature on community energy projects is full of good practices and processes that have contributed to project success; inclusion of multiple stakeholders, capacity-building, affordability, maintenance, demonstrations, thorough dissemination and education practices, local manufacturing and training centres, simultaneous implementation with employment-generating practices, the development of symbiotic services (irrigation, etc.). But not all of these will work in all contexts and what is vital to understanding which will work where and what other innovations might be needed cannot be achieved through top-down perceptual models based on finance and technology.
>
> *(Cloke et al., 2017, p. 270)*

Discussion of energy transitions need to be resituated from the technoeconomic into deliberative spaces accessible for the poor in the scenarios of coming decades of climate-changed livelihoods (Harriss-White, 2018). The evidence from studies on energy access for the poor in the Global South and technoeconomic projects of renewable energy development is that a great divide of perceptions separates energy modernity from off-grid. The way that energy developers see places as simply "where there is no infrastructure" has to be disrupted by a process of rendering energy relationships visible. Technology developers with knowledge of these unequal realities are indeed participants in democratic energy renewal, as are the neighbourhood decision makers who are translating "energy poverty" into a basis of mobilizing rights in cities, and in villages that can be transport linked and powered from local sources without profits from energy purchases being relocated elsewhere. Nonetheless, the difference between renewable energy, which merely adds more types of energy to the existing centralized, corporatized production and distribution systems, and truly alternative energy, which recognizes the need for profound change throughout the social energy system, remains substantial.

Notes

1 This chapter mobilizes a variety of project work undertaken under the umbrella of LCEDN. Principle web references include the following: www.lcedn.com/initiatives/category/USES-Network; www.lcedn.com/read; www.lcedn.com/song; www.lcedn.com/re4food; www.lcedn.com/samset.
2 This was from the Department of Energy and Climate Change, eager to complement the emerging UN Sustainable Energy for All program.
3 www.theminigridgame.org/.
4 www.lcedn.com/events/5th-international-conference-developments-renewable-energy-technology-icdret.
5 https://e4sv.org/about-us/.
6 See the LCEDN website: www.lcedn.com/blog/ethno-stuff. Ethnoengineering is "a method of defining and solving complex issues with constantly evolving deep experiential knowledge of the environment, without utilisation of modern mathematics, science and technology, relying on bottom-up management, practicing resourcefulness, and being contingent upon a holistic worldview" (Hess & Strobel, 2013, p. 58).
7 Of course, this is true not just in urban contexts. In rural areas, state offices are frequently institutions for everyday corruption. A Nepalese villager confided to one of the authors, "If you follow the law you will go hungry in this society."

References

Ahlborg, H., (2017). Towards a conceptualization of power in energy transitions. *Environmental Innovation and Societal Transitions, 25*, 122–141.

Appadurai, A. (2002). Deep democracy: Urban governmentality and the horizon of politics. *Public Culture, 14*(1), 21–47.

Bakke, G. (2016). *The grid: The fraying wires between Americans and our energy future*. Bloomsbury.

Batchelor, S., & Smith, J. (2014). Beyond literacy and knowledge: Energy proficiency for decentralised governance. *READ*. http://thereadproject.co.uk/wp-content/uploads/2014/04/Energy-Proficiency-for-Decentralised-Governance_Working-Paper-3-v29072014.pdf

Bonneuil, C., & Fressoz, J. (2017). *The shock of the anthropocene*. Verso.

Bridge, G. (2018). The map is not the territory: A sympathetic critique of energy research's spatial turn. *Energy Research and Social Science, 36*, 11–20.

Brown, E., Campbell, B., Cloke, J., To, L. S., Turner, B., & Wray, A. (2018). Low carbon energy and international development: From research impact to policymaking. *Contemporary Social Science, 13*(1), 112–127.

Brown, E., Cloke, J., & Harrison, J. (2015). *Governance, decentralisation and energy: A critical review of the key issues*. Working Paper 1. thereadproject.co.uk/?portfolio-view=116–2

Campbell, B., & Adamu, F. (2019). Energy on the move: Perspectives from the urban marginal. *DFID/ESRC Research for Policy and Practice*. www.energyonthemove.online
Campbell, B., Cloke, J., & Brown, E. (2016). Communities of energy. *Economic Anthropology, 3*, 133–144. https://doi.org/10.1002/sea2.12050
Castán Broto, V. (2019). *Urban energy landscapes*. Cambridge University Press.
Castán Broto, V., Stevens, L., Ackom, E., Tomei, J., Parikh, P., Bisaga, I., To, L. S., Kirshner, J., & Mulugetta, Y. (2017). A research agenda for a people-centered approach to energy access in the urbanizing global south. *Nature Energy, 2*, 776–779.
Chatterjee, P. (2004). *The politics of the governed*. Columbia University Press.
Cloke, J., Mohr, A., & Brown, E. (2017). Imagining renewable energy: Towards a social energy systems approach to community renewable energy projects in the Global South. *Energy Research & Social Science, 31*, 263–272. doi.org/10.1016/j.erss.2017.06.023
Das, V., & Walton, M. (2015). Political leadership and the urban poor: Local histories. *Current Anthropology, 56*(11), 44–54. https://doi.org/10.1086/682420
Davis, M. (2006). *Planet of slums*. Verso.
De Waal, A. (2018, November 21). *The reckless anthropocene*. Durham Castle Lecture.
Feldpausch-Parker, A. M., Sprain, L., Endres, D., & Peterson, T. R. (Eds.). (2019). *Energy democracy: A research agenda*. Frontiers Media. https://doi.org/10.3389/978-2-88963-197-1
Harriss-White, B. (2018). Science-policy interfaces in an era of global commodification. In C. Raj Kumar, S. Visvanathan, R. Sudarshan, & Y. S. R. Murthy (Eds.), *The future of Indian universities: Comparative and international perspectives*. Oxford University Press.
Harvey, P., & Knox, H. (2015). *Roads: An anthropology of infrastructure and expertise*. Cornell University Press.
Hess, J., & Strobel, J. (2013). Indigenous ways of doing: Synthesizing scholarly literature on ethno-engineering. *International Journal of Engineering, Social Justice and Peace, 2*, 55–80.
Hughes, D. (2017). *Energy without conscience: Oil climate change and complicity*. Duke University Press.
Ikejemba, E. C. X., Mpuan, P., Schuur, P., & Van Hillegersberg, J. (2017). The empirical reality & sustainable management failures of renewable energy projects in Sub-Saharan Africa (part 1 of 2). *Renewable Energy, 102*, 234–240.
Kaur, R. (2016). The innovative Indian: Common man and the politics of jugaad culture. *Contemporary South Asia, 24*(3), 313–327. https://doi.org/10.1080/09584935.2016.1214108
Lama, N. (n.d.). *Interview for 'energy on the move' ESRC/DFID project*.
Latour, B. (2017). *Facing gaia: Eight lectures on the new climatic regime*. Polity.
Lilliestam, J., & Hanger, S. (2016). Shades of green: Centralisation, decentralisation and controversy among European renewable electricity visions. *Energy Research & Social Science, 17*, 20–22. https://doi.org/10.1016/j.erss.2016.03.011
McNeill, J. R., & Engelke, P. (2014). *The great acceleration: An environmental history of the anthropocene since 1945*. Belknap Press.
Mitchell, T. (2011). *Carbon democracy: Political power in the age of oil*. Verso.
Moore, J. (Ed.). (2016). *Anthropocene or capitalocene: Nature, history and the crisis of capitalism*. Kairos.
Nathan, D., Shakya, I., Rengalakshmi, R., Manjula, M., Galkwad, S., & Kelkar, G. (2018). The value of rural women's labour in production and wood fuel use. *Economic and Political Weekly, 53*, 26–27.
Newell, P., & Phillips, J. (2016). Neoliberal energy transitions in the South: Kenyan experiences. *Geoforum, 74*, 39–48. http://dx.doi.org/10.1016/j.geoforum.2016.05.009
Nye, D. (1990). *Electrifying America: Social meanings of a new technology*. MIT Press.
Ockwell, D., Atela, J., Mbeva, K., Chengo, V., Byrne, R., Durrant, R., Kasprowikz, V., & Ely, A. (2019). Can pay-as-you-go, digitally enabled business models support sustainability transformations in developing countries? Outstanding questions and a theoretical basis for future research. *Sustainability, 11*, 1–21.
Perlman, J. (1975). Rio's favelas and the myth of marginality. *Politics and Society, 5*(2), 131–160.
Pueyo, A. (2018). What constrains renewable energy investment in Sub-Saharan Africa? A comparison of Kenya and Ghana. *World Development, 109*, 85–100. doi.org/10.1016/j.worlddev.2018.04.008
Redfield, P., & Robins, S. (2016). An index of waste: Humanitarian design, "dignified living" and the politics of infrastructure in Cape Town. *Anthropology Southern Africa, 39*(2), 145–162. http://doi.org/10.1080/23323256.2016.1172942
Scott, J. (2012). *Two cheers for anarchism*. Princeton University Press.
Silver, J., & Marvin, S. (2017). Powering sub-Saharan Africa's urban revolution: An energy transitions approach. *Urban Studies, 54*(4), 847–861. https://doi.org/10.1177/0042098016668105

Smits, M., & Bush, S. (2010). A light left in the dark: The practice and politics of pico-hydro power in the Lao PDR. *Energy Policy*, *38*, 116–127.

Solar Nano-Grids (SONG): An appropriate Solution for Meeting Community Energy Needs? (n.d.). *Low carbon energy for development network*. www.lcedn.com/song

Stirling, A. (2014). Transforming power: Social science and the politics of energy choices. *Energy Research and Social Science*, *1*, 83–95. http://dx.doi.org/10.1016/j.erss.2014.02.001

Supporting subSaharan Africa's Municipalities with Sustainable Energy Transitions. (n.d.). *Low carbon energy for development network*. www.lcedn.com/samset

Szulecki, K. (2018). Conceptualizing energy democracy. *Environmental Politics*, *27*(1), 21–41. https://doi.org/10.1080/09644016.2017.1387294

UNDP & WHO. (2009). *The energy access situation in developing countries. A review on the least developed countries and Sub-Saharan Africa*. http://content.undp.org/go/cms-service/stream/asset/?asset_id=2205620

Vaghela, D. (2017, September). *Energy democracy in Myanmar: Scaling up renewable energy social entrepreneurs* [conference presentation], UK Low Carbon Energy Development Network (LCEDN) Conference. www.lcedn.com/sites/default/files/files/Vaghela(1).pdf

36

ENERGY DEMOCRACY IN PRACTICE

Centering energy sovereignty in rural communities and Tribal Nations

Douglas Bessette, Chelsea Schelly, Laura Schmitt Olabisi, Valoree S. Gagnon, Andrew Fiss, Kristin L. Arola, Elise Matz, Rebecca Ong and Kathleen E. Halvorsen

Energy sovereignty and energy democracy

Energy democracy refers to an emergent social movement aimed at restructuring sociotechnological system regimes and reimagining energy politics in a way that links social justice, economic equity, and renewable energy transitions (Burke & Stephens, 2017). A focus on energy sovereignty and the rights of communities and individuals to make their own choices regarding the forms, scales, and sources of energy, as well as the patterning and organization of energy production and consumption, is key to realizing energy democracy (Friends of the Earth International, 2006; Laldjebaev & Sovacool, 2015; Paradis et al., 2009). Sovereignty holds different meanings for different types of communities. For nontribal communities, sovereignty is the ability to make community-scale decisions about issues like renewable energy development. For Tribal Nations, sovereign nation status is inherent and denotes a nation-to-nation relationship with the United States federal government (Bronin, 2016). Tribal Nations, in particular, are increasingly focused on energy sovereignty as an assertion of self-determination, a mechanism for community autonomy and development (Powell, 2015; Royster, 2008; Stefanelli et al., 2019), as well as an environmentally sustainable option for living in accordance with community values regarding relations among humans and nonhuman beings (McDonald & Pearce, 2013; Tsosie, 2013).

Current energy policy in the United States not only fails to prioritize energy sovereignty but often works against it, as energy systems are principally designed and deployed by policy makers and business leaders at the federal, state, and regional levels with little opportunity for household- or community-level input or decision making (Schelly et al., 2020). Real-world deployment of community-owned renewable projects is rare in the United States due to both the centralized energy grid and the entrenched fossil-based energy systems working along political lines to contest such projects (Breetz et al., 2018; Geels, 2014; Newell & Paterson, 1998). Very little research and no existing frameworks comprehensively consider the trade-offs between community-owned projects or the potential for hybridized combinations of systems or transition policies (Burke & Stephens, 2017). There is currently limited understanding of

 DOI: 10.4324/9780429402302-42

how this information might impact renewable energy technology deployment decisions for a particular community.

The concepts of energy democracy and energy sovereignty are interrelated. According to Burke and Stephens (2017), energy democracy arose from popular movements aimed at addressing climate and economic injustice, limiting fossil fuel use, and transitioning to renewable energy. Energy democracy has since become a broad social and cultural framework that acknowledges how energy systems impact society and reframes energy system change as an opportunity to actively redistribute political and economic power (Stephens, 2019). Various groups and organizations in the United States and Europe have explicitly used energy democracy as a central theme of discourse on energy and climate change since 2012 (Burke & Stephens, 2017), while groups elsewhere have used the terms "energy justice" (Jenkins et al., 2016) and "energy sovereignty" (Broto et al., 2018; Laldjebaev & Sovacool, 2015). Broto et al. (2018) write that energy sovereignty is more "emancipatory" in that it provides for "postcolonial analysis" and recognizes the "autonomy and self-determination of people in framing energy decisions that affect them," including the frames used to evaluate those decisions (pp. 648, 653). Burke and Stephens (2017) add that all three terms—energy justice, energy sovereignty, and energy democracy—must become more meaningful among actors and activists, and all actors must become more unified in their pursuit of more equitable distributions of political and economic power and greater community control of energy.

A key purpose of both energy sovereignty and energy democracy is to inform and facilitate a movement toward community-owned renewable energy projects. Examining that movement and determining what tools and resources are necessary to support it is the principal goal of the Michigan Community Anishinaabe and Rural Energy Sovereignty (MICARES) research team. MICARES is a collection of scholars and practitioners from Tribal Nations and colleges, universities, and community organizations in Michigan who recently began a five-year project to lay the foundation for a convergent transdisciplinary field of study and practice within the context of energy systems. This new field of study, sociotechnological systems transitions (STST), recognizes that technology is always deployed in a specific sociopolitical and technical context, which in many cases determines whether and how that technology is adopted, as well as whose interests it may serve (Stephens et al., 2008). This linked sociotechnological context is dynamic; technology affects the organization of social systems, and these social systems in turn shape technology development, adoption, adaptation, and dissemination.

Widespread system transitions require simultaneous evaluation of social, regulatory, and technical issues to effectively implement changes. Barriers can come from multiple directions, and creative solutions are often identified through cross-pollination of ideas. Such processes require not only a transdisciplinary team of social and natural scientists, engineers, stakeholder representatives, and nonacademic practitioners—of which the MICARES team is comprised—but also a means for pursuing deep integration and convergence, or the shared communication, language, and mental models necessary to generate breakthrough insights and solutions (Davidson, 2020; Dunn et al., 2019).

Energy democracy also requires a sustained commitment to collaborative and transdisciplinary research, as well as normative commitments, an empirical research agenda, and the practices and processes necessary to support energy systems transitions. The MICARES approach adopts each of these tenets, integrating theory and praxis in community-based research collaborations, participatory research, and communication strategies, along with trust building, productive, and equitable knowledge sharing (Halvorsen et al., 2019) and evidence-based lessons for nurturing transdisciplinary communities of practice (Cundill et al., 2015). Prioritizing community values in renewable energy deployment not only supports goals of energy sovereignty and democracy but is also critical in defining the technological solutions that contribute to a societal renewable energy transition.

In this chapter, we describe the methodological approach adopted by the MICARES team, specifically how our Tribal Nation partners' Anishinaabeg teachings of the medicine wheel, facilitates not only energy sovereignty and sociotechnological systems transitions but also the energy democracy movement. We also provide an initial investigation of how our approach could be operationalized in four communities informing MICARES's work.

The medicine wheel as a conceptual framework

At the heart of the MICARES project is a conceptual framework (see Figure 36.1) based on our Tribal Nation partners' Anishinaabeg teachings of the medicine wheel or sacred hoop symbol. This framework is significant not only because it is foundational and interwoven throughout our approach but because it acknowledges that MICARES's primary universities, Michigan Technological University (MTU) and Michigan State University (MSU), occupy the ancestral, traditional, and contemporary lands of the Anishinaabeg Three Fires Confederacy. The Anishinaabeg are one of the largest Indigenous groups in North America with nearly 150 different bands living throughout their original homeland in the present-day United States and Canada. Together known as the Three Fires Confederacy, the Anishinaabeg are currently identified by various names: the Chippewa, Ojibway, Ojibwe, or Ojibwa, as well as the Ottawa or Odawa, and Potawatomi or Bodewadomi. All of these names refer to the Anishinaabe people, the larger group who originated from the Great Salt Water (Atlantic Ocean) on the eastern shores of North America and migrated to the Great Lakes region from approximately 900–1400 CE. (Benton-Banai, 1988). These names originate with the Anishinaabeg, names they use to describe themselves that reflect specific relationships and responsibilities that the people have to their homelands and to one another. Slight variations in pronunciation and spelling reflect European understandings of the Native language they encountered, names that were written by differing European groups with contact and settlement. Throughout this chapter, we have chosen to use different spellings of the names depending on the place and people we are writing about, or the term chosen by particular scholars, because it is important to use the spellings that the Anishinaabeg chose to use for themselves (Younging, 2018). One of our Tribal Nation partners, the Keweenaw Bay Indian Community (KBIC) of the Lake Superior Band of Chippewa Indians, uses "Ojibwa"; another partner, the Sault Ste. Marie Tribe of Chippewa Indians, uses "Ojibwe." Additionally, the term "Chippewa" is also used because it is the legal spelling and used in the construction and ratification of treaties, reflecting the written understanding of those who negotiated treaties with the Anishinaabeg.

While we rely on Anishinaabeg teachings to inform our framework, it is important to note that Indigenous peoples employ varying versions of the medicine wheel as an interconnected system of teachings relating to the seasons, directions, elements, colors, and, overall, the cyclical nature of life (Johnston, 1990). As it has for millennia, the medicine wheel reflects the philosophy and practice of being in balance between and among time, space, and all beings. Illustrated in the framework are four equitable directions, with new beginnings originating in the East (e.g., sunrise, spring), transitioning clockwise to the South, and so on. As a whole and across time and space, the medicine wheel is inclusive of all interaction, all knowledge, and all virtues while aiming to sustain balance through continuous transitions of all kinds. The framework instantiates how human society is in a permanent state of transition. For the Ojibwe, the medicine wheel does not end, as a flower does not end, but instead transforms into a memory,

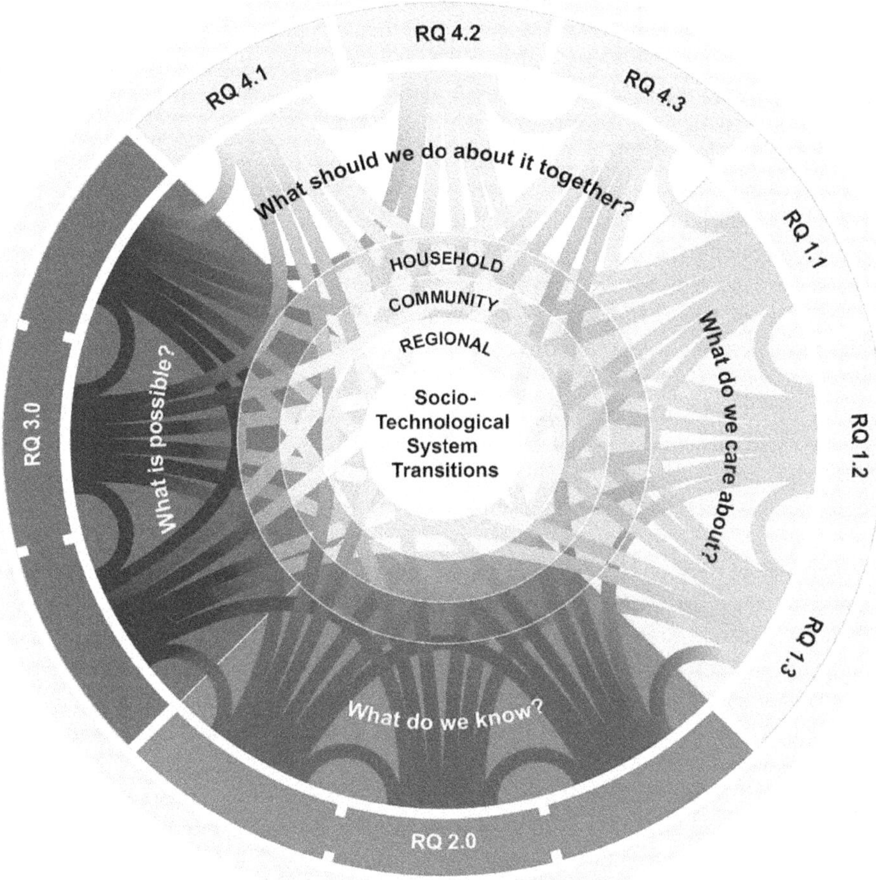

Figure 36.1 *MICARES conceptual framework.* This framework informs the methodological approach of the MICARES team and is based on our Tribal Nation partners' Anishinaabeg teachings of the medicine wheel or sacred hoop symbol.

organic material, nutrients, and energy, and the possibility of renewal (Johnston, 1990). In other words, the medicine wheel both reifies a practice and philosophy of research with, by, and as Indigenous peoples (Wehipeihana, 2008) and simultaneously indicates new beginnings, a process of discovery, and continuous transformation of the research and the researchers (Absolon, 2012).

The medicine wheel with its associated teachings also interweave the engagement between the MICARES team and Indigenous and rural community partners, centering our focus on energy sovereignty and aiding navigation of a respectful and ethical path through renewable energy and sociotechnological system change. In order to focus and track our movement along that path, we developed four associated research questions. These questions not only structure the research but can also be used to generate and demonstrate best practices for others to pursue.

The wheel and questions thus not only guide our partnerships and research with one another and Tribal and non-Tribal communities, but also provide balance, based on a relational coming together of voices, communities, and ecosystems. The conceptual framework also allows for an approach across and between three scales and dominions of decision-making control (household, community, and regional). The four questions appear as a cycle on the medicine wheel, beginning in the east, as described here:

1 *What do we care about? (East)* What are our core values, whether individually, as a Tribal Nation, as a community, or as a society? How do these values influence social choices regarding sociotechnological system research, design, policy, innovation, and implementation?
2 *What do we know? (South)* What is the current state of knowledge related to system transitions? What are the gaps in knowledge? What are the known risks, barriers, and opportunities to successfully navigating this transition?
3 *What is possible? (West)* What are the skills, resources, and technologies that can help facilitate successful transitions? Are there ways to turn risks and barriers into opportunities?
4 *What should we do about it together? (North)* How can we ethically and respectfully integrate knowledge, resources, and capabilities to facilitate sociotechnological transition, while simultaneously protecting our core values?

Reflecting on these research questions, the medicine wheel framework highlights that the process of systems transition is iterative, not linear, and does not end but continues to adapt to changes in society and our environment. One benefit of a cyclical framework for research and systems change is that it provides an opportunity for growth and strengthening of convergence between stakeholders. The medicine wheel framework and research questions also guide development of our community engagement protocol, which focuses on listening to, understanding, and visualizing shared community objectives and values, while at the same time, identifying both technical and social requirements for community energy system change and developing pathways to proceed ethically through transition in multiple communities.

Both energy sovereignty and democracy require learning from and helping to communicate best practices between communities at varying points and rates of transition. Both those that have faced unique—and many times colonized—endowments of natural, human and financial resources and renewable energy potential. By design, the economic and regulatory framework of energy delivery leaves very little decision-making control to communities. Aiming to center community knowledge, practices, and priorities for the research process, initial conversations with communities emphasize the position of researchers as learners (Carr & Halvorsen, 2001; Halvorsen, 2003). The epistemological diversity of the MICARES research team can then be a powerful tool for understanding the many factors that are brought to bear on how partner communities currently receive and consume energy, identifying opportunities to achieve transition in culturally appropriate and desirable ways.

To facilitate sharing, MICARES intentionally merges participatory community-based cultural mapping (Gagnon & Ravindran, 2014; Strang, 2010) and modeling, scholarship on sociocultural, biophysical, and technical factors, and community praxis. Additionally, the team has already begun to codevelop a community-focused web portal that serves as both a data repository and home for decision support tools (DSTs). Providing effective decision support is key to MICARES's work and the promotion of energy sovereignty and democracy overall. DSTs can facilitate multiscalar renewable energy transitions by characterizing how community-created energy strategies perform across shared visions and values, while incorporating social, regulatory and technical risks, barriers and opportunities, and spatial maps. Additionally, by incorporating tenets of multicriteria decision analysis, DSTs encourage creative thinking and a more explicit analysis of trade-offs and impacts across both systems and spaces (Bessette & Arvai, 2018). The strategies, visions, and values may vary across and between communities, but effective DSTs empower communities to determine the parameters.

While the overarching framework of MICARES helps to create a cohesive and unified vision for the work, the methods used to pursue this work are organized into subteams, following best practices established through other long-term transdisciplinary projects (Carr & Halvorsen, 2001; Halvorsen, 2003). Each subteam works independently, yet also in concert with one another, applying and building expertise and experience within teams, while regularly sharing with and learning from other subteams, practitioners, and experts in other disciplines and communities through regular meetings. This integrated and cyclical process of listening, sharing, and learning not only builds trust between and within communities and the MICARES team, but creates a mutual learning process through which we and our communities, our inquiry, and our approach undergo transformation (Absolon, 2012).

Communities in transition

In this section, we describe four of our case study communities, each at a different point and place in time and space, each experiencing unique challenges of and opportunities for transition.

Traverse City, Michigan

Despite only having approximately 15,000 residents, Traverse City is the largest community in Michigan's northern Lower Peninsula. The community has significant social and economic capital. The majority of the city's residents are politically aligned with and participate(d) in the same type of movements that spawned the energy democracy movement, and the city is home to several prominent nonprofit organizations that aggressively advocate for renewable energy and climate change action.

Traverse City may be especially well suited to pursue a renewable energy transition. In 2016, Traverse City's elected commission committed to sourcing 100% of the energy used for city operations from renewable energy sources by 2020. Two years later, the city's municipal utility, Traverse City Light and Power (TCLP) made a separate but similar commitment to achieving 40% renewables in its portfolio by 2025 and 100% renewables by 2040, making Traverse City the first Michigan city to commit to using 100% renewable energy on a community-wide basis rather than just in municipal buildings. TCLP also installed the state's first utility-scale wind turbine in 1996 and recently supported a utility-scale solar installation on the property beneath and adjacent to the wind turbine.

With its municipal utility, the city also maintains significant control over its energy. TCLP's priority is to serve its customers; in contrast, investor-owned utilities (IOUs), which provide electricity to 72% of US electric customers (Randolph, 2019), operate as for-profit entities. Relative to municipal utilities, IOUs have demonstrated a far greater commitment to the current fossil-fueled energy system and represent a significant technical, economic, and political barrier to increasing energy sovereignty (Geels, 2014; Newell & Paterson, 1998). TCLP, in contrast, is beholden to community interests and priorities, resulting in collaborative efforts such as a voluntary opt-in green pricing program, expanded solar energy projects across the state, and reductions in overall energy use. Application of the medicine wheel framework demonstrates that the community has shared collective values associated with environmental stewardship and responsibility and a shared condition of possibility in the structure of their public utility; these shared values and conditions of possibility shape their status as a community with a renewable energy commitment.

In spite of the many successes, Traverse City also faces challenges to its renewable energy transitions. Investing in green power purchasing and extralocal renewable energy development fails to enhance Traverse City's energy resilience, as neither develops community-controlled energy systems—something a community-scale microgrid would allow. These strategies also transfer the impacts of energy production to other communities, ensuring that residents outside of Traverse City, who tend to lack the legal, financial, and political capital to resist such development, must endure the externalities of renewable energy production. These externalities are considerable and include impacts to the visual landscape, rural values and ecosystems, disruption to farm production, and increased road traffic and noise from both the construction and operation of utility-scale energy systems.

Traverse City may be considered a case study of a successful renewable energy transition, yet application of the medicine wheel framework draws attention to the impact of divergent values within a community and how questions of action (what should we do about it together?) are necessarily predicated by the scope of inclusion (the "we" deciding collectively what to do). The city is a leader in terms of their commitment to achieving 100% renewable energy use in their community and in their utility. While their shared community values and the structure of their public utility both contribute to making this possible, questions remain regarding the extent to which this transition will enhance energy democracy or energy sovereignty in the community and beyond. Some of these questions involve the collective decisions the community makes, which will help determine whether its renewable energy used is physically located in the community, visible to community residents and providing benefits to community members. Also, as the community's shared values become more diverse when "community" is defined more broadly to include additional rural communities in the region, these more divergent values may require careful navigation and attention to align collective decisions about a renewable energy transition with energy democracy principles. In particular, the principles of inclusivity and engagement with broad and diverse publics.

Keweenaw Bay Indian Community

The Keweenaw Bay Indian Community (KBIC) is an Anishinaabe Ojibwa Tribal Nation, the oldest federally recognized tribe retaining the largest land base in the State of Michigan. Located 30 miles south of MTU, KBIC is successor in interest of the L'Anse and Ontonagon Bands of Lake Superior Chippewa Indians, signatories to the 1842 Treaty with the Chippewa (7 Stat., 591, 1842) and the 1854 Treaty with the Chippewa (10 Stat., 1109, 1854). In 2008, KBIC established the Committee for Alternative and Renewable Energy (CARE), which serves as an advisory council for energy decision making in the community. As part of their commitment to recognizing good relations among all humans and nonhumans, KBIC committed to sovereignty and sustainability in relations with the land and the water, including thoughtful engagement and planning regarding energy (KBIC, 2020).

KBIC and CARE continue to provide leadership in the renewable energy transition, particularly in solar energy installations and energy efficiency programs. Both have been awarded multiple federal grants for their energy transition projects and in 2018 won a Green Champions Award from Indian Health Services for a solar project installed at a local health clinic and residential treatment facilities, which saved the community approximately $29,000 a year (Vissers, 2018). The community's success in raising funds for renewable energy projects and in sustaining interest and engagement in a renewable energy transition demonstrates the link between the

Tribe's shared core values, resources, and capabilities. Engagement with KBIC also demonstrates that community values that are not directly linked to energy, such as values associated with human, ecological, and community health, can be motivations for pursuing renewable energy transitions. This demonstrates that consideration of core values (starting in the East) need not center on values associated with energy to inform actions associated with energy (meeting in the North)—a point that may benefit those interested in pursuing energy democracy.

KBIC also faces challenges in pursuing a comprehensive energy transition. The L'Anse Indian Reservation is served by multiple electric utilities: municipal, cooperative and IOU. This complex set of utility actors, which exist outside of KBIC's direct control, limit not only the community's ability to install renewable energy systems that connect to the electrical grid but also the economic compensation they receive for that generation. KBIC also lacks the financial resources to maintain permanent staffing or office support specifically focused on energy. This results in energy transition work being done in association with other projects and goals such as health or the environment more broadly. Although being part of other projects is not inherently negative, this leads to inconsistent capacity, sporadic attention, and reduced funding dedicated to renewable energy resources and increased concern about how this transition serves other community goals and values. Federal funding structures also promote piecemeal rather than comprehensive, strategies, so the community often navigates challenges associated with integrating multiple projects without coherent or consistent funding structures.

The conditions of KBIC demonstrate that possibilities and pathways may only indirectly connect to energy but may impact energy democracy and energy decision making nonetheless. Conditions of possibility in KBIC are complex, given their multiple competing demands for resources and engagement, as well as the multiple electric utilities operating in KBIC and the multiple, diverse funding mechanisms and program opportunities available to the community. Approaching this complexity through the medicine wheel framework, the MICARES project will focus on learning, in particular by listening to leaders and community members at KBIC to better understand how policy and formal capacities in energy expertise can be aligned with the community's priority interests and needs. The KBIC also practices both energy sovereignty and democracy utilizing the Seven Generations principle, which honors past ancestors' actions and teachings and protects the ability of future generations to sustain their identity as an Ojibwa people. Centering the intergenerational aspect of transitions is a crucial objective of the medicine wheel framework.

Sault Ste. Marie Tribe of Chippewa Indians

The Sault Ste. Marie Tribe of Chippewa Indians provides services to members residing within a seven-county service area in the eastern part of Michigan's Upper Peninsula. The Tribe is a leader in engagement in sustainability and resilience in tribal practice, planning, and advocacy, and their shared values around renewable energy development center on cultural, economic, and climate resilience. Recently, the Sault Ste. Marie Tribe's strategy to pursue a renewable energy transition involved a staged approach with the ultimate goal of net zero purchased energy, beginning with demand reduction, followed by pursuit of renewable energy systems sized and located to meet demand. The White House formally recognized this approach in 2014, earning the Tribe designation as a Climate Action Champion Community (DOE, 2014). This work has involved energy audits, investigating renewable energy development, energy efficiency programming, and renewable energy investment whenever possible.

The commitments and efforts of the Sault Ste. Marie Tribe of Chippewa Indians reveal several structural barriers for Tribal Nations aiming to pursue renewable energy transitions.

One involves the matching requirements embedded in federal granting programs like the ones offered by the US Department of Energy (DOE). These required matches can be upwards of 50% of the requested amount. Tribal Nation communities often cannot provide this level of funding. Additionally, DOE grants often fund either preliminary studies and analyses, or shovel-ready projects, but not the engineering design projects essential to bridging analyses and shovel-ready projects. This barrier is exacerbated by a lack of essential engineering expertise within the Tribal Nation and its broader geographical region.

Conflicts with environmental regulations also arise, specifically with wetlands, which are abundant in this region and form the majority of the lands relegated to Tribal peoples under the mission system. These regulations are particularly difficult to navigate when lands are defined by Tribal jurisdiction within multiple, overlapping regulatory regimes. Furthermore, all Tribal Nations face a challenge associated with renewable energy investment funds. The current federal investment tax credit is only available to entities with taxable status, which requires that Tribal Nations—and municipalities—who do not have taxable status enter into relationships with private development companies in order to apply this investment credit to renewable energy development. This both complicates planning and development and limits the ability to use renewable energy to enhance community-level energy sovereignty.

Applying the medicine wheel framework reveals that the Sault Ste. Marie Tribe of Chippewa Indians is positioned very similarly to KBIC; the shared community values associated with health and well-being for people and the Earth guide energy decision making, even if indirectly. However, this Tribal Nation explicitly utilizes more internal professional resources to advance its renewable energy transition, making energy democracy and decisions about how to pursue it more explicit. The community's challenges are associated with the difficulties in coordinating across its large, sparsely populated, geographical spaces served by multiple electric utilities. As is the case of KBIC, energy transition decisions are most typically pursued when in the service of other, more discrete community values and priorities. The MICARES project will focus on listening to learn from the experiences of the Sault Ste. Marie Tribe of Chippewa Indians and use targeted analysis to find ways to navigate these challenges while contributing to community energy democracy and sovereignty.

Gratiot County, Michigan

Located 240 miles due south of Sault Ste. Marie, Gratiot County is an agricultural community in the center of Michigan's Lower Peninsula. In the late 1970s, the county and its municipalities, amid declining oil, mining, and automobile manufacturing sectors and increasing unemployment, incorporated Greater Gratiot Development (GGD) to promote county-wide economic development (Rynne et al., 2011). Over the course of three decades, GGD worked to foster collaboration across the county and pursue countywide objectives, ultimately leading to residents initiating a countywide master planning process in 2008 and development of the Gratiot Regional Excellence and Transformation (GREAT) Plan—the "first countywide, locally developed, shared, adopted, and implemented plan in Michigan" (Rynne et al., 2011, p. 17).

Alongside the GREAT planning effort, a wind developer began investigating the county's wind potential and grid capacity, as well as building political and community support for wind energy development. Wind energy was introduced as a potential economic development strategy to GREAT's planners, as was the need to develop a county-wide wind energy ordinance. The latter led to a widespread education effort aided by MSU Extension, an information-gathering process, and public town hall meetings, the last of which was a public hearing of over 400 residents in 2010 (Rynne et al., 2011). Ultimately the GREAT process led to a unanimous vote by representatives from four townships to adopt the special use permit, and the first Gratiot wind farm was initiated. Today the county is home to over 320 utility-scale wind turbines, with the majority of those turbines owned and operated by either one of two IOUs, Consumers Energy and DTE Energy.

Per the GGD (2020), the economic benefits from wind development have been considerable: over $500 million invested since 2012, with county government, townships, and schools realizing $36.8 million in new tax revenue. Three hundred and fifty families now receive land payments, and 100 FTE positions were created by a wind investment–funded Agribusiness Park development. Finally, GGD (2020) claims that electrical transmission upgrades and four new substations have increased power reliability across the Mid-Michigan area and that production agriculture has been ensured "through careful placement of the wind turbines."

Applying the medicine wheel framework demonstrates the importance of identifying the specific shared values that spurred transitions in Gratiot, as well as acknowledging the considerable possibilities that accompanied an existing grid structure already suitable for turbines, and ample grid capacity resulting from the county's history of heavy manufacturing. Additional resources included a $65,000 grant through a sustainable communities program intended to kick-start master planning processes and a champion for collaboration in the GGD's initial director, who was open to and helped promote wind energy as a potential economic development opportunity (APA, 2011). Whereas utility-scale development and wind power may not be considered a boon to energy democracy or sovereignty elsewhere in Michigan—particularly in the Upper Peninsula and increasingly in the Thumb region—including Gratiot as a MICARES case study community provides an opportunity to investigate best practices in identifying shared values and to distinguish between the professional, infrastructural, and financial resources that make transitions possible. As in the Traverse City case, the MICARES project will investigate the extent to which Gratiot's transition contributes to energy sovereignty and democracy by centering discussions on values, diversity, possibilities, and potentials and by recognizing the complexity of defining community in relation to energy systems.

The interconnectedness of our research and communities

These four communities face unique challenges and contribute differently to energy sovereignty and democracy. They range in levels of access to technical expertise, necessary funding mechanisms, local control over utilities, renewable energy potential, and the complexity of their particular regulatory regimes. All demonstrate enduring commitments to climate and energy leadership and inclusive democratic participation.

MICARES's goals are to understand what is required to increase the rate and success of the implementation of novel renewable energy technology in communities such as these, while also improving their quality of life, protecting core values, and increasing energy sovereignty. Regarding both community control and decision making over renewable energy systems—as well as the nation-to-nation relationship with the US federal government—energy sovereignty remains the goal of both rural communities in Michigan and Tribal Nations. MICARES works not only to learn about the unique identities of Michigan communities and their technical, policy, and societal considerations, but also to help understand what it takes for communities to imagine and prepare for the renewable energy futures that best align with their values. Already our case study communities have provided insight regarding the energy sovereignty ramifications of defining "community" more broadly, the shared values that are indirectly driving energy transitions, and the influence of existing resources in opening new pathways. In full, by centering energy sovereignty in rural communities and Tribal Nations and sharing our core values, knowledges, possibilities, and pathways, MICARES offers a model of energy democracy in practice. Moreover, working with Indigenous communities, the MICARES group follows the medicine wheel framework of relationality, appreciating the interconnectedness of our research teams between and among not only human communities but also broader space and time. The wheel, after all, does not end but comes back to renewal and ultimately seeks transformation. Our work, though pursuing five-year goals, likewise aims to inspire further research, participation, and action.

Acknowledgments

The work in this chapter was supported by National Science Foundation Convergence Grant #1934346 "GCR: Collaborative Research: Socio-Technological System Transitions: Michigan Community and Anishinaabe Renewable Energy Sovereignty."

References

7 Stat., 591. (1842). In *treaty with the Chippewa. Proclamation, March 23, 1843*. Government Printing Office.

10 Stat., 1109. (1854). *Treaty with the Chippewa. Proclamation, January 29, 1855*. Government Printing Office.

Absolon, K. (2012). *Kaandossiwin: How we come to know*. Fernwood Publishing.

APA. (2011). *Wind energy case study*. American Planning Association.

Benton-Banai, E. (1988). *The Mishomis book: The voice of the Ojibway*. Indian Country Communications.

Bessette, D. L., & Arvai, J. L. (2018). Engaging attribute tradeoffs in clean energy portfolio development. *Energy Policy*, *115*, 221–229. https://doi.org/10.1016/j.enpol.2018.01.021

Breetz, H., Mildenberger, M., & Stokes, L. (2018). The political logics of clean energy transitions. *Business and Politics*, *20*(4), 492–522. https://doi.org/10.1017/bap.2018.14

Bronin, S. C. (2016). The promise and perils of renewable energy on tribal lands. In S. Krakoff & E. Rosser (Eds.), *Tribes, land, and the environment* (pp. 103–117). Routledge.

Broto, V. C., Baptista, I., Kirshner, J., Smith, S., & Alves, S. N. (2018). Energy justice and sustainability transitions in Mozambique. *Applied Energy*, *228*, 645–655. https://doi.org/10.1016/j.apenergy.2018.06.057

Burke, M. J., & Stephens, J. C. (2017). Energy democracy: Goals and policy instruments for sociotechnical transitions. *Energy Research & Social Science*, *33*, 35–48. https://doi.org/10.1016/j.erss.2017.09.024

Carr, D. S., & Halvorsen, K. (2001). An evaluation of three democratic, community-based approaches to citizen participation: Surveys, conversations with community groups, and community dinners. *Society & Natural Resources*, *14*(2), 107–126. https://doi.org/10.1080/089419201300000526

Cundill, G., Roux, D. J., & Parker, J. N. (2015). Nurturing communities of practice for transdisciplinary research. *Ecology and Society*, *20*(2), 22.

Davidson, E. J. (2020, January 14). *Self-evaluation for convergence science: Making the important measurable (not the measurable important)* [conference presentation]. Facilitated for NSF.

DOE. (2014). *Climate action champions: Sault STE*. Marie Tribe of Chippewa Indians. www.energy.gov/policy/climate-action-champions-sault-ste-marie-tribe-chippewa-indians-mi

Dunn, J. L., Knowlton, J. L., Handler, R. M., Pischke, E. C., Halvorsen, K. E., Mesa-Jurado, M. A., Selfa, T. L., Flaspohler, D. J., Licata, J., Mata-Zayas, E. E., Medeiros, R., Moseley, C., Nielson, E. A., Picasso Risso, V. D., Sacramento-Rivero, J. C., de Souza, T., Vazquez Navarrete, C. J., & Basiliko, N. (2019). Lessons from the transdisciplinary, international BIOPIRE project. In K. E. Halvorsen, C. Schelly, R. M. Handler, E. C. Pischke, & J. L. Knowlton (Eds.), *A research agenda for environmental management* (pp. 107–120). Edward Elgar Publishing.

Friends of the Earth International. (2006, September). Friends of the earth international. *Abuja Declaration, Resolution of Friends of the Earth International Conference on Climate Change*, 28–29.

Gagnon, V. S., & Ravindran, E. (2014). *Ethnography and oral history in the Keweenaw Bay Indian community: Collecting oral histories as a social experience with tribal narrators*. [conference presentation]. Oral History Association Conference.

Geels, F. W. (2014). Regime resistance against low-carbon transitions: Introducing politics and power into the multi-level perspective. *Theory, Culture & Society*, *31*(5), 21–40. https://doi.org/10.1177/0263276414531627

GGD. (2020). *Wind development*. https://ggdi.gratiot.org/wind-alternative-energy/

Halvorsen, K. E. (2003). Assessing the effects of public participation. *Public Administration Review*, *63*(5), 535–543. https://doi.org/10.1111/1540-6210.00317

Halvorsen, K. E., Knowlton, J. L., Schelly, C., Handler, R. M., & Pischke, E. C. (2019). Introduction to a research agenda for environmental management—through transdisciplinary, social science-rich environmental governance research. In K. E. Halvorsen, C. Schelly, R. M. Handler, E. C. Pischke, & J. L. Knowlton (Eds.), *A research agenda for environmental management* (pp. 3–7). Edward Elgar Publishing.

Jenkins, K., McCauley, D., Heffron, R., Stephan, H., & Rehner, R. (2016). Energy justice: A conceptual review. *Energy Research & Social Science*, *11*, 174–182. https://doi.org/10.1016/j.erss.2015.10.004

Johnston, B. (1990). *Ojibway heritage*. University of Nebraska Press.

KBIC. (2020). *Energy sustainability & sovereignty*. Keweenaw Bay Indian Community Natural Resources Department. http://nrd.kbic-nsn.gov/energy

Laldjebaev, M., & Sovacool, B. K. (2015). Energy security, poverty, and sovereignty: Complex interlinkages and compelling implications. In L. Guruswamy (Ed.), *International energy and poverty* (pp. 121–136). Routledge.

McDonald, N. C., & Pearce, J. M. (2013). Community voices: Perspectives on renewable energy in Nunavut. *Arctic*, 94–104.

Newell, P., & Paterson, M. (1998). A climate for business: Global warming, the state and capital. *Review of International Political Economy*, *5*(4), 679–703. https://doi.org/10.1080/096922998347426

Paradis, G., Yokey, J., & LeBeau, T. (2009). *Seneca Nation of Indians Energy Organizational Planning*. Red Mountain Energy Partners. https://doi.org/10.951271.

Powell, D. E. (2015). The rainbow is our sovereignty: Rethinking the politics of energy on the Navajo Nation. *Journal of Political Ecology*, *22*(1), 53–78. https://doi.org/10.2458/v22i1.21078

Randolph, K. (2019, August 19). EIA: Investor-owned utilities served 72 percent of US electricity customers in 2017. *Daily Energy Insider*. https://dailyenergyinsider.com/news/21198-eia-investor-owned-utilities-served-72-percent-of-us-electricity-customers-in-2017/

Royster, J. V. (2008). Practical sovereignty, political sovereignty, and the Indian tribal energy development and self-determination act. *The Lewis & Clark Law Review*, *12*, 1065.

Rynne, S., Flowers, L., Lantz, E., & Heller, E. (2011). *Planning for wind energy*. www.planning.org/research/wind/

Schelly, C., Bessette, D., Brosemer, K., Gagnon, V., Arola, K. L., Fiss, A., Pearce, J. M., & Halvorsen, K. E. (2020). Energy policy for energy sovereignty: Can policy tools enhance energy sovereignty? *Solar Energy*, *205*, 109–112. https://doi.org/10.1016/j.solener.2020.05.056

Stefanelli, R. D., Walker, C., Kornelsen, D., Lewis, D., Martin, D. H., Masuda, J., Richmond, C. A. M., Root, E., Neufeld, H. T., & Castleden, H. (2019). Renewable energy and energy autonomy: How Indigenous peoples in Canada are shaping an energy future. *Environmental Reviews*, *27*(1), 95–105. https://doi.org/10.1139/er-2018-0024

Stephens, J. C. (2019). Energy democracy: Redistributing power to the people through renewable transformation. *Environment: Science and Policy for Sustainable Development*, *61*(2), 4–13. https://doi.org/10.1080/00139157.2019.1564212

Stephens, J. C., Wilson, E. J., & Peterson, T. R. (2008). Socio-political evaluation of energy deployment (SPEED): An integrated research framework analyzing energy technology deployment. *Technological Forecasting and Social Change*, *75*(8), 1224–1246. https://doi.org/10.1016/j.techfore.2007.12.003

Strang, V. (2010). Mapping histories: Cultural landscapes and walkabout methods. In I. Vaccaro, E. A. Smith, & S. Aswani (Eds.), *Environmental social sciences: Methods and research design* (pp. 132–156). Cambridge University Press.

Tsosie, R. (2013). Climate change and indigenous peoples: Comparative models of sovereignty. In R. S. Abate & E. A. Kronk Warner (Eds.), *Climate change and indigenous peoples*. Edward Elgar Publishing.

Vissers, J. (2018). KBIC earns award for use of solar energy. *Daily Mining Gazette*. www.mininggazette.com/news/2018/09/kbic-earns-award-for-use-of-solar-energy/

Wehipeihana, N. (2008). Indigenous evaluation: A strategic objective of the Australasian evaluation society. *Evaluation Journal of Australasia*, *8*(1), 40–44. https://doi.org/10.1177/1035719X0800800106

Younging, G. (2018). *Elements of indigenous style*. Brush Education.

37
PART VI RESPONSE

Danielle Endres and Stephanie L. Gomez

Energy democracy is not one thing; rather it is a composition of theories, practices, tenets, and more. In this part of the handbook, chapters highlight various on-the-ground energy democracies in practice, which demonstrate a range of ways of working toward energy democracy across scales and contexts. Throughout this part of the handbook, each chapter highlights a case study that offers important lessons for energy democracy and provides important ideas about research going forward. In the remainder of this response, we will first reflect on themes that span the four chapters and then offer future directions for research.

Themes

Chapters in this section particularly tap into the power and participation elements of the justice, participation, power, and technology framework (see Chapter 1). In terms of power, each chapter wrestles with who has power, how power is deployed, and the ways that power redistribution can be used to further democratic ideals. McKasy and Yeo (Chapter 33) discuss the ways that carbon-neutral pledges are a way of transferring power from the federal government to a diverse community of local stakeholders. They point out that when the federal government refuses to lead on climate change, carbon-neutral pledges allow local communities to empower themselves and hold themselves accountable for climate change action. Pérez-Lugo et al. (Chapter 34) also discuss the transfer of power away from large universities that would normally act paternalistically to "help" communities without knowing what communities need. Instead, Pérez-Lugo et al. advocate for broad coalitions that include professors, students, experts, local officials, community members, and organizations. These coalitions avoid the top-down approach to power that often leads to less democratic outcomes. Campbell et al. (Chapter 35) also use their case study to highlight the importance of power redistribution. While they advocate for increased renewable energy solutions in the Global South, they also caution us to remember that renewable energy alone is not *inherently* democratic and that we need to be careful to make sure that new systems of energy production also adhere to principles of decentralized, local control. Bessette et al. (Chapter 36) also advocate for local control and highlight power redistribution in a microcontext, discussing how it uniquely functions in various communities. Ultimately, each chapter advocates for decentralized power and demonstrates the democratic possibilities

 DOI: 10.4324/9780429402302-43

that open up when communities are empowered to make decisions about their own energy production and consumption.

This sense of empowerment is only possible with local participation. Every chapter in this section discusses the importance of participation, including the participation of citizens in a particular local community (Chapters 33–36), professors and other professionals and experts (Chapter 34), and organizations that help facilitate greater local control (Chapters 34–36). Participation, then, is also about who is included and who is excluded in decision making. Importantly, these chapters all reject the assumption that local communities have nothing to say or contribute to energy decision making. Instead, they all start from the assumption that community members know what their communities need, hold forms of expertise, and want to hold themselves accountable for creating strategies and solutions unique to their locale. As Pérez-Lugo et al. (Chapter 34) astutely point out, communities are not simply empty vessels waiting for expert knowledge about energy systems to be poured into them; instead, each community has its own norms, practices, and conditions that need to be honored and valued. Each chapter points to the transformative possibilities that open up when communities are empowered to make decisions about their own energy systems.

Necessarily, the practice of empowering local communities also often means bringing together coalitions—sometimes across unlikely groups. For instance, McKasy and Yeo (Chapter 33) discuss the importance of more highly populated cities taking into account and working with rural cities in the creation of carbon-neutral pledges, which they admit is rarely accomplished in on-the-ground energy democracy work. Pérez-Lugo et al. (Chapter 34) discuss the formation of the National Institute for Energy and Island Sustainability (INESI), which not only connects all 11 campuses of the University of Puerto Rico system but also works in a true transdisciplinary fashion. INESI interrupts the common disciplinary hierarchy that places engineering departments across universities in charge of energy issues by including the social sciences and humanities, which are often excluded from these discussions. Moreover, INESI also coordinates with government agencies, municipalities, local officials, artists, community psychologists, and other grassroots and community organizations. By building these coalitions between groups that would normally not be in conversation with one another, Pérez-Lugo et al. argue that greater democracy and justice can be achieved around energy issues.

Campbell et al. (Chapter 35) discusses another coalition—the UK Low Carbon Energy for Development Network (LCEDN)—which is connected with a variety of other communities of practice, including NGOs, businesses, and policy communities. As a broad coalition, their focus is on using critical participatory approaches and engaging with grassroots and municipal initiatives and policies, thus broadening their coalition even further. Like Pérez-Lugo et al., Campbell et al. demonstrate that by bringing together members from a variety of sectors, more equitable solutions for energy production and consumption can be found.

Finally, Bessette et al. (Chapter 36) discuss Michigan Community Anishinaabe and Rural Energy Sovereignty (MICARES), which is a coalition that consists of scholars and practitioners from Tribal Nations and colleges, universities, and community organizations in the state of Michigan. Bringing together scholars, practitioners, and communities groups, as Pérez-Lugo et al. note, is key for building a coalition that can achieve real, practical results in increasing energy democracy on a local scale. Ultimately, these chapters show us that energy democracy is only possible when diverse groups, which are often unused to working with one another, come together to empower local, on-the-ground work.

Future directions

The chapters in this section highlight four embodied and emplaced case studies of energy democracies in action. While the examples in these chapters range from local cities to international, they do not cover all of the many ongoing energy democracies. With the exception of Campbell et al.'s chapter focused on the interplay between the UK (Global North) and countries in the Global South (see Chapter 35), the remainder of the examples come from the United States and Indigenous Nations in North America (which may, in part, reflect that the editors of this handbook are from the United States). As researchers continue to analyze case studies, we hope to see more geographic diversity in case studies. Of course, the wider literature on energy democracy demonstrates a range of examples from South America, Southeast Asia, and Africa (Bloem et al., 2021; Cantarero, 2020; Chávez, 2018; Delina, 2018; MacEwen & Evensen, 2021). Furthermore, other parts of this handbook highlight international case studies of energy democracies in Thailand (Chapter 24), Belarus (Chapter 27), and Latin America (Chapter 29). Yet there remains a preponderance of studies into energy democracies in North American and Western European contexts. In addition to viewing case studies of on-the-ground energy democracies as essential to the continued development of this research area, we call for efforts to further internationalize our research on energy democracy. Doing so may particularly help this research area grapple with the power dynamics and systems of oppression that characterize relations between the Global South and Global North and energy democracies in nondemocratic regimes.

Building from Pérez-Lugo et al.'s case study of the INESI's innovative transdisciplinary response to energy transition in Puerto Rico, a second future direction for ED research is to pursue more engaged transdisciplinary collaborations between energy democracy movements and universities. If energy transition is, as Pérez-Lugo et al. suggest, a wicked problem, then it will require not only the resources of academic researchers across many disciplines but also engagement with communities and case studies of energy democracy movements and transitions. As was argued in the introduction to this handbook (Chapter 1), an energy democracy research agenda has to be engaged with the communities that make up the movement.

In closing, the chapters in this section deliberately take us into the grassroots of the growing energy democracy movement. They highlight embodied and emplaced on-the-ground actions that can not only contribute to understanding energy democracy as a goal but also present lessons for emergent localized energy democracies. Thinking of the energy democracy research agenda as a composition (see Chapter 1), a crucial composition is that between the energy democracy movements and energy democracy research. Rather than seeking a merging or synthesis of the ED movement and research, as a composition, the parts remain, while also being composed in a series of compositions that connect movements with research. Through this composition, we see a future with many energy democracies.

References

Bloem, S., Swilling, M., & Koranteng, K. (2021). Taking energy democracy to the streets: Socio-technical learning, institutional dynamism, and integration in South African community energy projects. *Energy Research & Social Science*, *72*(101906), 1–14. https://doi.org/10.1016/j.erss.2020.101906

Cantarero, M. M. V. (2020). Of renewable energy, energy democracy, and sustainable development: A roadmap to accelerate the energy transition in developing countries. *Energy Research & Social Science*, *70*(101716), 1–15. https://doi.org/10.1016/j.erss.2020.101716

Chávez, D. (2018). *Energy democracy and public ownership: What can Britain learn from Latin America?* www.tni.org/en/article/energy-democracy-and-public-ownership
Delina, L. L. (2018). Can energy democracy thrive in a non-democracy. *Frontiers in Environmental Science, 6*(5), 1–5. https://doi.org/10.3389/fenvs.2018.00005
MacEwen, M., & Evensen, D. (2021). Mind the gap: Accounting for equitable participation and energy democracy in Kenya. *Energy Research & Social Science, 71*, 101843. https://doi.org/10.1016/j.erss.2020.101843

38
CONCLUSION
The future of energy democracies

Danielle Endres, Andrea M. Feldpausch-Parker, and Tarla Rai Peterson

Energy transition is undoubtedly one of human society's most acute global sociotechnical problems. In the midst of the climate crisis and its ongoing disproportionate impacts on marginalized, underrepresented, and underresourced communities, there is an urgent need to incorporate the tenets of democracy, including justice, equity, and participation, into the crucial decisions people make about energy in the next several years. There is an urgent need to explicitly interrogate the relations of power—particularly those emanating from the Global North—that maintain status quo fossil fuel–dependent energy systems and inequities, lest we simply replace fossil fuels with renewables without rethinking and shifting the balance of power in energy systems. Energy democracy (ED)—as both an ideal and a movement made of a variety of on-the-ground embodied and emplaced struggles across the planet—is the push to democratize ongoing and impending energy transitions, to bring the tenets of environmental justice into energy transition, and to challenge dominant narratives that energy transition is primarily technical and economic. Watching the ED movement unfold over the past several years has inspired us to bring more scholarly attention to this struggle in the hopes that simultaneous amplification of both activist and academic efforts for a just and equitable transition can disrupt the status quo and influence radical changes in how we conceive and implement energy systems.

ED is a composition that puts together energy and democracy (Chapter 1); the result is not synthesizing them into each other but viewing them in an ongoing process of composing anew in different contexts as *energy democracies* (Chilvers & Pallet, 2018; Latour, 2010). Just as the composition lens can apply to the ED movement, it also applies to emergent ED scholarship. This handbook represents a web of compositions that theorize, document, analyze, and innovate at the intersections of energy technology, justice, participation, and power.

As noted in Chapter 1, our central purpose with this handbook is to knit together a particular suite of contributions to the understanding and practice of ED. Although each chapter maintains its unique identity, all contribute to an inspiring symphony composed from varying arrangements of justice, participation, power, and technology. The handbook's internal diversity strengthens it in numerous ways. First, by bringing together researchers working in a wide variety of intellectual traditions and then encouraging them to make relevant aspects of those traditions extraordinarily transparent, the volume provides an opportunity for readers to compare and contrast the different kinds of learning that emanate from these traditions. That comparison is made even more fruitful when layered onto the diverse political contexts for each

DOI: 10.4324/9780429402302-44

study. Secondly, frequently highlighting opportunities for transdisciplinary networked research (Sprain et al., 2010) or research that compels iterative collaboration between scholars and activists, invites readers to reflect on ways they might build greater mutuality into their own scholarship and advocacy. While there is no guarantee that such reflection will result in more socially engaged scholarship or more intellectually informed activism, the invitation may initiate deeper conversations across the gulf between theory and practice.

As with other handbooks, this volume also explains key concepts and identifies a wide spectrum of the actors involved. Because ED is a relatively contemporary composition, linking the topic of energy with that of democracy requires significant attention. This challenge leads to another critical contribution of the handbook: investigating the diverse and sometimes disputed values, functions, and governance sites that may enable democratic renderings of energy systems. Finally, all contributors have sought to balance the urge to reimagine and innovate energy systems as democratic processes that prioritize justice and equity, with the pragmatic concerns common to all students of sociotechnical change. For ED, chief among those concerns is how to most effectively navigate path dependencies (Fouquet, 2016) that lock society into sociotechnical models that have thoroughly naturalized centralized decision making.

In the remainder of this conclusion, we will return to the justice, participation, power, and technology framework introduced in Chapter 1 to discuss how the entries in the handbook relate to it. Then we offer some areas of future research and conclude by inviting more and new voices and perspectives into the conversation toward a robust scholarly engagement with energy democracy.

Justice, participation, power, and technology

As stated in Chapter 1, although the handbook acknowledges a multiplicity of definitions for ED, one unifying theme is that "decision making [about energy systems is] characterized by a kind of equality among the participants at an essential stage of the collective decision making" (Christiano, 2018). This constraint complements Jasanoff's (2005) argument that critical analysis of civic epistemologies embedded in society should be a prerequisite to transforming relations between governance practices and technology. Mouffe's (2000) concepts of *democratic paradox* and *agonism* further contextualize our approach to energy democracy as a sociotechnical process grounded in disputation that produces frequent and always temporary junctures for developing and implementing policy. As an extension of these theoretical positions, the JPPT framework sees justice as a fundamental human right, power as a preexisting condition of political relations, participation as a means for enabling justice and negotiating power, and technology as a means for enabling the previous three components.

The chapters in the first part of the handbook highlight how the JPPT framework operates across the broad range of possible scales for energy democracy research and practice. The analyses in Chapters 3 and 5 operate at very different scales, with the former analyzing ED as it is imagined and instantiated across the multiple nations of the European Union and Chapter 5 analyzing ED as it is imagined and instantiated at a scale of Indigenous governance within what is now called the United States. Despite their scalar difference, however, they bring similar critical orientations, presenting ED as redirection of energy transformations toward the related problems of justice and power. They evaluate the relative success or failure of both participation and technology according to their functionality in contributing to more just (fair and equitable) ways of organizing power relations among citizens of the entity in question. While recognizing important differences, including the legal status of the EU as a supranational institution and indigenous nations within the United States as domestic, dependent, but sovereign entities, both analyses highlight disputations over sovereignty that severely limit democratic practice, while also identifying ruptures that may allow ED organizers and advocates to change current hegemonic configurations and contribute greater justice. Chapters 4 and 6 examine ED from a comparative perspective. Lenhart and coauthors (Chapter 4) contrast the ways electric cooperatives operate in Europe and the United States. They demonstrate that different operational philosophies influence the sense of community identity, key regulatory structures, the likelihood that decision making will function collectively, what kinds of market logics dominate system operation, and the varying levels of interdependence between energy and other sectors. While they note some similarities, they also note significant differences in the possibilities for broadening participation and restructuring the highly centralized decision making inherited from previous energy models. In Chapter 6, Goyal and Howlett use a multiple streams framework drawn from political science to tease out relationships between institutions, means, and eventual outcomes of efforts to build ED. By temporarily separating the streams of technology, problem, policy, and politics, they are able to evaluate ED opportunities across a broad range of spatial and temporal scales.

Chapters in the second part of the book focus on discourse and the work it does to frame understandings of energy. Because discourse is so intimately linked with enactment of power, these chapters hone in directly on the power component of the JPPT model. Chapter 9, which analyzes the language of energy security, demonstrates how this term disables people from imagining many potentially powerful means of democratizing energy systems, simply on the remote possibility that they could pose a security risk. Demski and Becker, however, push beyond this assertion to explore tensions internal to the language of energy security, arguing that the same intensity that initially appears to block ED also offers opportunities to broaden the scope of ED. In Chapter 10, Cozen takes a similar approach to energy poverty, arguing that current framing of access to commercialized energy as a basic human right simply strengthens the influence of powerful elites and further solidifies a Western perspective that already privileges individuals over communities. By critiquing a discursive process that limits people from imagining collective responses to energy system transition, Cozen hopes to open previously unimagined paths for democratized energy futures that are enabled by digitization but may remain elusive without an equally radical change in energy discourses. In Chapter 11, Batel reminds readers that technologies themselves are neither pro- nor antidemocracy. She demonstrates that, rather than being limited to fossil fuels, energy colonialism extends through all types of energy, including renewable energy. Her detailed descriptions of contemporary energy colonialism surrounding renewable energy demonstrate the importance of critically reviewing all energy transition initiatives in an attempt to avoid perpetuating colonial attitudes and practices. In Chapter 12, Schneider and Peeples intensify the critique of energy colonialism with their detailed analysis of the term "energy dominance." Their analysis clearly identifies ways that this terminology and its associated discourse validated the continuation and expansion of energy colonialism, while closing off opportunities for seriously considering issues associated with justice and power.

Parts III and IV bring together grassroots strategies of action for promoting ED and participatory principles that enable these strategies to succeed. In Chapter 15, Becker compares two strategies for developing ED within the institutional framework of Germany's energy transition. The strategy of remunicipalizing energy, which represents at least a partial rejection of neoliberal tenets urging privatization of public services, involves municipalities taking control of local energy systems. This often includes but is not necessarily limited to community ownership of infrastructure. A second strategy, one Becker notes is at least as common, is for previously unorganized citizens to initiate an energy cooperative. Becker notes that, while both strategies have strengths, it is important to realize that neither can guarantee that the reorganized hegemonic relations will produce justice. MacArthur and Tarhan (Chapter 16) narrow their focus to energy cooperatives at the same time they broaden their purview to multiple locations and a time span of centuries. They show that, while the organizational strategies that human societies have used as they struggle through sociotechnical transitions adapt as needed to fit new technologies, concerns about how to use those transitions as opportunities to develop more fair and equitable relations among citizens remain similarly complex and ill defined. After making the case that the energy sector remains patriarchal, Stephens and Allen (Chapter 17) point out how it perpetuates injustices long thought to have been corrected or at least hidden from view. They advocate for affirmative action to encourage women to take leadership positions within energy systems as an initial strategy for change. Burke (Chapter 18) combines specific strategies for change with democratic principles in his analysis of energy commoning, which he advocates as the best approach to ED theory and practice. Recognizing that his proposal requires a completely new (or old, depending on one's perspective) approach to social organization, Burke combines a description of specific strategies used by on-the-ground organizers with a careful explanation of philosophical grounding and organizing principles for both the practice of commoning and the tentatively imagined subject of that practice, the commons.

Those organizing principles, or in terms of the JPPT Framework, principles of participation, are explored in Part IV of the handbook. Kinsella (Chapter 21) uses nuclear energy to demonstrate how, even when faced with an industry that is heavily shielded from public view by both corporate and state interests, citizens have breached the barriers and figured out ways to participate in decisions about this highly controversial source of electricity. This chapter segues from strategic actions into participatory principles, because public participation in decisions about nuclear energy demonstrates how crucial it is for individual strategies to be linked together in ways that sometimes extend beyond the life of individual actors. Kinsella also points out that, although participatory principles form a sort of necessary infrastructure for organizing strategic actions, they do not guarantee success. For Barnes (Chapter 22), the challenge is to use the inevitable transformation of society's energy systems as an opportunity to produce *just* transitions, according to however the relevant citizens or stakeholders define justice. He sees the technological advances that permit *decarbonization*, *digitalization*, and *decentralization* as providing a fissure in an otherwise closed system and urges boisterous experimentation to democratize energy as much as possible. Delina (Chapter 24) takes another route to the discovery of participatory principles by developing a set of participatory principles based on observations of energy cooperatives in remote Thai villages that were both self-initiated and self-sustaining. One of the most striking aspects of these principles is that they are communally derived to focus on process rather than on any particular outcome. Scherhaufer reminds us in Chapter 23 that public acceptance should not necessarily be vilified. Scholars may need to be reminded that, however heady the air of commoning, decolonializing, and transformation may be, few of the venues that have enabled citizens to influence energy systems would have been available without an implicit promise that they would at least minimally promote public acceptance of some project or program.

Part V of the handbook includes chapters that apply an ED perspective to individual energy sources. Kasperski and Kuchinskaya (Chapter 27) begin this section with a case they found singularly lacking in ED values. They analyze a program designated as empowering victims of the massive technological failure of Chernobyl, demonstrating how this typically technocratic approach stripped all responsibility from the state and corporate entities and placed responsibility squarely on individuals whose bodies and land had been irreparably damaged. Finley-Brook's Chapter 29 demonstrates that the brutality of low-carbon energy is not limited to nuclear energy and provides powerful exemplars for Batel's (Chapter 11) contention that sociopolitical ideology and historical relations are more centrally connected with ED than are any specific energy technologies. Finley-Brook details the human rights violations accompanying large-scale dam projects in Central and South America and notes that such projects continue unabated in Asia and Africa. Despite the democratic potential of renewable energy, there are no guarantees that such potential will be achieved. On the other hand, Feldpausch-Parker and Batill-Bigler (Chapter 28) found that citizens of New York State (United States) were largely successful in attempts to ban hydraulic fracturing in their state, and the influence of their efforts appears to have spread to other potential sites. In an effort to avoid the double bind of pejorative characterization as being either outsiders or caring only about their own locale, the groups explicitly used terminology that focused attention on the need for justice and horizontal power relationships. Horton, Hernandez, and Peterson's (Chapter 30) study of Hawaii's energy transition also demonstrates hope for a democratic transition that seeks justice and equity. Although they found that energy professionals' discourse was highly technocratic, they also found evidence that these professionals were beginning to make implicit connections between the economic and technological factors they emphasized and the potential for a new energy system to promote justice and equity.

Chapters 33–36 in Part VI of the handbook emphasize the many ways that power and participation interact within the JPPT framework through case studies of campaigns and institutional collaborations. McKasy and Yeo (Chapter 33) use the device of carbon-neutral pledges to demonstrate the value of transferring political power to local communities. They demonstrate that the pledges have facilitated self-empowerment among communities that seek to become involved in mitigation of climate change. Pérez-Lugo and coauthors (Chapter 34) also argue that decentralizing decisions about energy transition provides opportunities for emergent leadership among participants in the coalitions that form around these opportunities, particularly in transdisciplinary collaborations between academia and local community organizations. Campbell and coauthors (Chapter 35) also emphasize the importance of decentralizing power. They accompany their advocacy for transitioning to more renewable energy in the Global South with another reminder that renewability does not make an energy technology democratic. At the same time, they urge supporters of ED to steadfastly push for decentralizing decision authority and providing realistic opportunities for local control. Bessette and coauthors (Chapter 36) also argue for local control through redistribution of political power within local contexts. Drawing from Indigenous knowledges of the medicine wheel, they highlight how power over energy decision making can be enhanced in Indigenous nations as well as in local non-Indigenous communities in Michigan. While at the beginning stages, this project holds great potential for decentralizing power and promoting justice and participation in communities.

Limitations, benefits, and future applications

As noted in the introduction to this handbook, the JPPT framework is a lens that allows researchers and practitioners alike to examine practices meant to enable energy transitions that value just, participatory, and equitable power relations. That said, like any lens, our framework enables analysis of some aspects of energy democracy and constrains others. In this section, we discuss the limitations as well as benefits to this framework for future scholarship.

First, let us address limitations. As Szulecki and Overland (2020) argue, it is no easy task to develop a framework that addresses all of the many energy democracies taking place globally. Their argument stems from the lack of an agreed-upon definition for ED, which they contend is why scholars should instead talk about understandings of energy democracy that are "broader than definitions or operationalisations" (Szulecki & Overland, 2020, p. 2). This type of thinking goes along with what Chilvers and Pallett (2018) argued two years prior when they called energy democracy ill defined and argued that we should instead talk about energy democracies "to account for the ways in which different forms of energy democracy and their publics are *made*, constructed, and co-produced" (p. 3; italics in original). When our team of scholars, including Dr. Leah Sprain, developed a previous iteration of the framework in 2017 prior to our Energy Democracy Symposium (Endres et al., 2017), we had yet to add the technology component to the framework and had not applied it outside of a North American context. Since this time, however, the framework has been used to evaluate European efforts at energy democracy as showcased in Carvalho and Horta's Chapter 3, allowing for a test in a different geographic location as well as at a different scale (i.e., subnational to international). Even with this broadened application, there have only been limited efforts to assess this framework in a Global South context, thus providing some future direction for us as well as others toward refining the framework for a variety of locations and situations.

Another limitation is that the JPPT framework does not follow more common conceptualizations of energy democracy that focus on economics, policy, politics, and technology (Kunze & Becker, 2014; Sorman et al., 2020). Having expanded our framework to include technological innovation (Chapter 1), we addressed one notable gap in our original framework. Although we argue that policy, politics, and economics play a role in energy transition, we made a conscious decision to instead incorporate these factors *through* the lens of justice, participation, and power. This decision stems from the concern that these factors, particularly economics and politics, too often "steer the ship" in scholarship and in practices of energy democracy, leading to perspectives that may occlude explicit discussion of justice, power, and participation. In our framework, policy, politics, and economics are a part of the greater context for multiple energy democracies, while justice, participation, and technology are cross-cutting relations of power. Our hope is that this lens offers a broader view of the sociotechnical aspects of energy systems beyond the economic and political.

Like any lens, the JPPT framework also offers numerous benefits for energy democracy scholarship. One benefit is that it makes explicit the ties between justice, power, and participation to energy systems and their related technologies. Because energy democracy is context specific (Chilvers & Pallett, 2018), it is valuable to acknowledge interdependencies between factors such as how power dynamics influence participatory processes, how higher levels of public participation can be used to address injustice and marginalization, and the role that technology and technological innovation play in achieving just transitions by providing a means for improving or recognizing limitations of technical energy systems. Thus, when the goal of energy democracy is to improve social *and* environmental conditions, examining these linkages among justice, participation, power, and technology can be enlightening for scholars and practitioners alike.

Another benefit to the JPPT framework is that it enables scholars and practitioners to turn a more critical eye to the role of technology within the goal of just and equitable transitions. The opportunities and limitations of particular energy technologies can vary by location, scale, and systems of governance. Technology is inherently influenced by social and cultural considerations, and the JPPT framework encourages seeing the sociocultural through lenses of justice, participation, and power. Multiple scholars in this book, most prominently in Chapters 10, 11, 21, and 29, have made it clear that no technology is inherently democratic, even renewables. This makes technology useful as a related concept to power, participation, and justice in energy democracy.

Future directions

The *Routledge Handbook of Energy Democracy* is not an ending but a beginning. Although this handbook builds from our collaborative engagement with the concept of energy democracy and represents a culmination of our efforts to build a research agenda and heuristic framework, it is also meant to spark new ideas, bring new voices to the conversation, and catalyze more research. ED research has seen tremendous growth over the last several years, yet it remains a nascent transdisciplinary area of research. The chapters in this handbook present the current state of the art of ED research across a range of theories, disciplines, empirical cases, scales, and practices that offer an opening for more engagement. Here we offer a series of future directions for ED research based on themes that came up in chapters across the six parts of the handbook.

First, we envision the ED research agenda as a "transdisciplinary networked process" (Sprain et al., 2010, p. 441). Sprain et al. define transdisciplinarity as "an approach to real world problems where collaboration reaches across academic disciplines and nonacademic sectors" (p. 442). As such, the future of ED research is not just interdisciplinary; it is transdisciplinary. It should include collaboration between academics, publics, and other stakeholders in the process of knowledge generation and composing solutions to the many interesting problems of the climate crisis, inequities, and energy transition. Sprain et al. (2010) propose thinking of transdisciplinary research as a network of processes and interconnections that is less focused on research outcomes (such as journal articles and books) and more focused on processes that contribute to meaningful contributions to solving societal problems. Pérez-Lugo et al. (Chapter 34) provide a compelling example of how the National Institute for Energy and Island Sustainability (INESI) at the University of Puerto Rico eschewed the ivory tower model to create a network of scholars and community groups to address the energy crisis in Puerto Rico before, during, and after the catastrophic impact of two hurricanes in September of 2017. We need more transdisciplinary networks like this one for ED research to realize its potential to make a difference in ensuring a just, equitable, and democratic energy transition locally and globally. As ED research continues to grow, we call for more transdisciplinary research networks.

As a part of transdisciplinary research, a second future direction for ED research is to fully embrace it as a form of engaged research. While there are numerous definitions of engaged research, we begin with Sergio Sismondo's (2008) characterization of engaged STS research as "placing relations among science, technology, and public interests at the center of the research program" (p. 21). Yet we take this a step further to include the various forms of community-engaged research practices (including forms of Participatory Action Research) that center relationships between researchers and on-the-ground communities and produce research that reflects the needs of both academics and communities. Engaged research, like the broader category of transdisciplinary research, aligns with the composition framework introduced in Chapter 1 of the handbook. Latour (2010) offers the composition framework (an alternative to constructionist and deconstructionist approaches) as a way for scholars to engage in composing solutions to the problems they research. Initially envisioned as a way for scholars to respond to the climate crisis, the composition approach is valuable for approaching energy transition not just as a technical process but as a process that needs the input of social sciences and humanities researchers as well as communities with local expertise. As an engaged research program, ED research is aimed at composing just solutions to energy transition. We hope to see engaged research at the center of this research program in years to come.

Third, given that ED research is still emerging, much scholarly attention has been devoted to theorizing and conceptualizing the key tenets of ED across a variety of settings. This valuable work has helped to define ED, think about it on the local, regional, and global levels, and demarcate areas for future research. While there have been important case studies of ED, including the many represented in this handbook, one goal we had for this handbook was to build up empirical analyses of energy democracies in action. We argue that ED research benefits from an interplay between the theoretical and empirical. Theoretical engagement with ED benefits from comparisons with empirical examples. Likewise, case studies of embodied and emplaced ED movements across the spectrum of possibilities benefit from the lens that theoretical and conceptual work can provide. Because a range of frameworks are available for ED research, including the JTTP framework, we call for a phase of building our knowledge through a robust growth in case studies. In addition to building knowledge, a focus on case studies is necessary to build the sort of transdisciplinary networks just discussed.

The chapters in this handbook suggest some topics that merit further scholarly attention. One area of potential tension between the ED movement and ED research is the role of renewable energy technologies in energy transition. While the ED movement actively promotes a transition to renewables as the answer to the climate crisis, ED research notes that democratic decision making could be at odds with renewables. For example, a democratic process could result in a community deciding that coal, natural gas, or oil is best for their community, which could technically be called energy democracy. Endres and Johnson (Chapter 5) pose the difficult question of what to do when a democratic process does not result in promoting a sustainable energy transition. Furthermore, as discussed, renewables are not inherently democratic. Several of the chapters in this handbook highlight that renewables can still be implemented in non-democratic, unjust, and inequitable ways. Finley-Brook (Chapter 29), for example, highlights the energy injustices that have plagued hydroelectric power via dams in Latin America. Batel (Chapter 11), likewise, conceptualizes *renewable energy colonialism* across a variety of scales. While solar energy has a strong connection with the possibility of energy democracy, in part due to its hyperlocalized implementation on individual houses in local communities, wind, hydro, and even solar have the potential to be implemented within problematic status quo systems that are not democratic. As such, the fourth future direction for ED research is to devote attention to the relationship between ED and renewable energies. Democracy is a process that can be brought (or not) to decisions about a variety of energy technologies. We hope to see more theorization and analyses that help to break down the idea that renewables are inherently more democratic and to build on a set of approaches for ensuring that the implementation of renewables is democratic.

Fifth, ED research would benefit from more analysis of the tensions in scale of governance and implementation of changes in energy systems. ED appears to be more implementable at local levels because of the availability of forms of direct participation. As ED practices scale up, participation shifts to representative models that still make room for democratic process at multiple scales, but the dynamics change and become more complicated. Demski and Becker (Chapter 9), for instance, highlight the need for ED research to do more research at multiple scales beyond the local, noting that considerations of how national energy systems can be democratized will be essential if ED is to be scaled up beyond local instantiations. As Feldpausch-Parker argues in Chapter 2, conceptualizing energy transition via polycentric governance (Carlisle & Gruby, 2019) accounts for the complexity of multiple centers of participation in decision making and allows for analysis of how power circulates through and across these multiple sites. Although not a guarantee, polycentric governance may allow for scaling ED up from local levels and is therefore worthy of future research.

Conclusion

ED is still in process, both as a movement and a research area. This handbook provides a valuable window into the current state of the research area but even more so provides a series of openings for scholars and scholarship to engage in this ongoing conversation. We hope that this handbook offers a catalyst for even more growth in ED movements and scholarship. The many compositions enabled by ED movements and research are essential to addressing the ongoing climate crisis.

References

Carlisle, K., & Gruby, R. L. (2019). Polycentric systems of governance: A theoretical model for the commons. *Policy Studies Journal, 47*(4), 927–952. https://doi.org/10.1111/psj.12212

Chilvers, J., & Pallett, H. (2018). Energy democracies and publics in the making: A relational agenda for research and practice. *Frontiers in Communication, 3*, 1–16. https://doi.org/10.3389/fcomm.2018.00014

Christiano, T. (2018, Fall). Democracy. In E. N. Zalta (Ed.), *The Stanford encyclopedia of philosophy*. https://plato.stanford.edu/archives/fall2018/entries/democracy/

Endres, D., Sprain, L., Feldpausch-Parker, A. M., & Peterson, T. R. (2017, July 12). *Justice, participation, and power: A starting point for energy democracy*. Energy Democracy Symposium.

Fouquet, R. (2016). Path dependence in energy systems and economic development. *Nature Energy, 1*, 16098. https://doi.org/10.1038/nenergy.2016.98

Jasanoff, S. (2005). *Designs on nature: Science and democracy in Europe and the United States*. Princeton University Press.

Kunze, C., & Becker, S. (2014). *Energy democracy in Europe: A survey and outlook*. Rosa Luxemburg Foundation.

Latour, B. (2010). An attempt at a 'compositionist manifesto'. *New Literary History, 41*(3), 471–490. http://muse.jhu.edu/journals/new_literary_history/v041/41.3.latour.html

Mouffe, C. (2000). *The democratic paradox*. Verso.

Sismondo, S. (2008). Science and technology studies and an engaged program. In E. Hackett, O. Amsterdamska, M. Lynch, & J. Wajcman (Eds.), *The handbook of science and technology studies* (pp. 13–30). MIT Press.

Sorman, A. H., Turhan, E., & Rosas-Casals, M. (2020). Democratizing energy, energizing democracy: Central dimensions surfacing in the debate. *Frontiers in Energy Research, 8*, 1–7. https://doi.org/10.3389/fenrg.2020.499888

Sprain, L., Endres, D., & Petersen, T. R. (2010). Research as a transdisciplinary networked process: A metaphor for difference-making research. *Communication Monographs*, 77(4), 441–444. https://doi.org/10.1080/03637751.2010.523600

Szulecki, K., & Overland, I. (2020). Energy democracy as a process, an outcome and a goal: A conceptual review. *Energy Research & Social Science, 69*, 101768. https://doi.org/10.1016/j.erss.2020.101768

39
AFTERWORD
Energy democracy—episode 196 of Cultures of Energy Podcast

Dominic Boyer, Cymene Howe, Danielle Endres, Andrea M. Feldpausch-Parker and Tarla Rai Peterson

This afterword is an excerpt from the Cultures of Energy Podcast episode on Energy Democracy (episode 196). Hosted by Dominic Boyer and Cymene Howe, the Cultures of Energy Podcast "invites scholars, artists and activists to discuss the pressing energy and environmental issues of our times." It "always strives to show how the insights of energy and environmental humanities can help us to better understand and overcome the dilemmas we face" (http://culturesofenergy.com/podcast/). In this episode, Boyer and Howe interviewed handbook coeditors Danielle Endres, Andrea Feldpausch-Parker, and Tarla Rai Peterson. This interview took place in September 2019, while Endres, Feldpausch-Parker, and Peterson were at a writing retreat in Solitude Utah to work on the handbook. What follows is a transcribed version of the portion of the podcast episode in which Boyer and Howe engage in conversation with Endres, Feldpausch-Parker, and Peterson about the handbook, energy democracy, and more. We encourage all of you to both listen to the full episode—which includes reflections on Brexit, climate action marches, and an environmental humanities workshop—as well as the other episodes of this podcast. While not explicitly linked with the energy democracy movement or scholarly agenda, the Cultures of Energy podcast is a shining example of an engaged scholarly practice that contributes to the tenets of justice, participation, interrogating power, and democratizing conversations about energy in society. Boyer and Howe, scholars at Rice University, exemplify the sort of engaged scholarship that is needed in energy studies broadly and in energy democracy research more specifically.

Dominic: We have a really interesting conversation today with three scholars who are putting together a handbook of energy democracy, which I think is a really interesting project. So, we're talking to Andrea Feldpausch-Parker, we're talking to Danielle Endres, and we are talking to Tarla Peterson, all three of whom are at a writing retreat right now putting together this handbook. It's meant to be a handbook that spans both the kind of scholarly world of research on resurgent energy democratic movements and actions. We're talking about everything from kind of remunicipalization campaigns, to pipeline protests, to people who are trying to create their own microgrids, all of that good stuff; and kind of good energy citizenship; and then also, people who are actually on the ground involved in these movements grassroots. And so, energy democracy: let's hope it comes soon.

Cymene: Yeah. We need to have more of that.

DOI: 10.4324/9780429402302-45

Dominic: We are so thrilled to have with us today Andrea Feldpausch-Parker, Danielle Endres, and Tarla Rai Peterson who are with us to talk about their marvelous project on energy democracy. Thank you all three for joining us.

Andrea: Thank you.

Cymene: As Dominic just mentioned, the three of you—among others, colleagues, activists, and scholars—have been working on this concept and the problem and the potential, I'll say, of energy democracy. I thought it might be useful to begin by talking about what energy democracy looks like, in the present and potentially in the future—we can kind of unfold that as we go along. You described that energy democracy is, on the one hand, a kind of set of social movement principles, or an advocacy or activist position, but that it's also an opening for scholarly intervention, for research protocols, and ways of kind of seeing the world differently through the intellectual lens of energy democracy. So, I wanted to ask all three of you—or a couple of you, whoever wants to respond to the question—if you could kind of walk us through what energy democracy looks like as a social movement and what it looks like as an intellectual project.

Andrea: I think as a social movement, if you think about it, there's a lot of similarities between the environmental justice movement and energy democracy in that people want to have more of a say over energy systems and be less beholden to these systems where others are making the decisions for them to a degree that some of these decisions are affecting them in other ways, climate change being a large driver. Knowing that we have to switch our energy systems away from fossil fuels and move towards alternatives, how do we go about doing that, and how do we hold the energy sector accountable for making such a switch? Because it's not easy. Our energy grids are set up on these fossil fuel–dependent systems. There are people in places of power that want to maintain the systems. How do you go about making that transition, and making it a just transition to these other types of sources? So that we're not maintaining these same power dynamics, but that people have a greater say in how these energy systems are being implemented. That's kind of how I see it. You guys wanna chime in?

Tarla: One point that I think is important can be encapsulated in that term "prosumer." It's an old term, but it helps me to remember that the energy democracy movement is not so much about having a whole bunch of people completely off-grid, just in sort of a Libertarian mode. Instead, it's people who are involved in the energy system. And the reason you use the word "prosumer" as a short-hand is because these are people who are producers and consumers at the same time. And by producer, I don't mean that they're just somebody who puts a set of solar panels on their roof. Yes, that may be it, but there's also the notion of the production of knowledge about the system that enters into that role of prosumer. It's much more than just putting up a wind turbine or putting up solar panels on the roof. I think that's really important to help understand the depth of the movement and to differentiate it from the escape idea that sometimes people have when they think about moving to more community or locally owned and controlled. I think that control and power are still—social power, political power—are still really important issues. But it's about decentralizing and changing those configurations of power, not doing away with them.

Cymene: And it seems that in thinking through energy democracy as a sort of research program, or set of projects that we can imagine, there's a kind of collaborative element at work here too; because you really do talk about it as something that requires multidisciplinary attention that needs to be a sort of networked set of scholars who are kind of swapping ideas back and forth with one another. So, is that part of how the energy democracy, as research program or as an intellectual space, came to you? Was it through this kind of collective action that's involved in energy democracy as a political movement? Are those two intertwined, do you think?

Danielle: Absolutely, I think that collaboration is really at the core of the research program, and on two levels. One, the level that you just spoke about, which is collaboration interdisciplinarily. We all have backgrounds with environmental studies, environmental communication, where we recognize that many of these problems that we as a society are facing cannot be solved with one discipline alone. And so, we've always worked in those ways and felt that to tackle something like energy, we need to be collaborating across disciplines.

Secondarily, we also really highlight the collaboration potential between the energy democracy movement and academia. We describe this as an engaged research program, drawing from some of the scholarship across disciplines about how academic work can be in conversation with what practitioners are doing on-the-ground, so to speak. I think in terms of that aspect of the research program; there's kind of a mutually constitutive relationship between the academic work and the social movement work where we as academics are not just responding to what the energy democracy movement is doing. And [we are] thinking through theoretical resources, forms of analysis that can kind of contribute to the ways that the energy democracy movement is forming and doing their work. But also hopefully our research would then speak back to those that are on the ground doing this energy democracy work. And we're very careful to say that engaged for us doesn't mean extractive. We don't want to be a research agenda that just kind of takes from the energy democracy movement, and says, "Hey, this is a really interesting movement. We're studying it, we're getting academic gain out of it." We really envision it as more of a collaborative, conversational mode of engagement between the research agenda and the social movement.

Cymene: That's great.

Dominic: Yeah, it's a very inspiring project, and it seems to me so on point for where we are in the twenty-first century. I mean, one of the cleverest things that the twentieth-century energy system did was to find all the ways it could possibly manage to distance people from energy production and to make them think that energy use kind of didn't matter, it had no consequences. So, they moved the power plants out of town, or they put them next to poor communities who didn't have the kind of political clout to resist. And meanwhile, we were encouraged to just continuing growing our consumption of fuel, growing our consumption of electricity all across the twentieth century. And now we're living in this moment where that kind of alienated relationship between people's everyday lives and energy is really coming back to haunt us. So, this idea of reconnecting seems really powerful to me, and I just wanted to ask if any of you had particular projects in energy democracy that you found inspiring? Either ones that you've been involved in, or ones that you've been studying that you would like to kind of bring to the attention of our listeners?

Danielle: I'm just starting work on a project that's looking at energy democracy in relationship to an Indigenous environmental organization called *Honor the Earth*. They're a group out of Minnesota and work with some of the Indigenous Nations near and around there—both in the US and First Nations in Canada. They're doing this really amazing work where they're taking decision making about energy into their own hands while also asserting sovereignty, Indigenous sovereignty. Some of the strategies that they're using are that they will find funding and they will go to Indigenous Nations and help them install solar and wind or whatever energy source is appropriate for that particular Nation; they'll work with them to get that installed. Indigenous Nations are doubly impacted—both by lack of energy access often, but also the impacts of climate change will affect Indigenous communities disproportionately. So, this move to bring energy production to Indigenous Nations, I think, is a really awesome example of energy democracy. Along with that, they are sending Indigenous students to learn the skills of energy engineering and energy production, so that they can bring that back to their own communities.

One other example that I think is really cool is that, for example, with the Keystone XL issue and the Indigenous Water Protectors, [they] found that they couldn't have access to decision-making processes and public hearings, which is typical for a lot of public hearing processes that are more "Decide-Announce-Defend." So, they used Native American sovereignty to hold their own public hearings and gather input and have a say in that and then sent the results of that to the decision makers. So, I think those are really exciting examples of energy democracy practices. I'm just starting to work on that, but that's something I'm really excited to keep working on under this research program.

Cymene: Yeah, really important stuff. We've had a couple of shows on Standing Rock as well, maybe even three, I think.

Dominic: Do we want to give our other guests a chance to see if they want to weigh in on that one?

Cymene: Yeah, sure.

Andrea: In all honesty, we have some projects that we're wanting to start up and I am not going to give a lot of detail, just because these are ideas in the works. What we also want to focus on is energy democracy in places that are harder hit by—I wouldn't call them natural disasters anymore, I would call them anthropogenic disasters, such as hurricanes hitting places that don't necessarily have the means or the control over their own energy systems. And so, that's something that we've been playing around with because as we're seeing these hurricanes that are larger and nastier, and they're hitting places that don't have the ability to quickly recover due to lack of federal funding for recovery, and things like that. I think [about] how they are able to adapt? Are they even able to develop resilient systems? That's something that we're focusing on: these coastal or island communities that are taking the brunt of these storms but at the same time are held hostage to these energy systems that are not allowing them to become more resilient to such weather events.

Tarla: I want to just tag onto the same thing that Andrea was talking about because it seems to me that at this juncture, we have a great opportunity for research and practice to work together. Because often it appears, in a couple of the cases that we're starting to look at, as though it's not because in and of themselves people in a particular location are not "able" to recover from some disaster. Say, let's stick with hurricanes. Again, it's not a natural issue. And I think what research can do is lift that cloak off that pretends that they are naturally unable. And it just can open a lot of peoples' eyes to some of the causes of the inability. And some of those causes of the inability are the very same causes that Danielle was talking about, with being excluded from the decision-making processes. It's not about lack of understanding of the technology, it's not about lack of access to the technology directly; it's about lack of access to the decision-making venues, and that's very carefully structured. Sometimes as a researcher, you can delve into that and make a little more explicit statements than you can when you're actually on-the-ground and have to continue working and living in the situation. I think we can help each other and work together in some of those [situations]. That's part of the reason they're so interesting because we have an opportunity to actually help do something good.

Andrea: Well, Tarla and I and some other coauthors wrote a piece a while back in relation to Superstorm Sandy and a lot of the mechanisms that we have in place do not allow us to build the system better. Funding allows us to build back what was there previously. And so, we think: 'Okay this anthropogenic disaster occurred, let's try to make a more resilient system.' But we haven't figured out a way of actually making that happen. Even in places like New York State where there's definitely an acknowledgment that climate change is happening, we need to do something, there's still barriers to doing that. So, that's something else that we're running up against, is that we have to really work on restructuring our systems to accommodate for these needs instead of constantly being reactive. Because we need to adapt and our systems do need to change—a centralized grid is not working to our benefit, considering what we're going to be tackling in the future. So I think that these ideas are happening, but it's in pockets, it's occurring in specific places. I think that's why a lot of people, when they think energy democracy, they're thinking at the local level versus scaling it up. Oftentimes, we have to think of energy democracy in scaled up [ways] because of the systems that are involved or apply pressure to our abilities to change our energy infrastructure.

Cymene: I mean, this is a really important point that you're making that these energy infrastructures and the grooves of finance that allow for certain kinds of rebuilding, and prohibit other kinds of building, are deep grooves. These are long histories of repetitious bad planning and a lack of consequences that were visible to people; or a refusal and denial about those consequences. So this is really a key piece.

I wanted to come back to a bigger question, maybe even a philosophical question, about public participation because this is something that's really fundamental to energy democracy, as you're describing it. As Dominic pointed out a few minutes ago, one of the things that the twentieth-century energy infrastructures created was the cloaking of these energy infrastructures—to hide them, to put them outside the city walls, if you will; to make invisible, in many ways, the sources of that energy, whether it's electric or fuel; and the consequences of course as well. So the consequences are coming home to roost. And we're seeing that with climate change and anthropogenic disasters everywhere. But I still wonder if there's not still an ambivalence, or a lack of interest, on the part of the public to really dig into these questions of energy on-the-ground as they're being produced. And I say this because I teach a class called the "Social Life of Clean Energy" and I asked my students on the first day whether they have any idea of where their electricity comes from, for example, and they never do; there's a real—they're not unengaged students, but they have no idea, and in some ways maybe not so much interest. So, I wonder how we generate that interest in a kind of proactive way, rather than in the kind of responsive way that we see with the Standing Rock protests, which were super important, but it was kind of a response to the deleterious effects of these infrastructures on Indigenous land and affecting water systems. How do we get people involved in a real way, in a passionate way, about energy; not just in response but proactively?

Andrea: That's an excellent question. I feel like we do very much react—I mean, just being able to plug in and then power be there, right? We are not made aware of things like electricity until it's not working. And when we have massive brownouts or blackouts, due to some catastrophic weather event or something like that, it really kind of hits home that this system is not always going to be there.

Maybe I'm jaded at this point, but I feel like we very much are reactive in this way because it's almost like energy has become a new right, when you think about air and other things, because the lack of access to it means that we cannot function in our daily lives. And so, if we look at it that way and if we treat energy as a right, then that also means that it should be evenly distributed across the population, that it should be there at all times, things like that. So, I think looking at it that way causes us to put the spotlight on energy systems in the way that we put the spotlight on clean air or clean water—saying that this is something that should be universal. Maybe that's a way of forcing us to move from the reactive to the proactive in thinking about energy systems because it brings it to the forefront.

Tarla: First off, I want to go back to one thing you said about the fact that we really don't think, as human beings in our society, we do not think about our energy. I don't think this is just with individuals. For instance, even with the most "environmentally engaged" of students, or the most "socially engaged" of students, one of the experiences that I know some of us have had, is that people who also work with us—in studying or doing research or in environmental movements—never have thought, even now, about the energy system as being relevant to those questions. Or if you participate, as I do, in a lot of conservation organizations or organizations that are focused on conserving wild spaces, you have to still sell them on the idea that what we do with our energy matters and makes a difference. These are organizations made of sophisticated, educated people who actually are interested. Now, for some of them, all you have to do is the informing step because they're already on the same page. But, it demonstrates how strongly we have hidden energy systems from all of ourselves. And we've allowed that to happen. I think that opening that up is just crucial, it's so important.

And I also want to say, I don't think that our reactivity is unique to energy. I think we need to kind of be careful about that and be aware that that's typically how we as a society operate. Maybe it's more rhetorically sane for us to start exploring: how do we work with a society that does things reactively? How can we—instead of trying to change that very deep tendency—work within it? And I think we can.

Danielle: A little bit on the flipside to what Tarla's saying. I do also think that we as researchers are really well equipped not only to work with this kind of crisis reactive frame that society is really embedded in right now; but we're also empowered to reject that and look at the mundane and the everyday. In some work Tarla and I did on energy communication as an area of study, we really encouraged scholars to stop focusing on energy only when a crisis happened—only when a Hurricane Maria or a Superstorm Sandy happened—and instead, really focus on uncovering that invisibility that we've been talking about. How do we create research programs that really do focus on these mundane everyday interactions that we have with energy? And then we can bring that to the classroom; it's not a short-term easy fix, but over time as we really focus on those everyday interactions with energy, we can potentially start changing some of the ways that we think about it.

Dominic: I wanted to ask you to talk a little bit more about the democracy part of this. I think it's really an interesting complex issue. On the one hand, we have certain kinds of democratic institutions, as we've been saying, some of which might enable a greater visibility of energy, greater concern, greater citizen engagement around energy; and some of those might disable those engagements. We certainly do have a tendency in the world that there are a lot of republics out there masquerading as democracies—like the United States, for example, we've found out—in which really radical democratic potential is always being held in check by these systems and hierarchies of elite privilege and certain groups being benefited at the expense of others.

So, I'll just throw this out to anyone who wants to respond to me: where do you see, at the institutional level, the places of hope or the places where we should concentrate in terms of trying to work towards broadening energy democracy? And where are the institutions that you think we have to be really honest with ourselves that need to change or that we really need to kind of focus our critical attention on as scholars, as activists, as citizens, in order to get where we need to go?

Tarla: I think one place we need to focus our attention if you're in the United States, is the US DOE. The Department of Energy has so much clout and is thoroughly controlled by some of these incumbent actors and interests that have no reason to give up any of their power. And, tied [to] one of the points that Andrea made, even if you take the approach—which I'm not opposed to—that even if your regional and national system isn't democratic, you can make your little village democratic, I think that's great. But, your ability to develop a more just energy democracy at the level of your little village is directly influenced and limited by the US DOE if you're in the United States. I know that's a fact, and anybody who thinks about it and starts paying attention will also be aware of that. So, sometimes we have to focus our attention on changes at that larger level, the national level. What's been happening in Germany overall is a good illustration of how you can try to work the top-down and a bottom-up approach together. But it's also an illustration of the fact that there are continually going to be these incumbent actors who will try to step in and stop the changes. They will be successful, to a degree, but that doesn't mean you should stop working.

Dominic: Yeah, and I'll just say—because we happen to be in Germany right now and we took part in the climate strike last week and the march—there also is a lot of criticism in Germany about their own fading consensus not on moving in the right direction but on accelerating the program and hitting their targets on time and so forth. There was an interesting fellow named Hermann Scheer, who is a Social Democrat here who really got the German energy turn going. He really believed that actually renewables and decentralized energy could contribute to strengthening democracy because of the lack of these large infrastructures that span states and countries, which are controlled in a few places and peoples' supply everywhere can be cut off by decisions that are made in a few places. He felt that a decentralized energy system was a really important infrastructure for democratic renewal. And I'm wondering if that's an idea that y'all have thought about too and whether that's something you might be seeing in practice in some of the cases you're looking at?

Tarla: Yes.

Andrea: I would say it has the potential, it's not a guarantee though. As we're changing, there's greater potential because there's the possibility of insertion of new players, which changes the dynamic, it shakes it up, but it's not a given. There can be situations where the same players are still trying to push out new stakeholders or new players from decision making because they don't want to relinquish that decision-making power. At the same time, as we're seeing these insertions of these new players, there's new conceptualizations of how the energy grid should be structured. There is that potential; that's the hope in that we can [build] on that potential. But I also want to say, I don't think it's a guarantee. We are seeing it, however, but it's not a guarantee.

Tarla: There are no guarantees in this. Trying to hold off until you find the guaranteed approach is not the way to do it. I think experimentation is more the keyword.

Andrea: Yeah, I'd agree with that.

Danielle: Yeah, and I was just going to add that were actually having this debate just yesterday because we're here working on our volume and thinking about, are these more local, decentralized approaches more democratic? Or do we need fundamental, larger, democratic change on a national or state level? So, we don't know that one is necessarily better than the other. When you're thinking about radical, democratic, participatory engagement you always have to, in whatever context, you've always got to be vigilant about keeping that going and looking for places where stakeholders are being excluded or folks at the local or the national level are gaining too much power, and then excluding voices.

Tarla: I also think that one of the issues that's really important here—it's one that's basic to public participation in the interests of justice in a wide array of things—is simply participation fatigue. People get tired of fighting the good fight, if you will. This is one reason why developing networks where you can develop mutual support is really important because that fatigue is going to set in. We see that in some of these small energy cooperatives that are emerging or remunicipalization efforts. We see over time that if the same people, if a small group of people, continually has to bear the brunt of it by themselves, they simply get tired. So, that's one reason why building a network and also working both at the local level, and the national, and even international level is crucial to help build in some—I'm using a terrible set of mixed metaphors—kind of safety net for these new approaches. Then there's also the notion that it's really important for us—again I just think this is crucial—for us to talk about it as energies. We're talking as though energy was a thing, a single thing and it's not. I mean, they're not. Energies is encouraging. This is why the terminology or the grammar that we use when we talk is really important.

Cymene: Did you want to say more about the plural energies?

Tarla: One of the reasons that I think energies is important is because it's important for us to realize that sunlight, and the way we can use it, is very different from the oil that comes from hydrofracking. When we put it all together, it helps us make the whole process invisible. It's like making it into the black box where the people who are running it say, "Trust me. It all works, you don't need to know how it works." But if we talk about it in the plural, we have different energies. It's easy for me—especially in our neoliberal world today—to say: "Okay, I own the pour space where this oil is stored, or I can lease it from you and buy it, or whatever. Therefore, it belongs to my company." It's harder to do that if you're talking about sunlight or wind power. So, it's important, in my mind, to specify that we have energies. I don't think it's just the localization that's important in helping energy democracy to move forward. I also think that focusing the attention on sunshine is an important way to help people think differently. Instead of thinking of just the energy system as the property of *x* person, it's easier if you talk about the sunshine to also introduce a new model for what it means, what energy means. And this goes along with talking about it at a more mundane, day-to-day level, instead of just this crisis mode.

Cymene: Yeah, I think this is really a key point. Again, this is sort of where Hermann Scheer, I think, is very interested because the truth of sunlight and radiation from the sun is that it is the shortest supply chain possible.

Dominic: Right.

Cymene: It's a direct source and you've got no pipelines, whatsoever. I think what you're getting at is really key: that the materiality of these energy sources is very important. The qualities of them as a force, as a physical entity that is different from fossil fuels and other petroleum products. This is really key. And the potential that we see in energy transitions really is an opportunity. We tend to think of it as—and maybe it's seeing, as you're saying, sort of the negatives of the reactive responsive nature of human beings or maybe just, you know, white, Western, settler–colonial regimes of thought is to respond to negative outcomes—but really energy transition is an incredible opportunity for us to do our politics differently. This is something Dominic and I've been arguing for a few years now, in terms of looking at renewable energy. Instead of mapping our old political forms onto these new energy infrastructures and systems, why don't we use these energy systems and their potential as a way to rethink new forms of politics, new forms of relating, new forms of getting along? Because the potential is there. It's not a guarantee, as you say, but I think there really is a true potential. So, maybe as a final question?

Dominic: Sure.

Cymene: Yeah, I wanted to get to this question of energy governance sites. And you write that, of course, these energy government sites involve a range of actors, democratic values and democratic functions. I want to think through with you what sites of energy governance are: what they look like, where they are, how they're distributed. How do you imagine those sites? Because it could be really quite extensive: energy governance could be happening at so many different points along the graph or the scale. So, I wanted to invite you to share what—in your imaginary—what are these the energy governance sites and what is their critical importance?

Andrea: Oh, you stumped us with this one. I think it's such an excellent question. Again, we're at a writing retreat right now, the three of us working on a handbook on energy democracy. We're really battling with this idea—and maybe battling's the wrong metaphor to use—really trying to think about at what scales of governance do we act? If we think about it, when the Trump administration took over and the Obama administration had pushed all of this climate legislation at the end of his term and then the Trump administration came in and was trying to dismantle a lot of this, a lot of people were feeling like "Okay, game over." And that's not actually true. What it means is that somebody else has to pick up the slack. And we were discussing [this] just the other day: yes it slows down the process but action is still [possible] at levels of governance if we think about what can happen at the state, or what can happen in the United States but in the subnational context, or what can happen at a more localized context. We have opportunities there so that as one level of governance, for lack of a better word, fails to address these issues, we do have the ability to not fully rectify but to continue the effort forward at a different level. That said, when we think about where are the spaces where this [energy transition] actually needs to happen, it's still at the national and also international levels of governance.

Tarla: Well, in the United States right now, for example, we've got California suing the federal government having to do with clean air standards. So, there's an illustration of exactly what you were talking about, Andrea. I realize we're falling back on the existing structure. I think it's important to simultaneously work with the existing structures but you are also attempting to build new structures to even potentially replace some of these old structures, particularly those that are the most problematic.

Dominic: Right. And I think that this idea—and maybe this is inspired by some of your own interest in energy communication and your scholarly background—of creating a handbook, does anyone want to speak to that? You know, what some of the objectives are there, and what you're hoping to accomplish with the handbook itself?

Andrea: Yeah, I'll let Danielle take this one.

Danielle: I think that we really want to not take ownership over this idea of energy democracy as a research program. Like we were saying in the beginning, we're so sure that to tackle these issues there needs to be vast interdisciplinary collaboration and thinking about it, that a handbook, for us, was a way to start and continue conversations about what a research agenda in energy democracy can be and to highlight the diverse and interesting research that we've been discovering through our engagement with this idea of energy democracy. So, we really want to put this handbook out there as a way to invite lots and lots of scholars and practitioners and activists to think about this as a potential area for more research. Obviously, we could have said we're going to write a book on energy democracy and we're going to put forward our perspective and our theory of this. But we chose the handbook because we wanted a conversation. We wanted to really open an area of research and invite a lot of folks into that conversation.

Andrea: I also wanted to say that energy is not just a technical system. I think that oftentimes, it's treated as such. So, social scientists and members of the humanities have had to push against the fact that it's not just a technical system, it is very much a social system. We have the ability to look at it from multiple lenses and analyze that social aspect of energy. So, I think that we are truly attempting to broaden the discussion and push against this kind of keeping it in the technocracy, if you will.

Tarla: Also, the opportunity that is presented by a handbook is that it's going to have usable chunks that a friend of mine who teaches in the engineering college might use a couple of chunks with the people he's working with who are then going to go out and work for some of the utilities. And, they will have that background that they didn't have before and they'll take that into their more technocratically oriented organization. So for me, that's one of the appeals of doing this kind of work: you can try, anyway, to make it something so that if people don't want to take all the time and effort to look through all these possibilities, there are chunks that they can pull out.

Dominic: That's great, and we just want to thank you for doing all this work to create this handbook that obviously is going to be a great resource for so many people and probably a wonderful gathering point where people are going to get to find out about kindred spirits, so to speak, that they hadn't even heard about before. So, I think even producing something like this is going to be kind of a network builder, and hopefully create even more collaborations and so forth. We're also not, because we're polite, going to ask you when it's going to be done, or when it's coming out, or anything like that. You'll get it done on your own good time, and we'll be happy to wait for it. We do want to thank you so much for the work you're doing and also for taking the time just to talk through the project and your interests with us. It's been really a fascinating discussion.

Andrea: Well, thank you this has been an absolutely wonderful experience.

INDEX

Page numbers in *italics* indicate a figure, and page numbers in **bold** indicate a table on the corresponding page.

For Product Safety Concerns and Information please contact our EU representative GPSR@taylorandfrancis.com
Taylor & Francis Verlag GmbH, Kaufingerstraße 24, 80331 München, Germany

www.ingramcontent.com/pod-product-compliance
Ingram Content Group UK Ltd.
Pitfield, Milton Keynes, MK11 3LW, UK
UKHW050231200926
475702UK00020B/2977
9781032130521